Global Political Ecology

Global political ecology

The world is caught in the mesh of a series of environmental crises. So far attempts at resolving the deep basis of these have been superficial and disorganized. *Global Political Ecology* links the political economy of global capitalism with the political ecology of a series of environmental disasters and failed attempts at environmental policies.

This critical volume draws together contributions from 25 leading intellectuals in the field. It begins with an introductory chapter that introduces the readers to political ecology and summarises the book's main findings. The following seven sections cover topics on the political ecology of war and the disaster state; fuelling capitalism: energy scarcity and abundance; global governance of health, bodies, and genomics; the contradictions of global food; capital's marginal product: effluents, waste, and garbage; water as a commodity, human right, and power; the functions and dysfunctions of the global green economy; political ecology of the global climate; and carbon emissions.

This book contains accounts of the main currents of thought in each area that bring the topics completely up-to-date. The individual chapters contain a theoretical introduction linking in with the main themes of political ecology, as well as empirical information and case material. *Global Political Ecology* serves as a valuable reference for students interested in political ecology, environmental justice, and geography.

Richard Peet holds degrees from the London School of Economics (BSc (Econ)), the University of British Columbia (MA), and the University of California, Berkeley (PhD). He is currently Professor of Geography at Clark University, Worcester, MA. His interests are development, global policy regimes, power, theory and philosophy, political ecology, and the causes of financial crises. He is the author of 12 books, 100 articles, and 50 book reviews.

Paul Robbins holds a BA in Anthropology from the University of Wisconsin and an MA and PhD in Geography from Clark University. He is currently Professor and Director of the School of Geography and Development at the University of Arizona, where his work focuses on the power-laden relationships between individuals (homeowners, hunters, professional foresters), environmental actors (lawns, elk, mesquite trees), and the institutions that connect them.

Michael J. Watts is Professor of Geography, and Director of Development Studies at the University of California, Berkeley where he has taught for 30 years. His research has addressed a number of development issues, especially food security, resource development, and land reform in Africa, South Asia, and Vietnam. He has written extensively on the oil industry, especially in West Africa and the Gulf of Guinea. Watts was a Guggenheim fellow in 2003 and was awarded the Victoria Medal by the Royal Geographical Society in 2004.

Global Political Ecology

**Edited by Richard Peet,
Paul Robbins, and
Michael Watts**

Routledge
Taylor & Francis Group

LONDON AND NEW YORK

First published 2011
by Routledge
2 Park Square, Milton Park, Abingdon, Oxon, OX14 4RN

Simultaneously published in the USA and Canada
by Routledge
711 Third Avenue, New York, NY 10017

Routledge is an imprint of the Taylor & Francis Group, an informa company

Typeset in Times New Roman by
Keystroke, Tettenhall, Wolverhampton

Printed and bound in Great Britain by
CPI Antony Rowe, Chippenham, Wiltshire

British Library Cataloguing in Publication Data
A catalogue record for this book is available from the British Library

Library of Congress Cataloguing in Publication Data
Global political ecology / edited by Richard Peet, Paul Robbins, and Michael Watts.
p. cm.
1. Political ecology. I. Peet, Richard. II. Robbins, Paul, 1967- III. Watts, Michael.
JA75.8.G56 2010
304.2—dc22
2010012691

ISBN: 978–0–415–54814–4 (hbk)
ISBN: 978–0–415–54815–1 (pbk)
ISBN: 978–0–203–84224–9 (ebk)

Contents

List of figures

List of tables

List of images

Notes on contributors

Karen Bakker is Associate Professor and Director of the Program on Water Governance, University of British Columbia, Canada.

João Biehl is Susan Dod Brown Professor of Anthroplogy and Woodrow Wilson School Faculty Associate at Princeton University. He is also the Co-Director of Princeton's Program in Global Health and Health Policy.

Gavin Bridge is Reader in Economic Geography in the School of Environment and Development, University of Manchester, UK.

Bruce Braun teaches in the Department of Geography at the University of Minnesota, Minneapolis, USA.

Adam Bumpus is a Postdoctoral Research Fellow at the Centre for Sustainability and Social Innovation at the University of British Columbia, Canada.

Sally Eden is Reader in Geography at the University of Hull, UK.

Jody Emel is Professor of Geography and Director of the Graduate School of Geography at Clark University, USA.

D. Asher Ghertner is Lecturer in Human Geography in the Department of Geography and Environment at the London School of Economics.

Julie Guthman is Associate Professor of Community Studies at the University of California, Santa Cruz, USA.

Leigh Johnson is a doctoral candidate in the Department of Geography at the University of California, Berkeley, USA.

Jake Kosek is Assistant Professor, Department of Geography, University of California, Berkeley, USA.

Mazen Labban is Assistant Professor of Geography at the University of Miami, Florida, USA.

Diana Liverman is Professor of Geography and Development and Co-Director of the Institute of the Environment at the University of Arizona, USA.

Becky Mansfield is Associate Professor of Geography at the Ohio State University, USA.

Joseph Masco teaches in the Department of Anthropology at the University of Chicago, USA.

Lyla Mehta is a Research Fellow with the Institute of Development Studies at the University of Sussex, UK, and is also a Professor II at Noragric, Norway.

Kristin Mercer is Assistant Professor at the Department of Horticulture and Crop Science, Ohio State University, USA.

Sarah Moore is Assistant Professor in the School of Geography and Development at the University of Arizona, USA.

Harvey Neo is Assistant Professor at the Department of Geography, National University of Singapore.

Richard Peet is Professor of Geography in the Graduate School of Geography, Clark University, USA.

Nancy Peluso is Professor of Society and Environment in the Department of Environmental Science, Policy, and Management, University of California, Berkeley.

Paul Robbins is Professor in the School of Geography and Development at the University of Arizona, USA.

Peter Vandergeest is a member of the Department of Sociology at York University in Canada.

Joel Wainwright is Assistant Professor in the Department of Geography at the Ohio State University, USA.

Michael Watts is Chancellor's Professor in the Department of Geography and Director of African Studies, University of California, Berkeley, USA.

Preface

The book was put together as the United Nations Climate Change Conference was taking place in Copenhagen in December 2009. These were sad days of utter failure even to reach an ineffectual accord on slightly restricting carbon emissions. They brought the realization to us that many of the more pessimistic conclusions emerging from the field of political ecology over the last few decades were more the case than even we had thought. That environmental destruction was endemic to "liberal democracy" was not a revelation, therefore, but the possibility that rationality would prevail before environmental catastrophe claimed its many, usually poor, victims came to feel all the more remote. As the conference moved towards its inevitable failure, the idea dawned on us again that the existing political structure is incapable of solving the drastic problems caused by the underlying economic system with its over-consumptive way of life. The existing system is not only corrupt, it is also dangerously ineffective – incapable of effectively discussing, let alone solving, environmental problems that interact into crisis.

On the other hand, there is always a core of hope underlying any radical or progressive politics. For every piece of evidence for the expansive impulses of destruction that prevail in the world economy, there are countless cases of surprise, emerging worldwide possibilities, and new forms of ecology, economy, and community, ranging from squatters gardening in the brownfields of urban Kenya, to socially organized anti-toxins crusaders in Eastern Europe, to community sponsored agriculture sprouting across the United States. To make better room for these political ecologies of *the possible,* it remains essential to sort through the causes of environmental crises and clearly evaluate the kinds of political-economic transformation necessary for reaching ecological sanity. The authors assembled here follow an urge to criticize, in order to re-think and organize for a rational, sane, equitable society capable of non-destructive environmental relations. Hope amidst sobering challenge is the guiding theme of this book.

The authors would like to acknowledge help with the production of this book. The photograph that opens Chapter 1 is reproduced courtesy of Associated Press. Chapter 18 appears courtesy of Sage Publications.

Richard Peet thanks his students at Clark University for their enthusiastic and politically dedicated support over the last 40 years. And his family, Anna Peet,

Eric Peet, Lukas Klapatch, and James Peet, but especially Elaine Hartwick for her loving help.

Paul Robbins would like to thank student members of "The Collective," past and present, for providing the best ideas of the last many years; the School of Geography and Development at the University of Arizona, for providing a safe working space for critical science; and Sarah Moore, Sallie Marston, and J.P. Jones for intellectual partnership.

Michael Watts would like to thank Dana Gerber for research assistance; the Class of 63 endowment at the University of California, Berkeley; the Berkeley Workshop on Environmental Politics and its motley crew; and the support of the UC Berkeley Luce Foundation Project on Green Governance. He would also like to express his deep and abiding love to Mary Beth, Nan, and Ethan.

Richard Peet, Worcester
Paul Robbins, Tucson
Michael Watts, San Francisco
14 January 2010

1 Global nature

*Richard Peet, Paul Robbins,
and Michael Watts*

Introduction: global warming as paradigm

It is a striking image. A global capitalist whose personal wealth is rooted in an industry, air transportation, distinguished by its massive carbon footprint, and a Nobel prize winning US politician and former Vice-President, honored for his contributions in placing global climate change, and the scientific work of the Intergovernmental Panel on Climate Change (IPCC) in particular, on the global political agenda. Tossing the globe into the air, British tycoon Sir Richard Branson announced to the world in 2007 that he was offering a $25 million prize for the scientist who discovers a way of extracting greenhouse gases from the atmosphere

Image 1.1 Sir Richard Branson and Al Gore

– a challenge to find the world's first viable design to capture and remove carbon dioxide from the air. Big Science meets Big Business meets Big Politics. But the prize – known as the Virgin Earth Challenge – was immediately attacked by a leading climate scientist, Kevin Anderson, of the Tyndall Centre for Climate Change Research at Manchester University, who offered the following assessment of Sir Richard's philanthropy: "He's misguided, misinformed and potentially quite dangerous in making people think there is some great technological hope out there." Sir Richard, accused of rank hypocrisy for creating a prize based on the profits of a firm and an industry responsible for massive carbon releases, replied: "I could ground my airline today, but British Airways would simply take its place" (*The Guardian* February 7th 2007; http://www.guardian.co.uk/environment/2007/feb/ 10/theairlineindustry.climatechange). Well, as a Berkeley bumper stick it has it: "At least the war on the environment is going well."

The photograph is above all a *planetary image*, in its own way a bookend to the famous NASA planet earth photograph AS17–22727 taken during the final Apollo mission in 1972. It is a picturing, or rendering, of a certain sort of global nature, global politics and global science all at once. If the NASA image came to be the lodestar for the United Nations Convention on the Human Environment held in Stockholm in 1972, perhaps the Branson-Gore photography captured perfectly the sentiments of the December 2009 UN Climate Conference in Copenhagen (COP15). Copenhagen was obviously not the first global forum in which big science, big politics, and big business have joined forces to address the conundrum of growth without limits and capitalism's massive material wastes and detritus – the "externalities" associated with converting the land, ocean and atmosphere into a global dumping ground. But the invocation of planet earth and 1960s crisis thinking about the environment in the run up to Copenhagen is historically resonant. Released in 1972 in the same year as the Stockholm Earth Summit, the famous *Limits to Growth* report – penned by a quartet of MIT physicists, cyberneticians and business management theorists – represented the apotheosis of a form of crisis thinking driven by a deep Malthusianism. On offer was a powerful discourse offering the prospect of chaos and collapse rooted in demographically driven scarcity (the five key sub systems calibrated in their World3 computer model were world population, industrialization, pollution, food production and resource depletion).[1] The global modeling exercise of *Limits to Growth* proved to be flawed in all sorts of ways but with the vantage of hindsight we can now see that it was prescient. In genealogical terms, the sort of "limits modeling" of the 1960s and early 1970s reappears in the general atmospheric circulation models (GCMs) of the 1990s. As they gained standing and analytical power, the new wave of global climate change models, without which there would have been no Montreal or Kyoto Protocols or COP15, were draped in the language of crisis and apocalypse. As Iain Boal put it "at COP15 it would be fair to say that versions of a secularised neo-catastrophism will be the dominant paradigm among climate scientists and laity alike" (2009: 3).

Implicit in the science behind the global climate change debate – there are after all doubters and legitimate scientific differences which have doubtless been exaggerated in the popular imagination by the release of the now famous e-mails from University of East Anglia climate scientists – is a worldview somewhat at odds with the Darwinian orthodoxy of evolutionary gradualism (Weart 2004; Boal 2009). Climate could, and did of course, change historically, but for human occupation and livelihood this represented a deep historical time – the very *longue durée*. On offer now is something unimaginable until recently, namely abrupt and radical shifts. It is a science of planetary disaster demanding a response – political, policy, civic and business – of an equal and opposite magnitude and gravity. Here is Al Gore on the matter: "What we are facing is a planetary emergency. So some things you would never consider otherwise, it makes sense to consider." We heard this same rhetoric in the wake of 9/11. What might the planetary ecological crisis entail?

> For some, therefore, it means that a war on global warming must be declared, quite as draconian as the global war on terror. Are we not faced with inhabiting – once again – the rubble of a ruined world? For others, typically of a social democratic cast of mind, it means pinning hopes on human adaptability and resilience in the face of melting glaciers, the end of irrigated agriculture and a return to dry farming. For the governments, green NGOs, and those others with seats at the table hoping for a leaner, low-fat capitalism, it means nego-tiating some version of the neo-liberal deal. That is, haggling over the further commodification of the earth and its productions – vegetable, mineral and animal – and legislating limits and rights to pollute, to trade toxins, to crank up derivatives markets recently vilified as a sure sign of the excesses of casino capitalism.
>
> (Boal 2009: 5)

In a discursive sense, then, climate change as a planetary emergency mobilizes powerful actors around the threat of massive risks and uncertainties. It is rather like the War on Terror, Ebola or nuclear weaponry and is fully consistent with what has been called a "culture of fear" (see Glassner 2000). Planetary challenges, however they are assessed and weighed empirically, are capable of eliciting very different responses. Climate change after all could entail a serious and multi-lateral push toward a zero-carbon economy or a privatized and corporate push to synthetic chemistry, "clean fuel" and nuclear energy.

Global climate change – as science, policy and politics – reveals starkly the sorts of problems that a global political ecology – the subject matter of this book – must confront. One can start with IPCC itself as a sort of transnational scientific network operating too as an advocacy group on a public landscape populated by a significant corporate (and Republican Party in the US) presence of climate change deniers. The scientific consensus is that humans have changed the chemistry of the earth's atmosphere, primarily by altering the concentration of CO_2 from pre-industrial levels of 280 parts per million to its current (and rising) level of over 400 (we discuss this at greater length later). But the very idea of human-induced climate change was contested from the very moment, in the 1980s, when it became a respectable matter of science. Oreskes and Conway (2008) have shown how the

Marshall Institute (MI) in Washington DC played a key role in the denial industry long before ExxonMobil and other oil companies, and indeed the George W. Bush administration, joined the denier fray. Populated by a group of retired physicists, the MI was an archetypical Cold War think tank devoted to what they saw as exposing scientific uncertainty and skepticism. They cut their teeth on Reagan's Strategic Defense Initiative (SDI) and what they saw as unprincipled *scientific* opposition to it. From the 1980s onward MI was a powerful voice (with robust Republican Party connections) in denying a raft of "uncertainties": that smoking causes cancer, that pollution causes acid rain, that CFCs destroy ozone, and that green gas emissions cause global warming. Behind this was the view that all scientific knowledge revealing alleged ecological or health costs was in the service of central planning and socialism! One of MI's founders, Fred Singer, articulated the view that behind the scientific work for global warming lay a "hidden political agenda" against "the free market. . .capitalistic system" (quoted in Oreskes and Conway 2008: 77). Lahsen (2004) has suggested that the science of global climate change denial more generally was rooted in the "paranoid style" (the term is from Richard Hofstader) of American politics: science and environmentalism were out to get market fundamentalism. In a sense they were right of course. Capitalism would *have* to change if it were to seriously address its own impact on the planet, something that institutions like the Marshall Institute could never accept.

The production of particular sorts of knowledge to discredit scientific orthodoxy speaks to not only questions of how environmental knowledge is produced and legitimated, but also to what Robert Proctor and Lnda Schiebinger (2008) call "agnotology," namely the willful production of ignorance and scientific ambiguity. One part of this story has to do with the extent to which corporate capital not only represent themselves as particular sorts of actors. We are thinking of BP's re-branding itself as "Beyond Petroleum" or Chevron's media barrage on the company's role in the clean energy transition. But also the extent to which they have their own in-house science – both sponsored research of the sort undertaken by the tobacco companies in their infamous denial of the links between smoking and cancer, and in-house corporate research programs of their own, as in the case of risk and reinsurance industries financing their own climate modeling on hurricane risks. What sort of knowledge is produced, in other words, and its legitimacy and authority, are central to the ways in which global environmental problems become, or do not become, "problems" and how they are construed and composed. How transnational scientific networks produce consensus amidst such scientific and popular contention – how epistemic communities (Haas 1992) are created, sustained and mobilized – is central to the IPCC story. But for every case of corporate climate change denial there is probably an equally problematic set of epistemological questions about how science is "reframed" in speaking truth to power. The disclosures that University of East Anglia climate scientists played "tricks" in presenting their data to the public and policy makers is a case in point. In other words, it is striking not only how "knowledge has emerged as a salient theme in projects of environmental governance" (Jasanoff and Martello 2004: 336) but also how a purportedly global or universal science is at the same time a "situated knowledge" (Haraway 1989), situated with respect to power and situated with respect to local knowledge-power formations (Jasanoff and Martello 2004).

The Framework Convention of Climate Change (UNFCC) which the UN adopted at the Rio Earth Summit is now eighteen years old. Its basic mission was to achieve the stabilization of greenhouse gas concentrations in the atmosphere at a level that would prevent dangerous anthropogenic interference with the climate system. The Kyoto Protocol, signed in 1997, to come into force in 2005, was established to realize these goals. CO_2 emissions are now 30 percent higher than when the UNFCC was signed; atmospheric concentrations of CO_2 equivalents are currently 430 parts per million (Gautier 2008; IPCC 2000). At the current rate they could more than treble by the end of the century. In effect this would mean a 50 percent risk of global temperature increases of 5 degrees C (the average global temperature now for example is only 5 degrees C warmer than the last ice age: see the *Economist* December 5th 2009: 3). Kyoto was of course a failure, in large measure because of the non-signatory states like the US. It is due to expire in 2012, and implementing a new treaty is expected to take three years. Much therefore rested on the Copenhagen Conference held in December 2009. As the *Economist* put it "without a new global agreement there is not much chance of averting serious climate change" (December 9th 2009: 3). Earlier in 2009 a G8 meeting agreed that increase in global temperatures should be no more than 2 degrees C above pre-industrial levels. According to IPCC calculations, this would involve cutting global emissions to half their 1990 levels by 2050 (for richer countries, this means a reduction to 2 tons of carbon per head from levels above 10 in Europe and 24 in the US). IPCC figures suggest the global south must cut emissions by 25–40 percent by 2020 (IPCC 2000).

The prospect of a meaningful and robust COP15 agreement turned on how the US under President Obama (after eight years of resistance from the Bush administration) and China could deliver on emissions reductions, and also deliver to the global south in terms of its needs. The EU, after all, was committed to a 20 percent cut, rising to 30 percent if the rest of the world comes through with significant reductions. Two weeks before Copenhagen opened, on December 7th 2009, Obama offered a 17 percent cut on 2005 emissions by 2020 (this is the figure in the Waxman-Markey Bill, now the Clean Energy and Security Act, that passed the House of Representatives but not the Senate), well below the expected figure for the developed world. The likelihood of further concessions and dilution in the US Senate is seen to be inevitable, not least because the energy industry has been the largest spender in lobbying against the carbon caps promoted by the Bill. The prospects are made no better with the fallout from the financial crisis of 2008. There is now a new found skepticism about the extent to which a new $1 trillion market in carbon will be manipulated (carbon derivatives for example) by Wall Street. As the world's biggest emitter, China has been fighting with the US and others over reductions, but it has promised a 40–45 percent cut in carbon intensity by 2020, a figure below US expectations. Not least there is the larger problem of the needs of the global south: to honor their commitments, money must be promised (both for moral-ethical [the West has, after all, been dumping carbon for 200 years] and practical [capital shortage] reasons). China says the developed states must hand over 1 percent of GDP ($400 billion a year). According to the International Energy Agency, the 2 degrees C target will cost the world $1 trillion a year, half of which will need to be spent in the developing world. Where the money will come from, and whether commitments will be honored, is another question. Cash-strapped governments may look to various taxes and carbon pricing and this is unlikely to fill the massive monetary need. Others suggest mobilizing capital markets and the surpluses in sovereign wealth funds to invest in energy infrastructure. It all looks rather sketchy.

COP15 turned out to be a colossal failure in spite of the arrival, at the last hour, of President Obama, who was able to cobble together a loose agreement among the so-called BRIC states (Brazil, Russia, India and China). Prior to the final agreement, a leaked UN document shows that a huge gap remained between the amount of emissions cuts that nations have pledged and what is in fact needed to keep global temperature rise below 2°C – the level IPCC scientists say is a tipping point for runaway climate change. The conference fell apart dramatically: two weeks of delays, theatrics, walk-outs, and last-minute deal-making. The end result was a grudging agreement by the participants to "take note" of a pact shaped by five nations. A 12-paragraph final accord, was only a statement of intention, not a binding pledge to begin taking action on global warming. Robert C. Orr, United Nations assistant secretary general for policy and planning, said that virtually every country had signaled that it would back the accord. But, in practice, the delegates of the 193 countries that had gathered in Copenhagen departed with nothing like firm targets for mid- or long-term reductions of greenhouse gas emissions, without clear and proportionate funding, and without a deadline for concluding a binding treaty next year. The world stood on the precipice and then walked away. Not least, EU leadership was shown to be insignificant in the face of Chinese and US power.

The global climate change issue is an exemplary illustration of two new powerful discourses: what is now referred to as "global environmental governance" and "global sustainability" (as opposed to the earlier language of "environmentalism" during the 1970s). The proliferation of institutions, organizations, principles, norms and decision-making procedures – what are conventionally seen as an "international environmental regime" – is reflected in the explosive growth of inter-state treaties, on average sixteen a year since the 1972 Stockholm Conference on the Human Environment, and nineteen a year since Rio (Mitchell 2003). The treaties which developed in the wake of Stockholm focused on limiting specific sorts of pollutants (SO_2, NO_2), banning ozone depleting gases, protecting key species (regulating commercial whaling) and preserving endangered ecosystems (wetlands). As Paterson (2008: 105) points out, these forms of global governance "correspond to an era where capitalism was itself organized and governed through extensive planning, from tripartite corporatist management in many Western countries to the nationalization of many industries, and extensive multilateral management through the Bretton Woods system." But this model was to shift. By the 1980s the Brundtland Report (*Our Common Future*) enshrined sustainability as a political discourse which sought to address both the growing North–South conflicts (the relations between poverty and global sustainability) and the growing counter-revolution by firms and states to the idea of regulation, that is to say to attempt to install an environmental governance "compatible" with no limits to growth (the pre-condition of neoliberal capitalism).

Remnants of "managed" or "organized capitalism" – that is to say remnants of a Keynesian project – did endure. But by the 1990s, on the back of a major neoliberal push in the US, UK, Germany and through the multilateral development institutions like the IBRD and IMF, these relics rapidly disappeared. Deregulated markets, privatization of state owned industries, the deregulation of exchange rates and financial flows, assaults on corporate taxes and the welfare state were all buttressed by the power of global regulatory institutions and international agreements praying at the altar of free trade and robust property rights (for example the WTO, TRIPS, GATS). In a radically new ideological environment – variously dubbed Thatcherism Reaganism, Kohlism – environmental problems were now subject to the implacable logic of the markets, prices and capital flows. As Paterson put it (2008:107): "global environmental governance . . .became increasingly guided by an imperative less to *organize* and directly *manage* capitalism to pursue sustainability than to enable private sector actors to pursue their economic interests in ways which simultaneously promote sustainability." What began with state-directed command and control ended with governments creating markets for environmental goods and services and in subsidizing (that is to say providing) new incentives for green industries. Much of this new model of green governance was wrapped up with voluntary agreements for new practices and standards set by industry for itself (for example the ISO 14000 series on environmental management standards), and by rafts of international environmental NGOs, think tanks and foundations following the money into new forms of market-based green governance.

Global climate change has become, in other words, a theater for "governance through markets": government provides incentives and subsidies, and corporations establish their own (voluntary) standards. Global climate change policy and struggles over its shape and form must be rooted, then, in a very specific set of political economic changes over the last four decades, and in specific capitalist order (Bernstein 2001; Paterson 2008; Heynen *et al.* 2007). 2010 looks, in this regard, very different from 1970. Naomi Klein (2007) and George Monbiot (2006) make the point that in this ascendant neoliberal order – it is not at all clear in this regard that the financial crisis of 2008–9 has made a serious dent in neoliberalism's armor – even environmental calamity and reconstruction is a source of corporate profit and capitalist consolidation (so-called "disaster capitalism"). At the very least, in the shift from environmentalism to sustainability, Tim O'Riordan is surely right when he says that "we are nowhere near a business model for sustainability" (2008: 319).

The history of climate change governance has been, one might say, the success of global science knowledge production and knowledge mobilization but the uneven record of national and multilateral policy (and in some cases a total failure). Climate change *mitigation* was the touchstone of the IPCC findings, and this meant limiting emissions through three instruments: regulation, carbon pricing, and subsidies

(Gautier 2008). The entire debate – both in terms of assessing costs and policy solutions – has been dominated by economics (which is to say an internal debate within the profession about costs and benefits of climate change and its mitigation, and the deployments of markets and prices to ameliorate the inter-generation effects of global climate change). There are those economists – most famously the Copenhagen Consensus led by the infamous environmental skeptic Bjorn Lomborg (see http://www.lomborg.com/cool_it/) – which refutes the notion of any serious costs associated with climate change, but the conventional argument really resides in how the serious costs are calculated: which is to say how climate change and its environmental and economic impacts are to be valued and costed, and how prices are assigned over time in a cost-benefit framework. Virtually all of the climate change policy debate is then about the process (the means, methods and techniques) of commodifying nature, and creating markets in those parts of climates outputs (for example carbon) that can trade our way out of catastrophe.

Most economic analyses begin with the likeliest outcome – the apex of the probability curve – which is usually taken as the IPCC position: 2.8 degrees C over the next 100 years. A central question then becomes the discount rate: the rate at which future benefits and costs are discounted. The Stern Review (2006) used a rate close to zero. William Nordhaus of Yale University (http://nordhaus. econ.yale.edu/dice_mss_072407_all.pdf) argues for a 3 percent discount rate commensurate with "today's marketplace real estate and savings rates" which implies that benefits accrued in 25 years are worth half of their current value. Marvin Weitzman of Harvard (http://papers.ssrn.com/sol3/papers.cfm?abstract_id = 992873) argues that we need an insurance policy (2 percent of GDP per year) to insure against less likely threats (surprises in other words) but which cannot be ruled out. Others have suggested Stern's figure (roughly 1 percent of GDP) is much too low. The point is that the process and logic of costing, discounting and valuing – the heart of the economics of climate change – is contested and, perhaps appropriately, uncertain and risky.

In the current neoliberal order, the short term future depends, whatever the precise costing and price structure, upon the extent to the energy companies will drive the shift to a low-carbon economy, and the prevailing carbon prices (set through taxes or cap and trade). In the wake of the financial crisis and the boom and bust in oil prices, Big Oil, in spite of the rhetoric and media frenzy around clean technology and renewables, is moving very slowly toward low-carbon (see *The Economist* op cit., 2009). With gas prices at an all time low (and new sources of shale gas now available technologically), gas is being posed as the new "clean" fuel. Investment in clean energy in the last years has fallen catastrophically, and in spite of the availability of state subsidies and funds through various national stimulus packages (New Energy Finance, a consulting firm, estimates that green stimulus money globally so far adds up to $63 billion but only $24 billion was dispersed globally in 2009), private equity and venture capital seem uninterested. The oil majors in particular, and other energy companies, see fossil fuels as still cheap and profitable: the oil industry sees oil and gas as the source of their future for the next 35–50 years. Unless governments or consumers or both change radically in their behavior and commitments, the incentives will change little. Carbon prices look no different. The EU's Emissions Trading Scheme (ETS), which started in 2005, is the only large scale attempt to set a carbon price (EU countries receive national allocations parceled out to firms in five dirty industries). The ETS makes up the lion's share of the global market at $122 billion, but the price ($22 per ton) does not encourage much of an energy transition. In the US, Congress is proposing $12 a ton, which will not encourage any serious investment. Most experts believe that onshore wind energy needs a carbon price of $38, and solar cells of $196 per ton, respectively. Against this backdrop, the failure at Copenhagen – and the much vaunted transition to a low carbon economy – is even more telling.

The economics of global climate change *mitigation* has, in general, drowned out the need to think about what is now called climate change *adaptation*. In this arena the work of geographers and other social scientists has been central, raising the questions of how the possible burdens of climatic change, sea level rise, and possible catastrophic events are to be distributed geographically and in social and class terms (see Adger *et al.* 2009; Schipper and Burton 2009). It is already clear that the real, material burden of climate change will fall heavily on Africa (Toulmin 2009): with large numbers of poor and vulnerable rural and urban populations, deepening food insecurity, famine, degradation of livelihoods and resource-conflicts (precipitated by climate-induced reconfigurations in the resource base) are seen to be in the continent's immediate future. This is partly why the African delegations were so demonstrative at the Copenhagen meetings about resources being made available in order to honor their commitments to a problem they did little to create.

On its face the research on adaptation to climate change seems to build on insights drawn from political ecology (see Neuman 2004; Robbins 2004) – in particular that one must start from patterns of social, economic and political vulnerability of the poor and the sorts of entitlements they have to control and gain access to resources and ecosystem services. Yet so much of this work in practice is a recycling of an older sort of cultural ecology – systems theory dressed up as new institutionalism – in which there is much talk of adaptive capacity, resiliency and flexibility of local social systems, but almost no serious account of political economy and the operations of power. The best of this (for example, Agrawal and Perrin 2009) inventories the insights of political ecological work on agrarian societies in showing how rural communities adapt to climate change through mobility, storage, diversification, communal pooling and exchange by drawing upon social networks and their access to resources. An important illustration of green governance from below is the work published recently (2009) in *Proceedings of the National Academy of Sciences* in which Chhatre and Agrawal show across a large number of case studies how communities can often manage forest more effectively that either the market or the state (http://www.pnas.org/content/106/42/17667.abstract).

In the first study of its kind, which tracked the fate of 80 forests worldwide in 10 countries across Asia, Africa, and Latin America, over 15 years and under differing models of ownership and management, Chhatre and Agrawal of the University of Michigan conclude that "locals would also make a better job of managing common pastures, coastal fisheries and water supplies" (http://www. pnas.org/content/106/42/17667.abstract). Carbon storage potential, it turns out, especially improve when community organizations and their institutions incorporate local knowledge and decentralize decision making to restrict their consumption of forest product. All of this drives home the extent to which key non-state actors – communities, households, civic groups, NGOs, think tanks and social movements – have not only created a cacophony of voices and counter-discourses to the spectacular United Nations events, but provided an important if sometimes disorganized and unruly political counterweight to the forces of government and capital (see Hochstetler and Keck 2007 on networks; on climate change see http://climateactioncafe.wordpress.com/2008/10/28/towards-radical-critique-and-action-on-climate-change-politics-and-copenhagen-2009/).

But generally, in this vast industry of work on adaptation to climate change, critical social science, and hard edged political economy, is strikingly absent. The rough and tumble of actual struggles and the relations between households, communities and power state and corporate agents is missing. Instead, on offer, is a shopping list of "conditions" for adaptive governance, including "policy will," "coordination of stakeholders," "science," "common goals" and "creativity" (see for example Cole and O'Riordan 2008) rather than the complex political, cultural and social dynamics at work – that is to say what political ecology has stood for.

* * * * *

We have begun *Global Political Ecology* with this extended discussion of the political ecology of global warming because it is seems to us to embody, sub-stantively and theoretically, what is centrally at issue in this book and in the world of environmental analysis more generally. The first is the *planetary character of the ecological crisis*. A global sense of nature is not new of course – the emergence of a powerful environmental movement in the 1960s was closely tied to a sense of the future and fragility of the planet earth tied to a deep sense of Malthusian scarcity. But climate change has asserted the inescapability of processes which profoundly compromise the ability of capitalism to reproduce its conditions of production.

Second, and relatedly, climate change debate affirms *the centrality of expert knowledges* (and discourses) in giving shape to the definition of problems and solutions, and of the indisputable significance of transnational scientific mobilization coupled to a historically new raft of actors, norms, conventions and treaties – a sort of liberal international green regime – which provides a ground on which intense political struggles are now being waged (see Goodman *et al.* 2008). The very idea of expertise is, however, a contested issue, and there is a profusion of forms of situated knowledges which can, and should, be seen as "counter-discourses" – corporate deniers for example, or activist lawyers like Climate Justice (http://www. climatelaw.org/) – to the conventional theory and practice.

Third, the question of carbon emissions *roots global problems in the material world of basic provisioning systems, and in the energetic foundations of modernity itself*. This is certainly about patterns of dependency upon fossil fuels and the viability of hydro-carbon civilization, but it implicates directly the political economy of carbon-capitalism and the corporate and political power attached to control over and access to sources of energy. Regimes of global governance – that is to say states, community institutions, civil society groups, social movements, firms and multilateral organizations – must be understood as products of particular sorts of capitalist social orders. At this moment in history global nature confronts an assertive neoliberal capitalism which is simultaneously destroying (existing) and creating (new) commons through complex processes of dispossession, annihilation and creative destruction (Harvey 2005). All of this, at present, is shaped by a deep crisis of finance capital, a global recession and the unstable hegemonic powers exercised by the US, the EU and China.

Finally, global climate change reveals the ways in which global and local knowledge-power formations call attention to a very particular sort of global order, and *a particular sort of environmental rule or governmentality* (what some have called "environmentality" – see Li 2007). Hardt and Negri (2000, 2009) call this an imperial machine and they invoke through the work of Michel Foucault new kinds of sovereignty exercised through powerful systems for the identification, classification, and organization of knowledge and persons. In this sense global nature must be construed in terms of empire and new forms of sovereignty that mark a passage from "disciplinary society" to a "society of control." If disciplinary society turns on diffuse networks of social command that produce and regulate customs, habits and practices (sanctioning and prescribing the normal and the deviant), then the society of control is an intensification and generalization of disciplinary apparatuses which internally animate practices through and well beyond social institutions (Hardt and Negri 2009: 55). In this sense, the society of control operates through what Foucault (2008) calls biopower, the power to administer and produce life through the government of populations. Life, says Foucault, has become an object of power. The idea of green global governance to preserve life in the face of climate change or energy insecurity is an exemplary case of biopower at work. Green biopower or green governmentality is both a way of reproducing particular subjectivities (the subjects created through the expert knowledge and practice of sustainable governance), but also, as Hardt and Negri (2009: 57) show, of generating "life as resistance, of another power of life that strives toward another existence." One of the main purposes of this book is to show what sorts of subjectivities and practices have been produced at the intersection of neoliberal capitalism and sustainable development (what has been called "environmental neoliberalism" [see Heynen *et al.* 2007]) and what are the sites and sources of resistance to them.

We have organized the book around several broad substantive themes which throw into broad relief the sorts of questions – about science, knowledge, power, discourse, states, markets, political and social movements – that a critical political ecology must attend to. The first addresses the intersection of food, health and the body through detailed case studies of obesity, the livestock industry, and the marine crisis. The second we call capital's margins, and it addresses the political ecology of the ultra-poor and disenfranchised, in this case the slumworld of New Delhi, the social and ecological life of trash, and massive human displacement through dams. Our third theme addresses global governance particularly in regard to climate change by examining carbon trading and sequestration, the risk and insurance industries and climate change, and the politics of certification and "greenstamping." Another section turns to the intersection of ecology and security. Here the focus is on the relations between nature and militarism, the role of forests in insurgency, mutant ecologies in the US southwest, and finally the energy security. Finally several papers address the implications of the bio-molecular revolution for both the production of new natures and for the sorts of ways in which the genetic revolution is shaping both new sorts of rule and governance and new sorts of subjects.

How then can we understand the gravity of the current multiple and interlinked environmental crisis in substantive and theoretical terms? This question must address the relations between global nature and a global capitalism defined by what Perry Anderson (2002) has dubbed a "neoliberal grand slam." The "fluent vision" of the Right has no equivalent on the Left he concluded, and embedded liberalism is, he says, as remote as "Arian bishops"; neoliberalism rules undivided across the globe and is the most successful ideology in world history. How is this neoliberal hegemony related to the current ecological crisis? A second question must address the theoretical and conceptual toolkit that a critical political ecology can provide in both explicating the dynamics of socio-ecological systems and the sorts of changes and practices required if sustainable development is to be more than a pipedream. It is to these questions we now turn.

Global capital, global nature

As our *tour d'horizon* of global climate change revealed, the world is firmly caught – some would say trapped – in the mesh of multiple and interlinked environmental crises. The centrality of environment and sustainability talk in the new millennium, seemingly given a new credibility by the ascendancy of Barack Obama and his administration, immediately draws us back to the late 1960s, and the explosion of books predicting environmental collapse and systemic crisis, and the emergence of varied green politics – legislation, social movement, transnational institutions – focused on a new awareness of planet earth. With the power of hindsight we can now see that attempts at resolving the deep basis of the crises have been superficial and disorganized, and some profound environmental problems (global climate change among them) were either ignored or deferred in the face of the necessity, under capitalism, of continued economic growth. Once again then, propelled forward by the realities of massive changes in the world's climate, the talk of crisis, extinction, radical vulnerability and so on are back with us, and on the political agenda.

As a starting point, this book poses the question: what has happened since the 1960s that renders this moment distinctive and different? As we have seen, current ecological challenges stand in the wake of momentous technological and scientific changes, three decades of neoliberal economic and social policies, a massive global expansion of capitalist accumulation marked by deepening international and sub-national inequalities, and one of the deepest economic recessions in the history of modern capitalism that now render agreement by the "international community" virtually impossible. Second, what are the contours and structures of contemporary ecological threats – and the particular ways in which global and local biophysical and political economic processes interact – that must now be confronted? What are the political and other forms of institutions on offer to address the assault on the global and local commons? What, in relation to the dire history of the last four decades, is actually on offer in terms of clean or green technologies, green governance, green capitalism, and alternative fuels and energy sources? And not least, what is the conceptual and theoretical toolkit available to us to better grasp both the dynamics of the political ecological problems themselves and the forms of rule and systems of governance capable of addressing them?

Capitalism and alienation from nature

Waves of multiple global environmental crises break with particular ferocity on the shores of the popular imagination: destruction of the rain forests, the disappearance of species, pollution from carbon dioxide emissions, melting of the polar ice caps, the poisoning of the seas, the return of nuclear proliferation, global pandemics, massive oil leaks, and the threats of genetically modified organisms are regular staples of the mainstream media. As we have said, realizations about specific crises cohere into a singular acknowledgment that there is a universal environmental crisis, with the potential to become catastrophic: climate change. Most concerned people share these environmental concerns. But there are problems with such moments of realization. Environmental catastrophe "will" occur "sometime in the future," unless "we do something about it." What to do about the emerging crisis is but vaguely mentioned, and the collective "we" doing that "something" is even more cloudy and indeterminate. So, continual moments of realization take the form of perpetual periods of evasion – the history of near-accords recorded above. The basic causal issues – what are often called "driving forces" in the conventional literature of human dimensions of global environmental change – are rarely confronted, and when they are, they seem to fall back onto the old faithfuls: population growth, technological change, over-consumption or bad policy. We have so far been looking mainly at the discourse of environmental change. Now we want to deepen this into what used to be called "the basic, structural causes." And for this we turn to . . . Marx, Karl that is.

In his early work, influenced by Hegelian idealism (Marx 1959), and again later in *Capital* (Marx 1967, Volume 1), Marx speaks of a characteristic he calls "alienation," that he finds fundamental to capitalist culture. As capitalism develops, Marx argues, increasing socialization binds workers into a more extensive labor process that they do not collectively control. The socialized labor process loses its inherent meaning as: the social production of human existence, through the collective transformation of nature. The result is a severing of relations: among workers, and between workers and capitalist owners of production systems; between the individual and its species being; between producers and their products; and between producers and the environment on which continued existence nevertheless depends (Ollman 1976: Chapter 8). Alienation extends to a separation between living for the moment, and continuing to exist in the future. Alienation takes the social form, in capitalism, of competitive individuals each pursuing self-interested goals, yet with the whole economy, as optimistically assumed by Adam Smith (1937), acting rationally, as though guided by an "invisible hand." There is a dense and somewhat obscure passage in Marx's (1973: 243–245) *Grundrisse* criticizing this supposed Smithian elevation of competitive selfishness into a higher order of the common interest: Marx thinks instead that the common interest "proceeds as it were behind the back of these self-reflected particular interests, behind the back of one individual's interest in opposition to that of another."

More than this, the "common interest" decided in this selfish way becomes an alienated force controlling individuals rather than being controlled by them, so that they are forced by competition to do things they already know to be socially and environmentally destructive. And the price system, that supposedly signals the costs and consequences of production and consumption, works only for a limited part of the content of commodities, mainly the labor content and capital investment. Market prices do not represent social and environmental costs and long-term consequences at all. As a result, market systems are environmentally destructive and socially irresponsible. Yet, the most sophisticated, liberal political economists see carbon trading as the most compelling solution to climate warming – as though pricing and commodifying carbon can solve what commodity markets created in the first place. And further, extending the notion of separation, competition in markets makes short-run economic survival so difficult to secure that long-term care of the environment becomes a utopian luxury. Marx's insights into a capitalist system then unfolding have potential for understanding a capitalism we know only too well today. The theory of alienation and the critique of the market allow us to understand that something scarcely credible might indeed be happening: "normal" production and consumption destroy the natural environment, historical origin and material source of human existence, to the point of its collapse.

If, as Marx thought, alienated production renders human existence essentially meaningless, then still life must retain enough immediate purpose for human activity to continue. Marx himself examined this "meaningless purpose" in the following way: religion, he said, is the heart of a heartless world, the soul of soulless conditions; "It is the opium of the people" (Marx 1977). Re-phrasing this, in terms of environmental consciousness: the agony of destroying nature is relieved through environmental sanctimony – crying out against hurting a spiritualized Earth, prayers to Mother Nature. Yet renewing the ancient deification of Nature proves insufficient for providing meaning in a partly, intermittently secularized world. Production itself has to have some kind of meaning if its agents are to continue destroying the environment, every day. For its owners, the capitalist production system destroys the environment with immediate purpose: every moment of production, each product made, or service performed, has the clear intent of making profit. So nature is destroyed in the prior interest of profit. And profit means power, in its multiple forms of control over other people, its endowing high social status on entrepreneurs, its provision of an over-abundance of property on the "deserving few" – not much in the long-run history of meaning, perhaps, but enough to keep them trying hard right now. Then too, the making of profit has long had a conscience in the form of philanthropy, especially in Calvinist cultures – "environmental defense funds" and "green investment" in this case. Ownership has its rewards, especially in the purchase of sophisticated excuses, through endowed research.

It is the armies of the poor that pose the problem for the meaning of environmental destruction by Capitalist production. Why should those who do not own the system, nor benefit from it as much, collaborate in the destruction of nature? Well, there is the sheer necessity of labor to consider – the brute fact that work has to be performed for others, under conditions not of the worker's choosing, for a wage to be earned, and families supported – what the Italian Marxist Antonio Gramsci (1971) would call the domination of life by the real necessity of production. Yet in the "advanced societies," where damage to environment reaches its highest amounts, domination by utter necessity (fulfilling real material needs) fades in significance before the real purpose of production. For in these societies there is the even-more compelling necessity: that everyone must vastly over-consume to support the over-production that keeps economies growing – what Gramsci (1971) called "hegemonic control" over consciousness resulting from the imposition of over-consumption. Domination engenders a work ethic; hegemony entails a compulsion to consume; and each has a somewhat distinct environmental ethic. Under domination, production retains its link with necessity, and the mass realization of economic and environmental crisis remains possible, yet is masked by the drive for development. Under hegemony, production is artificially linked with necessity, and mass realizations of the real causes of anything, including environmental crises, are lost to the pleasure principle. Under hegemony, the pleasure of consumption substitutes for the meaningless of existence, yet again provides enough in the way of immediate excuse for ruining nature. Or, referring back to the notion of short-term memory, and driving this further, one of over-consumption's many delights might be the rush to consume like crazy while time remains.

In sum, analyzing the nexus of production and consumption, in its modern capitalist guise, to seems to us to be the indispensable starting point for understanding the basic causes of the destruction of the global environment, in terms of physical, material causes, in terms of legitimating beliefs, and in terms at last of forming real, fundamental solutions. Not that the "actually existing socialism" did any better. As Judith Shapiro (2001) shows for Maoist China, the traditional Chinese ideal of "harmony between heaven and humans" was abrogated in favor of Mao's insistence that "Man Must Conquer Nature" often with disastrous consequences for people and the natural environment.

The social production of environmental destruction

The destruction of nature is not primarily an ethical issue that can be cured through moral resolve to live simpler and re-cycle more. It does not begin in discourse, either. Instead the ecology of destruction results from an alienated form of the production of human existence, one that is not democratically controlled, that is organized indirectly through markets, that is based in the self-interested pursuit of profit, and that has to grow to survive. If we want to understand what is happening to the environment, we have to understand the origins, development, structure and dynamics of capitalism: its systematic imperatives.[2]

How can we grasp the contours of these imperatives? Briefly, and again following Marx (1967) and Marxist theory, capitalism is a system in which capital is invested to buy waged labor for the purpose of making commodities for sale in markets. In Marx's version of the classical theory of value, labor has the power to produce not only valuable commodities or services, for which it receives a wage, but also a surplus of value over and above the value already expended in the creation of the laborer. When owners of money (capitalists) control the conditions under which labor makes commodities, by controlling factories, offices, etc. (the means of production), surplus value can be expropriated (or taken) from the real producers of value (human workers), to form the profit that is the real purpose of capitalist production – Marx calls this "exploitation." Yet, under market conditions, any individual capitalist has to produce commodities at prices regulated by inter-capitalist competition. This forces even the environmentally, socially concerned capitalist to produce at the lowest cost, regardless of the "external" consequences. Capitalist development is an utterly contradictory, violent process essentially because of the contradictory nature of its defining social relations – exploitation and competition. Marx conceptualized capitalist development as socially unjust (the benefits were unevenly distributed), geographically and temporally uneven (occurring more at some places and some times than others), expansionary (invading and controlling societies all over the world) and full of crises (recessions and depressions) periodically necessary for restoring the conditions of profitability destroyed by fierce competition (Marx 1967; Harvey 1982; Becker 1977; Weeks 1981).

This exploitative, competitive system originates from the destruction of older social systems, feudalism immediately in Western Europe, and the gradual formation, during historical processes full of class struggles, of a new kind of production system – hence early capitalism witnessed two class struggles, capitalist versus feudal nobility, and capitalist versus peasants and workers. In particular, these struggles involve removing peasants from ownership of land, or rights to their own means of production, ejecting them into the labor market as owners merely of their own persons ("individual liberty"), and forcing them, by threat of starvation, to sell their value-creating capacity to capital for a wage. Similar processes of "accumulation through dispossession" accompany the spread of capitalism to this day (Harvey 2003). Rather than the early capitalists coming from the feudal ruling class, they were originally commercial farmers or small manufacturers (artisans, craftspersons), investing their savings, but also borrowing capital derived from "primitive accumulation" on a world scale, that is to say through the exploitation of societies dominated by historically changing forms of imperialism. In the work of Jason Moore (2000) ecological crisis originated in the sixteenth-century transition to capitalism in a "metabolic rift" – a progressively deepening rupture in the nutrient cycling between the country and the city. While initially the capitalist manufacturing economy might be thought to merely extend the earlier natural economy, in the sense that textile machines were made of wood, the energy used to drive them was provided by water power, and the cotton and wool raw materials came from a largely un-mechanized agriculture, drastic changes were in store for the environment with the steam engine

Industrial capitalism

We can make good use here of E.A. Wrigley's (1988; 2000) notion of change from an earlier "organic economy" to a later "energy-based mineral economy." In organic economies the ultimate source of all wealth was the land, or the conversion of the sun's energy through photosynthesis by crops and animals. Nearly all the motive power driving production was derived from organic sources – human and animal muscle, supplemented by wind and water, with heat provided by burning wood. Economic growth was conditioned by this universal dependence on organic raw materials and as Wrigley (1988: 29) argues, "organic economies were subject to negative feedback in that the very process of growth set in train changes that made further growth additionally difficult because of declining marginal returns." With the industrial revolution of the nineteenth century, growth was of a new type in a mineral-based energy economy, freed by the extensive use of coal from the limitations to growth inherent in the earlier phase. This mineral-based, energy economy was subject to positive feedback effects, in that each step made the next easier to take. The crucial point came when workers who previously used hand-held tools turned to machinery powered at first by water, and then by steam produced by burning coal, and then electricity generated by burning all kinds of fossil fuels. Many of the biggest, modern industries, Wrigley points out are freed from dependence on animal or vegetable raw materials; capital goods are constructed mainly from metal, concrete, and bricks; consumer durables are made from metal or plastics. The point for the environment, he adds, is that the supply of mineral ores, clays, oil, and coal, the raw materials from which many products are manufactured, is not unlimited, while converting such materials for human use entails expending huge quantities of energy, with polluting results. So the move away from an exclusively organic economy was a sine qua non of achieving a capacity for exponential growth and massive environmental damage (Wrigley 2000: 139).

Looking at the evidence we can see what Wrigley means. The British economy was eight times larger in 1900 than it had been in 1800, industrial production increased thirteen times, and coal production seventeen times (Mitchell 1988: 247–249; 431–432; 822). The industrial revolution, in other words, produced modern societies utterly different from anything that had existed before. This steam-powered, metal-mechanized, railroad-connected industrial revolution of the nineteenth century produced the satanic mill, smoke-belching scenes that pervade our environmental memories. So in Britain, to take but one indicator of environmental effect, carbon dioxide produced in one main form, as emissions from burning fossil fuels, increased seventeen times during the nineteenth century, from 7.3 million tons of carbon to 114.6 million tons (CDIAC 2009).

The problem with such accounts, however, is the tendency towards technological determinism in the sense that bigger resource-intensive, polluting systems were made inevitable by their superior productivity – W.W. Rostow's (1960) universal stages of growth all over again. Similar formulations of technological determinism appear in some of Marx's writings too. But as Marx's ideas matured through the excruciating task of writing *Capital*, the most thorough critical analysis of capitalism of his time and perhaps ours, there is a change in emphasis towards social relations and social struggle as main causes. So, for Marx, competition is the external, coercive force compelling capitalism towards perpetual technological revolution. As Marx and Engels said:

> The bourgeoisie cannot exist without constantly revolutionising the instruments of production, and thereby the relations of production, and with them the whole relations of society. Conservation of the old modes of production in unaltered form, was, on the contrary, the first condition of existence for all earlier industrial classes.
>
> (Marx and Engels 1969: 115)

In other words, the energy-intensive, mechanized, resource-eating, polluting industrial economy that developed in the nineteenth century came from competition among capitalists, and among capitalist economies, as with Britain, Germany and the United States. Capitalists had to mechanize in order to survive, and mechanization meant energy-intensification. Yet the practical, "efficient," competitive rationality used every day in capitalism reverses the social and environmental rationality needed to sustain continued social existence in the longer run. The difference between an historical analysis based in Marx and other accounts is the emphasis on social relations rather than technological inevitability. All this was protected, aided and abetted by the supposedly *laissez faire* liberal state, meaning in reality a state that left capitalist enterprises alone as much as possible, in terms certainly of environmental regulation, but acted on their collective behalf externally in terms of imperial expansion and colonial control. The Marxian explanation for imperialism is different from other, purely political theories. The internal contradictions of capitalism are seen as being resolved through what David Harvey (1982) calls a "spatial fix" – external expansion into societies that were converted into markets, or providers of food and raw materials, without regard for cultures, environments, or the previous economies. It also produced ecological disasters wherever it touched down (in the "New World" for example).

Fordism

Even so, there were limits on economic growth and resource use even under classical (West European) industrial capitalism. The limits to growth were not set primarily by technological or (as yet) resource limitations, but by the same competition that propelled capitalist efficiency. For "efficiency" meant limiting the wages of industrial workers, with the consequence of lack of domestic demand for the products being made, a dearth relieved intermittently by exports to an under-developing, de-industrialized, peripheralized "Third World." How this contra-diction was resolved is best described by the theories of the neo-Marxist Regulation School (Boyer 1990). The regulation school sees market forces as essentially anarchic (i.e. the catastrophic collision of millions of selfish actions "coordinated" by a hand that is invisible because it is not there) and emphasizes the complementary functions of social, cultural and political mechanisms, like collective identities, common norms and modes of calculation, in guiding continued capital accumu-lation. The regulation school theorizes society in terms of development models, their parts and transformations: regimes of accumulation (basically periods of development) describe the main production-consumption relationships; modes of regulation describe the cultural habits and institutional rules related to each period of capitalist development. Capital accumulation is stabilized by modes of regulation made up from the laws, institutions, social mores, customs and hegemonies that collectively create the institutional conditions for long-run profit-making (Lipietz 1985;1986;1987; Aglietta 1979).

What the Regulation School calls "Fordism" (a term originally coined by Gramsci) was pioneered by Henry Ford in the immediate pre-World War I years, became generalized in the US from the 1920s onwards. Ford linked two innovations: the semi-automatic assembly-line; and a doubling of the prevailing wage. The expansion in productivity from the assembly line was counterbalanced by an equally massive growth in consumption, first by well paid (and increasingly unionized) wage earners in the automobile industry, later by many other sectors of the population. Fordism consisted of domestic *mass production* with a range of institutions and policies supporting mass consumption, including stabilizing *economic policies* and *Keynesian* demand management that generated national demand and social stability; it also included a class compromise or social contract entailing job stability and wages that could comfortably support families, leading to broadly shared prosperity – rising incomes were linked to national productivity from the late 1940s to the early 1970s. This provided a social logic for capitalist development that worked well in terms of resolving the previous social limits to growth – incomes that have to be spent on consumption, and that increase with the productivity of commodity output, are fundamentally necessary to complete the virtuous circle of mass production/mass consumption. Fordism was generalized in the capitalist social formations of the center countries after World War II. Economic growth of 4 percent a year and more that lasted until the crises of the 1970s. The model Fordist economy in the United States economy (measured in terms of real GDP) was essentially the same size in the 1930s as it had been at the end of the nineteenth century, when it had already overtaken Britain to become the biggest economy in the world. It then tripled in size between 1940 and 1980 (Historical Statistics of the United States table Ca9–19) or rephrasing this in environmental terms, US emissions of carbon dioxide, that were already at the level of 500 million tons of carbon a year in 1940, more than doubled to 1980 when 1300 million tons were emitted (CDIAC, 2009). The revolution in production had already created huge industrial complexes on the landscape. The revolution in consumption swelled these huge economies into giants, gobbling resources at one end and polluting air, water and soil at the other.

Even so, resistance to Keynesian Fordism among leftist political and social movements in the 1960s and 1970s extended through the anti-Vietnam war and Civil Rights movement to broader cultural critiques of consumption and environmental destruction. Thus the first Earth Day in 1970 saw participation by 20 million people. Such was the popular mood of the time that the notion of extending Keynesian regulation into a broader framework of state intervention that included environmental management came to be held by people of just about every political persuasion. Even, by conservative parties – the US Environmental Protection Agency, formed in 1970, was signed into law by Richard Nixon, a Republican President. For a few years, amidst the rampant consumption, and perhaps because of it, the possibility seemed to exist for transformative change, in the new social movements of the First World countries, including a huge and growing environmental movement, and in radical social and political movements among Third World peoples.

As we saw in our account of global climate change, the managed capitalism and environmentalism of the 1960s and 1970s was radically refigured as the political mood shifted, drastically, suddenly, disastrously in the mid- to late 1970s. The conventional explanation is that Keynesian regulation of the economy entered into crisis characterized by stagflation – high rates of inflation coinciding with high rates of unemployment – although the solution was a new, military Keynesianism of the so-called Star Wars years of the 1980s. Rather, the mid-late 1970s witnessed a secular shift in political-economic opinion in all the capitals of the West. Business reacted to a Keynesian welfare state that they thought had gone too far: income had flowed down to the poorest people, instead of up to them; and the state had tolerated, even mollycoddled, student and worker protestors, including the early environmental movement. There were plenty of places, like the Trilateral Commission or the Business Roundtable, where these "disturbing tendencies" were discussed. But the extent of elite reaction, and its commonality of themes, indicates a broad consensus occurring through simultaneous realizations by thousands of increasingly like-minded patriotic, conservative people. Then, too, capitalism was changing. Production was re-orienting towards high-technology methods and products. Globalization increased the intensity of competition. Finance capital was on the ascendancy – no longer outdated notions of investments made for life in trusted, established companies, but more investments made for a few days, maybe a few minutes, even a few seconds, in activities that once had been the purview of disreputable gamblers. The late 1970s and early 1980s counter-revolution made rightist commitment not only acceptable, but even necessary for policy formation – it took a right-wing intellect to formulate a right-wing policy. The counter-revolution positioned hundreds of think tanks at the center of policy formation. Think tanks have remained there ever since. So the Right won the interpretive war of words against the political culture of "the 1960s," and all that meant in terms of protest against war, imperialism and environmental destruction. In came politicians like Margaret Thatcher in Britain and Ronald Reagan in the US and a political economics called neoliberalism (Harvey 2005; Peet 2007).

Neoliberalism and finance capitalism

Neoliberalism revives late nineteenth-century, free-trade classical Liberalism, under the assumption that markets should rule internally and states intervene externally. Internally, Neoliberalism employs monetarist economics under the conceptual belief that macroeconomic problems, like inflation and debt, derive from excessive government spending (fiscal deficits). The nation state withdraws from macroeconomic management except in times of deep economic or political crisis. But also the notion of regulation of the economy by the state becomes anathema all over the world in what came to be known as the Washington Consensus (Williamson 1990). Instead regulatory power over economies is displaced upwards to the international institutions (IMF, World Bank, G7/8/20) within a "global community" dominated by the US, Western Europe, and Japan. "Structural adjustment" – a set of neoliberal policies forced on countries by the IMF and World Bank – re-enforces neoliberal political-economic policies everywhere. While regional variations in speed of adoption and level of commitment, persist, the neoliberal regime responded positively to the globalization of economy, society, and culture of the late twentieth century. Indeed neoliberalism helped organize globalization that benefits a newly re-emergent, super-wealthy, financial-capitalist class, mainly living in the leading Western countries, especially the US, but operating transnationally in terms of investment activity (Peet 2009). Neoliberal globalization resulted in the de-industrialization of the First World, and the industrialization of parts of the Third World – Brazil, South Korea, China, India – and therefore a huge upsurge in emissions in a spectacular globalization of environmental destruction. China's carbon dioxide emissions from burning fossil fuels amounted to 407 million tons of carbon in 1980 and 1,665 million tons in 2006; India's went from 95 million tons in 1980 to 411 million tons (CDIAC, 2009). And yet, under neoliberalism we find state regulation of development, and its relations with the environment, of diminished significance due to changing beliefs about government, markets and policies.

At the same time, capitalism changed in form towards global financial capitalism, meaning that finance is the leading fraction of capital. Finance normally operates on a global scale, and governments and global governance institutions are integral parts of that capital – so neoliberalism might more accurately be interpreted not so much as state withdrawal, as state re-direction in a kind of Keynesianism for the elite. In *A Brief History of Neoliberalism*, David Harvey (2005: 31–38) argues that ownership (shareholders) and management (CEOs especially) of capitalist enterprises have fused together as upper management is paid with stock options. Increasing the price of the stock becomes the objective of corporate operations. All this, Harvey says, is connected to a burst of activity in an increasingly unregulated, and rapidly globalized, financial sector in a process he describes as "the financialization of everything," meaning the control by finance of all other areas of the economy. Nation states, individually (as with the US) and collectively (as with the G7/8/20), have to support financial institutions and the integrity of the financial system, for that is what keeps their economies going. The tremendous economic power of the new entrepreneurial-financial-political class enables vast influence over the political process.

Essentially this kind of global power is exercised by controlling access to the biggest capital accumulations in the world and directing flows of capital in various forms – as equity purchases, bond sales, direct investment, etc. – to places and users that are approved by the financial analytic structure of the Wall Street banks and investment firms. Control over investment capital and technical expertise like this give finance capital and its banking representatives tremendous power – over policy making, over economies, over employment and income, over advertising and image-production. . .over everything. Production, consumption, economy and the use of environments are subject to a more removed, more abstract calculus of power, in which ability to contribute to short-term financial profit becomes the main concern, and long term consequences are not so much ignored as glossed over through ideological incorporation ("we are all environmentalists now"). And when the contradictions of global finance capitalism moved the system into crisis as we saw in 2008–2010, the state comes to the rescue of capital, the resurrection of continued growth is the urgent necessity, while the environment is the "necessary" sacrifice. Instead, the problems that capitalism periodically encounters can be solved through the market mechanisms (carbon trading) that critics say causes them.

Globalization of this neoliberal, financial kind means that economic growth rates slow down in the de-industrialized center, and increase rapidly, to rates of 8–10 percent a year, in some peripheral industrializing countries. China's economy grew 14-fold between 1980 and 2006, to the equivalent of a GDP of \$4.4 trillion, and India's economy grew sixfold to \$1.2 trillion (IMF 2009) with carbon dioxide emissions quadrupling in both countries (CDIAC 2009). Much of this production and pollution is connected to consumption in the First World – 40 percent of China's product is exported, and 20 percent of India's, while both economies have become dramatically more export-oriented. So we have seen the globalization of an economy, centered still on serving consumption in the high-income countries. This has led to an intensification of the globalization of pollution, as evidenced from carbon dioxide emissions. In 2006 global fossil-fuel carbon emissions amounted to 8230 million metric tons of carbon. In global terms, since 1751, 329 billion tons of carbon have been released to the atmosphere from the burning of fossil fuels and cement production, half of these emissions happening since the mid-1970s (Figure 1.1) when it was already becoming known that greenhouse gases were causing global warming (Schneider 1976). The point is that environmental pollution is driven by economic necessity under capitalism. Within the existing political-economic context, drastically decreasing pollution can only be brought about by economic recession. Thus between 2008 and 2009 there was a decline of 5.9 percent in global carbon dioxide emissions from burning fossil fuels. This was brought about by a decline of 2.5 percent in global GDP, a decline of 11.5 percent in the manufacturing production index, and a reduction of 40 percent in raw steel production (EIA 2009). It is politically impossible for parties or governments to suggest, in effect, that the necessary price of ending environmental destruction is social and economic calamity. Again the "solution" is to displace discussion "upwards" from the national scale to the international. Rather than setting up powerful institutions, as with the Bretton Woods agreement on regulating the global economy, upward displacement

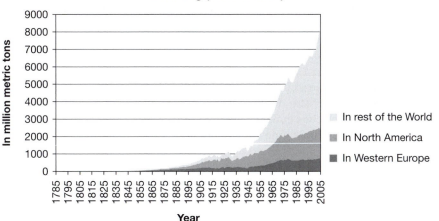

Global Carbon Dioxide Emissions from Fossil Fuel Burnikng (1785 to 2005)

Figure 1.1 Global fossil fuel carbon emissions

in the environmental discourse necessarily takes the form of UN conferences, "Earth summits" and non-enforceable Protocols. Economic necessity produces endless political evasion on the environment, rather than "Obama's lack of resolve."

To summarize, capitalism and its historical transformations is a *starting point* for any account of the destruction of nature (what we, following Marx, called alienation from nature). Global political ecology, in other words, emphasizes global political economy as a main causal theme. We have provided a very simple period-ization – industrialization, consumption, globalization, financialization – driven by "internal" social, economic and political contradictions rather than "external" contradictions with environments. A global political ecology must invest these forgotten externalized moments in the social production of existence with a sense of urgency, and within a political position that does not accept the continuation of capitalist economic growth, in more or less the sense we have known it so far. Political ecology is predicated on an ecologically conceptualized view of politics: it is attentive to the hard edges of capitalist accumulation and global flows of labor, capital and information, but also attuned to the complex operations of power-knowledge, as we shall see below, all within a system prone to political-economic crisis.

Towards a critical global political ecology

The destructive potential of these spectacular and crisis-ridden trends in the global economy play themselves out in many unexpected ways, once set loose in actual ecosystems around the world, where people, microbes, plants, and animals interact through systems that are both dense and complex. Over the past twenty years, these very *political ecologies* have become increasingly well understood, though they remain subject to intense debate and scrutiny.

Political ecology, as a field of investigation and a form of criticism that explores these outcomes, traces its lineage to the 1970s and 1980s, when analysts became impatient with the largely apolitical forms of explanation that saw environmental problems as a reflection of population growth, inappropriate technology, or poor management (see Blaikie 1985; Watts 1983; see also Neumann 2004 and Robbins 2004 for a history of the field). The set of available explanations and methods for investigation were painfully circumscribed. For many environmentalists, most notably, the core problem driving reckless human behavior was seen as one of values. Improve the ethical register of people, and better stewardship of the earth would follow. Better environmental thinking – deep ecological values – would precede better environmental doing (Naess 1973). While this is instinctively attractive, many political ecological observers who reflected on the "hard-wired" tendencies of both multi-national companies and consumers around the world came to suspect that *ideas* were insufficient to change history. A material system must be addressed first.

Conversely, for market enthusiasts, environmental destruction was typically identified as the failure to fully free the economy from constraints put on it by environmentalists, state regulations, or unions. Capitalism unleashed, it was assumed, could remedy environmental problems. It did not matter that such problems were precisely caused by processes of accumulation inherent in markets. Political ecologists became skeptical of market environmental optimism in a world where markets had failed to protect forests, oceans, and air for decades, and had likely hastened their decline (Nevins and Peluso 2008; Zimmerer and Bassett 2003). The commodification of nature to save it seemed contradictory at best.

A burgeoning body of research and writing arose in political ecology as a response to precisely these kinds of assumptions and formed an alternative set of explanations, based on observations around the world. Defined as a combination of "the concerns of ecology and a broadly defined political economy" (Blaikie and Brookfield 1987: 17), the field of political ecology coalesced around investigations into: first, the impact of capitalist development on the environment; second, the social and political implications of environmental protection, conservation, and management; third, the political economy of the way new natures (species, landscapes, and ecosystems) are produced. The field's empirical discoveries have done much to undermine simple assumptions about environmental change.

Political ecology of environmental degradation

Take the classic problem of soil degradation. Around the world, the quality and quantity of priceless topsoils, on which global food production depends, have long been acknowledged to be in a state of potentially catastrophic decline. With declines in soil fertility – resulting from declines of key nutrients and organic matter – come declines in agricultural yield. This is potentially followed by regional food crises, or at least localized economic problems for farmers growing for turbulent markets. For many years, soil conservation became the watchword of environment and development (Eckholm 1976; Brown and Wolf 1984), and has remained a central trope of environmental concern (Montgomery 2007). International development assistance went to farmers to instruct them in improved methods, to "educate" them concerning "proper" use of the land, and to support them to undertake sometimes costly conservation techniques. Farmers, it was largely assumed, were the cause, and therefore the solution, to a major environmental problem (Grossman 1997).

Political ecologists, however, are keen to discover *why* farmers behave the way they do. What logics compel farmers to use land in a heavy, exploitative fashion? Work on the ground with producers reveals that most producers are well aware of the dwindling capacity of their land and dread the high levels of extractive production they place on the resource base. It has long since been demonstrated instead that, under conditions where production margins are slim, prices are low, labor availability is strained, and the capacities of households are economically limited, producers often mine soils (Blaikie 1985; Edwards 1995; Zimmerer, 1993). The process of commodification – how differing sorts of households and individuals are linked into markets – shape the process of what farmers and local resource mangers can do. The logic of the market shapes the imperatives of producers, and so into the land: the rural poor may find themselves in a position that they must work harder and longer and exploit their land resources to the maximum as prices fall (Watts 1983), what has been called the "simple reproduction squeeze." As we shall see in this volume, this logic of production also leads to the increased intensification of livestock production, resulting in the brutal factory farms mushrooming across the world, as Emel and Neo demonstrate forcefully (see Chapter 3).

This should not, by any means, lead to the assumption that every farmer involved in any market will inevitably destroy their soil base. On the contrary, many farmers find ways to maintain the land and their livelihood (Benjaminsen 2001). Where markets are generous (usually a temporary condition – commodity prices have a tendency to rise and fall precipitously!), capital is often available to reinvest in the environment, to rest the land, or to subsidize or maintain soil nutrients. The critical lesson is rather that where massive and widespread soil degradation has happened on farmlands or rangelands, the producers involved are typically responding to excruciatingly forceful pressures, often dictated by the brutal logics inherent in global markets.

Much the same has been demonstrated repeatedly for degradation of grazing lands. Heavy stocking of ranges may result from anarchic grazing systems and so lead to soil poverty (Runge 1981), an argument made most famous in Hardin's (1968) classic essay on "The Tragedy of the Commons." Even so, culturally rich and historically deep common property management systems have allowed sustained grazing in many parts of the world, especially where tradition societies and hybrid economies persist (Jodha 1987; Agrawal 2001).

Similar observations have been made about a wealth of related environmental problems. The earth is currently going through the highest rate of extinction in the history of the human species. Biodiversity decline in rainforests is most notably acute, since these areas are so dense in species (consider that entomologist E.O. Wilson once famously observed forty-three species of ant in a single Peruvian rainforest tree; Hölldobler and Wilson 1990) and rapaciously undergoing deforestation. This decline in forest diversity is associated not simply with farming, but with the kind of large-scale extractive cultivation of plantation-based commodities (e.g. bananas) that are built into the unfair terms of trade between tropical countries and wealthy global consumers (Vandermeer and Perfecto 2005). Similarly, as Mansfield points out in this volume (Chapter 4), many fisheries have been driven to the edge of collapse, but not as is often assumed because of anarchic small fishers competing for the "last catch." Instead, the globalization of the production system, and the increasing influence of large-scale industrial production systems has mercilessly taxed the oceans. Resources can be managed collectively, but the pressures of global markets and owners pushes these resources towards a state of crisis.

Conversely, neoliberal logics hold that such resources cannot be managed collectively but must be owned privately. Such logic has been used to propel new rounds of primitive accumulation (see Harvey 2005): the acquisition of historically collective and public resources by firms for exclusive control and production. In some cases this involves an extension of the rainforest frontier and dispossessing indigenous or poor communities; in other instances, technological innovation has opened up a new scramble for resources (new frontiers such as deepwater oil and gas exploitation) in remote and previously unexplored environments such as the Arctic and the ocean floor (Redclift 2006); in still further cases, resources like water have been commodified and their provision privatized – see Chapter 16 by Bakker. The repetitive nature of primitive accumulation – what Harvey calls accumulation by dispossession – marks off particular epochs which are especially destructive environmentally for old and new commons alike (Boyle 2008, Bollier 2003, Parry 2007).

In other words, political ecology has repeatedly shown what we argued previously: that environmental degradation is not an unfortunate accident under advanced capitalism, it is instead a part of the logic of that economic system. Environmental degradation is a consistent symptom of various logics and trajectories of accumulation – and the deadly operations of markets – worked out on the land and for specific resources, as most evidently in the case of oil, which Bridge surveys in detail (see Chapter 14). The outcomes of these privatizations are necessarily complex, moreover, and as Bakker shows here (see Chapter 16) the privatization of water presents particularly complicated challenges both for companies that attempt to control and harness the commodity as well as governments and states that seek better services and greater efficiencies by doing so.

Political ecology of environmental conservation

With that in mind, it would seem that efforts to stem the major environmental problems of our time would best be addressed by going to the heart of the problem, the typically perverse driving engines of industrial capitalism, economic growth, and the uneven power of different players contending over the use and management of natural systems. Political ecological work has revealed, however, that many efforts at conservation, environmental protection, and ecological amelioration – whether in protection of endangered species, threatened ecosystems, or degraded air and waterways – have been inattentive to these underlying forces and have instead drawn upon dated, indeed frequently colonial, models of environmental management (Adams and Mulligan 2003).

Political ecology takes as a second thread of investigation, therefore, the impacts, logics, and operations of conservation and environmental protection itself. The central concern here is the way that conservation, purportedly an effort to create better conditions in the world, can frequently be a mechanism for (or more cynically a "cover" for) powerful players to actually seize control of resources and landscapes, and the flow of value that issues from these. Most prominently, global efforts for the protection of wildlife through the creation of national parks has received great scrutiny in the field. While certainly the catastrophic decline of key species around the world is a matter of universal concern, it is notable that a generation of efforts to save animals (e.g. African lions, Indian tigers, Chinese pandas, etc.) have largely been built around the "fortress" model, where urban elites call for the enclosure of lands long used and occupied by rural, indigenous, and local people, all in the name of protection (Brockington 2002).

It is frequently the case, moreover, that people living in and around the wildlife of concern are not only not the central existential threat to wildlife (relative to urban growth, for example, or state corruption through poaching), but that their landscape management practices are precisely ones that have created habitat or protected such species in the past. The political ecological history of parks around the world, including some of the key foundational ones in the United States (i.e. Yellowstone National Park) reveals the hidden exclusion, violence, and seizure that sits at the core of what might otherwise be viewed as unmitigated actions of environmental "good" (Jacoby 2001). As Neumann (1998) has described for the case of Tanzania, moreover, local communities often fight back against such impositions, breaking park rules in an effort to reclaim land they understood as effectively stolen. In many cases the governance of protected areas and species resembles the militarization and intensive surveillance of ecosystems.

It is certainly true that contemporary conservation has gone through a period of innovation and decolonization (Adams and Mulligan 2003), in which new models have been developed that are more inclusive of local needs, geared towards consideration of livelihoods, and increasingly aware of the role of power in creating undesirable conservation outcomes. The resulting forms of governance are increasingly localized and decentralized as a result. But even these changes, political ecology suggests, demand scrutiny, since "a potential pitfall of governance at this level is their influence and entrenchment of sharp social inequalities that operate in such local milieus as villages and communities" (Zimmerer 2006: 11). Furthermore, they typically side-step the underlying issues of political economy by focusing on institutional fixes that do not challenge critical drivers of biodiversity decline, the global market itself.

Li (2007) has explored these questions through the lens of environmental rule in Indonesia, exploring the power of a liberal "will to improve," understood as a two-century-long project to secure the welfare of populations, but rooted in a historically complex situation of government practice, operating within the jagged rhythms of capitalist accumulation. Li is especially concerned with the ways government programmers draw boundaries around, and "render technical," aspects of landscape, conservation and livelihood. Simultaneously, she demonstrates how these practices have limits, imposed by the contradictions between improvement and sovereign power, and between the rationalities and practices of government and their ability to actually regulate dynamics social relations. These open up the terrain of "contestation and debate between people with different interests and claims" (Li 2007: 270).

Focusing explicitly on politics surrounding a national park, she has shown how technical conservation efforts served to screen out marginal households among recipient communities, a process which produced limited development benefits and encouraged community radicalization. In one case a Free Farmers Forum emerges from a century of failed improvement; in another, highland villagers reject the park and efforts by the Nature Conservancy. In all of this, local politics turns on the contradiction of a form of rule as trusteeship in which agents with power are ultimately unprepared to relinquish their authority, however much it is draped in the rhetoric and discourse of participation and empowerment.

This basic political ecological insight, that environmental conservation in its many forms, for better or worse, is basically a form of environmental control (green rule), has been extended to a range of other problems. Carbon offsetting most notably, as Bumpus and Liverman point out in this volume (Chapter 10), while an innovative conservation technique for potentially addressing global carbon, also represents a form territorial control – merely a "spatial fix" for capitalism's ongoing ferocious growth. It is also a technique that extends the "financialization of everything" – as noted above – by building and extending markets for carbon, only deepening the problematic role of capitalism in environmental management. Similarly, as Eden discusses in this volume (Chapter 8), "greenstamping" certain commodities (most notably timber) to assure their sustainable production and harvesting, while a laudably clever way to extend conservation goals in a market environment of runaway consumption, essentially concedes the logics and power relations that already exist in those markets, making them a poor challenge to the crisis of sustainability in global capitalism. Arguments and analyses in political ecology, therefore, are as concerned with the modalities of green protection – how forms of rule are instantiated that both produce subjectivities and environmental outcomes – as much as the way in which it is abused.

Political ecology of environmental production

It is clear, however, that the environments of the world around us are increasingly, and to some degree have always been, the product of human activity and trans-formation. The proliferation of genetically modified organisms (GMOs), for example, is a case where altogether new forms of life are being innovated by human beings. Once set loose into the environment, these are further interacting with plants, animals, and people, in ways that are sometimes unexpected, though always with implications for people's livelihoods, for the fate of surrounding biodiversity, and for power and economy.

Such innovations have a deep history of course. The domestication of animals and plants has a 12,000-year legacy. The movement of plants, animals, and microbes, moreover – whether domesticated or not – is a critical part of the political economic history of the earth, as in the case of the conquest of the Americas, where cattle, smallpox, and fodder grasses spread across the landscape with European people (Crosby 1986), accompanied by joint stock corporations, plantation agriculture, and capitalist class relations. New ecologies are always proliferating, therefore, but the political economic drivers and implications of these demand careful attention.

The third area of political ecological interrogation is into the political and economic implications of such new ecologies, environments, and species. In this regard, political ecology takes as its core understanding of the environment that nature is *produced*. Following geographer Neil Smith (1996), this does not mean that all nature is controlled by people, only that – at this point – it is effectively impossible to imagine an environment where human activities and industries are not a core component of emerging ecosystems (how else, after all, can we imagine the earth's climate in an era of global change?). With this in mind, we can further ask how *specific* environmental conditions are produced and how these are or are not entangled with the tendencies of global capitalism outlined above: accumulation, growth, and crisis. In the production of nature we can see how nature itself – through harvesting of genetic diversity for new medicinal patents, for example, or the creation of nano-biotechnological living engines – becomes a strategy for accumulation (Smith 2007). These new natures frequently have capitalist goals and logics bound up within them.

Several of the chapters in this book address these questions directly. Bruce Braun (Chapter 18) addresses how the molecularization of life has brought us to a new moment in the history of biopolitics, one in which bodies are understood in terms of their genetic inheritance. Here, the management of risk is individualized and, concomitantly, the make-up of our bodies, and not just their conduct, becomes the subject of "technologies of self." But this is not the only way in which the molecularization of life has been apprehended. For every story in the U.S. media that speaks breathlessly of advances in stem cell research and gene therapy, or that worry over the "post-human" futures these might usher into being, we find two or three other stories that speak ominously of migrating birds and backyard chickens, which mix together Vietnamese peasants, influenza viruses and homeland security. This conjunction of biopolitics and geopolitics, shows the way the molecularized body and the problem of biosecurity are linked.

Wainwright and Mercer (Chapter 19) tell a related story, tracing the travels of maize, a plant species that came out of Mexico as an ancient domesticate, only to return as a "transnational transgene" after genetic modification. Objections to genetically modified organisms are many and distinct, but they all point to simultaneous threats to environment and livelihood. The evaluation also stresses the difficulty of adjudicating disputes about the impact of GMOs given divergent assumptions and necessarily normative stakes. Given the produced nature of this ecology, however, ecological and agroecological science can only take us so far, and the critical problem of survival for Mexican producers and native genetic diversity requires a more comprehensive and explicitly political and economic evaluation.

Admitting that nature is produced opens further political opportunities, nonetheless. While powerful actors are at work producing new natures, it is at least possible, within political ecology, to ask "how, and to what ends, *alternative natures* might be produced" (Smith 1996). Viewing the often discouraging record of natures produced by the powerful, therefore, may open a view onto new natures based on other economic and social logics, ones that are more sustainable, democratic, or desirable.

Working people around the world have produced nature by establishing gardens in the city, for example, often seizing and occupying city territory and turning it to productive use. Such was the case where African American communities in Midwest U.S. cities "greened" the industrial urban environment around them at the turn of the century, in spite of widespread disenfranchisement and state under-investment in their communities (Moore 2007). Beyond urban contexts, working rural people have conserved floral, faunal, and avian biodiversity through the creation of artificial forests in pursuit of livelihoods in sites across Africa (Fairhead and Leach 1996) and India (Ranganathan *et al.* 2008).

In a very different context, Moore shows (Chapter 6) that the most fundamentally produced ecology of all may be garbage, the flowing streams of which circle the globe both as an unwanted hazardous byproduct of capitalist development and as a commodity. Her analysis, moreover, stresses the way garbage becomes both a resource for the world's poor and a political lever to gain access to state development investment. This case dramatically reminds us not only that all contemporary ecologies are produced, but also that even the most apparently anthropogenic (human-caused) objects are ecological, bound into complex life-systems, along with their inevitable attendant politics.

The central themes of political ecology, in sum, include: first, the grounding of environmental degradation in the trajectories of accumulation and the operations of market-based power; second, the intertwining of environmental conservation with struggles over environmental control; and third, the ongoing emergence of new ecologies, developing from human productive activity, with implications both for environmental destruction as well as for creative environmental alternatives. The book makes the point that a full accounting of environmental degradation must powerfully link ecological process to poverty and local political struggle (see Chapter 7 by Ghertner on the slums of New Delhi and Chapter 17 by Mehta on dam displacements) but also to the highest levels and concentrations of state and corporate power and wealth (see Chapter 9 by Johnson on the insurance industry and Chapter 15 by Labban on energy security).

Emerging problems in political ecology

To support such an analysis, recent research in political ecology has explored a trio of key ideas and inquiries, all of which are in evidence in this volume. First among these are *questions of power and forms of rule/governance*. Power over nature and society is exercised, it is suggested, not only through complex forms of social control and hegemony but also normative ideology, and governmentality.

Second, political ecological analysis has maintained a *sensitivity to representation, both as a set of discourses and as a field of practice*. Knowledge of environmental problems is not always immediately available and unproblematic. Rather, our categories, priorities, and interpretations are mediated by complex systems of discourse that frame problems and focus the scope of how we imagine them. Hegemonic control of knowledge about environmental crises themselves (from global warming to Katrina) is a fundamental part of the political economy of nature.

Finally, recent political ecological scholarship has attended to questions of *expertise and democracy*. Science and science-based inquiry are both essential to solving environmental problems, but these are also historically problematic parts of those problems. Science is a critical and rigorous way for environmental conditions and change to be known, measured, and described. Conversely because environmental science is always embedded in the larger context in which it is produced and employed, and because "scientific" accounts tend to exclude or marginalize other critical contextual ways of knowing, science can be highly partial, reductionist, and instrumental in achieving and maintaining political control over nature. We seek to emphasize that rule, discourse/representation and expertise/ knowledge operate in conjunction and in complex configurations when nature is at stake.

Green governance

Environmental problems (and their possible solutions) are inevitably entangled with questions of power and governance. To enclose common resources, as we described previously, for example, it is essential that there is a mechanism that justifies and enforces control, as where "property" is given the force of law so that "private" owners and the state have the *power* to enforce exclusion. Power has also come to be understood, however, in more complex and subtle ways, as where individuals come to obey or take for granted "property" laws, effectively internalizing certain forms of control and authority as normal and natural. Political ecology, as a result, has come to consider and reconsider the diverse nature of power and authority in differing environmental contexts (Robbins 1998).

Sovereign environmental power

Power in environmental management is most crudely and commonly understood as the capacity of a polity or state to control the actions of people (or organizations or firms) within its jurisdiction, what theorist Michel Foucault referred to as "sovereign" power (Foucault 1980, see especially pages 95–96). This capacity to dominate or subjugate is obviously essential to the promulgation of environmental problems as well as to the control of environmental degradation through regulation. Consider, for example, the monopoly of force required to take control of large areas of land for surface mining, to exclude traditional or nearby inhabitants, and to enforce exclusive rights to exploit the land through large-scale construction of open pits, the removal of mountain tops, or the saturation of the land with acids for in-situ leaching of minerals. Such power is further extended by stifling or controlling resistance to the health and ecosystem costs such development entails, either through the legal protection of the rights to exploit the land, or more dramatically through the collaboration of state force to put down or silence opposition. Consider, for example, the rapacious destruction of Appalachian landscapes and communities through wholesale mountaintop removal mining, leaving toxic environments and worked-over communities in its wake, with little room for community resistance or legal recourse (Burns 2007). This is raw sovereign environmental power in its crudest and commonest form around the world.

Needless to say, such socio-environmental force has historically been confronted by its corollary: popular or state power to restrict, control, or exclude environmentally or socially destructive practices. The roots of formal modern environmental regulation lie in such forms of power, as where a spate of environmental laws in the United States in the late twentieth century challenged the power of corporations to emit air and water pollution or to dump hazardous wastes indiscriminately, all practices common throughout the two centuries prior (Colten and Skinner 1996). The roots of many if not all of these reforms developed from local and regional resistance, leading most recently to anti-toxins campaigns fought in the name of "environmental justice" (Szasz 1994).

Of course, the exercise of sovereign environmental or "green" power of this kind, itself holds contradictions. The monopoly of power that allowed the United States government to establish a wilderness preserve at Yellowstone National Park in 1872 – an action that later allowed the maintenance and revival of buffalo, elk and wolf populations throughout the west – came with a set of violent exclusions, including the elimination of the Native Americans in the region, which included the removal of perhaps a dozen native nations from the area to resource-impoverished reservations. It also required the forceful removal of traditional white subsistence land users (Jacoby 2001). The creation of a wilderness at Yellowstone was necessarily an act of social and environmental subjugation, for better and for worse. In this sense, political ecology has long been about the machinations and contradictions of raw *sovereign* power.

Environmental subjects and environmentality

The landscapes of political ecology have shifted dramatically in recent years, both in terms of the forms of power exercised in pursuit of ecological control and resistance, and also in the conceptual tools at our disposal to understand such power. Specifically, political ecology is increasingly attendant to the way power is exercised *within* – rather than over – individuals, communities and societies. As people come to understand themselves, regulate their activities, and help oversee the actions of others, they are not merely the objects of external force, but are themselves embodied power. Foucault observes, "individuals are the vehicles of power, not its points of application" (Foucault 1980: 98) and further suggests that government (or sovereign) power depends upon the extension of the state itself through internalization and acceptance of individuals as state subjects, a condition he refers to as "governmentality" (Foucault 1991).

In environmental politics, these forms of governmentalized control – or forms of green rule – are evident in countless examples, ranging from the quirky to the profoundly problematic. Consider in an innocent example, the way recycling, once a foreign and somewhat awkward social practice, has become fully normalized behavior. This has allowed an enormous quantity of metal and paper to be removed from the traditional waste stream (Ackerman 1997). It also has perverse effects, of course, insofar as the effort of the average homeowner to separate, haul, and donate aluminum and paper through recycling is effectively a gift of their labor to companies that receive artificially cheap resource inputs. The key issue, however, is that such practices are normal, naturalized, and socially scrutinized. Neighbors, after all, know who is not participating in roadside recycling programs and children are drilled from a young age with training in the "Three Rs": reduce, reuse, recycle. Recycling – no matter how potentially ecologically beneficial – is socially regulated, therefore, and represents a form of "soft" or obedient environmental power.

Such internalized environmental practices cut in different ways, moreover. Consider that the maintenance of lawn yard landscapes through the use of environmentally harmful lawn chemicals and high-input power mowers in many countries is predicated upon a normative and socially enforced environmental aesthetic, that effectively associates good citizenship with environmentally harmful activities. Such behaviors benefit the corporate entities that produce, package, and market the goods and services that maintain such an aesthetic, but it cannot be said that these companies *forced* anyone, in any simple way, to act as they do. Rather, the exercise of power is enacted internally, through the production of a certain kind of "subject," whose identity as a good citizen is associated with a set of specific environmental activities (Robbins 2007).

Kosek's book *Understories* (2006) provides another illustration of how forests (in this case the US southwest) are classified, organized and ruled in a way that is intended to produce particular sorts of subjects (including Smokey the Bear!) and property relations. Yet, at the very moment that forests are declining as local sources of revenue and employment, they become the basis for powerful (yet different) sorts of insurgent consciousness and practice among both Hispanic and white rancher communities.

The implications of this way of thinking are wide-ranging, especially as we examine how environmental attitudes and practices change over time. Arun Agrawal, in his book *Environmentality* (2006), has observed the tendency over time of communities who self-govern resources to change their attitudes about the environment and slowly internalize responsibility for governing nature. In his case (research in India), communities that long opposed colonial and government control of forests – going so far as to set them on fire rather than see them dominated by an external sovereign state – eventually come to fully accept the mantle of protecting forests on behalf of the state, changing the way they think about forests, but also their own views about themselves. Agrawal refers to this transformation as a case of "environmentality" where people's identities, activities, and attitudes come to internalize previously external norms or mandates.

The two chapters here (5 and 12) by Biehl and by Peluso and Vandergeest shows these processes at work. In the latter, forests become a theatre for contending sorts of political projects loosely held together around the idea of nation building. In the same way that Kosek describes how forests are vested in ways that relate to local struggles, so, in the case of Malaysia and other southeast Asian states, forests are central to national identifications and struggles. Biehl explores the impact of the AIDS treatment rollout throughout the country's government, health systems and personal lives in Brazil. In charting the lives of poor patients before and after they had access to ARVs, he reveals the real-life outcomes of novel national, inter-national, and corporate policies, with the arts of government that accompany econo-mic globalization and the remaking of people as market segments (specifically, therapeutic markets).

These issues are explored too by Braun in his chapter (see Chapter 18) who explores the idea (see Rose 2007) that the molecular revolution entails a shift within the biopolitical regimes of modernity, from political rationalities directed toward the management of risk at the level of populations, to the individual management of the genetic risks peculiar to one's own body, or what Rose calls "ethopolitics."

Environmental representation and practice

Accepting that control of the environment is, at least in part, also about the exercise of power within people, their identities, and their practices, political ecology is also increasingly concerned with how we come to know about the environment, how it is defined and categorized, and how environmental problems are represented, acknowledged, and defined. This concern with the so-called "construction" of nature does not come at the expense of an understanding of the environment as real, concrete, and material; rather it accepts the inevitability of our partial knowledge of the environment and the way human knowledge of the environment can be interpreted, controlled, and indeed manipulated. Political ecologies are always, in part, about what and how we know about the environment, therefore (Hajer 1997; Leach and Mearns 1996).

Typically, such knowledges are produced and conveyed through "discourses," a term taken to mean: "frameworks that embrace particular combinations of narratives, concepts, ideologies, and signifying practices" (Barnes and Duncan 1992: 8). From this point of view, framing environmental problems is something people *do*, a set of practices like making maps, writing newspaper articles, giving speeches, sending emails, displaying photographs, selling advertising, posting flyers, posting blogs, or telling stories at a bar. These framings always depend on a variety of diverse pre-existing ideas, categories, images, and words, but the way these separate pieces come together into persistent, stable, and consistently repeated and expected ways of thinking about and describing problems, distinguishes them from discourses. Discourses can profoundly situate and control how we think about environmental crises and what we do (or do not do) about them (Adger *et al.* 2001).

And these discourses are rarely ones we each individually dream up out of thin air. Instead, they are typically constituted from clusters of well-cemented concepts that circulate through the global media and through common understanding. This makes unthinking them very hard indeed, since it is difficult to imagine outside the categories already at your disposal.

What is natural?

Most obviously, the relegation of an environmental event to a status of "natural" is a common form of representation. Consider the very idea of a "natural disaster," most notably, where a violent or destructive outcome is understood as natural and therefore somewhat inevitable. In the 2008 earthquake in Sichuan, China, for example, perhaps 10,000 schoolrooms collapsed (Yardley 2008; Zhang and Jin 2008), making children horribly overrepresented amongst the fatalities of that event. Earthquake deaths were in this way precisely stratified across the already existing fault-lines of growing class stratification in China, especially in the provisioning of services and state resources for the rural poor. In what sense are those deaths the product of a "natural" disaster instead of a systematic political and economic underdevelopment of critical infrastructure – the horrific results of the earthquake in Haiti in early 1010 being a case in point (Soper 1995; Castree and Braun 1998)?

In the heat wave that blistered Chicago in the summer of 1995, nearly 500 people died from heat-related causes as the morgue overflowed with bodies, ambulance services were paralyzed, and the city's water and energy infrastructure shut down. But the distribution of deaths was by no means equal, with elderly and African American people killed in dramatic disproportion, and men far more likely to have died than women. Such differences reflect deeply stratified and structured differences within racist and ageist capitalism, in terms of access to resources, social capital, and infrastructure, among other things, and therefore underline the very politicized nature of the disaster's outcome. But more than this, because of the way the event was covered by journalists and reflected in the national and international media (to the degree that it was at all) as a "natural" tragedy, one free from precisely these fundamentally unjust misdistributions of vulnerability and opportunity, the heat wave was both depoliticized and utilized to secure expert power at the expense of meaningful understanding. As Eric Klinenberg (2002: 23) explains in his analysis of the heat wave:

> Journalistic, scientific, and political institutions benefit from their symbolic power to create and to impose as universal and universally applicable a common set of standards and categories, such as natural disaster and heat-related death, that become the legitimate frames (or organizing concepts) for making sense of an unexpected situation. . . . Examining the ways in which

features of the catastrophe were brought to light or concealed helps to make visible the systems of symbolic production that structured the public understandings of the disaster.

A political ecology of such events must attend not only to the very real political fault-lines across which vulnerability is distributed (for rural children in Sichuan or urban African Americans in Chicago) but also draw attention to the way any characterization of such outcomes as "natural" is itself a dangerous form of representation, which erases the very political nature of environmental crises, yet may coincidentally produce other sorts of politicized nature.

Joseph Masco speaks directly to the nature question in his chapter (see Chapter 13) which examines the Manhattan project and the atomic bomb testing, and its legacies, in the U.S. southwest. He seeks to extend our theorization of the complexity of nature-culture forms via the concept of "mutation" (i.e. when the ionization of an atom changes the genetic coding of a cell, producing a new reproductive outcome). Understanding the cultural effects of atmospheric nuclear testing requires an investigation into the different conceptions of nature that inform local forms of knowledge. Debates and practices involving new "species" logics in the nuclear age examine how the pursuit of security through military technoscience has raised questions about the structural integrity of plants, animals, and people, revealing what Masco calls mutant ecologies.

Diverse and very political narratives

The political ecology of environmental representation is even more complex than this, however, and often entails extremely subtle ways of categorizing or naming natural objects or conditions, and narrating events or problems. By so framing environmental crises, people set in motion and concretize understandings of the world that are both politically effective but also, conversely, often dangerously constraining.

Consider for example, the crisis of clear-cutting of forests in the Pacific Northwest of North America. Here, an apparently obvious environmental problem has come to be narrated in ways that are often politically problematic. As shown elsewhere, in Braun's (2002) analysis of this crisis, the environmental community, in pursuit of protection of the forests of the region, dramatically narrated apocalyptic representations of forest destruction through pictures and words, showing maps and images of unbroken forest in the early twentiethth century to denuded landscapes of the present. The power of these images effectively led to galvanizing support against a rapacious forestry industry that, despite its claims of plantation and stewardship, has unquestionably helped to transform the region in undesirable ways. Braun's examination of the mapping and rhetoric of this story, however, reveals that the effort is rooted in an insistence on radically distinguishing "natural" (and therefore desirable) forests from "modified" (and therefore unnatural) ones. This narrative's dependence upon deeply colonial images and ideas of the forest, leads to the exclusion of a range of people and human activities, especially indigenous people. Re-narrating the conditions of the forest, on the other hand, might make it possible to imagine better and more inclusive methods of control and new opportunities for people to live in and around forests.

This is similar to Helmreich's (2009) recent account of how revolutions in genomics, bioinformatics, remote sensing have pressed marine biologists to construct the sea in a different way: its nature now resides in marine microbes which in turn becomes objects of debate as regards the origins of life, global climate change and biotechnology.

Put in plain terms: arguments over the apparently "given" facts and categories of ecology, are always also arguments over social and political control of nature. The lens through which environmental problems are constituted and projected inevitably assigns specific causations and empowers and disempowers different actors. Behind every story of environmental crisis, therefore, is a narrative of political and social control.

Myriad case examples from this volume suggest the importance and stubborn pervasiveness of environmental representations. Guthman's chapter notes that in drawing attention to the "obesity crisis" in terms that draw on notions of moral citizenship, for example (see Chapter 2), activists have unquestionably produced political momentum for public health. On the other hand, these narratives precisely obscure the political economic forces at work on people's food habits and activity. Labban (see Chapter 15) shows how, in the wake of 9/11, discourses of empire, war and terror are welded together in powerful ways around the notion of "energy security." These issues are also part of a larger debate over the ways in which the environment becomes a geo-strategic question: the environment can cause conflict or generate "eco-terrorism" and ecology becomes a ground on which militarism and security is to be conducted (see Dalby 2002; Peluso and Watts 2005). Ghertner (see Chapter 7) offers another kind of example in his account of how "green aesthetics" provide a compelling legal and political discourse rooted in middle class concerns about cleanliness and health to push forward massive slum clearance in New Delhi (see also Davis 2006 on the political ecology of the slumworld). He calls attention to this "green speak" not only to suggest that the government has been straightforwardly "greenwashing" environmentally deleterious projects, although this is true in many cases, but rather to reveal how it acquired expanded epistemological authority in the context of new political-economic and govern-mental imperatives.

Environmental and green narratives – from the always-changing fields of ecology, hydrology, climatology, and biology but also from the religious and non-scientific secular realms – delimit and direct social and political imperatives and opportunities. This presents some problems for coming to terms with science, however, whose role in the production, alleviation, and explanation of environ-mental crises is another central emerging concern in political ecology. As our account of the global warming debate revealed, who speaks and the sites of knowledge production are central to the legitimacy and the hegemony of particular sorts of narratives and what passes as conventional wisdom.

Science and society

The rigorous application of science to environmental problems is a critical component of political ecology, since measuring and tracking environmental change requires techniques and methods that can reliably identify the causes of undesirable and unequal impacts, externalities, and costs. To explore the ecological costs of privatizing forests, for example, we require an ability to track forest cover change, transformations of ecological structure, and altered biodiversity and hydrology. Without the ability to track fish populations and abundance, or to map it across the diverse fisheries and ownership systems of the world's oceans, in another case, it is impossible to explore the implications of privatization, global treaties, or regional innovations in fisheries management. Science is fundamental to political ecology.

And yet science is itself a highly problematic global political enterprise. Science is never conducted entirely separately from the global political and economic forces that make it possible. The exclusive role of science as an adjudicator of environmental conditions or "truths" has historically led to the marginalization of different ways of knowing and explaining the world, putting undue influence and power in the hands of technical experts. These simultaneous realities – the inevitably political character of science as well as its apparently non-political status – hold implications for global political ecologies.

One pathbreaking study in this area by Michael Goldmann (2005) examines the relations between a powerful global organization (the World Bank) and its particular fusion of knowledge-power in going "green." Here Goldman takes the reader through the normalization process of environmental knowledge and practice at the Bank. He begins with the fact that the Bank is a massive producer of "green" knowledge and then provides a rigorous account of the environmental research and project cycle. The devil, as Goldman shows, is in the details, in that opportunities for data collection within projects are small (driven by all manner of pressures placed upon project managers) and that staff training for environmental assessment – and the monitoring of environmental performance – is impacted by an array of external forces. These include (i) the dominance of cost benefit analysis, (ii) the project donor's liability, and (iii) the enormous import for project development and management of any environmental assessment (121ff.). Goldman demonstrates the overwhelming extent to which Bank knowledge only refers to Bank knowledge ("narcissism," p.131) and that while the Bank prides itself on scientific rigor, numbers and no-bullshit analysis of an empirical sort, it fails to meet even the most basic academic standards. Nobody believes the national environmental assessments, 60 percent of the projects had no baseline surveys; and most everything on Africa seems to be little more than "airy proclamations" (p.132). At the end of the day, says a Bank official, "we are selling a product" (p.133).

Johnson's chapter in this volume (Chapter 9) is an exemplary case in this vein. It investigates the dynamics of knowledge production and value creation emerging as the risk industry seeks ways to measure and manage the impacts of climate change. Given the extreme visibility and rhetorical power of global warming impacts, she examines how the insurance industry both responds to and shapes the world of risk in relation to corporate profitability. She shows how as both climate risks *and* our knowledge of them grow, so will the market for new insurance and risk-management products, thus reproducing the industry's own conditions of existence. This heavily influences which sorts of risks companies and governments feel the need to plan for, or to ignore.

The politics of the laboratory and the field

Similarly, despite the obvious advantages of science to achieve crucial knowledge (i.e. the replicability of its methods, the rigorous review of its findings, and the careful annunciation of its assumptions), the practice of science has long been known and demonstrated not to occur in a social and political vacuum. Within their own communities, scientists respond to social incentives to pursue certain questions and not others, leading to paradigms and "group think" (Merton 1973; Kuhn 1970). More dramatically, scientists bring into their craft their own previously held convictions and ideas, all of which are formed within social and political worlds they inhabit; the history of ideas in any field, from primatology (Haraway 1989) to physics (Traweek 1988), are influenced by the social realties of scientists themselves, in terms of their life history, gender, and class, but also in terms of the historical preoccupations and political systems in which their labs or fieldwork are constructed and funded.

Context – the way political and economic context influences scientific inquiry – holds perhaps the most implications for the political ecology of global environmental problems. Consider, for example, how long-held racist assumptions about poor stewardship by Arabs led to the persistence of colonial-era assumptions about relatively recent North African desertification, for which little meaningful evidence actually exists (Davis 2007).

Such political economies of knowledge extend into the modern economy as well. As Kaushik Sunder Rajan (2006) has argued, new genetic knowledge and the increased ability to manipulate genes has transformed capitalism in fundamental ways, making personalized medicine possible, for example, and transforming the way risk and health are priced and managed in markets. Equally significantly, however, the basic research conducted in labs involved in advancing this science – producing "DNA chips" for example that help to determine people's genetic predispositions – emerges and operates precisely within the context of these emerging markets for genetic products. One cannot possibly explain the influence of new sciences and technologies on health and economics without acknowledging the influence of those economic landscapes on the development of these forms of knowledge to begin with.

Political ecology is increasingly sensitive; therefore, to the problem that the science we use to apprehend political and economic impacts on the environment (the earth, the body, and the whole natural world in which these are enmeshed) is itself a product of both political economy and the changing environment in which it is practiced.

The problem of expertise and local knowledge

Finally there is the question of knowledge and expertise. Given the critical importance, both historically and at present, of science in the characterization and management of the natural world, it is essential to ask what forms of knowledge, authority, and acquisition of information are shunted aside in the imposition of scientific management (Forsyth 2003; Robbins 1998). This problem has been a traditional concern in human-environment research, which has long acknowledged both the efficacy of traditional and local forms of environmental knowledge and the way the utilization and deployment of science has led to an actual overall *loss of knowledge*, coincident with a *loss of power* for the traditional communities that developed and maintained those knowledges (Berkes 1999).

In many cases, the hegemony of scientific expertise, often tied to narrow and instrumental management goals, is in part an explanation, therefore, for social and environmental change, as for example, in the case of north African desertification mentioned above. When modern range science has been introduced in countries like Morocco, Diana Davis has observed, local knowledge of range conditions and variability have been discounted, leading to stocking decisions and settlement policies that have worsened instead of improved grazing conditions and the productivity of the land (Davis 2005). Nathan Sayre (2006) has documented similar processes at work, under differing political and institutional conditions, in the US rangelands of the southwest. The history of environmental justice conflicts has similarly been marked by a discounting of local health opinions, typically articulated by women from marginalized class and race communities, leading to inaction and further exacerbation of the negative health effects of harmful industrial siting (Seager 1996).

Kosek's chapter (see Chapter 11) on the bee crisis points to all of these issues at work, and to the deep complexity of science, political economy, politics and context. He shows how that the current state of the honeybee is undeniably dismal, experiencing considerable decline in populations even before recent reports of "colony collapse disorder." Global environmental changes have been devastating, whether the intensification of industrial agriculture, toxic pollution, climate change, loss of habitat, or the spread of disease and parasites. Few researchers, however, pose the more fundamental question: How has the modern bee come into existence in a way that has made it vulnerable to new threats? In fact, the largest funding for bee research and bio-engineering during the Bush administration was by military intelligence and weapons research agencies who hope to harness and develop bees' abilities as part of the "war on terror." His chapter therefore emphasizes the fundamentally political ecology of scientific knowledge and practice.

At bottom then, it is not science as a specific and useful form of knowledge production that is questioned in political ecology. Instead, it is the way specific forms of ecological knowledge are selected and validated, the way environmental problems are narrated and structured, and what assumptions and practices becomes normal and internalized for people. In this way, the emerging concerns in political ecology – subjectivity, representation, and knowledge – are interlinked.

Understanding and resisting the way critical environmental problems are produced and promulgated in global capitalism depends on knowing and grappling with how people internalize, narrate, and explain the world around them.

Conclusion

> The promise is that again and again, from garbage, the scattered feathers, the ashes and broken bones, something new and beautiful may be born.
>
> (John Berger)

Our account of the failure at Copenhagen – if it indeed is paradigmatic for other sorts of profound global environmental crises – points to a deep vein of intellectual and political pessimism. The need for rapid policy change, coupled with the demands for far reaching transformative action at the sub-national level, and for a meaningful degree of international co-operation and consensus globally, render the failures at Copenhagen all the more debilitating. Political ecology pushes us toward this point by highlighting the fundamental contradictions between the logic of global, and other forms, of capitalism and the very idea of reaching ecological resiliency and sustainability. At base, political ecology says that we need a set of interventions and frameworks capable of laying the groundwork upon which we in the North, and the South too in different ways, must change our *whole way of life*. This is not to suggest a "return to the Stone Age." It is simply to suggest that tinkering around the edges of capitalism will not do, will not make sufficient difference. And social transformation must happen, moreover, within a time frame – our lifetimes – that is intimidating in its urgency. Evidence, nonetheless, continues to point to human capacity for collectivity, cooperation, and transformation, especially amidst the kinds of disasters and crises predicted by a runaway global political economy. As Rebecca Solnit has observed, it is precisely during catastrophe that people appear to act with the most unforeseen sociability and collectivity. Indeed, it is arguably the breakdown of stifling restraint that frees the possibilities of social and political transformation:

> The possibility of paradise hovers on the cusp of coming into being, so much so that it takes powerful forces to keep such a paradise at bay. . .If paradise now arises in hell, it's because in the suspension of the usual order and the failure of most systems, we are free to live and act another way. . .The positive emotions that arise in. . .unpromising circumstances demonstrate that social ties and meaningful work are deeply desired, readily improvised, and intensely rewarding. The very structure of our economy and society prevent these goals from being achieved.
>
> (Solnit 2009: 29)

So, what cynics dismiss as utopian dreaming is actually the sober consideration of necessary social response. Hence, much of *Global Political Ecology* is normative and prescriptive, pointing toward a raft of modalities (movements, technologies, policies and practices) from which we may draw strength. Even in the depth of darkness, we must all take heart from the late Edward Said's (1983: 247) observation about the human condition: the unstoppable predilection for alternatives.

Notes

1   Donnella Meadows, Jørgen Randers, and Dennis Meadows updated and expanded the original version in *Beyond the Limits* in 1993 – a 20-year update on the original material. The most recent updated version was published in 2004 by Chelsea Green Publishing Company and Earthscan under the name *Limits to Growth: The 30-Year Update*. In 2008 Graham Turner at the Commonwealth Scientific and Industrial Research Organization (CSIRO) in Australia published a Working Paper entiled "A Comparison of 'The Limits to Growth' with Thirty Years of Reality" which compared the past thirty years of reality with the predictions made in 1972. He found that changes in industrial production, food production and pollution are all in line with the book's predictions of "economic and societal collapse in the 21st century" (see http://www.csiro.au/files/files/plje.pdf).

2   This is a pretty tall order we realize. We recommend reading Marx's *Capital* (1967) or one of excellent political-economic histories of capitalism (for example Hobsbawm 1979).

References

Ackerman, F. 1997. *Why do we recycle?* Washington DC: Island Press.

Adams, W.M. and M. Mulligan, eds., 2003. *Decolonizing development: Strategies for conservation in a post-colonial era*. London: Earthscan.

Adger, W.N., T A. Benjaminsen, K. Brown, and H. Svarstad 2001. Advancing a political ecology of global environmental discourses. *Development and Change* 32 (4): 681–715.

Adger, Neil, Irene Lorenzoni, and Karen O'Brien, eds, 2009. *Adapting to climate change*. London: Cambridge University Press.

Aglietta, M. 1979. *A theory of capitalist regulation*. London: New Left Books.

Agrawal, A. 2001. Common property institutions and sustainable governance of resources. *World Development* 29 (10): 1649–1672.

Agrawal, A. 2006. *Environmentality*. Durham, NC: Duke University Press.

Agrawal, Arun and Nicholas Perrin, 2009. Climate adaptation, local institutions and livestock, in Neil Adger, Irene Lorenzoni, and Karen O'Brien, eds, 2009 *Adapting to climate change*. London: Cambridge University Press, pp. 350–368.

Anderson, P. 2000. "Renewals," *New Left Review*, 1, 5–24

Barnes, T.J. and J.S. Duncan 1992. Introduction: Writing worlds, in T.J. Barnes and J.S. Duncan, eds., *Writing worlds: Discourse, text, and metaphor in the representation of landscape*, pp. 1–17. New York: Routledge.

Becker, J. 1977. *Marxian political economy*. Cambridge: Cambridge University Press.

Benjaminsen, T.A. 2001. The population-agriculture-environment nexus in the Malian cotton zone. *Global Environmental Change-Human And Policy Dimensions* 11 (4): 283–295.

Berkes, F. 1999. *Sacred ecology: Traditional ecological knowledge and resource management*. Philadelphia: Taylor & Francis.

Bernstein, S. 2001 *The compromise of liberal environmentalism*. New York: Columbia University Press.

Blaikie, P. 1985. *The political economy of soil erosion in developing countries*. New York: Longman Scientific and Technical.

—— and H. Brookfield 1987. *Land degradation and society*. London and New York: Methuen.

Boal, Iain 2009. *Climate, globe, capital*. Berkeley: RETORT.

Bollier, D. 2003 *Silent theft*. London: Routledge,

Boyer, R. 1990. *The regulation school*. New York: Columbia University Press

Boyle, J. 2008 *The public domain*. New Haven, CT: Yale University Press.

Braun, B. 2002. *The intemperate rainforest: Nature, culture, and power on Canada's west coast*. Minneapolis: University of Minnesota Press.

Brockington, D. 2002. *Fortress conservation: The preservation of Mkomazi Game Reserve, Tanzania*. Bloomington: Indiana University Press.

Brown, L.R. and E.C. Wolf 1984. *Soil erosion: Quiet crisis in the world economy*. Washington, DC: Worldwatch Institute.

Burns, S.S. 2007. *Bringing down the mountains: The impact of mountaintop removal surface coal mining on southern West Virginia communities, 1970–2004*. Morgantown: West Virginia University Press.

Castree, N. and B. Braun 1998. The construction of nature and the nature of construction: Analytical and political tools for building survivable futures, in B. Braun and N. Castree, eds., *Remaking reality: Nature at the millenium*, pp. 3–42. New York: Routledge.

CDIAC (Carbon Dioxide Information Analysis Center), 2009. Available online at http://cdiac.ornl.gov/trends/emis/meth_reg.html

Chhatre, Ashwini and Arun Agrawal 2009. Trade-offs and synergies between carbon storage and livelihood benefits from forest commons. *Proceedings of the National Academy of Science*, October 20th #106/42, pp. 17667–19770.

Cole, S. and T. O'Riordan 2008. Adaptive governance for a changing coastline, in N. Adger *et al.*, eds., *Adapting to climate change*. Cambridge: Cambridge University Press.

Colten, C.E. and P.N. Skinner 1996. *The road to love canal: Managing industrial waste before the EPA*. Austin: University of Texas Press.

Crosby, A.W. 1986. *Ecological imperialism: The biological expansion of Europe, 900–1900*. Cambridge: Cambridge University Press.

Dalby, S. 2002. *Environmental security*. Minneapolis: University of Minnesota Press.

Davis, D.K. 2005. Indigenous knowledge and the desertification debate: Problematising expert knowledge in North Africa. *Geoforum* 36 (4): 509–524.

—— 2007. *Resurrecting the granary of Rome: Environmental history and French colonial expansion in North Africa*. Athens: Ohio University Press.

Davis, M. 2006. *Planet of slums*. London: Verso.

Eckholm, E.P. 1976. *Losing ground: Environmental stress and world food prospects*. New York: W.W. Norton.

Edwards, D. 1995. *Small farmers and the protection of the watersheds: The experience of Jamaica since the 1950s*. Kingston, Jamaica: Canoe Press, University of the West Indies.

EIA 2009. Emissions of greenhouse gases. Report Number: DOE/EIA-0573(2008). Available online at http://www.eia.doe.gov/oiaf/1605/ggrpt/index.html.

Fairhead, J. and M. Leach 1996. *Misreading the African landscape: Society and ecology in a forest-savanna mosaic*. Cambridge: Cambridge University Press.

Forsyth, T. 2003. *Critical political ecology : the politics of environmental science*. London: Routledge.

Foucault, M. 1980. Two Lectures, in C. Gordon, ed. *Power/knowledge: Selected Interviews and other Writings 1972–1977*. New York: Pantheon.

—— 1991. Governmentality, in G. Burchell, C. Gordon and P. Miller, eds., *The Foucault effect: Studies in governmentality*, pp. 87–104. London: Harvester.

—— 2008. *The birth of biopolitics*. London: Palgrave.

Gautier, Catherine 2008. *Oil, water and climate*. Cambridge: Cambridge University Press.

Glassner, B. 2000. *Culture of fear*. New York: Basic Books.

Goldman, M. 2005. *Imperial nature: The World Bank and struggles for social justice in the age of globalization*. New Haven, CT: Yale University Press.

Goodman, David, Maxwell Boykoff, and Kyle Everd, eds., 2008 *Contentious geographies*. Aldershot: Ashgate.

Gramsci, A. 1971. *Prison notebooks*. London: Lawrence and Wishart,

Grossman, L.S. 1997. Soil conservation, political ecology, and technological change on St Vincent. *Geographical Review* 87 (3): 353–374.

Haas, P. 1992. *Saving the Mediterranean*. New York: Columbia University Press.

Hajer, M.A. 1997. *The politics of environmental discourse: Ecological modernization and the policy process*. Oxford: Oxford University Press.

Haraway, D. 1989. *Primate visions: Gender, race, and nature in the world of modern science*. New York: Routledge.

Hardin, G. 1968. The tragedy of the commons. *Science* 162: 1243–1248.

Hardt, Michael and Antonio Negri 2000. *Empire*. Cambridge, MA: Harvard University Press.

—— 2009 *Commonwealth*. Cambridge, MA: Harvard University Press.

Harvey, D. 1982. *The limits to capital* Oxford: Blackwell

—— 2003. *The new imperialism* Oxford: Oxford University Press

—— 2005. *A brief history of neoliberalism* Oxford: University Press

Helmreich, Stefan 2009. *Alien ocean*. Berkeley: University of California Press

Heynen, Nik, J. McCarthy, W.S. Prudham, and P. Robbins, eds., 2007. *Neoliberal environments*. London: Routledge.

Historical Statistics of the United States Millennial Edition http://hsus.cambridge.org/HSUSWeb/toc/hsusHome.do

Hobsbawm, E.H. 1962. *The age of capital*. London: Signet.

Hochstetler, Kathryn and Mimi Keck 2007. *Greening Brazil*. Durham, NC: Duke University Press.

Hölldobler, B. and E.O. Wilson 1990. *The ants*. Cambridge, MA: Belknap.

IMF 2009. *World Economic Outlook Database*. http://www.imf.org/external/pubs/ft/weo/2009/01/weodata/index.aspx

IPCC 2000. *Emissions scenarios*. Cambridge: Cambridge University Press.

Jacoby, K. 2001. *Crimes against nature: Squatters, poachers, thieves, and the hidden history of American conservation*. Berkeley: University of California Press.

Jasanoff, Sheila and Marybeth Martello 2004. *Earthly politics*. Cambridge, MA: MIT Press.

Jodha, N.S. 1987. A case study of the degradation of common property resources in India, in P. Blaikie and H. Brookfield, eds., *Land degradation and society*, pp.186–205. London: Routledge.

Juhasz, Antonia 2008. *The tyranny of oil*. New York: Morrow.

Klein, Naomi 2007. *Shock doctrine*. New York: Picador.

Klinenberg, E. 2002. *Heat wave: A social autopsy of disaster in Chicago*. Chicago: University of Chicago Press.

Kosek, Jake 2006. *Understories*. Durham, NC: Duke University Press.

Kuhn, T.S. 1970. *The structure of scientific revolutions*. Chicago: University of Chicago Press.

Lahsen, Myanna 2004. Transnational Locals, in Sheila Jasanoff and Marybeth Martello, eds., *Earthly Politics*. Cambridge, MA: MIT Press, pp.151–172.

Leach, M. and R. Mearns 1996. *The lie of the land: Challenging received wisdom on the African environment*. Portsmouth, NH: Heinemann.

Li, Tania 2007. *The will to improve.* Durham, NC: Duke University Press.

Lipietz, A. 1985. *The enchanted world.* London: Verso.

—— 1986. New tendencies in the international division of labor: Regimes of accumulation and modes of regulatio in A. Scott and M. Storper, eds., *Production, Work, Territory,* pp. 16–40. Boston, MA: Allen and Unwin.

—— 1987. *Mirages and miracles.* London: Verso.

Marx, K. 1959. *Economic and political manuscripts of 1844.* Moscow: Progress Publishers.

—— 1967. *Capital* 3 vols. New York: International Publishers.

—— 1973. *Grundrisse.* London: Penguin.

—— 1977. *Critique of Hegel's "Philosophy of right"* Cambridge: Cambridge University Press.

—— and F. Engels 1969. Manifesto of the Communist party, in Marx and Engels *Selected Works* Vol. 1: 98–137. Moscow: Progress Publishers.

Merton, R.K. 1973. *The sociology of science.* Chicago: University of Chicago Press.

Mitchell, B.R. 1988. *British historical statistics.* Cambridge: Cambridge University Press.

Mitchell, T. 2003. *Rule of experts.* Berkeley: University of California Press.

Monbiot, G. 2006. *Heat.* Toronto: Doubleday.

Montgomery, D.R. 2007. *Dirt: The erosion of civilizations.* Berkeley: University of California Press.

Moore, J.W. 2000. Environmental crises and the metabolic rift in world-historical perspective. *Organization and Environment* 13, 2: 123–157.

Moore, S. 2007. Forgotten roots of the green city: Subsistence gardening in Columbus, Ohio, 1900–1940. *Urban Geography* 27 (2): 174–192.

Naess, A. 1973. The shallow and the deep, long-range ecology movement: A summary. *Inquiry* 16: 95–100.

Neumann, R.P. 1998. *Imposing wilderness: Struggles over livelihood and nature preservation in Africa.* Berkeley: University of California Press.

Neumann, Rod 2004. *Making political ecology.* London: Hodder.

Nevins, Joseph and Nancy Peluso, eds., 2008. *Taking Southeast Asia to market.* Ithaca, NY: Cornell University Press.

Nicholson-Cole, Sophie and Tim O'Riordan 2009. Adaptive governance for a changing coastline, in Neil Adger, Irene Lorenzoni, and Karen O'Brien, eds., 2009 *Adapting to climate change,* pp. 368–383. London: Cambridge University Press.

Ollman, B. 1976. *Alienation: Marx's conception of man in capitalist society*, second edition, Cambridge: Cambridge University Press.

Oreskes, Naomi and Erik Conway 2008. Challenging knowledge, in Robert Proctor and Linda Schiebinger, eds., *Agnotology*, pp. 55–89. Stanford, CA: Stanford University Press,

O'Riordan, Tim 2008. Reflections on pathways to sustainability, in Neil Adger and Andrew Jordan, eds., *Governing sustainability*, pp. 307–328. London: Cambridge University Press.

Parry, B. 2007. *Trading the genome.* New York: Columbia University Press.

Paterson, Mathew 2008. Global governance for sustainable capitalism? in Neil Adger and Andrew Jordan, eds., *Governing sustainability*, pp. 99–122. London: Cambridge University Press.

Peet, R. 2007. *Geography of power.* London: Zed Books.

—— 2009. *Unholy trinity: The IMF, World Bank and WTO*, second edition, London: Zed Books.

Peluso, Nancy and Michael Watts, eds., 2005. *Violent environments.* Ithaca, NY: Cornell University Press.

Proctor, Robert and Linda Schiebinger, eds., 2008 *Agnotology.* Stanford, CA: Stanford University Press.

Ranganathan, J., R.J.R. Daniels, M.D.S. Chandran, P.R. Ehrlich, and G.C. Daily 2008. Sustaining biodiversity in ancient tropical countryside. *Proceedings of the National Academy of Sciences of the United States of America* 105 (46): 17852–17854.

Rajan, K.S. 2006. *Biocapital: The constitution of postgenomic life.* Durham, NC: Duke University Press.

Redclift, Michael 2006. *Frontiers.* Cambridge, MA: MIT Press.

Robbins, P. 1998 *Authority and environment: Institutional landscapes in Rajasthan, India.* Annals of the Association of American Geographers 88 (3): 410–435.

—— 2000. The practical politics of knowing: State environmental knowledge and local political economy. *Economic Geography* 76 (2): 126–144.

—— 2004. *Political ecology: A critical introduction.* New York: Blackwell.

—— 2007. *Lawn people: How grasses, weeds, and chemicals make us who we are.* Philadelphia: Temple University Press.

Rose, N. 2007. *The politics of life itself.* Princeton, NJ: Princeton University Press.

Rostow, W.W. 1960. *The stages of economic growth: A non-communist manifesto.* Cambridge: Cambridge University Press.

Runge, C.F. 1981. Common property externalities: Isolation, assurance, and resource depletion in a traditional grazing context. *American Agricultural Economics Association* 63: 595–606.

Said, Edward 1983. *The world, the text and the critic.* Cambridge, MA: Harvard University Press.

Sayre, N. 2006. *Ranching, endangered species and the urbanization in the southwest.* Tuscon: University of Arizona Press.

Seager, J. 1996. "Hysterical housewives" and other mad women: Grassroots environmental organizing in the United States, in D. Rocheleau, B. Thomas-Slayter and E. Wangari, eds., *Feminist political ecology: Global issues and local experiences*, pp. 271–283. New York: Routledge.

Schipper, Lisa and Ian Burton, eds., 2009. *Adaptation to climate change.* London: Earthscan.

Schneider, S.H. 1976. *The genesis strategy: Climate and global survival.* New York: Plenum.

Shapiro, J. 2001. *Mao's war against nature: Politics and the environment in revolutionary China.* Cambridge: Cambridge University Press.

Smith, A. 1937. *The weath of nations.* New York: Modern Library.

Smith, N. 1996. The production of nature, in G. Robertson, M. Mash, L. Tickner, J. Bird, B. Curtis and T. Putnam, eds., *FutureNatural: Nature/Science/Culture*, pp. 35–54. New York: Routledge.

—— 2007. Nature as accumulation strategy. *Socialist Register* 43: 16–34.

Solnit, Rebecca 2009. *A paradise built in hell.* New York: Viking.

Soper, K. 1995. *What is nature?* Oxford and Cambridge: Blackwell Publishers.

Stern, N. 2006. *The economics of climate change.* Cambridge: Cambridge University Press. Available at: http://www.occ.gov.uk/activities/stern.htm.

Szasz, A. 1994. *Ecopopulism: Toxic waste and the movement for environmental justice.* Minneapolis: University of Minnesota Press.

Toulmin, C. 2009. *Climate change in Africa.* London: Zed Press.

Traweek, S. 1988. *Beamtimes and lifetimes: The world of high energy physics.* Cambridge, MA: Harvard University Press.

US Energy Information Administration 2009. Short-term energy outlook supplement:

understanding the decline in carbon dioxide emissions in 2009. Available online at http://www.eia.doe.gov/emeu/steo/pub/special/pdf/2009_sp_06.pdf.

Vandermeer, J. and I. Perfecto 2005. *Breakfast of biodiversity: The truth about rainforest destruction*, second edition, Oakland, CA: Food First.

Watts, M.J. 1983. On the poverty of theory: Natural hazards research in context, in K. Hewitt (ed.) *Interpretations of calamity*. Boston, MA: Allen and Unwin, pp. 231–262.

Weart, S. 2004. *The discovery of global warming*. Cambridge, MA: Harvard University Press.

Weeks, J. 1981. *Capital and exploitation* London: Edward Arnold.

Williamson, J., ed., 1990. *Latin American adjustment: How much has happened?* Washington, DC: Institute for International Economics.

Wrigley, E.A. 1988, *Continuity, chance and change: The character of the industrial revolution in England*. Cambridge: Cambridge University Press.

Wrigley, E.A. 2000 The divergence of England: The growth of the English economy in the seventeenth and eighteenth centuries. *Transactions of the Royal Historical Society*, 6th series, X: 117–141.

Yardley, J. 2008. Chinese are left to ask why schools crumbled. *New York Times*, May 25, page 1.

Zhang, M.Z. and Y.J. Jin 2008. Building damage in Dujiangyan during Wenchuan earthquake. *Earthquake engineering and engineering vibration* 7 (3): 263–269.

Zimmerer, K.S. 1993. Soil erosion and labor shortages in the Andes with special reference to Bolivia, 1953–1959: Implications for conservation-with-development. *World Development* 21 (10): 1659–1675.

—— 2006. Geographical perspectives on globalization and environmental issues, in K. Zimmerer, ed., *Globalization and new geographies of conservation*, pp.1–43. Chicago: University of Chicago Press.

—— and T. J. Bassett, eds., 2003. *Political ecology: An integrative approach to geography and environment-development studies*. New York: Guilford.

Part I

Food, health, and the body: political ecology of sustainability

2   Excess consumption or over-production?: US farm policy, global warming, and the bizarre attribution of obesity

Julie Guthman

In April 2008 researchers at the London School of Hygiene and Tropical Medicine released a study which "showed" that obese people contribute more harmful gases to the planet than thin people.[1] Using the presumption that obese people eat more and drive more than thin people, they effectively deduced that the extra fuel devoted to producing and distributing food for the obese, as well as that spent on transporting them, was causing global warming. As stated by one of the principal investigators, Phil Edwards, "The main message is staying thin. It's good for you, and it's good for the planet."

This was not the first time obesity was linked to global warming – or environmental degradation and resource depletion more broadly. In a 2006 article in the journal *Agriculture and Human Values* entitled "Luxus consumption: Wasting food resources through overeating" (Blair and Sobal 2006), the authors described several calculations they had made to ascertain "the impact of eating on the ecosystem" based on current estimates of obesity rates in the US. They stated that the 4.5 kg of extra fat each person is carrying, for a total of 9.9 trillion kcals nation wide, would be released as CO_2 at our death (p. 65). They also argued that the 600 per day per capita increase of calories made "available" between 1983 and 2000, of which they estimated 400 calories were eaten and 200 wasted, was using an additional 0.36 hectares of land per capita, for a total of 100 million hectares going to produce this excess food (p. 67). They concluded by pointing to the utility of "luxus consumption" as a concept that has great potential to motivate and offer students, in particular, a link between over-consumption and environmental degradation (p. 71).

For those skeptical of the way that the rhetoric of obesity is being deployed these days, as is this author, these studies appear to be driven more by funding opportunities that encourage the invocation of millennial issues (e.g. obesity *and* global warming) than serious endeavors at explanation. Among other problems, the researchers assume that obesity is fundamentally a consequence of excessive energy intake relative to energy expenditure (calories in-calories out), a presumption which, in fact, has not been well-established in obesity science (Gard and Wright 2005). Simply put, the link between "excess" eating and body fat is not as simple as it is portrayed. Even those more convinced of the verity of the energy balance equation might struggle with the causal sequence the study suggests. Public health professionals, for example, tend to see obesity as a reflection of the "toxic environment" rather than the cause of it (e.g. Brownell 2004). Importantly, though, Brownell and others like him tend to refer to the ubiquity of fast, junky food and dearth of exercise opportunities in the *built* environment, rather than other potential environmental causes of obesity, such as the widespread use of agricultural and household pesticides which disrupt endocrinal systems (Newbold *et al.* 2008). As such, he shares with the researchers above a tendency to draw attention to the consumption behaviors of individuals.

Casting individual consumption practices as the source of public health and environmental problems, even those as complex as global warming, demonstrates the persistence of Malthusian thinking. The heart of Thomas Malthus' argument, penned in 1798, is a claim that unchecked population growth would outstrip food production. The claims above, then, are more of the *neo*-Malthusian sort, which is to say that the scope of concern is widened from food production to environmental problems and resource shortages more broadly, while the point of causation is reoriented from over-population to over-consumption. And in parallel with neo-Malthusian discourses that have blamed individual food producers at the point of production for environmental degradation, due to their lack of knowledge or negligence, here blame is attributed to individual food consumers who appear to be uneducated to or negligent of the profundity of their eating impact on either their bodies or their carbon footprints. As (I assume) the introduction of this volume discusses, the field of political ecology was in part animated by these sorts of claims and looked for explanations outside individual ignorance or negligence, such as the role of states and markets in surplus extraction or that of industrial malfeasance in pollution.

Still, the aspect of the argument that most motivates this chapter is its assumption about the economics of food production. For, in naming over-eating as a cause of global warming it would seem to suggest that excessive consumption is what drives food production and distribution. This formulation could not be more wrong. As scholars of agrarian political economy have shown many times over, the problems with the international food economy stems from excess production, particularly in the world's wealthier countries. Many forces, historical, structural, and technical, have converged to create a quite persistent logic of over-production of food, and the environmentally deleterious use of energy and chemicals are more effects of this logic rather than causes. Given emerging evidence that obesity may stem from the many environmental pollutants associated with intensive, industrial agriculture, it is profoundly ironic that fat people are being blamed for its effects. My objective in this chapter is to make this case, highlighting the relationship between food over-production, environmental degradation, and, paradoxically, food insecurity. Therefore, I will give special focus to the historical evolution of US farm and food policy, which has been especially significant in generating this inter-related set of problems. As I will show, technical and political efforts to subvert this logic have often exacerbated it, and have thus become inextricable with the US's larger geopolitical ambitions. I will thus suggest that if anything, the focus on obesity detracts from the larger social injustices and ecological concerns currently at stake in the way food is produced and distributed.

Agriculture and the state: a lesson in agrarian political economy

To understand the origin of contemporary US farm and food policy and its broader functionality within the context of its larger political and economic ambitions, it is first important to consider the underlying dynamics of agrarian capitalism, a discussion which begins with the farmer and a set of abstract propositions. Farmers, as many scholars of agrarian political economy have noted, tend to be price takers – meaning they take what they can get. In a climate of routine over-supply, as has characterized much of what goes on in basic commodity farming, high profits only come when someone else's misfortune reduces such supply through pests, drought, and other "acts of God." The problem of routine over-supply can largely be attributed to two fundamental tendencies of agricultural production, one related to its basis in land and the other related to its end in food (Fine 1994; Kautsky 1988). Regarding the former, those with already existing access to the most critical means of production – land – are loath to give it up. In wanting to hold onto their land, farmers are notorious for not fully operating in accordance with market signals. Getting some return is sometimes better than none, especially with a crop already planted, so farmers will often harvest and sell their crop no matter what the price consequences. This gives tremendous power to the buyers of farm products, such as grain traders, to set prices, a historical development that Kautsky had already noted when he wrote *The Agrarian Question* in 1898. Regarding the latter, markets for food do not expand with supply. The problem, known as Engel's Law, is that as individual income increases, people do not generally buy or consume more food proportionate to those increases, as they might DVDs or automobiles, although they do indeed buy different food as income increases. All else being equal, commodity farmers (i.e. those that grow and sell basic, undifferentiated crops such as grain) can only persist so long as price-takers. Eventually they have to pay the mortgage and/or borrow money to plant next year's crops. Without returns from the previous years, they go out of business.

Given a clear and almost invariable propensity for price competition, farmers find other ways to stay in business. One approach, referred to as *extensification*, involves expanding farming enterprises, by bringing more land into production. In the modern world such opportunities are limited unless farmers obtain land that other farmers have lost. Another approach, referred to as *intensification*, involves improving the productivity of and/or value reaped from land already in production. Intensification has generally meant farmers' adoption of various yield-enhancing technologies such as higher-yielding varietals, heavy fertility inputs, pesticides to reduce crop loss, and labor-saving technologies such as tractors and combines. Intensification is precisely the basis of increased farm productivity in the last sixty years – and also the cause of many of its environmental problems. A third approach is to move to higher-value crops, from say, wheat to oranges, as did farmers in California in the 1880s. Yet, once one farmer innovates in one of these latter two ways and reaps unusually high profits, others catch on and prices fall, making all gains temporary. This is the classic treadmill of production, which, among other things, makes farming very prone to systematic over-production and boom-bust cycles (Cochrane 1993).

Thus far this discussion has begged the role of the state, yet the state is no minor player in this predictable dynamic. Feudal states rose and fell through the tough balancing act of appropriating grain from the peasantry for both military development and lavish consumption, for which they provided military protection for their charges only in the service of more appropriation. Over-appropriation led to starvation and/or rebellion. Since the industrial revolution, ensuring a stable food supply has come to be a key source of legitimacy for both capitalist and socialist states alike. While some governments are so embedded with or indebted to others as to still depend on food imports (e.g. S. Korea and the US), many governments strive to have a stable agrarian sector.

For industrializing countries, making food affordable has often been part of the political bargain that ensures a productive and complacent labor force. Affordable food is part of what scholars call the social wage – the overall package of basic goods and services supplied through either direct wages or public goods and entitlements that allow for the reproduction of the labor force (Sen and Drèze 1989). These sorts of public goods provide a subsidy of sorts to business interests, which no longer need to pay the full costs of reproducing their labor force. Food that is sold below the cost of producing is thus such a subsidy. Cheap food policies were central to industrial development in many countries, including the former USSR which fed the cities at considerable costs to agrarian producers. India's fair price shops provided an important social safety net until its economy was neoliberalized in the 1990s. In addition, food production can provide employment and income for sizable numbers of the population – sometimes the vast majority. These days rural income mitigates destabilizing rural-to-urban migration.

Yet, a cheap and stable food supply is not the only rationale for the food and farm policy of modern states. Increasingly, the rationale for particular agricultural policies is to encourage agricultural *exports*, which can provide an important source of state revenues, through direct and indirect taxes. As such, governments with the capacity to do so (of which there are fewer all the time) want to encourage exports, while discouraging imports, which cut into employment at home. They may limit imports through protectionist mechanisms such as tariffs (taxes on incoming goods), and quotas (quantity limits on incoming goods). To make exports competitive on the world market, however, they have to be bought more cheaply than elsewhere. Export subsidies, where states pay domestic farmers a premium on world market prices, as well as indirect subsidies to agriculture such as public sector research and extension, infrastructural development, and cheap labor policies, are thus also protectionist. Farmers who do not have to pay the "full cost" of producing grain are not as vulnerable to price fluctuations. In short, there are good reasons to keep farmers in business and the surest ways to keep them in business are to moderate over-production, to make them super-competitive through technological innovation, or to open new markets. These are precisely the strategies the US government applied to its farm sector, albeit in different ways at different times. As I also will show, these fixes effectively extended the problem and created a lot of environmental and public health damage along the way, yet were done with very different aims from satisfying the desires of fat people for more food.

US Farming before the New Deal

It is arguable that US farming was bound toward problems of over-production, given US expansionist tendencies coupled with generous resource policies. Through various distributions of the public domain, most famously the 1865 Homestead Act, white farmers obtained access to land at rock bottom prices, notwithstanding the bamboozlement by some land speculators (White 1991). These land giveaways themselves had geopolitical underpinnings. The rationale of such hand-outs was not only to sow Jeffersonian democracy but also to further continental aspirations of (white) settlement from "sea to shining sea," no matter what or who stood in the way (Limerick 1987).

Contrary to highly romanticized notions of US agrarianism, farming was commercialized almost from its inception. By 1820 farms were beginning to specialize and commercialize (Danhof 1969); by 1860 regional crop specialties were firmly in place: corn/sorghum grown in lower Midwest, wheat in the upper Midwest, and, of course, cotton and tobacco, much bound for UK, in the south (Cochrane 1993; Post 1982). A system of marketing through granaries encouraged farmers to specialize in one crop or two. Early on, farmers were beholden to the quality controls and price setting of the granaries, and were thus made to compete on the basis of productivity (output per unit of land) (Cronon 1991). Regional specializations emerged in livestock production as well, with hog production centered in the mid-Atlantic states and cattle raised in the Great Plains.

Early US economic growth owed much to the grain economy and cattle boom. As Page and Walker (1991) have argued, agro-industry provided a springboard for more generalized capitalist growth. Midwest cities developed around agro-industry, in both upstream and downstream support. That is, much of the early industry was in farm inputs (tools, seeds) and outputs (granaries, processing, transport, meatpacking). Crucially, the canals and then railroads that developed to move goods to market provided both investment opportunity and opened up new markets. The relatively equitable ways in which land and other resources had been distributed to people (whites) in the Midwest helped fuel a home market for the goods being produced (Post 1982). Expansion, in these ways, helped absorb both surplus production and surplus capital (investment funds).

Yet, the US was not the only place in the world growing grain. As food scholars Friedmann and McMichael (1989) have written, an extensive, trade-oriented, world food system existed before 1914. Through the turn of the century, the British and other European powers had continued to pursue an imperial strategy of colonization to provide industrial inputs and food stuffs to their rapidly industrializing economies. Grain colonies besides the US included Canada, Prussia/Armenia, and Australia. The problem with such rapid replication was one of price competition. World grain prices collapsed in the 1870s, signaling the first instability in this already globalized economy (Cochrane 1993). After a brief recovery in the 1880s fueled in part by the cattle boom in the US, agricultural prices continued to fall even though the growth of capitalism in the cities was hale and hearty, most centrally through the development of the railroads and other heavy industry. In fact, expansion of the railroad, and the beginnings of refrigeration, actually opened up new markets in the world of "fresh" commodities. The strategic alliance of the California orange growers cooperative (Sunkist) with the Southern Pacific Railroad that allowed oranges to be shipped all over the US was a beacon of the role transportation capital would play to encourage more production.

In the US, the ability to extensify in agriculture other than through consolidation (mergers and acquisitions) was effectively terminated with the closing of the frontier in 1890. At that time the federal government declared that it would no longer give away land, and most arable land in fact had been effectively privatized or otherwise purposefully preserved in the public domain for conservation and other interests (Limerick 1987; White 1991). The proliferation of steam-powered farm machinery the 1890s allowed a wave of intensifying instead, even among family farmers. Whereas earlier wheat farms required hired labor for the arduous work of plowing, harvesting, and threshing, farms could now be run solely with family labor, supported with tractors and combines (themselves products of industrialization) (Friedmann 1978). Yet this too contributed to overproduction. Pent-up frustration led to an agrarian revolt, as farmers organized to contest the price-gouging railroads and to establish cooperative marketing arrangements to prop up poor prices. Nevertheless agricultural prices continued to plummet until 1897, the trough of a depression (Cochrane 1993).

Marking the beginning of the American century, the US embarked on its own imperial project in the Philippines, Oceania, and Latin America, to shore up resources and expand markets. Midwest farmers supported this expansion in hopes that Latin America would provide a market for wheat surplus (Trubowitz 1998). Even without these markets, the farm sector recovered nicely after 1897. Much of the glut was absorbed by new immigration when hundreds of thousands of Jews, Irish, and Italians arrived in the cities to provide a labor force for America's industrial revolution. New technologies increased farm productivity, yet not at the pace of population growth, so farm prices were high (Cochrane 1993). World War I proved an even more prosperous period for the US farm sector, when the disruptions of war abroad heightened exports.

Working in parallel, companies such as United Fruit were developing plantations and railroad transportation throughout Latin America and the Caribbean, to bring bananas and other tropical fruit into to the US (Striffler and Moberg 2003).

The decline in foreign demand after the war led to yet another glut of agricultural production in the 1920s. The Capper-Volstead Act of 1922, which exempted farmers from anti-trust legislation and thus allowed them to cooperate in producing, handling and marketing their products, did little to stem the tide of declining farm prices that were first to signal the great slump of the 1930s (Cochrane 1993). The "dustbowl" in the American southeast was a symbol of much that was lacking in agricultural policy. Without production controls or orderly marketing, that is, farmers would produce no matter what the conditions. In this case, the soil erosion that precipitated the dust storms rested squarely with high land prices, which had forced farmers to adopt technologies and over-produce to pay their rents and mortgages (Worster 1979). The economic plight of this group of tenant farmers was eventually saved by the war industry. In the meantime, pervasive hunger was not for lack of food or too high food prices. As Poppendieck (1986) chronicles in *Breadlines Knee Deep in Wheat*, unemployed urban consumers and rural farmers alike lacked income to buy the food they needed, presaging the paradox of plenty and want that has resulted from US farm policy.

New Deal farm and food policy

As the history described thus far shows, extensification and intensification, both means by which farmers attempted to increase farm income, exacerbated problems of poor crop prices and thus contributed to chronic instabilities. Current US farm policy, originally designed to minimize these boom-bust cycles, owes much to the New Deal. The gist of New Deal farm policy was enhanced government spending to restore farm prices and hence farmer incomes. Specifically, the Agricultural Adjustment Acts of 1933 and 1938 entailed government loans that would allow farmers to store commodities rather than market them, so not to glut the market. The loan program provided a minimum price support, because if market prices fell below the set rate, farmers would then put excess grain in storage; if the crops were never sold, the government effectively bought them (Cochrane 1993). At first, the adjustment acts applied to "basic commodities" e.g. wheat, corn, cotton, rice, tobacco, peanuts; eventually price supports were extended to other commodities including soy, hogs, and dairy products.

Subsidizing certain commodities put growers of these crops on a much stronger footing politically. These farmers could stay in business when otherwise competition would prevail. Their profitability – government supported – helped support organizations such as the Farm Bureau that became an effective lobby for large farm interests. The subsidy system thus gave rise to a powerful farm bloc, which continued to advocate for more of these types of programs – including guaranteed prices (McConnell 1953). The specifics of the programs shifted over years but the basic mechanism to support prices became production controls, and most production controls took the form of acreage restrictions. In other words, it was these sorts of programs that were later characterized as paying farmers not to grow crops. But, as is the case with many policies, these policies produced unintended consequences. Acreage restrictions effectively encouraged growers to intensify production on the ground they had, playing yet again into the cycle of over-production.

Yet, it was not only direct subsidies that encouraged over-production. The price-taking position of farmers proved a recurring motivation to adopt the newest technologies that would yield, well, more yield. There was lots of money to be made in selling these technologies. New technologies thus played a huge part in the major intensification in agriculture that characterized the Post-WWII period. Following its successful use to control malaria and typhus during World War II, DDT was soon thereafter introduced to crop production, and the use of agro-chemicals to control pests proliferated. The widespread adoption of hybrid corn in the 1940s (albeit invented twenty years prior) was one of the most critical for this story. To be sure, hybrid varieties allowed for enormous yield increases in corn, and lower, but still significant increases for other grain crops. Hybridization had other effects, as well. Since hybrid varietals do not "breed true," farmers committed to such varietals were forced to return each year to the seed company to buy more – and they often become reliant on other chemical inputs designed to work with hybrid varietals (Kloppenburg 2005). Hybridization thus hastened appropriationism, referring to processes where industry seizes processes once part of farm production and sells them back as inputs (Goodman *et al.* 1987).[2] Meanwhile, the development of the inter-state highway system following the war, along with the availability of cheap domestic oil from Texas and California, also encouraged high petroleum energy forms of food production and transportation.

Even before the neoliberal attack on regulation in the 1980s, not much was done to thwart the use of toxic substances in crop protection and food processing. The 1958 Delaney clause of the Food and Drug Administration (FDA) had ordered zero tolerance for those food substances deemed to be carcinogenic, yet very few such substances were actually banned from use following that amendment. Saccharin was the most notable exception, banned when it was found to cause cancer in laboratory rats in the early 1970s, but was put back on the market in 1977. It was not until the publication of Rachel Carson's *Silent Spring* in 1962, which condemned DDT and other synthetic pesticides for their carcinogenic properties, that the environmental externalities of intensive crop production received much attention at all. Following that, the Environmental Protection Agency was established in 1970 and was charged with regulating agricultural chemicals. Even then, the EPA had only limited successes with keeping dangerous agro-chemicals off the market, with the 1972 prohibition of several chlorinated hydrocarbon pesticides such as DDT being the most notable.

What Goodman *et al.* (1987) called "substitutionism" also played a large part in chronic over-production. Through the colonial period, the heyday of which ended at World War I, the colonized world had become a major exporter of key food and fiber materials: rubber, tropical oils, ground nuts, and sugar, along with those not easily substituted desirables that are still widely traded: coffee, tea, bananas, and cocoa. In the face of political instabilities, early twentieth-century US food manufacturers were looking for cheaper, more reliable sources of these food items and the US government encouraged development of internal supplies (Friedmann 1993). Commodities that could be grown in temperate climates substituted for these tropical imports. Beet sugar first replaced cane sugar, much before high fructose corn syrup derived from corn became the most prevalent sweetener in processed foods. Corn, safflower, and rapeseed (canola) oils substituted for palm and coconut oils, and dairy-produced butter, as well. At some point or another virtually all of these import substitutes were subsidized through farm support programs, as so-called strategic commodities. Many have since become the primary source for processing aids in mass-produced products. Lecithin, a derivative of soybeans, for example, is a widely-used emulsifier.

Another critical piece of the story is the symbiotic relationship that developed between excess grain production and a significantly more intensified livestock sector (Friedmann 1993). Rather than land-extensive grazing, grain feeding satisfied two problems: competition from other uses for ranching land and insufficient demand for the grain that was being produced. Voluminous meat consumption became a regular feature of American life, made affordable through subsidies to grain (and water) as well as the scale economies of more industrialized ways of managing livestock.

During this period, the US government found additional ways to address chronic food surpluses while simultaneously addressing issues of poverty, with entitlement programs such as school meals (1946) and food stamps (1964). Policy that served multiple ends was not limited to domestic use, however. Chronic food surpluses were put to strategic use abroad, as well (Garst and Barry 1991). Most famously, Public Law 480, AKA Food for Peace, instituted in 1954 allowed the US government to dispose of crop surpluses through direct aid, barter (for strategic raw materials), and concessionary sales. As put by then President Eisenhower, PL 480's purpose was to "lay the basis for a permanent expansion of our exports of agricultural products with lasting benefits to ourselves and peoples of other lands" (USAID 2004). Never intended for charity alone, but rather to increase the consumption of US agricultural commodities and improve foreign relations, the law proved to be an invaluable weapon for extracting political and military concessions (McMichael 2004). For example, Egypt became one of the largest recipients of US food aid in dollars upon its post-1973 accord with Israel (Dethier and Funk 1987).

Of course, the enhanced markets that both domestic and international food aid provided for farmers encouraged them to produce even more. A blip in the late 1960s and early 1970s rendered the circumstances by which even more over-planting became thinkable. *Life* magazine photos of starving children in Biafra had already kindled the idea of world food shortage. The clincher, though, was the sale of massive amounts of grain to the former Soviet Union during the détente years. The 30 million metric tons of grain sold between 1972 and 1973 represented three quarters of all commercially traded grain in the world at that time (Friedmann 1993). Almost instantly, US grain supplies shifted from surpluses to grain scarcity; this led to soaring prices for meat and grain alike. US farmers, typically responsive to price signals and egged on by then Secretary of Agriculture, Earl Butz, did as they were told and planted "from fencerow to fencerow." In doing so, they brought marginal land back into production – the very land they had been encouraged to set aside through previous policies. And to plant and market this grain, they mortgaged themselves to the hilt. When the grain sales terminated, grains yet again glutted the markets. Predictably, prices plummeted, and farmers no longer had the income to pay their mortgages. A major farm crisis ensued in the 1980s.

Post-1980 food and farm policy

In terms of farm policy, the answer to the 1980s crisis was yet more production controls as well as renewed emphasis on exports: i.e. not much new under the sun, despite declining political support for the national farm program. The production controls took the form of incentive programs to set aside acreage, such as the

Conservation Reserve Program, which began in 1985 and provided payments to farmers who agreed to grow only soil-conserving plants, such as grasses, and not to harvest or graze except in limited circumstances. Only the 1996 Farm Bill was a serious effort to dramatically reform farm policy to be more in keeping with the neoliberal agenda that Clinton, more than any other President, heartily pursued. Officially called the Federal Agricultural Improvement and Reform (FAIR) Act of 1996, it was trumpeted as the Freedom to Farm bill. The objective was to "decouple" payments to farmers from traditional commodity payments. Basically, farmers who had produced subsidized crops were allowed to shift to crops they wanted to grow, irrespective of planting history, and grow on acreage they wanted to grow on. In return they received fixed "market transition payments" that were eventually supposed to be phased out (Orden *et al.* 1999). Yet, the 2002 Farm Bill (under George W Bush) reinstated the countercyclical payments that keep farmers in business when prices fall below production costs. Not incidentally, it also allowed Conservation Reserve land to be used to grow grasses for biofuel, meaning that farmers would still receive their payments for participating in CRP while selling a harvestable crop. That biofuel has been championed as an alternative to oil dependency, but is not expected to reduce CO_2 emissions, casts further doubt on the role of excess eating in contributing to global warming. As the bioethanol boom makes clear, much of the excess production of strategic commodities is not for domestic food at all. Much is exported, and goes to feed grains· such as soy and corn, or non-food crops – and often so for geopolitical ends. It also makes clear that all fixes are temporary. After causing record-breaking corn prices in 2008, the biofuel boom began to cool, driven in part by international competition as more farmland around the world was converted to biofuels.

Nevertheless, as countercyclical programs, crop subsides became costly for the federal government, both fiscally and politically. The US treasury spent over $11 billion in commodity program payments in 2006, even with high market prices (Environmental Working Group 2009). Furthermore, approximately two-thirds of farmers did not receive any direct payments during that period, while the top 10 percent of payees received 60 percent of the payments (Cook and Campbell 2009). Why, then, did farm subsidies persist? Increasingly they were justified on the need to encourage agricultural exports. No longer, that is, did exports serve only to prop up prices at home, while exacting political concessions; they became important sources of revenue in their own right in the face of a growing balance of payment problem (Orden *et al.* 1999). These subsidies effectively enabled the US to out-compete third world producers (for instance, African cotton producers) on the world market. And so they persist, despite the fact that the US is out of compliance with the Agreement on Agriculture, negotiated in the Uruguay round of the GATT and enforced through the WTO. The power of the WTO to enforce trade agreements through allowing unilateral retaliation has been thwarted by a government willing to take the hit on certain cases because those subsidies are so powerful geo-politically. As such, US crop subsidies have had serious repercussions in relation to food sovereignty, referring to the rights and ability for so-called sovereign nations to produce their own food and eliminate reliance on politically volatile and often costly imports (McMichael 2000). Subsidized corn, shipped to Mexico, put many a Mexican corn farmer out of business and thus contributed to the flow of migrants.

Not only the subsidy system, but the insufficiency or inefficacy of public health and environmental regulation continued to play a major role in over-production. In fact, just as regulation in this arena was getting some traction, the Reagan administration came into power and did much work to delegitimize such regulation in the interest of business profitability. After the successful ban of DDT, efforts to review other agricultural chemicals were stalled, and so the much more toxic, but with shorter half-life, organophosphates stayed on the market. The Food Quality Protection Act, passed in 1996 under Clinton, actually overturned the Delaney clause, notwithstanding that it was under this act that methyl parathion was banned. Importantly, the FPPA did not address pesticides, herbicides, and fungicides associated with endocrine disruption at all, yet many widely used agricultural chemicals such as Malathion, Aldicarb, Lindane, and Atrazine are believed to be endocrine disruptors (Solomon 2000). They contribute to over-production because substantial restrictions on the use of these chemicals would likely make the prolific production of crops that rely on them less lucrative. And, as endocrine disruptors, they appear to have links with obesity – a piece of the equation that has yet to make it into the public discussion.

Still, crop production itself is not the end of the story; the economic importance of food processing, distribution and sales has become much more pronounced in the last thirty years. So to further make the case that fat consumers are not the perpetrators of the problem, we also need to know more about the role of the food industry, and how it has contributed to overproduction, food insecurity, and environmental degradation.

Beyond the farm

One sure trend since 1980 is the intense consolidation of those industries that surround the farm: the seed and chemical companies that supply farm inputs, the grain buyers, meat packers, and food processors that purchase farm commodities and then ship that food around the world. These are some of the largest corporations in the world: Monsanto, Pioneer Hybrid, Unilever, Conagra, Cargill, Tyson, Dole. The 1980s saw major mergers in this sector, and restructuring has since gone unabated. Heffernan and Constance (1994) attribute this first wave of consolidations to the decreased anti-trust enforcement during the Reagan years. Since then these firms have become quite strategic in buying and selling different brands to shore up market advantages but such market power ensures that farmers get very little (Heffernan 1998). Arguably, it is the companies who purchase crops that most support the subsidy system today, because subsidies allow them to pay even less to farmers.

The impact of these large buying firms on livestock production is particularly pronounced. For example, FDR's promise of a chicken in every pot was far surpassed, and distorted, by a chicken industry that took a no-holds barred approach to cheapening chicken, beginning in the 1960s. Its success began with a regional shift to the American south to take advantage of already existing rural poverty that would make for willing farmers and laborers. There, Tyson, Purdue and the other big chicken "integrators" began to contract the grow-out phase of chicken production to economically marginal family farms. Working with breeders, they produced chickens that would grow faster and plumper, but were also more prone to diseases such as salmonella. Since the ever-shrinking (current 38-day) period from chick to broiler is the most biologically risky for the already over-bred chickens, the economically marginal farmers were made to comply with highly specified production standards. The integrators set the prices in accordance with needs of retailers and food service corporations such as KFC, McDonalds, and Wal-Mart, and thus these farmers became price-takers in every possible way (Boyd and Watts 1997).

Intensification in the beef industry was brought to a new level with de-skilled slaughterhouse practices, as depicted by Schlosser in *Fast Food Nation* (2001). In the 1980s IBP's (now Tyson Foods) success in making dramatic cost cuts per unit forced virtually all meat processors to adopt similar practices. IBP also put the squeeze on ranchers, who either had to sell to IBP at lower rates than they previously received or go out of business. In effect, low prices to ranchers forced many into feed lot husbandry, making beef a particularly efficient protein source on *a per acre basis*, but at an enormous costs to the environment, public health, and animal welfare. For example, Confined Animal Feedlot Operations (CAFOS) are known to release enormous amounts of methane, a greenhouse gas, into the atmosphere. This would not have be possible if the costs of feedlot husbandry had not been effectively "externalized" by lax regulation of the meat industry.

The tendency toward substitutionism is the one of particular interest in relation to the question of obesity. As discussed earlier, substitutionism referred to import substitution and the replacement of agricultural products with synthetic inputs. Substitutionsim's biggest successes were first in fiber and industrial materials: nylon for cotton, latex for rubber. However, Goodman *et al.* also theorized it as a process by which industrial (processing) activity accounts for a rising proportion of value-added in food production. Firms would want to maximize substitution, as it is where profits are to be made. Today, much of what goes on in the area of substitutionism is to make food cheaper than the real deal. Consider, for instance, the shift from fruit juice to fruit drinks. The way substitutionism works is not only to change out sugar for high fructose corn syrup, but also artificial flavorings for fruit extract. Substitutionism in fact uses all sorts of micro-ingredients to impart particular food qualities and enhanced shelf life. This is where the development of generic processing additives such as wheat gluten, lecithin, guar gum, "natural" flavors, and so on come in. They cheaply (and hence profitably) add all sorts of qualities: from mouthfeel to sweetness to sometimes orgiastic flavor. A bag of Cheetos today has far more micro-ingredients than one of twenty years ago and, of course, hugely more intense flavors. These are qualities that are imparted by design, as is the particular mix of flavors that make it very difficult to eat just one once that package is open. And it is precisely these qualities that have led many to draw links between the availability of snack food and obesity.

By the same token, and crucially, substitutionism is the process that has allowed the development of diet foods. These days, many diet products work specifically to thwart the metabolization of food calories into body fats. Splenda (or sucralose), a low calorie sugar substitute, is ten times less dense, but 600 times sweeter than sugar. Very little of Splenda's sweet component – sucralose – is metabolized. The very few calories it does contain come from dextrose or maltodextrin filler. Olestra, a fat substitute designed to provide mouthfeel without caloric intake, literally passes through the digestive tract without being absorbed. Of course there are limits to products that thwart metabolic function, as the anal leakage and vitamin depletion associated with Olestra so vividly conjures up. Even with record-setting complaints about the side effects of Olestra, the FDA's primary intervention was to require that snack food containing Olestra state on the label that "Olestra may cause loose stools and abdominal cramping." After one peer reviewed study showed few side effects, in 2003 the FDA removed all labeling requirements for Olestra. Surely, if the same food system that produces "fattening" food also produces non-digestible or noxious food, with little regulatory restrictions, we have to re-think the relationship between food availability, consumption and body size.

Finally, this last period saw a huge expansion of the global food trade, especially in high value goods such as fresh fruits and vegetables, shrimp, and coffee. Much of this growth stemmed from post-colonial necessity. Namely, the failed project of development left many countries in debt, and export production was touted – no, forced upon indebted countries in compliance with the conditionalities of structural adjustment loans (McMichael 2004). Shifting from domestic food crops to internationally bound cash crops had deleterious effects on both food security and the environment in many places in the developing world. For instance, shrimp, once a luxury item, was able to become a regular feature of Sizzler and all-you-can-eat venues at bargain prices because of the development of shrimp farming all over the Pacific Rim. The ecological damage to mangroves has had clear livelihood effects on coastal fishers in those export areas, just as unfolding price competition has ruined the livelihoods of shrimpers who obtain their sources wildly. Yet, it is the ecological costs of transporting these high value crops that bears special scrutiny. By the early 90s global cool chains, made up of linked systems of refrigerated ships, trucks and airplanes, were moving fresh fruits and vegetables, shrimp and chicken, and cut flowers around the world (Friedland 1994). Surely, if the same system uses enormous amounts of fossil fuel to cool and move what are considered "healthy foods," we have to re-think the relationship between obesity and global warming.

Toward lipo-fuels?

In contrast to the neo-Malthusian (and, frankly, bizarre) idea that the over-consumption of food is causing environmental problems such as global warming, this chapter has endeavored to locate environmental problems (and food insecurity) in the over-production of food. Highlighting key moments in the American century (roughly 1880 to the present), I have specifically tried to show that US food policy has been guided by expansionist efforts at home and abroad – to expand and secure markets – and through policies that encourage high yielding crops and cheapened food, to produce as much food as possible. I have further argued that the post-1980 period often associated with neoliberalism was less a break with this pattern than an exacerbation of it. Even while the US continued to urge the liberalization of farm and food sectors abroad, US farm policy remained highly protectionist and thus at odds with neoliberal free trade ideology.

Food safety and environmental regulation, however, which was already weak, became even weaker; with agency mandates (and capacities) geared toward non-enforcement of regulations seen as unfriendly to business. With less oversight on farm inputs and additives, food sectors began to engage in unrelenting sub-stitutionism and appropriationism, which have both had tremendous effect in lowering the cost of food and in creating the public health and environmental externalities associated with global warming. Meanwhile, in the service of promoting free trade (and expanding export revenues), as well as its own sectoral fortunes, the global procurement and shipping industry has become a huge part of the food economy, as well.

Yet, for all that is at stake in the current food system, there seems to be intense concern with obesity, specifically understood as a problem of over-eating. Even those who take up the perversities of the US farm subsidy system (e.g. Pollan 2006) assume a rather linear and simple relationship between junk food availability and fatness, which, among other things, does an injustice to those with non-normative bodies, and in important respects plays into Malthusian discourses. That over-consumption is so readily accepted as a driver of the food system and its effects on global warming says as much about the poorly understood political ecology of obesity as it does about the persistence of Malthusian discourse. For, both environmentalists and food system critics have largely ignored the possible relationship between environmental toxins and obesity. That some of the possible toxins have clear links with crop production alone suggests the need to turn away from the current obsession with individual consumption habits and body sizes and at least engage more deeply with a set of farm, food safety, food trade, and environmental policies that appear deeply unethical. These policies are a result of political choices, not consumption choices, and political choices that require much more attention to the broader injustices that the food system rests on and perhaps less attention to who is eating what. If we are really interested in reducing fossil fuel dependence and solving the problem of obesity, why not simply mine bodies for excess body fat (say, lipofuels) so we could kill two birds with one stone? It could even be done locally. Dear readers, this is a joke, yet one that brings light to how bizarre the links are between obesity and global warming.

Notes

1   The term "obesity" is a medicalized notion of fatness or large size and is deeply problematic. Since space does not permit me to elaborate this point, I will use the term throughout this paper, albeit reluctantly.
2   The Ford and Rockefeller foundations played an enormous role in encouraging the adoption of such varietals around the world, in the name of improving farm output, and

also to subvert socialism, and even more capital-friendly agrarian reform around the world (Ross 1998).

References

Blair, Dorothy and Jeffery Sobal. 2006. Luxus consumption: Wasting food resources through overeating. *Agriculture and Human Values* 23: 63–74.

Boyd, William and Michael J. Watts. 1997. Agro-industrial just-in-time: The chicken industry and postwar American capitalism. In *Globalising Food: Agrarian Questions and Global Restructuring*, edited by D. Goodman and M. J. Watts. London: Routledge.

Brownell, Kelly D. 2004. *Food Fight: The Inside Story of the Food Industry, America's Obesity Crisis, and What We Can Do About it*. New York: McGraw-Hill.

Cochrane, Willard W. 1993. *The Development of American Agriculture*. Minneapolis: University of Minnesota Press.

Cook, Ken and Chris Campbell. 2009. *Amidst Record 2007 Crop Prices and Farm Income Washington Delivers $5 Billion In Subsidies* 2009 [cited January 23 2009]. Available online at http://farm.ewg.org/farm/dp_text.php.

Cronon, William. 1991. *Nature's Metropolis*. New York: W.W. Norton.

Danhof, Clarence. 1969. *Changes in Agriculture: The Northern United States, 1820–1870*. Cambridge, MA: Harvard University Press.

Dethier, Jean-Jacques and Kathy Funk. 1987. The language of food: PL 480 in Egypt. *MERIP Middle East Report* 145: 22–28.

Environmental Working Group. 2009. *EWG Farm Subsidy Database* 2009 [cited March 6 2009]. Available onlline at http://farm.ewg.org/farm/regiondetail.php?fips = 00000& summlevel = 2.

Fine, Ben. 1994. Towards a political economy of food. *Review of International Political Economy* 1 (3): 519–545.

Friedland, William H. 1994. The global fresh fruit and vegetable system: An industrial organization analysis. In *The Global Restructuring of Agro-Food Systems*, edited by P. McMichael. Ithaca, NY: Cornell University Press.

Friedmann, Harriet. 1978. World market, state, and family farm: Social bases of household production in the era of wage labor. *Comparative Studies in Society and History* 20 (4): 545–586.

—— 1993. The political economy of food. *New Left Review* 197: 29–57.

Friedmann, Harriet and Philip McMichael. 1989. Agriculture and the state system: The rise and decline of national agricultures, 1870 to the present. *Sociologia Ruralis* 29 (2): 93–117.

Gard, Michael and Jan Wright. 2005. *The Obesity Epidemic: Science, Morality, and Ideology*. London: Routledge.

Garst, Rachel and Tom Barry. 1991. *Feeding the Crisis: U.S. Food Aid and Farm Policy in Central America*. Lincoln: University of Nebraska Press.

Goodman, David, Bernardo Sorj, and John Wilkinson. 1987. *From Farming to Biotechnology*. Oxford: Basil Blackwell.

Heffernan, William. 1998. Agriculture and monopoly capital. *Monthly Review* 50 (3): 46–59.

Heffernan, William D. and Douglas H. Constance. 1994. Transnational corporations and the globalization of the food system. In *From Columbus to ConAgra*, edited by A. Bonanno, L. Busch, W. Friedland, L. Gouveia and E. Mingione. Kansas City: University Press of Kansas.

Kautsky, Karl. 1988. *The Agrarian Question*. London: Zwan Press. Original edition, 1899.

Kloppenburg, Jack. 2005. *First the Seed: The Political Economy of Plant Biotechnology*. Madison: University of Wisconsin Press.

Limerick, Patricia. 1987. *The Legacy of Conquest: The Unbroken Past of the American West*. New York: Norton.

McConnell, Grant. 1953. *The Decline of Agrarian Democracy*. Berkeley: University of California Press.

McMichael, Philip. 2000. The power of food. *Agriculture and Human Values* 17: 21–33.

—— 2004. *Development and Social Change*. 3rd ed. Thousand Oaks, CA: Sage.

Newbold, Retha, Elizabeth Padilla-Banks, Wendy Jefferson, and Jerrold Heindel. 2008. Effects of endocrine disruptors on obesity. *International Journal of Andrology* 31: 201–208.

Orden, David, Robert Paarlberg, and Terry Roe. 1999. *Policy Reform in American Agriculture*. Chicago: University of Chicago Press.

Page, Brian and Richard Walker. 1991. From settlement to Fordism: The agro-industrial revolution in the American midwest. *Economic Geography* 67: 281–315.

Pollan, Michael. 2006. *The Omnivore's Dilemma: A Natural History of Four Meals*. New York: Penguin.

Poppendieck, Janet. 1986. *Breadlines Knee Deep in Wheat: Food Assistance in the Great Depression*. New Brunswick, NJ: Rutgers University Press.

Post, Charles. 1982. The American road to capitalism. *New Left Review* 133: 30–51.

Ross, Eric B. 1998. *The Malthus Factor: Poverty, Politics, and Population in Capitalist Development*. London: Zed Books.

Schlosser, Eric. 2001. *Fast Food Nation: The Dark Side of the American Meal*. Boston: Houghton Mifflin.

Sen, Amartya and Jean Drèze. 1989. *Hunger and Public Action*. New York: Oxford University Press.

Solomon, Gina. 2000. Pesticides and human health: A resource for health care professionals. Available online at http://www.psr-la.org/files/pesticides_and_human_health_solomon. pdf. Santa Monica, CA: Physicians for Social Responsibility and Californians for Pesticide Reform.

Striffler, Steve and Mark Moberg, eds., 2003. *Banana Wars: Power, Production, and History in the Americas*. Durham, NC: Duke University Press.

Trubowitz, Peter. 1998. *Defining the National Interest: Conflict and Change in American Foreign Policy*. Chicago: University of Chicago.

USAID. 2004. Fifty years of food for peace: The history of America's food aid. Available online at www.usaid.gov/our_work/humanitarian_assistance/ffp/50th/history.html (accessed October 30, 2007).

White, Richard. 1991. *"It's Your Misfortune and None of My Own": A New History of the American West*. Norman: University of Oklahoma Press.

Worster, Donald. 1979. *Dust Bowl: The Southern Plains in the 1930s*. New York: Oxford University Press.

3   Killing for profit: global livestock industries and their socio-ecological implications

Jody Emel and Harvey Neo

Smithfield Foods, the largest and most profitable pork processor in the world, killed 27 million hogs last year. That's a number worth considering. A slaughter-weight hog is 50 percent heavier than a person. The logistical challenge of processing that many pigs each year is roughly equivalent to butchering and boxing the entire human populations of New York, Los Angeles, Chicago, Houston, Philadelphia, Phoenix, San Antonio, San Diego, Dallas, San Jose, Detroit, Indianapolis, Jacksonville, San Francisco, Columbus, Austin, Memphis, Baltimore, Fort Worth, Charlotte, El Paso, Milwaukee, Seattle, Boston, Denver, Louisville, Washington, D.C., Nashville, Las Vegas, Portland, Oklahoma City and Tucson.

(Jeff Teitz, *Rollingstone* 2008)

Introduction

In a report released in 2009, Greenpeace (2009) stated that the cattle sector in the Amazon is the single largest driver of global deforestation, responsible for 14 percent of world's annual deforestation and 80 percent of all deforestation in the Amazon. The United Nations Food and Agriculture Organization's 2006 study, *Livestock's Long Shadow*, revealed that the livestock sector releases 18 percent of greenhouse gas emissions (measured in CO_2 equivalent) – more than the transportation sector (FAO 2006). The pollution, animal cruelty, and worker abuse resulting from factory farming are increasingly covered by the popular press, as illustrated by this chapter's opening quotation. The rise of the global livestock industry has clear socio-political roots, and distinct ethical-environmental ramifications. It demands closer scrutiny. The multifarious consequences of the global livestock industry are best viewed in the broad lens of political ecology, for it is an industry that "combines the concerns of ecology and a broadly defined political economy" (Blaikie and Brookfield 1987: 17). Our focus will be on the livestock industry, in its industrial form where mass production is predicated on: the concentration of live animals (pigs, poultry or cattle), manure, and urine to small spaces where feed is brought in (concentrated animal feeding operations (CAFOs or factory farms)); and operations involving large-scale deforestation, irrigation, and improved genetics for "grass-fed" production of cattle. While mainstream discourses on factory farming and mass produced "grass-fed" beef have hitherto centered on the protection of human health and welfare in general, not least because of recent health crises like mad cow disease and avian flu, there are significant repercussions to workers, animals and the environment as well. To put it simply, the global livestock industry is political, social and ecological (Walker 2005; 2007). This chapter outlines the key economic and ecological issues, particularly relative to factory farms; as well as the implications for human and animal well being. Using various cases, with a specific focus on pigs, this chapter will articulate a political ecology of the global livestock industry and critique the attempts (or the lack thereof) to minimize its negative ramifications.

The chapter is divided into three sections. Following this short introduction, the nature and development of the global livestock industry is briefly outlined. The second, substantive, section elaborates on the various ecological issues confronting production of animals for food. The third provides more detail about the pig sector of the global livestock industry. And the conclusion reflects on the possibility of enhancing workers' and animal welfare, and ecological implications, despite the unceasing expansion of factory farming.

Intensification of the global livestock industry and factory farming

The "Livestock Revolution," which began in the 1970s, on the wave of increasing average income in developing countries, has persisted unabated. Globally, farmers produced 276 million tons of chicken, pork, beef and other types of meats in 2006, four times more than in 1961 (Halweil and Nierenberg 2008: 61). The top three most popular meats in the world, by tons consumed, are (in descending order) pork, poultry and beef. Together, they represent 93 percent of global meat output. Such a dramatic increase in demand and supply is not possible without concomitant changes in the way meat is produced.

The defining feature of the contemporary meat industry is its unceasing concentration and intensification – fewer but bigger farms or factories, with more specialization of feed and other inputs, and fewer farm workers. For example, in the United States alone, the number of pig farms decreased drastically from 2 million in 1950 to 73,600 in 2005 while the production of pigs in the same period rose from 80 million to 100 million. This concentration is accompanied by other key developments in the livestock industry. First, the production of meat has increasingly relied on contract farming, where different farms are contracted, by larger meat packing companies, to rear livestock at specific stages of the animals' growth. Second, there has been a growing market consolidation of the top meat producers, particularly in developed economies, through expansion, mergers and acquisitions. For example, in the United States, several leading companies now control most of the supply of meat in the country. In 2005, the top three beef packers in the United States controlled more than 80 percent of the market, while the pork packing industry was 64 percent controlled by four companies, up from 40 percent in 1990 (Hendrickson and Heffernan 2007). Some of these leading meat companies include Tyson Foods, which saw sales of $26.9 billion in financial year 2008 (Tyson

Foods 2008); and Smithfield Foods, which netted sales of $11.3 billion in the same period (Smithfield Foods 2008). More generally, food conglomerates have continually achieved high profits in the midst of global hunger. GRAIN, an international NGO that promotes the sustainable management and use of agricultural biodiversity, reveals that the top four fertilizer corporations in the world (Potash Corp from Canada; Norwegian Yara; Sinochem from China and US-based Mosaic) saw their 2007 profits increased 44 percent–41 percent from 2006 (GRAIN 2008).

For leading meat companies and other investment firms too, the search for bigger markets to exploit and bigger profits to reap has extended beyond national boundaries in the past decade. The Brazilian beef packing company JBS has since 2007 bought out several leading meat packing companies in the US, and is now the biggest beef processor in the world. Brazil is now the biggest exporter of beef by volume (Australia is by value) and has the biggest cattle herd (some 200 million head). Goldman Sachs in the same year bought out the largest pig producer in China for USD 252 million. China's case is particularly instructive. Although China is the world's biggest consumer of pork and accounts for between 45–48 percent of the world's pork production each year (*Pig International* 2007: 12), its pig industry is relatively undeveloped. Data released by the Chinese agriculture ministry in 2002 put the number of pig "farms" in China at an astounding 105,367,514 (*Pig International* 2005: 11). For the most part, such farms produced just 2–3 pigs for sale each year. As recently as 2002, only 4,132 farms produced more than 3,000 pigs a year (*Pig International* 2005: 11). By 2005, farms which produce less than 100 pigs a year still make up 70 percent of total output (Yin 2006: 22). In other words, China represents a gold mine for global meat processors to capture.

Besides China, other post-socialist countries in Central and Eastern Europe, which are transitioning to a more market-based economy, offer immense possibilities for investment. Smithfield, the largest pig producing company in the world, has in the past 20 years made several investments in Eastern Europe. In fact, Smithfield can be found in countries such as Spain, Poland, Romania, United Kingdom and Mexico. The spread of these global meat producers has meant that more meat is being produced by fewer farms in fewer places. This in turn results in the daily transportation of millions of animals from one end of the world to another – an unsurprising phenomenon because, for the most part, many countries are not self-sufficient in meeting their demand for meat.

The changes in the intensity of the livestock sector reflect the economic logic of a Fordist regime and produce significant social-political and environmental ramifications too. In such a regime, production tasks are divided into minute detail, and goods are mass produced. In fact Henry Ford credited his idea for car assembly to swine disassembly in slaughterhouses in Chicago (Pew 2009). For the livestock industry, mass production has led to environmental ramifications at scales that were unheard of as recent as 30 years ago. The production of meat has continued to be predicated upon increasing productivity and standardization. For ease of transportation, slaughtering, packaging and consumers' perennial demand for health and convenience, livestock animals in "modern farms" are reared to precise requirements (Ukes 1998). In many cases, the animals are owned by the "processors" or slaughterhouse owners from birth to death – contract farmers are essentially hired hands. Increasing numbers of farmers are adapting to such "modern" production and organizational methods (e.g. a contract system of farming and highly mechanized and circumscribed modes of production) introduced by established meat companies. Beyond that, farmers are also more likely to accept such changes as beneficial to themselves given, for example, the proclivity of the "market" for "standardized meat." Commodification and intensification are thus presented as a "natural" development and essential for survival to many less developed livestock producers. In his study of the Greek poultry system, Labrianidis (1995: 206) argues that while there is extreme flexibility in production (in terms of varying the quantity of meat produced) afforded by an intensified, subcontracting-based system, it is only flexibility accrued to the big processors. To sustain a system that "does not lead to an upgrading of work force," previously independent producers essentially become cheap, albeit land-owning, labor for the principal firm. More importantly, such flexibility is often made possible by contract farmers drawing from their underpaid familial networks.

Intensification processes also involve the "economic colonization of the rural areas". . .a colonization which should be resisted by rural people so as to "preserve their priceless rural culture. . .and to pursue a different strategy of "sustainable" rural development" (Ikerd 2003: 34–38; see also Whatmore 1994 for a European perspective). Anthropologist Walter Goldschmidt concurs. Reflecting on the intensification and concentration of the rural pig industry in Iowa, he noted that the "sense of community, the ideals of mutuality and the social value of civility" are eroded by the changing systems of production (Goldschmidt 1998: 185). Further, the welfare of the workers employed in big factory farms is another concern, as is the plight of animals in the midst of such economic transformations. To illuminate these concerns, we will use the hog industry as our case study, following a more generalized description of the ecological impacts of the intensifying livestock industry.

Ecological impacts of livestock production and consumption

Industrial livestock production is one of the most significant generators of ecological impacts at the global, regional, and local scales. Flooding the global markets with cheap meat, milk, and eggs has huge implications for biogeochemical cycles and land cover change. Twenty-five percent of the earth's surface is managed grazing, making it the biggest category of land use (Asner *et al.* 2004). Thirty-four percent of the world's cropland is dedicated to producing feed for livestock. Counting grazing land, as well as lands in feed crop production, the livestock sector occupies 30 percent of the ice-free terrestrial surface of the planet (FAO 2006: 4). Transformation of forest and grassland into range lands and fodder or grain crops is occurring at alarming rates, especially in South America (McAlpine *et al.* 2009). Pastures and feedcrops account for a 70 percent decrease in forested land in the Amazon (FAO 2006). These land use changes generate carbon dioxide emissions, alter biodiversity and hydrologic cycles, and produce new pollutants. Concentrated livestock production produces its own set of hazards to people and the environment, including serious nitrogen and phosphorous pollution of water resources, new viruses, and exotic new drug-laced pollutants. The FAO's comprehensive study of the ecological effects of livestock production globally generated considerable press commentary because the authors found global livestock production responsible for 18 per cent of greenhouse gas emissions, making it the single greatest anthropogenic source (Food and Agriculture Organization 2006). The Pew Foundation followed FAO's ground breaking study with its own expert-led examination of the industry in the United States. Its conclusion: "The present system of producing food animals in the United States is not sustainable and presents an unacceptable level of risk to public health and damage to the environment, as well as unnecessary harm to the animals we raise for food" (2009: viii).

The primary types of ecological impacts we consider in this chapter include GHG emissions, water pollution and supply issues, air pollution, and biodiversity impacts. These impacts differ according to the type of livestock considered, the type of production system involved (extensive or intensive), and the geography of the production sites (rural/urban, arid/temperate/tropical, coastal/interior, etc.). While we are not able to consider all of the impacts, we briefly examine the broad impacts that are detailed in other sources. We discuss the implications of these impacts and briefly consider the mitigation measures proposed. Of course, it is difficult to limit ourselves to the ecological impacts, because the intensification and expansion of this industry has many social, economic and ethical impacts that should be addressed simultaneously. The failure to combine the social, economic, ethical, and environmental impacts produces an ineffective and contradictory set of solutions.

GHG emissions

The livestock industry is responsible for generating between 4.6 and 7.1 billion tons of greenhouse gases each year, or between 15 and 24 percent of total global GHG emissions measured as CO_2 equivalents (Fiala 2008). This includes 9 percent of anthropogenic CO_2 emissions, 37 percent of methane (with 23 times the global warming potential (GWP) of CO_2) and 65 percent of anthropogenic nitrous oxide (with 296 times the global warming potential of CO_2) (FAO 2006). Livestock accounts for some 64 percent of global anthropogenic emissions of ammonia, and processing may be a significant source of high GWP gases (e.g. HFCs) as well as of CO_2 (FAO 2006). GHG emissions are most pronounced from deforestation, enteric fermentation, and manure. Cattle production accounts for most of the deforestation and much of the enteric fermentation. Pig and cattle production account for most of the methane produced from manure. These emissions are expected to grow rapidly as "demand" for meat and dairy products doubles over the next 50 years (FAO 2006), and will continue to increase despite further intensification (Fiala 2008). According to Subak (1999) and Fiala (2008), producing 1 kg of beef has a similar impact on the environment in terms of CO_2 as 6.2 gallons of gasoline, or driving 160 miles in a mid-size American car.

Other air pollutants

Concentrated animal feeding operations (CAFOs) are big producers of air pollutants. Major air-polluting gases include hydrogen sulfide, ammonia, airborne particulate matter and VOCs (Sneeringen 2009; Hoff *et al.* 2002). Particulate matter includes fecal matter, skin cells, feed materials, and the products of microbial action on feces and feed materials (Heederick *et al.* 2007). These are linked to respiratory infections, infant respiratory distress syndrome, perinatal disorders and spontaneous abortion. Sneeringer (2009) found that a doubling of livestock numbers in an area was significantly correlated with a 7.4 percent increase in infant mortality with damage to the fetus being the most likely promoter. In addition, acidification and nitrogen deposition arises from ammonia volatilization in the soil after deposition, a large part of which derives from animal excreta. This can produce forest die back and possibly other impacts although these are relatively understudied (Food and Agriculture Organization 2006: 83). Almost no consistent air pollution emission data exist for concentrated livestock production facilities, making ecological, public health and other implications virtually impossible to estimate or prevent (Sneeringer 2009; Heederick *et al.* 2007).

Water use and pollution

Global water use by the livestock industry is estimated at 16.2 km^3 per year, with cattle using the largest quantity (11.4 km^3) (see Table 4.4 in FAO 2006: 131). Drinking and servicing requirements represent only 0.6 percent of all freshwater use, but when added through the food chain, estimated water requirements, still quite simplified to include only product processing (slaughterhouses and tanneries) and feed production, exceed 8 percent of the global human water use. Feed production constitutes the largest portion: 7 percent of total human use. Estimates of groundwater depletion from feed production total 15 percent of global water depleted annually (ibid. 167). Local depletion of groundwater aquifers from livestock production is prominent on the Southern High Plains of the United States and in parts of India, China and Botswana (Brooks *et al.* 2000; Food and Agriculture Organization 2006). Seaboard Farms (one of the largest pork producers in the U.S.) in the Oklahoma Panhandle is primarily responsible for over three feet of decline in the nonrenewable High Plains Aquifer from 2001–2006 (OWRB 2007).

Water quality issues derive from both extensive and intensive livestock waste production. Animal waste can harm water quality through surface runoff, leaching into soils and groundwater, direct discharges, and spills. The more intensive the production process, the worse the brew of chemicals discharged. Nutrients (N and P primarily), sediment (erosion), pesticides, antibiotics, heavy metals, chemical disinfectants, and pharmaceuticals such as hormones are primary constituents of intensively produced animal facilities. Hormones, used to enhance growth, include testosterone, progesterone, oestradiol, zeranol, trenbolone, and melengestrol (the latter three are synthetic). Recombinant Bovine Somatotropin (rBGH) is the most prevalent hormone used to stimulate milk in cows.[1] Endocrin disruption in humans is a hormone-based concern, and aquatic systems are quite vulnerable (Raloff 2002). Even under extensive production modes, streams and groundwater can be polluted by nitrogen from excreta.

Land application of animal manure can lead to the accumulation of heavy metals and phosphorus in soil. And dairy soil has been shown to be a reservoir of multi-drug-resistant bacteria in the transmission of infectious disease from farm animals to humans (Burgos *et al.* 2005). While metal accumulation in soil is context dependent, Ni, Co, and Cr are detected in most samples of swine slurry (Suresh *et al.* 2009). Also, pesticides used in feed stock production pose threats to soil and water quality, as do fertilizers.

Biodiversity

Livestock production affects biodiversity in a number of ways, depending upon the mode of production and cultural context. Wholesale deforestation impacts a vast array of plant and animal species. Pollution affects terrestrial and aquatic species. Some of the biggest fishkill catastrophes in US history occurred from pig facility lagoon ruptures during large storms, e.g. a billion fish were killed in a North Carolina river in 1991 – bulldozers were used to clean them off beaches. Some 150 miles of Missouri's streams were polluted from swine CAFOs, causing 61 fish kills and killing more than 500,000 fish. Buffalo Lake National Wildlife Refuge in Texas experienced large fish kills in the 1960s and 1970s due to field run-off and discharges from cattle feedlots. Eutrophication may be the biggest threat to biodiversity (on the North Carolina Coastal Plain alone an estimated 124,000 metric tons of nitrogen and 29,000 metric tons of phosphorus are generated annually by livestock) in areas where cyanobacteria blooms and outbreaks of avian botulism and avian cholera also occur (Schwarz *et al.* 2004). Concentrations of hormones fed to animals may have great implications for the health of aquatic organisms, as research shows that even low-level exposure to select hormones can illicit deleterious effects in aquatic species (Kolpin *et al.* 2002).

The pig industry

The development of the pig industry is a suitable case to ground the preceding discussion. Pig meat is the biggest category of global production in weight (over 100 million tons) although continuing increases have been thwarted by rapidly spreading diseases. A typical pig CAFO in the US might consist of several aluminum buildings with no windows, housing tens of thousands of pigs. The pigs, smarter than dogs, with behavioral and social impulses of their own, are forced to live their lives out in crowded pens with slatted floors, standing and sleeping in their own waste (much the same as feedlot cattle). They produce prodigious amounts of waste (each adult pig produces ten times more than a human) and require very few workers, especially in the new computerized facilities. Yarger (1996, cited in Horwitz 1998: 42), described the modern industrialized system of rearing pigs as such:

If you aren't familiar with the confinement method of raising hogs, picture a warehouse, not a barn, housing animals who never see the outdoors. They live in individual pens inside buildings where the feed and water is completely mechanized and the need for human labor slight . . . From an animal rights perspective, confinement is inhumane and unnatural. The animals often experience crippling because of the metal or concrete floors, and sow's legs eventually break down under the stress of being forced to overproduce piglets . . . The piglets can be grown stacked on top of each other in crates. These unnatural conditions breed numerous diseases in the animals and necessitate dependency on antibiotics and sulfa drugs; how these drugs affect humans who consume the meat is unknown.

Whereas on mixed use farms, the manure of pigs is used to fertilize the soil for crops to supplement or even supplant inorganic fertilizers (OECD 2003), this course of action is untenable in mega-farms because "large operations are more likely than smaller operations to have an insufficient land base for utilizing manure nutrients" (Thu 2003:17). A typical pig CAFO of 100,000 animals can generate more waste than a city of 1 million people (Kennedy 2005). This means that the management of manure will be "driven by lowering disposal costs rather than optimizing the nutrient needs of crops and pasture" (OECD 2003: 30). Manure thus becomes a form of waste rather than a potential resource, and waste from pig facilities is particularly onerous. It may contain not only the usual excess nutrients but pathogens, trace elements, antibiotics, and hormones. A study of wetlands nearby waste lagoons in Nebraska found abundant cyanobacteria and incrocystin toxins. Tetracycline, macrolide, and diterpene antibiotics were detected in lagoon and canal sediment and water samples, as were concentrations of 17-B estradiol and testosterone. The latter exceeded toxicity thresholds for aquatic life. The magnitude of the problem is illustrated by the fact that feed mills in the US that add typical microbials and other drugs must be registered with the US Food and Drug Administration as a "drug establishment."

Pig farms also emit strong odors which can degrade the quality of life of residents nearby. As smells are more tangible and immediately apparent than water pollution to local residents, odor complaint is one of the significant public issues faced by the pig (and poultry) industry (Paton 2003). Neighbors have also reported staph infections that will not heal, as well as groundwater pollution, psychological damage, infant mortality, and multiple other respiratory problems.

Because the conditions are so dreadful (processing plants have almost 100 percent turnover annually), workers have to be recruited from desperate populations of illegal immigrants and even those who have yet to migrate (Cooper 2007). Nevertheless, the promise of "economic development" persuades government officials and some community members to sponsor or advocate for intensified animal production facilities. In three mid-western US states (Kansas, Oklahoma and Missouri), laws on the books prohibiting out of state corporate farming were lifted at the explicit demands of industry lobbies representing the interests of pig producing corporations (Seaboard and Premium Standard Farms) (Williams 2006).

Farmers had won prohibitions on corporate farms in eight Great Plains states since the 1970s. But the slogan, "pigs, poultry or prisons" denotes the choices facing many communities in rural America.

The desire to be more productive is a deep seated one, even in far flung places. In 2006, the local government in the city of Jinghong, Southern Yunnan, introduced a livestock rearing scheme, ostensibly to help marginalized ethnic minorities increase their household incomes. The scheme proved to be a modest success for all four of the villages that first adopted it, allowing them to rear a specialized indigenous breed of pigs (the Small Ear Pig) in a small scale and environmentally friendly manner (Neo and Chen 2009). However, tantalized by the prospect of profiting from the discerning tastes of increasingly affluent Chinese urbanites, the local agriculture officials began to draw up plans to "modernize" the production of pigs for themselves and are actively seeking foreign investors. As one official explains:

> Even for indigenous breeds like the Small Ear Pig, we have to try to compete with the bigger farms. The market is there and it is a big one. We must grab the chance to push more out to consumers. We have to think big and use modern technology.

> Interview with Chinese agricultural official 2008

The generally lax environmental standards and weak enforcement, coupled with an almost non-existent concern for animal welfare, means that any attempts to "modernize" livestock production in China will tend to result in significant negative impacts to animals and the environment. Moreover, the modernization process, should it materialize, would mean that the small scale marginalized farmers will be pushed aside.

Amongst other things, "modernization," especially when it is coupled with intensification and the subcontracting system, robs individual farmers of their independence and forces a complete change in the ways the latter "make their own decisions and accept the responsibilities for the impacts of their decisions on the land and on other people" (Ikerd 2003: 30). Clearly, the restructuring of the meat industry, particularly when wrought by foreign investors in less "modern" places, has real impacts on the local community and the social lives of the farmers. In Poland, whereas the meat industry was the focal point of community life in the socialist era (for example, spearheading commune social activities), foreign investors in post-socialism have generally found it difficult to replicate those kinds of ties with the local community, despite persuasive public statements on corporate social responsibility. Moreover, resistance towards these investors wanes as farmers eventually take to the organizational and production methods of the corporations; particularly when those who oppose are forced, by the "market" or otherwise, to leave the industry altogether (Micek *et al.* 2009). Nonetheless, in other places, protests against such factory farming and its environmental and social-political consequences have continued to gather strength with citizens fighting to keep these enterprises out.

Within the United States, a large number of local communities have been prevented from legislating or zoning against hog facilities (for example, in Missouri and Pennsylvania). Local community rights were sacrificed to what politicians from a distance saw to be "economic development." Many of those leading the resistance were farmers (Johnson 2003). Premium Standard Farms sued Lincoln Township in Missouri for $7.9 million after community leaders passed zoning preventing their facilities from locating there. These facilities have become environmental justice targets because of their perseverance in attempting to locate within poorer spaces – like the Rosebud (Sicangu Lakota) Reservation and the Oklahoma panhandle. Hog facilities are much more likely to be found near poor and non-white communities (Wing *et al.* 2000).

The debate over animal welfare is another way in which conflict over the pig industry manifests itself, within and beyond the local community. The key concern here, as with other livestock farming, is a universal one that objects to the distressing conditions in which pigs are born, reared and slaughtered for consumption in intensive barn systems. Responding to such criticisms, Britain has adopted five key principles of "freedom" which is considered to be a global standard in ensuring the welfare of livestock. The five principles are freedom from: malnutrition; thermal and physical discomfort; injury and disease; fear and stress; and freedom to express most normal patterns of behavior (http://www.fawc.org.uk/freedoms. htm). Such principles, because of their expressed universality, demand to be adopted everywhere – regardless of any recourse or excuse to place-based socio-cultural "exceptionalism." In other words, it is not acceptable that particular places reject these principles because their "culture," "history" or "context" is different.

Scholarly works on the abuse and ill-treatment of livestock remain scarce, largely due to the difficulty in penetrating the inner workings of factory farms (but see Marcus 2005; Eisnitz 1997). Producers and their political sponsors make it difficult to even photograph facilities in the US. The "Animal Enterprise Terrorism Act," passed in 2006, names anyone a terrorist who crosses state lines

> for the purpose of damaging or interfering with the operations of an animal enterprise; and in connection with such purpose (A) intentionally damages or causes the loss of any real or personal property . . . (B) intentionally places a person in reasonable fear . . . (C) conspires or attempts to do so.

Non-violent physical obstruction could earn someone 18 months in jail. It is important to note that a large part of the animal welfare problem is the result of too few workers in these facilities, operating under tight schedules with poor training, getting paid minimum wages. In slaughterhouses the line used to be slower, with fewer animals butchered per minute. Workers in these facilities are themselves treated as "animals," and have, in some cases, been whistle blowers on the cruelty to the other "animals."

Global livestock industry

Clearly, any attempt to regulate just one dimension of the industrial or concentrated animal factory system is insufficient, because the system is wanting in many respects.

In the US CAFO regulation is viewed as an unmitigated disaster by most environmental, public health and animal welfare groups. CAFO rules proposed in 1999 to bring states into compliance with the Clean Water Act were suspended on the first day of the Bush administration in 2001. A lawsuit by the Waterkeeper's Alliance, headed by Robert Kennedy Jr., actually resulted in suspension of the 2003 rules, which were judged so weak that they actually expanded the right to pollute. In 2008 CAFO rules came out which give even more away to corporate producers, promulgated during the last days of the Bush administration. The National Resources Defense Council called them a "Halloween Trick of the Bush Administration: Treat to Factory Farms" (NRDC http://enewsusa.blogspot.com/2008/11/industry-supports-new-cafo-rules.html). There are even fewer environmental regulatory controls in other countries. In 2007, it was estimated that China had 14,000 factory farms of which 90 percent were entirely uncontrolled (Ellis 2007). Ellis estimates that only about 5 percent of animal waste is treated in China.

Such weak regulation is apparent in other places too. For example, the Helsinki Commission (an intergovernmental group that works to protect the Baltic Sea comprising the nations of Denmark, Estonia, Finland, Germany, Latvia, Poland, Russia and Sweden) notes that:

> Industrial animal farming seeks out competitive advantage by shifting the costs of production onto individual farmers and the environment. Smithfield Foods, a world leader in such practices, pays only a marginal fee for its place in the Polish countryside yet at the same time leaves its own unique mark – and noises and smells. Even more serious are the impacts in neighboring village where the ammonia emissions from the pig farms have been rising drastically.
>
> Helsinki Commission 2004

Indeed, large corporations are adept in getting public funds to spearhead their expansion ambitions. When Smithfield took over Animex Ltd of Poland in 1999, it managed to obtain US$ 100,000,000 from the European Bank for Reconstruction and Development (EBRD) and Rabobank Poland to ostensibly modernize the pork industry in Poland. While EBRD insists on an "environmental mandate" for its loans, Smithfield eventually used the money to finance the construction of new industrial pig farms, indicating that the regulatory and governance environment is generally weak and ineffectual. This turn of events has led the Central and Eastern Europe Bankwatch Network, an international NGO, to comment that:

> What makes this situation even more scandalous is that the EBRD loan, officially intended to develop meat processing, has helped to bring about rapid growth in industrial animal farms throughout Poland. Through public farms, therefore, small farms disappear, food quality deteriorates and the environment is transformed.
>
> Helsinki Commission 2004

A weak and oftentimes inconsistent regulatory regime in Malaysia's pig industry, has led to a general reluctance on the part of the pig farmers to improve the environmental performance of their farms (Neo 2009). Thus, what is needed is not over-regulation or minimal regulation but clear, consistent, enforceable and environmentally-sound regulation.

Citizens have had to work very hard to get governments to regulate these facilities. In general, governments have cooperated with corporations and farmers' alliances. Because the US federal government has been slow to regulate, states and local governments have taken the initiative. North Carolina became the first US state to ban the construction or expansion of new lagoons and sprayfields in 2007. California, Arizona, Colorado and Florida have banned the use of gestation crates for sows – crates that would not allow them to turn around. California also banned the use of layer hen cages. In general, the EU and northern European governments are much more proactive.

On the industry side, most of the efforts to "fix" environmental impacts have focused on greening their image. The biggest efforts involve the agricultural industrial scientific complex in efforts to capture the gases and to re-engineer the animal or the feed. Methane efficiencies can be achieved through productivity increase from hormone treatment for eliciting and prolonging lactation, through use of anabolic steroids (progesterone and testosterone), an increase in reproductive performance (artificial insemination), alteration of ruminant flora, systematic breeding to reduce methane. Many corporations are trumpeting their biogas burners which capture the gases and use the fuel to produce electricity. Others are joining carbon exchanges.

Recognizing the benefits to health, animals, and environment from consuming less meat, NGOs, public health programs, and local governments are pushing for meatless days. Ghent is the first city we know of to institute municipal meatless days in order to reduce their meat footprint. Lower meat consumption to reduce cancer rates (especially colorectal cancer) is encouraged by the World Cancer Research Fund and the American Cancer Institute for Cancer Research. Their most recent report calls for public resources to be expended on such programs because "[m]arket economies are not designed to protect public health and cannot be relied on to do so."

Anti-microbial use in livestock is recognized as a significant problem by the World Health Organization and the American Medical Association. The World Health Organization recognized "There is clear evidence of the human health consequences resulting from non-human use of antimicrobials" (WHO 2003). Some of this use has been banned in Europe but in the US, it is reported that 70 percent of antibiotics are used for non-therapeutic use in livestock production (Mellon *et al.* 2001). Animal welfare movements are achieving levels of consumer education resulting in the regulatory redefinition of production modes in parts of Europe and the US.

Conclusion

Overall, the increasing globalized nature of meat production presents several environmental-health and social-welfare challenges in the immediate future. In a

2005 report, the United Nations Food Agriculture Organization highlights that livestock production traded across international borders has increased from 4 percent in the early 1980s to 10 percent in 2005. Among other things, the report states that globalized markets are exclusive, in that only some producers meet the requirement to access them. Moreover, safety and quality requirements can become non-tariff barriers, and prices in general for meat might increase due to costs of packaging and quality control (FAO 2005: no pagination). There are also likely to be continual distinct environmental externalities where poultry and pig exporters take advantage of economies of scale to create large industrial units with potential problems of waste management and threat to the quality of local water supplies (FAO 2005: no pagination). International meat trade also increases the scope and risks of zoonotic diseases.

The search for ever lower costs of production and ever greater profits has meant that local people, workers, animals, and the environment get squeezed and squeezed some more. Intensified production thus does not provide "economic development" (Bonnano and Constance 2006; Fink 1998). Some hog facilities are 100 percent automated. Others employ only a few workers, and turnover is high because it is difficult to work under those conditions with animals that are miserable. Slaughterhouse turnover is sometimes 100 per cent per year. Corporate representatives go into Mexico looking for workers who will come to the plants. Furthermore, these facilities do not purchase many local inputs. Feed and other inputs come from contracted sites, as do the animals themselves. Owners are not local (for example, Seaboard – with farms all over the Midwest – is owned by a family in Newton, MA, USA).

The producers are clearly not going to be the solution to this problem of enormous ecological and human health (not to mention animal welfare) impact. The problem can only be solved by consumers choosing meat from other methods of production, choosing meatless alternatives to meat proteins, and pushing governments to take seriously their regulative requirements in protecting the environment and public health. Meat is tied up with notions of masculinity and the "good life" in many cultures. It has become an ingrained entitlement for many, an indicator of class equality (where there is none). Consumers seem deaf to the cries of local people stuck in environmental justice confrontations, much like those communities battling mine location. Consumers and activists need to push for reform of farm subsidy programs, the funding of small producers, anti-trust legislation to reduce vertical integration, protection of local communities, worker's rights, and anti-corporate farming regulation. Lower meat consumption is absolutely requisite to improving animal welfare, reducing biodiversity loss, and reducing GHG emissions. The growing intensity of public institutions asking for social change regarding consumption, however, suggests that more people are becoming aware of the enormous ecological footprint of the industry, in addition to the worker, farmer, and animal welfare issues. As one NGO representative has remarked, "the meat industry is broken from start to finish." It is incumbent upon us all to fix it.

Notes

1   Hormone use in livestock is the subject of multiple WTO battles between Canada, the US and the EU. Despite the controversy over its use, the US Food and Drug Administration has refused to permit labeling of non-rBGH milk products. It is outlawed in the EU.

References

Asner, G.P., Townsend, A.R., Bustamante, M.M.C., Nardoto, G.B. and Olander, L.O. (2004) "Pasture degradation in the Central Amazon: Linking carbon and nutrient dynamics with remote sensing." *Global Change Biology* 10(5): 844–862.

Associated Press (2009) "Energy from pig slurry helps fight climate change."Available online at http://www.google.com/hostednews/ap/article/ALeqM5gjMI86Hl5pTmHFlm DK8POHxUvsjgD98HCRBG0.

Blaikie, P. and Brookfield, H. (1987) *Land degradation and society*, London and New York: Methuen and Co. Ltd.

Boehlje, M (1995) "Vertical coordination and structural change in the pork industry: discussion," *American Journal of Agricultural Economics*, 74: 1225–1228.

Bonnano, A. and Constance, D. (2006) "Corporations and the state in the global era: The case of Seaboard Farms and Texas," *Rural Sociology* 71(1): 58–74.

Brooks, E., Emel, J., Robbins, P. and Jokisch, B. (2000) *The llano estacado of the U.S. Southern High Plains: Environmental transformation and the prospect for sustainability*. New York: United Nations University Press.

Burgos, J., Ellington, B., and Varela, M. (2005) "Presence of multidrug-resistant enteric bacteria in dairy farm topsoil," *J. Dairy Science* 88: 1391–1398.

Cooper, M. (2007) "Lockdown in Greeley: How immigration raids terrorized a Colorado town," *The Nation* 284(8): 11–15.

Eisnitz, G. A. (1997) *Slaughterhouse: The shocking story of greed, neglect, and inhumane treatment inside the U.S. meat industry*, New York: Prometheus Press.

Ellis, L. (2007) *Environmental health and China's concentrated feeding operations. Research brief prepared for China environment forum*. Washington, DC.

Fiala, N. (2008) "Meeting the demand: an estimation of potential greenhouse gas emissions from meat production," *Ecological Economics*, 67(3): 412–419.

Fink, D. (1998) *Cutting into the meatpacking line: Workers and change in the rural Midwest*, Chapel Hill and London: University of North Carolina Press.

Food and Agriculture Organization (2005) *The globalizing livestock sector: Impact of changing markets*. Food and Agriculture Organization of the United Nations. Available online at http://www.fao.org/docrep/meeting/009/j4196e.htm.

—— (2006) *Livestock's long shadow: environmental issues and options*. Food and Agriculture Organization of the United Nations. Available online at http://www.fao.org/ docrep/010/a0701e/a0701e00.HTM.

Furuseth, O.J. (1997) "Restructuring of hog farming in North Carolina: Explosion and implosion," *The Professional Geographer*, 49(4): 391–403.

Goldschmidt, W. (1998) "The urbanization of rural America" in K.M. Thu and E.P. Durrenberger (1998) (eds.) *Pigs, profits and rural communities*, Albany: State University of New York Press.

GRAIN (2008) *Making a killing from hunger*. Available online at http://www.grain.org/ articles/?id = 39.

Greenpeace (2009) *Slaughtering the Amazon*, Greenpeace International, downloadable from: http://www.greenpeace.org/international/press/reports/slaughtering-the-amazon

Halweil, B. and Nierenberg, D. (2008) "Meat and seafood: The most costly ingredients in the global diet," in L. Starke (ed.) *State of the World 2008: Innovations for a Sustainable Economy,* pp. 61–74. New York: W.W. Norton.

Heederick, D., Sigsgaard, T., Thorne, P., Kline, J., Avery, R., Bonlokke, J., Chrischilles, E., Dosman, J., Duchaine, C., Kirkhorn, S., Kulhanova, K., and Merchant, J. (2007) "Health effects of airborne exposures from concentrated animal feeding operations." *Environmental Health Perspectives*, 115: 298–302.

Helsinki Commission (2004) "Assessing the environmental impact of industrial pig farms: Polish green groups open up debate at the international level." Available online at http://www.ccb.se/pdf/CCB%20Press03-03-04%20Helsinki.pdf.

Hendrickson, M. and Heffernan, W. (2007) *Concentration in agricultural markets*. Available online at http://www.nfu.org/wp-content/2007-heffernanreport.pdf.

Hovorka, A.J. (2006) "The No. 1 ladies' poultry farm: A feminist political ecology of urban agriculture in Botswana," *Gender, Place and Culture*, 13(3): 207–225.

Hoff, S., Hornbuckle, K., Thorne, P., Bundy, D., and O'Shaughnessy, P. (2002) "Emissions and community exposures from CAFOs," In Iowa State University and the University of Iowa Study Group, *Iowa Concentrated Animal Feeding Operations Air Quality Study*, pp. 45–85.

Horwitz, P. (1998) *Hog ties: Pigs, manure and morality in American culture*, New York: St Martin's Press.

Huang, H. and Miller, G.Y. (2006) "Citizens' complaints, regulatory violations and their implications for swine operations in Illinois," *Review of Agricultural Economics*, 28(1): 89–110.

Ikerd, J (2003) "Corporate livestock production: Implications for rural North America" in A.M. Ervin, C. Holtslander, D. Qualman, and R. Sawa (eds.) *Beyond factory farming*, Ottawa: Canadian Centre for Policy Alternatives.

Johnson, C. (2003) *Raising a stink: The struggle over factory hog farms in Nebraska,* Lincoln: University of Nebraska Press.

Kennedy, Robert, F. Jr. (2005) *Crimes against nature*, New York: Haper Perennial.

Kolpin, D., Furlong E., Meyer, M., Thurman, E., Zaugg, S., Barber, L., and Buxton, H. (2002) "Pharmaceuticals, hormones, and other organic wastewater contaminants in US streams, 1999–2000: A national reconnaissance," *Environmental Science and Technology* 36(6): 1202–1211.

Labrianidis, L. (1995) "Flexibility in production through subcontracting: The case of the poultry meat industry in Greece," *Environment and Planning A* 27: 193–209.

Lyford, C. and Hicks, T. (2001) "The environment and pork production: The Oklahoma industry at a crossroads," *Review of Agricultural Economics* 23(1): 265–274.

McAlpine, C.A., Etter, A., Fearnside, P.M., Seabrook, L., Laurance, C. F. (2009) "Increasing consumption of beef as driver of regional and global change: A call for policy action based upon evidence from Queensland (Australia), Colombia and Brazil," *Global Environmental Change* 19(1): 21–33.

Marcus, E. (2005) *Meat market: Animals, ethics and money*, Boston, MA: Brio Press.

Matisziw, T.C. and Hipple, J.D. (2001) "Spatial clustering and state/county legislation: The case of hog production in Missouri," *Regional Studies*, 35(8): 719–730.

Mellon, M., Benbrook, C. and Benbrook, K. (2001) *Hogging it: Estimates of microbial use in livestock*. Cambridge, MA: Union of Concerned Scientists.

Micek, G., Górecki, J. and Neo, H. (2009) "Relations: Firm and local milieu in the context of foreign direct investment in the Polish pig industry" in Z. Górki and A. Zborowskiego (eds.) *Society and Agriculture*, pp. 297–309, Krakow, Poland: Instytut Geografii i Gospodarki Przestrzennej (in Polish).

Neo, H. (2009) "Institutions, cultural politics and the destabilizing of the Malaysian pig industry," *Geoforum*, 40(2): 260–268.

Neo, H. and Chen, L.H. (2009) "Household income diversification and the production of local meat: The prospect of small scale pig farming in Southern Yunnan, China," *Area*, 41(3): 300–309.

Nierenberg, D. (2005) *Happier meals: Rethinking the global meat industry*, Worldwatch Paper 171, Washington, D.C.: Worldwatch Institute.

Novek, J. (2003) "Intensive hog farming in Manitoba: Transnational treadmills and local conflicts," *The Canadian Review of Sociology and Anthropology* 40: 3–26.

Oklahoma Water Resources Board (2007) *Oklahoma Comprehensive Water Plan 2007 Status Report*. Available online at http://www.owrb.ok.gov/supply/ocwp/pdf_ocwp/WaterPlanUpdate/OCWPStatusReport2007.pdf

OECD (2003) *Agriculture, trade and the environment: The pig sector*. Paris: Organization for Economic Co-operation and Development.

Paton, B. (2003) "The smell of intensive pig production on the Canadian prairies" in A.M. Ervin, C. Holtslander, D. Qualman and R. Sawa (eds.) *Beyond factory farming*, Ottawa: Canadian Center for Policy Alternatives.

Pew Commission on Industrial Farm Animal Production (2009) *Putting meat on the table: Industrial farm animal production in America*. Available online at http://www.ncifap.org/.

Pig International (2005) "Profiling China," *Pig International*, 35(1): 12–13.

—— (2007) "China's biggest for pork," *Pig International*, 37(1): 12–19.

Raloff J. 2002. "Hormones: Here's the beef." *Sci News* 161: 10–12.

Rikoon, J.S. (2006) "Wild horses and the political ecology of nature restoration in the Missouri Ozarks," *Geoforum*, 37(2): 200–211.

Roe, B., Irwin, E.G. and Sharp, J.S. (2002) "Pigs in space: Modeling the spatial structure of hog production in traditional and nontraditional production regions," *American Journal of Agricultural Economics*, 84: 259–278.

Savard, M. and Bohman, M. (2003) "Impacts of trade, environmental and agricultural policies in the North American hog/pork industry on water quality," *Journal of Policy Modelling*, 25(1): 77–84.

Schwarz, M.S., Echols, K.R., Wolcott, M.J. and Nelson, K.J (2004) *Environmental contaminants associated with a swine concentrated animal feeding operation and implications for Mcmurtrey National Wildlife Refuge*. USFWS DEC ID: 6N45 FFS 200060006.

Smithfield Foods (2008) *Annual Report 2008*. Available online at http://investors.smithfield foods.com/financials.cfm.

Sneeringer, S. (2009) "Does animal feeding operation pollution hurt public health? A national longitudinal study of health externalities identified by geographic shifts in livestock production," *American Journal of Agricultural Economics*, 91(1): 124–137.

Storper, M. (1989) "The transition to flexible specialization in the US film industry: External economies, the division of labour and the crossing of industrial divides," *Cambridge Journal of Economics*, 13: 273–305.

Subak, S. (1999) "Global environmental costs of beef production." *Ecological Economics* 30: 79–91.

Suresh, A., Choi, H., Oh, D., and Moon, O. (2009) "Prediction of the nutrients value and biochemical characteristics of swine slurry by measurement of EC-Electrical conductivity," *Bioresource Technology* 100: 4683–4689.

Tan-Mullins, M. (2007) "The state and its agencies in coastal resources management: The political ecology of fisheries management in Pattani, Southern Thailand," *Singapore Journal of Tropical Geography*, 28(3): 348–361.

Teitz, J. (2008) "Boss Hog," *Rollingstone*, http://www.rollingstone.com/news/story/2172 7641/boss_hog/; visited 11/16/09.

Thu, K.M. and Durrenberger, E.P. (1998) (eds.) *Pigs, profits and rural communities*, Albany: State University of New York Press.

Tyson Foods, (2008) *Tyson Fact Book 2008*, downloadable from: http://www.tyson.com/ Corporate/PressRoom/docs/FY08_Fact_Book_FINAL.pdf

Ukes, F.M. (1998) "Building a better pig: Fat profits in lean meat" in J. Wolch and J. Emel (eds.) *Animal Geographies*, London, New York: Verso.

Walker, P.A. (2005) "Political ecology: Where is the ecology?," *Progress in Human Geography*, 29(1): 73–82.

—— (2007) "Political ecology: Where is the politics?," *Progress in Human Geography*, 31(3): 363–669.

Whatmore, S. (1994) "Global agro-food complexes and the refashioning of rural Europe" in N.Thrift and A.S. Amin (eds.) *Globalization, institutions and regional development in Europe*, London: Oxford University Press.

Williams, H. (2006) "Fighting corporate swine," *Politics and Society*, 34(3): 369–398.

Wing, S., Cole, D. and Grant, G. (2000) "Environmental injustice in North Carolina's hog industry," *Environmental health perspectives* 108(3): 225–231.

World Health Organization (2003) *Nutrition and the prevention of chronic diseases*. WHO Technical Report 916. Geneva: WHO.

Yin, J. (2006) "China's pig profile in numbers," *Pig International*, 2006 (September): 22–23.

4 "Modern" industrial fisheries and the crisis of overfishing

Becky Mansfield

Until late in the twentieth century, many people thought that the world's oceans were so big and fish so numerous that human activity could never have any substantial impact. What is clear now is that people have profoundly affected the world's oceans both directly and indirectly. This chapter focuses on how people's efforts to capture fish and shellfish have caused rapid declines all over the world in the abundance of many species and in the mix of species. For example, fisheries scientists recently estimated that over the past 50 years the global biomass of large predatory fish – such as tuna and swordfish – has declined by 90 percent, and that the diversity of these fish has declined 10–50 percent (Myers and Worm 2003; Worm *et al.* 2005). The decline of fish populations is often particularly hard on poor coastal communities – in both the global North and South – where many people depend on fishing (and fishing related industries, such as boat building and fish processing) for food and employment. The crisis of overfishing, then, has both environmental and socio-economic dimensions: overfishing is a problem for fish, their ecosystems, and people that depend on them.

After defining overfishing, the heart of the chapter explains why overfishing happens, arguing that it is caused by industrialization of fisheries for economic development. While every case is somewhat different – the decline of Pacific salmon is different from the decline of Atlantic cod, for example (Weber 2002) – it is clear that the main cause of overfishing is the rapid growth of fishing and seafood processing since World War II. The chapter discusses five features of the industrialization of fisheries. First is the huge scale of much fishing today: large vessels, staggering nets and fishing lines, advanced fish-finding technology, and very large seafood firms. Second, there are now global commodity chains that provide relatively wealthy consumers of the global North with a vast array of fresh fish. Third, government policies have *encouraged* industrialization of fisheries, in the name of economic development and modernization. From the US to Ghana to the World Bank, individual governments and intergovernmental agencies have not only treated fish primarily as economic resources, but have urged fishers to catch and sell more fish, and enticed them to do so with financial incentives, technical assistance, and the like. Fourth, industrial fisheries have tended to displace small-scale and artisanal fisheries, which tend to be more equitable and environmental friendly. Fifth is that, as a capital-intensive industry, the fish industry faces an inherent "contradiction" that arises because firms depend on the environment to provide necessary resources (the fish!), but – especially with competitive pressures to reduce costs – they actively avoid paying the full costs of protecting the environment on which they depend. In sum, harmful industrial fishing is the purposeful outcome of ongoing efforts to foster a western, capitalist model of development, and this capitalist model of development brings with it new pressures to continue to expand fishing effort even if this leads to degrading the very resource on which the industry depends.

The chapter also shows that the dominant explanation for overfishing is misleading and, in fact, is part of the problem, because it encourages further industrialization as the solution. The dominant explanation pivots on the seemingly apolitical idea of "the tragedy of the commons," which suggests that degradation in fisheries is inevitable as long as fisheries are treated as a "commons" rather than as private property, because in a commons no one has the incentive to conserve. This explanation ignores a host of important features of contemporary fisheries, including the vast differences between small-scale and industrial fishing, the many examples of successful management of fishing commons, and the numerous factors that influence fishing decisions. These factors indicate that individual rationality in specific property regimes is not the underlying problem. Policies based on this dominant explanation encourage capital-intensive fisheries (as opposed to labor-intensive ones), consolidated among fewer and fewer firms – in the name of efficiency, modern economic development, and market incentives. In other words, policies based on the dominant explanation tend to encourage increased industrialization of fisheries. Therefore, these policies are part of the problem – for both ocean ecosystems and poor people – not the solution. This also shows that even while dominant explanations appear apolitical, they are highly political, in that they lend support to certain outcomes and groups of people over others.

Global overfishing: definitions and evidence

What is overfishing, and what evidence shows that it exists? For this chapter, the term "overfishing" refers to a situation in which fishing substantially reduces the abundance of a population of fish; this then causes a variety of broader ecological and socio-economic changes. Fishing can reduce a population not just by killing many fish, but by reducing the abundance of breeding adults, so that they are unable to reproduce quickly enough to replenish the population. Overfishing can lead to changes in a local or regional ecosystem; for example when predatory fish are removed, smaller herbivorous fish may increase in abundance, restructuring the entire food web and making recovery of the predatory species less likely (Frank *et al.* 2005). Overfishing of some species can also reduce marine biodiversity, which undermines the resilience of marine ecosystems and can lead to collapse of additional fish populations (Worm *et al.* 2006). All of these biological and ecological changes have socio-economic dimensions as well; for example, it may take more effort (time, technology) to catch the same quantity of fish, fish may become more expensive, or desired fish may no longer be available.

One set of evidence for overfishing is the collapse of a variety of individual fisheries around the world. Such "crashes" occur when catch levels in a fishery decline due to changes in the abundance of the fish (rather than because people stopped trying to catch them); in other words, crashes occur when fishing is halted or dramatically reduced because there is no longer enough fish to sustain catch at previous levels. One of the most well-known examples is the collapse of the cod fisheries on the Georges and Grand Banks in the Atlantic Ocean off the coasts of Canada and the United States (Kurlansky 1997; Pauly and Maclean 2003; Weber 2002). Atlantic cod are infamous for their former abundance – so thick early colonists claimed they blocked ships – but in the late 1980s and early 1990s, these fisheries were declared severely overfished and were closed by both the US and Canadian governments. Closure of the Grand Banks fishery was especially devastating in Newfoundland, where 30,000 people were put out of work all at once in 1992. These fisheries have not recovered – even as some fishers (at the urging of local and national governments) have moved on to fish for other species, such as monkfish (a deep sea fish), that were formerly undesirable but now are themselves overfished. It is important to note that crashes such as that in the cod fishery suggest that overfishing had been occurring for a long time. In the short to medium term, overfishing can be masked by increased effort or improved fish finding technology that allow fishing to continue even as abundance of fish plummets.

Information about global fisheries suggests that overfishing is not just a series of isolated events, but is quite widespread. The Food and Agriculture Organization of the United Nations (FAO) is the main intergovernmental organization that collects information on fisheries and provides fishery development assistance around the world. The FAO now concludes that almost *one third* of fish stocks today are overfished (the FAO refers to them as overexploited, depleted, or recovering), while half of fish stocks are fully exploited, meaning that any expansion would lead to overfishing (FAO 2008b). In other words, all told, 80 percent of fish stocks globally are fully or over exploited. Fisheries researchers have also noted the extent to which fishing activity has expanded spatially in the past fifty years. Fisheries for large ocean-going predators (e.g. tuna, billfish) covered most of the oceans by the 1980s (leading to the reductions noted in the chapter's introduction), and fisheries that target bottom-dwelling fish (e.g. cods, flatfish, lobster) now cover the world's continental shelves to a depth of 200 meters (Myers and Worm 2003; Pauly *et al.* 2003; Worm *et al.* 2005).

This information indicates the extent of the overfishing problem, and the limited options for a quick economic fix. Were overfishing limited to a few isolated cases, these would be localized ecological and socio-economic tragedies of environmental degradation and local hardship. For the fishing industry overall – and especially larger, more mobile firms – such localized problems would not constitute a larger crisis, for they could simply move on to other places and other species. What current data suggest, however, is that this is no longer possible. Fisheries that are not already overfished are fully exploited; the global fishing industry has already moved from place to place and from species to species. This situation is reflected in FAO information regarding global fish production, which shows that today people globally produce around 140 million metric tons (over 300 billion pounds) of seafood a year – seven times as much as they did in 1950, when total global production was just 20 million metric tons; further, total production continues to climb every year (FAO 2008b). This information might seem to suggest that, in fact, fisheries are quite healthy: they are large and still growing. What these aggregate numbers mask, however, is that after rising for decades, the global fish *catch* leveled off in the late 1980s and early 1990s, at around 90–95 million metric tons per year (FAO 1998, 2000, 2008b). Capture fisheries are not growing – although, in aggregate, neither have they declined.

The remainder of global production – and all growth in production – comes from aquaculture (also known as fish farming) in which fish are raised instead of captured. Aquaculture is now the fastest growing animal food sector in the world, growing at an annual rate of almost 7 percent since 1970, so that now aquaculture provides more than a third of total volume of fish (and almost a half of fish produced for human consumption) (FAO 2008b). Farming fish represents a potential fix for fish firms looking for a way out of the crisis of overfishing – yet it is important to note that aquaculture also *contributes* to the crisis in a variety of ways (Mansfield forthcoming). For example, intensive aquaculture depends on wild fisheries to provide feed for farmed fish. It may destroy or pollute local habitats, and it drives down prices for key species such as salmon and shrimp (thus deepening the crisis for fishers trying to make a living on the wild versions of these species), and in some cases it introduces new chemicals into fish that may be harmful to human health. It seems then, that while aquaculture is becoming increasingly important in the global seafood business, it is not in itself a solution to widespread problems in fisheries.

Explaining overfishing: industrialization of fisheries for modern economic development

Today's crisis of overfishing is caused by industrialization of fisheries, since the 1950s, as an engine for capitalist economic development. This section discusses five features of industrialization that together explain why overfishing is happening on the scale it is today: first, the massive scale of fisheries today, second, the flow of fish from South to North, third, government policies for modernizing fisheries, fourth, the threat industrial fisheries pose to small-scale fisheries, and fifth, pressures to overfish faced by capitalist, industrial fishing.

Big boats and big business

One of the most striking features of contemporary fisheries is the staggering size and sophistication of available technology. It is this industrial revolution that fueled the incredible growth in the global catch of fisheries from the 1950s to the 1980s – when, as discussed above, catch leveled off. While the FAO (2008b) defines as "industrial" any fishing vessel over about 24 meters (75 feet) in length, a vessel that size would appear small compared to the largest vessels, which are over 130 meters (400 feet) long and can stay at sea for over a year (FAO 2008a). There is a variety of industrial fishing methods, including trawling (using a long net pulled behind a vessel), purse-seining (using a net to surround a school of fish), and longlines (fishing lines up to tens of kilometers long with thousands of hooks) (FAO 2009). Iconic of industrialization are the factory vessels that not only capture fish but have processing facilities on board. These factory vessels were invented in the early 1950s by European countries (led by the United Kingdom) as part of a strategy of development for war recovery (Standal 2008). These vessels were widely adopted in the 1960s by many major fishing nation-states, such as Norway, Japan, and the Soviet Union, and in the 1970s by the United States. Today industrial vessels (as defined by the FAO) are found in countries of all regions of the world, though they comprise a higher proportion of the vessels in Europe, North America, and Latin America than in Asia or Africa (FAO 2008b). Further, industrial vessels are only possible because of a range of other technological developments, including advanced refrigeration, hydraulic machinery to haul gigantic nets and lines, and fish-finding technologies such as sonar and satellite guidance systems (FAO 2008a).

Another key dimension of industrialization is that seafood – fishing, processing, marketing, etc. – is now big business. This is true not just because of the volume of fish that is caught, or its total value, which is over US$90 billion (FAO 2008b). It is also true because a few, large fishing firms from countries such as Japan, Russia, Norway, Thailand, and the United States dominate the world of commercial seafood. For example, the world's largest fishery for human consumption is that for Alaska pollock, which is found across the northern Pacific Ocean (other fisheries are for fish meal and oil used in animal feed and fertilizer). Annual catch of Alaska pollock is close to 3 million tons (FAO 2008b), about half of which is caught in US waters off Alaska (NMFS 2007). This entire amount, around 1.5 million tons, is caught by about 120 vessels, including 21 factory trawlers owned by just five firms (e.g. Trident Seafood); the other 100 vessels deliver their catch to just eight onshore processors, which are largely owned either by the same companies that own the factory trawlers or by large Japanese fish firms (e.g. UniSea, which is owned by Nippon Suisan Kaisha) (NMFS 2009; Mansfield 2004b). This fish is then used in a variety of industrial preparations – you will almost never see pollock on the menu as itself. Instead, pollock is one of the main species used in the fish sticks and fried fish fillets that are ubiquitous in grocery store freezer aisles and fast food restaurants; pollock is also one of the main species used in "surimi," a fish paste used to make imitation crab legs and other imitation products (Mansfield 2003). In other words, pollock is very much an industrial product: it is caught in vast quantities by a small number of vessels owned by very large firms, and it is mass-produced and sold by large food chains.

Consumption in the global north

Part of what makes seafood so profitable for these large firms is that fish are world travelers: much of the fish caught in industrial fisheries is consumed not by the poor, but by relatively wealthy consumers of the global North. The volume of seafood traded internationally is large and growing – and the majority of it is ending up in North America, Japan, and the European Union. Almost 40 percent of seafood production, worth almost US$90 billion, enters into international trade (data in this paragraph from FAO 2008b). Even adjusted for inflation, this is more than double the volume and value of seafood traded twenty years earlier, in the mid-1980s. The top ten exporting (producing) countries include countries from the global North and South, while in contrast, the top ten importers (consumers) are all in the North except for China – which imports many fish to process and re-export them to the North. Looking beyond these "top ten," in terms of value about 75 percent of fish exports from the South are destined for the North, and about 80 percent of imports in the North are from the South; indeed, Japan, the USA, and the EU account for 72 percent of total import value. In addition, the South provides 70 percent of world exports of non-food fish – that is, the fish meal and oil that are used in animal feed (for farmed fish, livestock, pets) and fertilizer.

What all this means is that seafood is coming to be like many other products from timber to toys: it is produced in the South and consumed in the North. It remains true that fish is an important source of protein for poor people in coastal communities around the world; for example, recent estimates suggest that fish provides about 20 percent of protein in developing countries (Béné *et al.* 2007). But these data on trade contradict common claims that demand for fish is driven by "population growth" (e.g. FAO 2008b: 164), which locates the problem in the global South (where the populations of many countries are still rising). Rather, the flow of fish from the South to the North contains a simple lesson: blame for overfishing cannot be divided equally among all people or all places. Just as it is important to understand differences between industrial and small scale fisheries, it is important to understand differences in who benefits from industrial fishing. A disproportionate share of the world's fish catch is ultimately destined for wealthy countries of the global North.

Industrial fisheries = "modern economic development"

Explaining the rise of industrial fishing is impossible without understanding the role of fisheries development policy. While industrialization might seem to be the inevitable outcome of a seemingly natural process of economic development, in fact industrialization had to be both envisioned and fostered. Fisheries have been targeted by both national and international governmental bodies (e.g. the FAO, World Bank) as an engine for regional or national economic development – for example as a resource for isolated regions with few economic options, or as a source of foreign exchange earnings for poor countries. But it is not any and all fishing that is encouraged for economic development. Rather, fisheries development has followed the model of "modernization" applied in other areas as well, such as agriculture and manufacturing. In this model, small-scale, labor-intensive fishing for subsistence and local markets is seen as irrational and inefficient, and therefore as part of the problem. Development means replacing these fisheries with "modern," capital-intensive industrial fishing that can generate the highest profits.

The most prominent form of government fisheries development assistance is subsidies, or funds for governments used for everything from building and outfitting vessels to port development to marketing fish. Even today, governments worldwide contribute about US$16 billion to increasing fishing capacity, and another US$4–8 billion in fuel subsidies (Sumaila and Pauly 2006). But governments have not just provided funds for fisheries development; they also have been central to envisioning fisheries as capital-intensive enterprises that can fuel economic development. One illustration of this larger role of government policy is the development of the fisheries along the west coast of the United States, including the fishery for pollock discussed earlier (Mansfield 2001b, 2001a). Historically, fishers in this region targeted near-shore species such as salmon and crab. In the 1960s, Japanese and Soviet factory trawlers started to target offshore species such as pollock. Then in the 1970s US government decided to embark on a program of what it called "Americanization" of these "underutilized" species, by which it meant developing a new industrial US fishery to capture these fish "for the benefit of the nation." Subsidies were part of this fisheries development program, but it also involved a complex mix of new laws, nation-to-nation negotiations, new business models, and even cultural work to make Americanization a more general goal and to make new fish products desirable to consumers. The result of this comprehensive "modern-ization" program is a fleet of large trawlers and factory trawlers that targets not only pollock but a variety of other offshore species. Despite the fact that these fisheries have been tightly managed, by the late 1980s these fisheries were considered to be at capacity (both ecologically and economically), and several species (not including pollock) have been overfished.

Another, very different illustration of the role of fisheries development policy is provided by the many countries of the South that invite distant water fishing fleets from countries of the North into their waters in exchange for financial compensation and bilateral aid. These arrangements are quite prevalent in some parts of the world. For example, on the basis of these longstanding arrangements Western Africa has been called "the fish basket" of Europe (Alder and Sumaila 2004). And the world's largest and most valuable tuna fishery is located in the western and central Pacific Ocean, where it is caught not by fleets from Pacific Island countries, but by fleets from Japan, the USA, the EU, and Australia (Petersen 2002, 2003). The fees paid for these rights to access fish are often very low, often because the receiving countries are not able to bargain effectively, given their dependence on aid from the countries doing the fishing. People studying these colonial-style arrangements have concluded that distant water fishing – which is itself subsidized in the home country – competes with local fisheries, contributes to overfishing, undermines local fishery development, exposes poorer countries to financial risk, and, ulti-mately, hinders economic development while increasing environmental degradation (Alder and Sumaila 2004; Petersen 2002, 2003). It seems then, that these distant water fishing arrangements are exemplary of how industrial fishing has been encouraged in the name of economic development (and consumption by the wealthy) – and of how such industrial economic development leads to further economic marginalization of the poor and degradation of the natural environment.

Industrial vs. small-scale fishing

In the name of modernization and economic development, policy makers have encouraged industrial fishing to replace small-scale and artisanal fishing. Yet, evidence suggests that it is small-scale fisheries that appear to offer a variety of environmental and economic benefits. There are currently just over two million motorized vessels worldwide; of these only 10 percent are longer than 12 meters in length, and less than 25 thousand (just over 1 percent) are industrial vessels (FAO 2008b). Because many people fish with non-motorized vessels such as canoes, this means that industrial vessels (again, those over 24 meters) account for much less than 1 percent of total vessels worldwide. From these numbers, it might be easy to conclude that it is small vessels that are "overpopulated," and that poor fishers with small boats must be the culprit in overfishing. This is implied in the commonly repeated phrase that "too many boats are chasing too few fish." But this attention to simple numbers ignores vast differences among kinds of fishing, such as those compiled by Daniel Pauly, a fisheries biologist who has become famous for tolling the warning bell regarding industrial overfishing. While he does not provide a precise definition of "large scale" and "small scale," his comparison is quite informative (Pauly 2006):

1. *Catch*: Large-scale fisheries capture half the annual catch for human consumption (30 million tons annually) but almost all the fish caught for fishmeal and oil (20–30 million tons). Large-scale fisheries also produce somewhere between eight and 20 million tons of "bycatch" (bycatch is unwanted fish that are then discarded dead). Small-scale fisheries, on the other hand, account for the other half of annual catch for human consumption, with almost no catch for industrial uses or bycatch.
2. *Employment*: While capturing about half of fish catch for human consumption, small-scale fisheries employ 24 times as many people as do large-scale fisheries (12 million vs. a half million). For each US$1 million invested in vessels, large-scale fisheries employ only 5–30 people, while small-scale fisheries employ 500–4,000.
3. *Fuel Use*: Large-scale fishing uses almost 40 million tons of fuel, while small-scale fishing uses just 5 million tons. Looked at in terms of how much fish you get for your fuel, large-scale fisheries catch just one to two tons of fish for every ton of fuel, whereas small-scale fisheries catch four to eight tons of fish for the same ton of fuel.

Information such as this cautions us to ask more questions when faced with raw numbers regarding "too many" of anything, whether people or fishing vessels. We must be careful to ask not just "how many," but "what are the differences among them." Even a simple distinction between large- and small-scale fisheries (which ignores large differences within these categories) suggests that labor-intensive, small-scale fishing can make important contributions to providing food and employment to coastal regions worldwide, and can do so with much less fuel and less intensive technology than does capital-intensive, industrial fishing.

This information aligns with new research on small-scale fisheries, which reverses two common assumptions. The first erroneous assumption is that poor fishers are poor because they are fishers; this assumption leads to recommendations that these people should be something else (like factory workers) and that fishing should be entirely industrial. Instead, it turns out that people fish because they are poor – in other words, fishing provides unique opportunities for *alleviating* poverty (rather than making it worse), and small-scale fisheries should be encouraged rather than undermined (Allison and Horemans 2006; Béné 2003; Béné *et al.* 2007). Second, people assume that because of their poverty, poor fishers have no other choice but to deplete fisheries to the point of overfishing. Instead, small-scale fisheries are turning out to be an important *model* for the future of fishing, because overall they are more efficient and *less* degrading than industrial fishing, and people in these fisheries are often very effective at managing their resources (Allison and Ellis 2001; Dyer and McGoodwin 1994; Pauly 2007). While it is certainly true that small-scale fisheries can degrade local environments under some conditions, it seems that blaming them for the majority of depletion is a diversion. Rather, industrial fisheries (and intensive fish farming) often compete directly with small-scale fisheries, for example by catching the same fish, disrupting ecological dynamics, or degrading habitats in ways that undermine local fisheries. In other words, what all this information suggests is that it is not poor fishers with their small boats who cause the majority of overfishing, but rather that these fishers are harmed by depletion they do not themselves create.

Contradictions of capitalism

So far, this chapter has explained overfishing as the outcome of industrial fishing happening the world over for the enjoyment of Northern consumers, all of which is envisioned and encouraged by governments in the name of fisheries development and foreign exchange earnings. It is crucial to recognize, then, that capitalist industrialization brings constant pressures for individual firms (big or small) to keep down costs. One of the main ways firms do this is by "externalizing" the costs of their impacts (including environmental, social, and health impacts), which means making the costs external to the firm itself – in other words, finding a way to make someone else pay those costs. In fisheries, this means that firms benefit from the environment – they profit from the fish – but they do not pay the full costs of the fisheries. Certainly this is the case when there are subsidies, but it is also so in less obvious ways. For example, fishing firms do not pay the full costs of fisheries management or for recovery when an area has been overfished. They do not pay when they destroy habitat or release pollutants. Industrial fleets do not pay when they undermine small-scale fisheries. Certainly this is unfair, and it is essential to understand the unequal distribution of who gets the benefits and who bears the costs when evaluating the successes and failures of a particular fishery (or fisheries in general).

Beyond immediate questions of fairness, it is also important to recognize the ways that this process of passing off the costs – of gaining benefits from fisheries without paying the full costs – represents what some scholars have called an inherent contradiction of capitalism (O'Connor 1998; for one discussion and application of this idea, see Bakker 2003). On the one hand capitalist firms depend on the environment to provide goods and services firms themselves cannot produce; on the other hand to profit and continue to grow they are under constant pressure to destroy (by externalizing costs) the very environment on which they depend. For fisheries this means that firms fundamentally depend on environmental resources they did not create – not just the fish, but the healthy ecosystems that support the fish – at the same time that they actively undermine those same environmental resources by removing fish and degrading habitats (Mansfield forthcoming). And once firms have made substantial financial investments, they have strong pressures to keep fishing, even if so doing is destructive. There are numerous ways firms, and even whole sectors, might try to overcome this contradiction. For example, they might try to apply more technology so they can find and catch fish even while they are declining – but this makes fish more expensive and leads to less profit and more overfishing. Or they might lobby for increased government subsidy – but not only does this represent a direct externalization of costs, it also leads to more rather than less overfishing. Or, as many seafood companies are doing, they might switch from fishing to fish farming – but this comes with its own pressures to externalize costs of pollution, habitat degradation, and so on. In other words, all of these efforts to escape the contradiction only exacerbate it: externalization of the problems of fisheries undermines the very resources on which fisheries depend.

In sum

Overfishing is caused by the dynamics among industrial technology, consumer markets, models of development, and capitalist relations to nature. Overfishing is the result of the massive industrialization of fisheries since the 1950s, which vastly expanded global capacity to catch fish. But technology alone is not the ultimate cause; rather fishing technology is part of a broader political – and cultural – economy of fishing that since the 1950s has focused on "modernizing" fisheries across the world. During this time, capital-intensive fishing that generates profits and foreign earnings by feeding Northern consumers has been prioritized over "traditional" small-scale and artisanal fishing for subsistence, local markets, and poverty alleviation. Overfishing, then, is not simply the result of technological capacity, but is also explained by the need to profit by externalizing costs – even if this means undermining the resources on which fisheries depend.

Not the "tragedy of the commons"

The explanation of overfishing in this chapter focuses on *fisheries development as a political process*. That is, fisheries development imposes a particular, culturally specific vision of what nature is, who should control it, how people should use it, and who should benefit. By industrializing fisheries following this Western model of modernity, fisheries development not only leads to overfishing, but it also intensifies socio-economic inequality. It benefits some groups of people, in particular wealthier fishers and fishing firms with access to capital for building and outfitting large vessels, as well as relatively well-off Northern consumers. And it makes things worse for others, especially poorer fishers (of the North as well as the South) who lose access to fisheries due to increasing costs and environmental degradation. In other words, modernized, industrial fisheries lead to both degradation and marginalization, each of which exacerbates the other.

The rest of this section shows that the dominant, mainstream, and seemingly *apolitical* explanation for overfishing conveniently overlooks all of this (see also Mansfield 2001a, 2004a, 2006). The dominant approach ignores all of these dynamics, instead explaining overfishing simply in terms of "the tragedy of the commons," which is based on the idea that individual decisions are determined by property rights (Gordon 1954; Hardin 1968). In this view, if individuals do not own a resource (such as fish) they have no interest in protecting it. This is not because they don't care, but because it is not profitable for them to do so: the individual cannot be sure that s/he will be the one to benefit, because, without ownership, someone else might come along and take whatever has been conserved. In other words, "rational" individuals are those who maximize their profits. The inverse argument is that ownership gives individuals control over access to resources, which ensures that the owner will be the one to benefit from conservation. In this view, then, private property provides incentives that match individual rationality to conservation goals. While this general argument has been applied to a wide range of resources (from trees to the internet), there is no arena in which the tragedy of the commons is more popular than in fisheries. References to "the tragedy of the commons" or to "incentives" and "rights-based" approaches, which are based on these underlying ideas about property and conservation, are ubiquitous in discussion of fisheries today. Long the view of mainstream fisheries economists (Gordon 1954; Hannesson 2004), examples also abound in the popular media (e.g. Easterbrook 2009; The Economist 2009: 17), in public policy from the Obama administration to the World Bank (NOAA 2009; World Bank 2009), and in leading scientific journals (Beddington *et al.* 2007; Costello *et al.* 2008). In this view, a lack of property rights is the problem and implementation of property rights is the solution. The most commonly referenced property right in fisheries today is some form of "catch share" or "transferable quota" system in which fishers own access to a specified share of the total fishery.

There are many things wrong with this explanation and the "solutions" to overfishing based on it. First, case study research around the world shows conclusively that the commons can be a *benefit* to conservation rather than the root of the problem (Berkes *et al.* 1989; Dyer and McGoodwin 1994; McCay and Acheson 1987; Rowe 2008). People can communicate with each other, cooperate, and have all sorts of explicit and implicit rules limiting who can use resources, when, in what ways, and so on. By showing that the commons is indeed a kind of property, not the same as open access, this evidence offers an important counter-balance to simplistic notions about common vs. private property.

Second, property-based explanations ignore the politics of fisheries development over the past 60 or so years. That is, these explanations ignore the political and cultural dimensions of fisheries development in which industrialization is a purposeful project based on Western notions of modernity and capitalist relations, as outlined in the previous section of the chapter. Instead, property-based explanations pretend that the explosion in fishing capacity in the late twentieth century "just happened" as the result of rational individual decision-making in an open-access situation (perhaps one that was encouraged by government policy, which is seen as distorting – rather than encouraging – capitalist markets). This assessment of the situation not only ignores a whole host of historical facts of the sort addressed in this chapter, but, given that it supposes a universal process, it also fails to explain why the explosion in capacity happened when and how it did. The explosion of fisheries in the twentieth century is not due to a particular property regime, but rather to the imposition of industrialization as a model of development.

Third, the problem is not just that the property-based explanations focus entirely on individual rationality while ignoring the politics of western, capitalist development. Rather, it is that by ignoring this politics mainstream analysts can pretend that "individual rationality" (defined as profit maximization) is a trait of human nature. In property-based explanations, individuals are assumed to be just like capitalist firms, in which profit is the primary motivating force. Closer attention shows that it is through this politics of western, capitalist development that many people *are forced to be* profit maximizing. For example, fishers forced into debt to keep their fishing operations alive must focus on profits. Indeed, while development specialists might pretend that being profit motivated is simply human nature, at the same time any sign of a lack of profit motive among resource users (fishers, but also farmers, hunters and gatherers, etc.) is seen as a sign of irrationality and "backwardness," and as something to be fixed. This is, in large part, what modernization entails – encouraging people to become the profit-maximizing individuals that help drive capitalist markets worldwide (Barnes 1988; Davis 1991; Feeny *et al.* 1996).

Finally, property-based approaches are a problem not just because they misdiagnose the problem, but because they propose solutions – such as "individual transferable quotas" or other sorts of "catch share" programs – that exacerbate the problems. Quota or share programs are a way of creating property rights not to the fish themselves (which is particularly difficult), but instead rights to access the fish; they generally take the form of providing some guaranteed right to a percentage of the total fishery. The first thing to note is that it is not clear that these property rights have any direct effect on how much fish is caught. Rather, it is a government authority that determines what the total catch will be (along with seasons and other regulatory measures), while the quota determines simply *who* will catch the fish (Mansfield 2004b, 2007). In other words, any environmental protection still comes from government authority, rather than from individual incentives to conserve provided by property rights. At the same time, by determining who will catch the fish, property rights in fisheries lead to increased inequality and increased industrialization of fisheries – the very thing that has caused problems in fisheries today (problems including overexploitation, ecological degradation, and decline of small-scale fisheries). The whole idea of property rights regimes in fisheries is to give some people access while excluding others. Quota programs can be designed to benefit different groups of people over time, but this does not negate the fact that (unless they aren't working as intended!) property-rights approaches provide the resource to some and take it away from others. Because quota permits become another expensive item that fishers must own in order to fish – the boat, the gear, and now the quota permit – in most cases those who are already better off will benefit the most (e.g. Mansfield 2007; Palsson and Helgason 1995). Those with access to capital will be able to buy quota permits and expand their operations, and those without will reduce the amount they fish, or stop altogether. Privatized quotas on their own do nothing to prevent overfishing, while they do much to encourage further consolidation of fishing into the hands of the wealthy, and therefore to increase inequality. In other words, quota programs encourage the further demise of small-scale fishing and intensification of industrial fishing, and do so in the name of conservation!

Conclusion

In conclusion, property-based explanations and solutions to the problem of overfishing should be seen as new chapter in the ongoing story of the politics of fisheries development that has been the focus of this chapter. In the mainstream view, the lack of private property in fisheries is seen as a sign that fisheries are traditional and backward (much as a lack of profit-maximization is a sign that small-scale fishers are irrational and backward). Using quota programs to enclose the oceans as private property is the latest means for turning fisheries into the modern, capitalist, industrial enterprise that has been envisioned and encouraged for decades. Because quota programs are rooted in notions of individual rationality and the necessity of private property, they are not only completely consistent with this vision of capitalist economic development, but in fact extend it in new ways.

But there is a fundamental problem with this vision, which is that dominant approaches to fisheries are only exacerbating the underlying problems driving overfishing today. This chapter has documented that the cause of overfishing is not a lack of property rights, but the massive and very purposeful industrialization of fisheries as a driver of capitalist economic development, which then leads to contradictory pressures to degrade the very environment on which fisheries depend. By encouraging consolidation of capital-intensive fisheries, property-based approaches to fisheries management only intensify the very sort of fishing that has created problems in the first place.

Despite these fatal problems, the tragedy of the commons remains popular as an explanatory framework. This is because it is so simple and because it blames all people equally. In so doing, it allows us to avoid thorny political questions, such as about who gets to make decisions, whose lives matter more, and who benefits from both using and conserving fish and the ecosystems that produce them. But by avoiding these political issues, property-based approaches show themselves to be *highly* political. They are part of a western, capitalist model of development that ignores history and politics by naturalizing overfishing as a problem of human nature that can be solved through capitalist markets. In the end they promote privatization as a way of further intensifying the market-relation in fisheries, and through that encourage increased industrial control of fishing. A better approach would be to promote the many small-scale fisheries that appear to be more equitable and environmentally friendly.

References

Alder, Jacqueline and Ussif Rashid Sumaila. 2004. Western Africa: a fish basket of Europe past and present. *Journal of Environment and Development* 13 (2): 156–178.

Allison, Edward H and Frank Ellis. 2001. The livelihoods approach and management of small-scale fisheries. *Marine Policy* 25: 377–388.

Allison, Edward H and Benoît Horemans. 2006. Putting the principles of the sustainable livelihoods approach into fisheries development policy and practice. *Marine Policy* 30: 757–766.

Bakker, Karen. 2003. *An Uncooperative Commodity: Privatizing Water in England and Wales*. Oxford: Oxford University Press.

Barnes, Trevor J. 1988. Rationality and relativism in economic geography: an interpretive review of the homo economicus assumption. *Progress in Human Geography* 12 (4): 473–496.

Beddington, J.R., D.J. Agnew, and C.W. Clark. 2007. Current problems in the management of marine fisheries. *Science* 316: 1713–716.

Béné, Christophe. 2003. When fishery rhymes with poverty: a first step beyond the old paradigm on poverty in small-scale fisheries. *World Development* 31 (6): 949–975.

Béné, C., G. Macfadyen and E.H. Allison. 2007. *Increasing the Contribution of Small-Scale Fisheries to Poverty Alleviation and Food Security, FAO Fisheries Technical Paper 481*. Rome: UN Food and Agriculture Organization.

Berkes, F., D. Feeny, B. J. McCay, and J. M. Acheson. 1989. The benefits of the commons. *Nature* 340: 91–93.

Costello, Christopher, Steven Gaines, and John Lynham. 2008. Can catch shares prevent fisheries collapse? *Science* 321: 1678–1681.

Davis, Anthony. 1991. Insidious rationalities: the institutionalisation of small boat fishing and the rise of the rapacious fisher. *Marine Anthropological Studies* 4 (1): 13–31.

Dyer, Christopher L. and James R. McGoodwin, eds. 1994. *Folk Management in the World's Fisheries: Lessons for Modern Fisheries Management*. Niwot, CO: University Press of Colorado.

Easterbrook, Gregg. 2009. Privatize the seas. *The Atlantic* 304 (1): 58.

FAO. 1998. *The State of World Fisheries and Aquaculture 1998*. Rome: Food and Agriculture Organization of the United Nations.

——. 2000. *The State of World Fisheries and Aquaculture 2000*. Rome: Food and Agriculture Organization of the United Nations.

——. 2008a. *Fisheries Capture Technology*. Food and Agriculture Organization of the United Nations [cited August 21 2009]. Available from http://www.fao.org/fishery/topic/3384/en.

——. 2008b. *The State of World Fisheries and Aquaculture 2008*. Rome: Food and Agriculture Organization of the United Nations.

———. 2009. *Fishing Gear Type Fact Sheets*. Food and Agriculture Organization of the United Nations, Fisheries and Aquaculture Department [cited August 21 2009]. Available from http://www.fao.org/fishery/geartype/search/en.

Feeny, David, Susan Hanna, and Arthur F. McEvoy. 1996. Questioning the assumptions of the "Tragedy of the Commons" model of fisheries. *Land Economics* 72 (2): 187–205.

Frank, Kenneth T., Brian Petrie, Jae S. Choi, and William C. Leggett. 2005. Trophic cascades in a formerly cod-dominated ecosystem. *Science* 308 (10 June): 1621–1623.

Gordon, H. Scott. 1954. The economic theory of a common-property resource: the fishery. *The Journal of Political Economy* 62 (2): 124–142.

Hannesson, Rognvaldur. 2004. *The Privatization of the Oceans*. Cambridge, MA: MIT Press.

Hardin, Garrett. 1968. The tragedy of the commons. *Science* 162: 1243–1248.

Kurlansky, Mark. 1997. *Cod: A Biography of the Fish that Changed the World*. New York: Penguin.

McCay, Bonnie J. and James M. Acheson. 1987. *The Question of the Commons: The Culture and Ecology of Communal Resources*. Tucson, AZ: The University of Arizona Press.

Mansfield, Becky. 2001a. Property regime or development policy? Explaining growth in the US Pacific groundfish fishery. *Professional Geographer* 53 (3): 384–397.

———. 2001b. Thinking through scale: the role of state governance in globalizing North Pacific fisheries. *Environment and Planning A* 33: 1807–1827.

———. 2003. Spatializing globalization: a "geography of quality" in the seafood industry. *Economic Geography* 79 (1): 1–16.

———. 2004a. Neoliberalism in the oceans: "rationalization," property rights, and the commons question. *Geoforum* 35 (3): 313–326.

———. 2004b. Rules of privatization: contradictions in neoliberal regulation of North Pacific fisheries. *Annals of the Association of American Geographers* 94 (3): 565–584.

———. 2006. Assessing market-based environmental policy using a case study of North Pacific fisheries. *Global Environmental Change* 16: 29–39.

———. 2007. Articulation between neoliberal and state-oriented environmental regulation: fisheries privatization and endangered species protection. *Environment and Planning A* 39: 1926–1942.

———. forthcoming. Is fish health food or poison? Farmed fish and the material production of un/healthful nature. *Antipode*.

Myers, Ransom A and Boris Worm. 2003. Rapid worldwide depletion of predatory fish communities. *Nature* 423 (15 May): 280–283.

NMFS. 2007. *Fisheries of the United States 2006*. Silver Spring, MD: National Marine Fisheries Service.

———. 2009. *American Fisheries Act (AFA) Pollock Fisheries Management*. National Marine Fisheries Service, Alaska Regional Office [cited August 21 2009]. Available online at http://www.fakr.noaa.gov/sustainablefisheries/afa/afa_sf.htm.

NOAA. 2009. *Press Release: NOAA Announces Catch Share Task Force Members*. National Oceanic and Atmospheric Administration [cited August 21 2009]. Available from http://www.nmfs.noaa.gov/sfa/domes_fish/catchshare/index.htm.

O'Connor, James. 1998. *Natural Causes: Essays in Ecological Marxism*. New York: Guilford Press.

Palsson, Gisli and Agnar Helgason. 1995. Figuring fish and measuring men: the individual transferable quota system in the Icelandic cod fishery. *Ocean and Coastal Management* 28: 117–146.

Pauly, Daniel. 2006. Major trends in small-scale marine fisheries, with emphasis on developing countries, and some implications for the social sciences. *MAST* 4 (2): 7–22.

——. 2007. Small but mighty. *Conservation* 8 (3): 25.

—— and Jay Maclean. 2003. *In a Perfect Ocean: The State of Fisheries and Ecosystems in the North Atlantic Ocean*. Washington, DC: Island Press.

——, Jackie Alder, Elena Bennett, Villy Christensen, Peter Tyedmers, and Reg Watson. 2003. The future for fisheries. *Science* 302: 1359–1361.

Petersen, Elizabeth. 2002. Economic policy, institutions and fisheries development in the Pacific. *Marine Policy* 26: 315–324.

——. 2003. The catch in trading fishing access for foreign aid. *Marine Policy* 27: 219–228.

Rowe, Jonathan. 2008. The parallel economy of the commons. In *State of the World 2008: Innovations for a Sustainable Economy*, edited by L. Starke. New York: W.W. Norton.

Standal, Dag. 2008. The rise and fall of factory trawlers: an eclectic approach. *Marine Policy* 32: 326–332.

Sumaila, Ussif Rashid and Daniel Pauly. 2006. *Catching More Bait: A Bottom-up Re-estimation of Global Fisheries Subsidies*. Vol. 14, *Fisheries Centre Research Reports*. Vancouver: Fisheries Centre, University of British Columbia.

The Economist. 2009. *A Special Report on the Sea* Vol. January 3, 2009. London: The Economist.

Weber, Michael L. 2002. *From Abundance to Scarcity: A History of US Marine Fisheries Policy*. Washington, DC: Island Press.

World Bank. 2009. *Sunken Billions: The Economic Justification for Fisheries Reform*. Washington, DC: The World Bank and the FAO.

Worm, Boris, Edward B. Barbier, Nicola Beaumont, J. Emmett Duffy, Carl Folke, Benjamin S. Halpern, Jeremy B.C. Jackson, Heike K. Lotze, Fiorenza Micheli, Stephen R. Palumbi, Enric Sala, Kimberley A. Selkoe, John J. Stachowicz, and Reg Watson. 2006. Impacts of biodiversity loss on ocean ecosystem services. *Science* 314: 787–790.

Worm, Boris, Marcel Sandow, Andreas Oschlies, Heike K. Lotze, and Ransom A. Myers. 2005. Global patterns of predator diversity in the open oceans. *Science* 309: 1365–1369.

5 When people come first: beyond technical and theoretical quick-fixes in global health

João Biehl

"It is the financial part of life that tortures me"

I begin with a poem by João Cabral de Melo Neto (2005) on the people of Northeastern Brazil, one of the poorest regions in Latin America. João Cabral writes of people who are one with that inhospitable environment, yet with a unique fluidity that creates potential. The poet grew up there, and it is there that I will take you in this chapter:

> And from this indigent river,
> this blood-mud that meanders
> with its almost static march
> through sclerosis and cement
> and from the people who stagnate
> in the river's mucus,
> entire lives rotting
> one by one to death,
> you can learn that the human being
> is always the best measure,
> and that the measure of the human
> is not death but life.

Life is in transit. This was certainly true for Evangivaldo. "What a joy you give me by coming back," the 38 year-old man beamed as he saw me and photographer Torben Eskerod in December 2001 at Caasah, a community-run AIDS hospice in Salvador. Considered by many "the African heart of Brazil," Salvador is the capital of the state of Bahia. It has an estimated population of 2.5 million, with more than 40 percent of families – like Evangivaldo's – living below the poverty line. I could barely recognize him. But the stark visual side effects of AIDS therapies were the least of Evangivaldo's concerns. "Today I woke up anguished. We have no gas to cook with."

We were happy to help him out and told Evangivaldo that we had been trying to reach him for three days but had the wrong address. "I already had to move four times. The neighbors discovered that we have AIDS," he said. "When it was just Fátima and me we could improvise things, but now that we have a child it is another

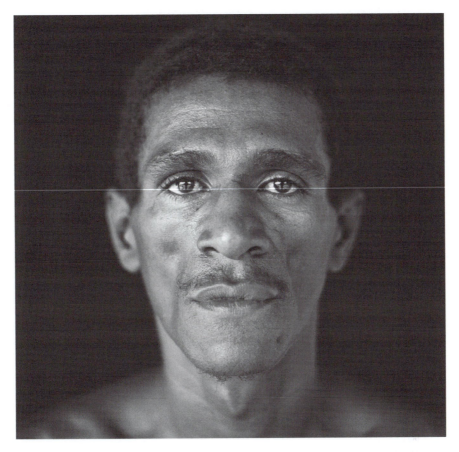

Image 5.1 Evangivaldo, 2001

matter. We can go hungry but Juliana cannot. I always take my ARVs [antiretroviral drugs], even if I just have coffee in my stomach." Evangivaldo explained that "we did not plan to have a child. The condom broke. But now that she is here, I see that this is what I wanted most in life. I thought I would die. . .but now I have a fruit of the earth." He paused and then added: "*It is the financial part of life that tortures me.*"

Via pharmaceuticals and at the mercy of a volatile economy, Evangivaldo and his loved ones lived in flux. Like millions of other poor AIDS patients worldwide who now have access to treatment, he struggled to move out of the stream of history and into a technologically prolonged life. Scavenging for resources and care, Evangivaldo conveyed desperate and extraordinary efforts to swerve and exceed constraints of all kinds. As he drove to singularize out of economic death, he also expressed world-altering desires. This chapter is about Evangivaldo and the social fields that the new people of AIDS invent and live by. Their drives and doings upset probabilities, bias estimates, and expand the limits of what can be known and acted on in the new world/market of global health.

Model policy

Brazil accounts for 43 percent of all HIV/AIDS cases in Latin America. An estimated 730,000 Brazilians were living with HIV/AIDS in 2007 – an adult prevalence of 0.6 percent (Figures 5.1, 5.2). For about a decade, incidence has hovered between 20 and 25 per 100,000 for men and between 10 and 15 per 100,000 for women. But social epidemiological studies show considerable heterogeneity in HIV infection rates, with large numbers infected among vulnerable groups, such as men who have sex with men, commercial sex workers and injecting drug users. Brazil is indeed known for its stark socio-economic inequalities and for its persistent development challenges. Yet, against all odds, Brazil invented a public way of treating AIDS.

In late 1996, groundbreaking legislation guaranteed universal access to anti-retroviral therapy (ART) (Figure 5.3). This policy resulted from potent rights-based social mobilization and novel public–private partnerships. The democratic Constitution of 1988 granted the right to health to all citizens and mandated the creation of a national healthcare system – AIDS activists were the first group to effectively equate this right to drug access. Some 200,000 Brazilians currently take antiretroviral drugs paid for by the government. The government managed to reduce treatment costs by promoting the production of generics. It also negotiated sub-stantial price reductions from pharmaceutical firms.

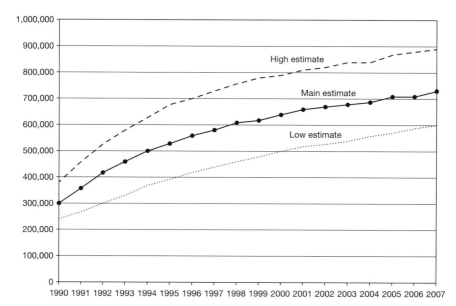

Figure 5.1 Number of people living with HIV, Brazil

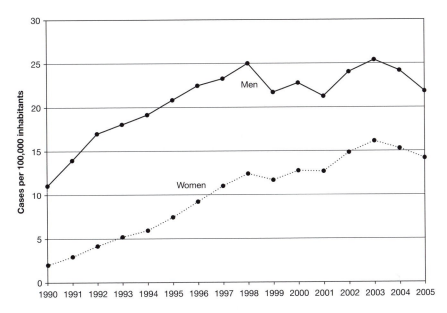

Figure 5.2 AIDS incidence, Brazil

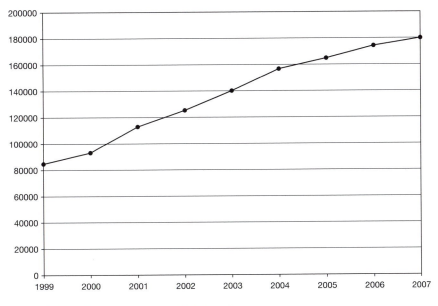

Figure 5.3 Number of patients on ART, Brazil

According to the Health Ministry both AIDS mortality and the use of AIDS-related hospital services fell by more than 50 percent (Figure 5.4). Perhaps even more impressive is the decline in mortality during the first year after diagnosis (Figure 5.5), signifying the transformation of HIV/AIDS from an acute to a chronic disease. Brazil's bold, multi-actor, and large-scale therapeutic response to AIDS has made history (Figure 5.6). It empirically challenged the economic and medical orthodoxies that treating AIDS in resource-poor settings was infeasible and that poor patients could not adhere to these complex drug regimens – as a result, Brazil has been a leader in the struggle to universalize access to AIDS therapies.

Yet, I wondered, what would be the effects of the universal treatment policy on the country's poorest and most marginalized citizens, among whom HIV/AIDS was spreading most rapidly. How would people such as Evangivaldo and Fátima transform a death sentence into a chronic disease? What social innovation could make such medical transformation possible?

Moving in the direction of the incomplete

For over ten years, I have explored the impact of the AIDS treatment rollout throughout the country's government, health systems, and personal lives. I interviewed policy makers and health professionals and carried out a long-term study of marginalized AIDS patients in Salvador. In charting the lives of poor patients before and after they had access to ARVs, I wanted to open a window into the real-life outcomes of novel national, international, and corporate policies (Biehl

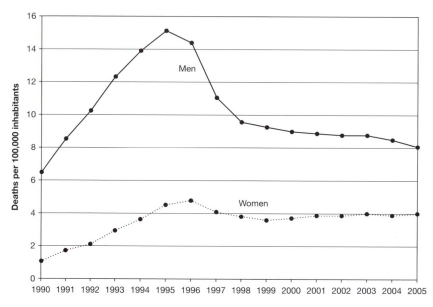

Figure 5.4 AIDS mortality rate, Brazil

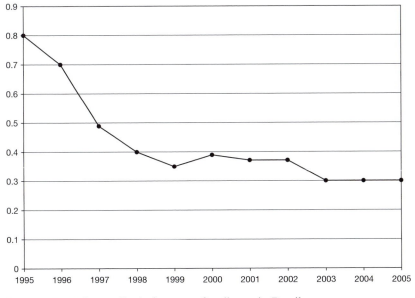

Figure 5.5 AIDS mortality in first year after diagnosis, Brazil

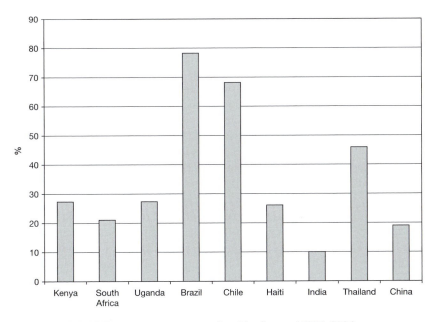

Figure 5.6 ART coverage among people with advanced HIV, 2006

2007). Broadly speaking, I have been concerned with the arts of government that accompany economic globalization and the remaking of people as market segments (specifically, therapeutic markets). How do citizen-consumers draw from government and make it resourceful as they negotiate the vagaries of the market and survival? When and under what conditions are marginalized people accounted for as population-subjects in new biomedical regimes?

In my ethnographic work, I also engaged nongovernmental and pharmaceutical communities as they took up the call for responsibility and care in the face of AIDS. Following the Brazilian lead, initiatives are being launched today, seeking to address AIDS therapeutically in places where treatments have been scarcely available. Whereas in the past, the field of international public health was dominated by multilateral and bilateral organizations, a complex matrix of partnerships (nongovernmental, philanthropic, industrial and governmental) has arisen and is shaping health interventions worldwide under the frame of security and humanitarianism (Fidler 2008) – the field of global AIDS treatment is paradigmatic of this trend.

Public–private partnerships in global health come in multiple forms, and they have diverse interests. Ranging from the Gates Foundation, to corporate drug donation programs, to PEPFAR (the U.S. President's Emergency Plan for AIDS Relief), to exemplary pilot projects such as those of Partners In Health in Haiti and Rwanda, these various actors have elastic relationships with each other. They set goals and new norms for institutional action and sometimes fill voids in places where national systems and markets are failing to address public health needs or have been absent altogether (Reich 2002). Whatever differences there are across corporate, activist, and public health agendas, the new rubric of "value" appears to reconcile these differences and folds them into an ethos of collective responsibility. Arguably, participants can become impervious to critique as they point to dire global health statistics and their non-optional duty to act (i.e. to partner, making treatment accessible and saving lives – see Sachs 2005; Singer 2009).

So far, few, if any, institutions are in place to monitor this burgeoning and somewhat disordered "public goods" field (Biehl 2008; Samsky 2009). In practice, the interests and concerns of donors, not recipients, tend to predominate, and the operations of international organizations tend to reinforce existing and unequal power relations between countries (Banerjee 2005, 2007; Epstein 2007; Ferguson 2006; Ramiah and Reich 2005). Moreover, initiatives are increasingly dominated by scientifically based measures of evaluation, revolving around natural experiments, randomized controlled trials, statistical significance, and cost-effectiveness (Duflo *et al.* 2008; Todd and Wolpin 2006) – a technical rhetoric aligned with the demand of funding organizations for technical solutions. Traditional public health initiatives are now slated in the category of "non-science" and this "scientific preoccupation" tends to overlook the on-the-ground dynamics of programs, assuming that they will work in other settings, replete with distinct institutions, practices and rationalities (Adams *et al.* 2008).

Indeed, much is side-stepped and remains unaccounted in this global form of experimentality and "post-politics" (Petryna 2009; Ecks 2005). How can donors be held accountable in the long-run, especially in this financially volatile time?

How do global health trends affect the role of governments and their human rights obligations? Moreover, how are other deadly diseases of poverty that have less political backing being dealt with? Which projections and value systems underscore policy-decisions and medical triage? Problems and questions that were not necessarily known in advance and that now have to be addressed as life-saving imperatives have been converted into pharmaceutical and new geopolitical capital.

In his recent book *Cold War, Deadly Fevers*, historian Marcos Cueto (2007) documents the story behind the Malaria Eradication Program that played a crucial role in Mexico's public health policy during the politically charged years of the Cold War era. While constantly keeping in view the campaign's international political implications, Cueto's detailed account of the way the eradication campaign unfolded on the ground leads him to unexpected anthropological terrains: he documents a profound disconnection between how the campaign was designed to work by the Rockefeller Foundation and elite national health experts, and the complex ways it was actually received by the indigenous residents of rural Mexico. In rural communities, many families simply refused to let the DDT sprayers into their homes, and there were cases when spontaneous protest even bordered on the edge of armed violence. After the first several years, even people who had complied with earlier rounds of DDT spraying angrily noted that it worked less effectively every time, and that many insects already seemed to be developing resistance and growing bigger instead of dying off.

It was in this charged historical moment that medical anthropology emerged as an applied science. Anthropologist Isabel Kelly, a former Berkeley student of George Foster, began collaborating with Héctor García Manzanedo and the Mexican Health Secretariat on rural projects in 1953. As the pair began researching how the malaria eradication program was being received in indigenous communities, they conceived their roles to be those of listeners and cultural brokers. Their report suggested many complex reasons why the program was not achieving its anticipated success, which stretched far beyond the already underestimated language barriers. For example, the medical anthropologists' report explored complex rotational housing patterns according to the seasons, meaning that families often seasonally abandoned the house that had been sprayed or preferred to simply sleep outside in the heat of summer. More fundamentally, indigenous communities often had their own healing systems and understandings of fever that coexisted uneasily with the public health information about malaria that the government distributed. And as the medical anthropologists ultimately pointed out, this environment of suspicion was underpinned by a fundamental difference in health priorities. In many communities, malaria was not conceived of as a major health problem or even a single disease, and many people in rural areas wondered why it was being addressed when their other more pressing health concerns were being ignored.

Cueto's complex portrait captures the fact that this collision between local values and international public health agendas was hardly just a fluke or footnote in the history of Malaria Eradication – it was a key reason why the campaign ultimately failed. Without paying attention to how this intervention became embedded in local economies and politics, national health officials often treated social resistance as a communications problem in a population that needed to be educated, instead of reflecting on the structure of the intervention itself. The implications of these realities run deep for our health policies today. In 2007, the Gates Foundation revived the failed campaign, pledging to eradicate malaria from the world. A year earlier, the World Health Organization once again approved the spraying of houses as an appropriate part of malaria eradication. As Cueto notes, pyrethroid-soaked bednets and pharmaceuticals have become the technical fixes of a supposedly "new era," the goal of malaria eradication now resurrected four decades after its original failure was declared in 1969 (see Bleakley 2009).

The fact is that magic-bullet approaches are increasingly the norm in global health – that is, the delivery of health technologies (usually new drugs or devices) that target one specific disease regardless of myriad societal, political and economic factors that influence health. Drawing from my study of the Brazilian therapeutic response to AIDS, this chapter explores the limits of the vertical-technical-fix approach in global health and the feasibility of "people-centered" initiatives. We need analytic frameworks and institutional capacities that move beyond the repetition of history and that focus on people: on-the-ground involvements that address the politics of both control *and* non-intervention, the fragmentation of efforts, the presence of heterogeneity, the personal and the interpersonal, people's inventiveness.

It is time to attribute to the people we study and describe the kinds of complexities we acknowledge in ourselves, and to bring these complexities into the picture of global health. Policy and popular accounts tend to cast people as helpless victims, over-determined by environment, history, and power, or as miraculous survivors who bear witness to the success of external aid. Details are suspended. Broken institutions, rifts that deepen, and larger political economies in which these lives unravel seem peripheral to both analysis and activism. In the social sciences, methods such as randomized trials have been hailed as magic bullets in the quest for scientific evidence and for keys to unlocking the mysteries of health and development. People are put into pre-conceived molds. The human populations that constitute the subjects of health and development studies are not just the source of problems. Their practical knowledge may well yield effective solutions – experiential knowledge all too readily disqualified by sponsors of technical fixes in the search for quick results.

People's everyday struggles and interpersonal dynamics exceed short-term experimental approaches and demand listening and long-term engagement (Biehl 2005; Scheper-Hughes 2008; Tsing 2004; Whitmarsh 2008). Anthropology's task in the field of global health is to produce different kinds of evidence, approaching bold challenges such as the pharmaceuticalization of health care delivery and crucial questions such as what happens to citizenship when politics is reduced to survival – with a deep and dynamic sense of local worlds (Petryna *et al.* 2006). The anthropologist demarcates uncharted territories and tracks people moving through them. In the field, the unexpected happens everyday and new causalities come into play. An openness to the surprising and the deployment of categories that are important in human experience can make our science more realistic and hopefully better.

"My politics is to see things humanly"

I first met Evangivaldo in 1997. Homeless and with contagious scabies, Evangivaldo had been sent from the AIDS ward of the state hospital to Caasah – one of the 500 Brazilian "houses of support" (*casas de apoio*) that helped poor AIDS patients navigate the precarious health system. Antiretrovirals were becoming available, but public institutions were barely functioning, and the government was increasingly outsourcing care to grassroots services. "I need to talk, to speak all truths," I remember Evangivaldo saying through the door of a room that quarantined him: "I have this sad psychosis in my head."

Evangivaldo's parents died when he was young, and he was raised by an uncle. As a teenager, he moved to Salvador in the early 1980s: "I carried many sacks of flour on my back to buy my first pair of sandals, he said.

"Later, I escorted prostitutes to the ships that docked here." He was struggling to belong: "There are people here who think that they are superior because of

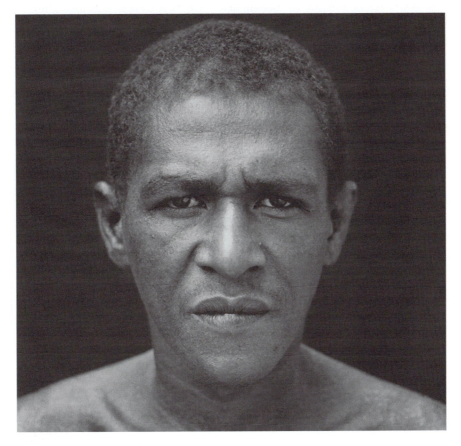

Image 5.2 Evangivaldo, 1997

the color of their skin, or because they have a doctor who likes them, or because they are in better health."

Caasah was founded in 1992, when a group of homeless AIDS patients squatted in an abandoned hospital formerly run by the Red Cross. Soon, Caasah became a nongovernmental organization (NGO) and began to receive funding from a World Bank loan disbursed through the Brazilian government. By the mid-1990s, the unruly patients in Caasah had been evicted and a smaller group underwent an intense program of resocialization run by psychologists and nurses.

"With time, we domesticated them," recalled Celeste Gomes, Caasah's director. "They had no knowledge whatsoever. We showed them the importance of using medication. Now they have this conscience, and they fight for their lives."

Evangivaldo was one of the few who got that chance. He and his fellow "AIDS citizens" (as many in Caasah called themselves) knew all too well that inequalities of power, ranging from economic destitution to racial discrimination, determined who had access to what services. They had to take up a new patient identity, and this newly learned ability to "accumulate" health at Caasah was also a highly competitive enterprise. "Did you ever see an AIDS patient in here hoping for the other's well-being?" Evangivaldo asked me. In fact, residents were constantly pointing out each other's faults and comparing clinical conditions. The other's misbehavior or sickness was a measure of their own health progress. Money was also at stake. Caasah was facilitating application for AIDS disability pensions, and priority was given to those residents who showed change. People kept to

Image 5.3 Caasah in the mid-1990s

Image 5.4 Caasah in the mid-1990s

themselves. As one patient told me: "One Luis has died and another has emerged. I got used to the medication. Medication is me now. Today people only die from AIDS if they want to." But Evangivaldo thought differently: "*My politics is to see things humanly*. The one who is strong now must help the weaker." After a year in Caasah, Fátima came into his life, he said. "As I got better I told her 'Like you, I have no family. I have nobody for me. Let's live together.' And she said 'yes.'"

In the meantime and what is outside biopower

There is no short cut to understanding how a technologically-prolonged life is achieved – be it in Brazil or in the growing number of other poor countries where AIDS is finally being treated through an unprecedented array of public- and private-sector initiatives. More than 25 million have died of AIDS to date, and an estimated 33 million people are living with HIV worldwide, about two thirds in Sub-Saharan Africa. Some 10 million people are in need of antiretroviral therapy. The battle for access has been hard-fought, and nearly 4 million are now on treatment in low and middle-income countries.

Global AIDS treatment rollouts rightly open the door to drug access, but they also exemplify the inadequacies of a magic bullet approach to health care. Drugs are ancillary to the full treatment of the disease. Alone, neither money nor drugs nor sophisticated pilot projects guarantee success. Healing, after all, is a multi-faceted concept, and "healing" is no more synonymous with "treatment" than "treatment" is with "drugs." Statistical strategies and profit motives hover above, by and large missing the interpersonal networks that link patients, doctors, and governments, which are especially important in resource-poor settings, where clinical infrastructures are not improving. AIDS death and HIV infection keeps growing among the destitute. An estimated 3 million people become newly infected each year. For them, HIV/AIDS is one tragedy among many others.

These realities are not reducible to the theories we bring to the field. Numerous anthropologists have been using Michel Foucault's formulation of biopower – how natural life has been taken as an object of modern politics (1990; 2007) – to assess emerging assemblages of technology, medicine, and governance, particularly in the face of HIV/AIDS (Nguyen 2005; Comaroff 2007; Fassin 2007; Robins 2006). Yet this influential biopolitical analytic – "making live and letting die" – deserves deeper probing as it might assume transcendent forms of power and homogeneous people and overly normalized populations (Foucault 1990, 2007). As Ian Hacking acknowledged in his essay on how new kinds of people can be "made up" by medical diagnostics: "my concern is philosophical and abstract. . .and [I] reflect too little on the ordinary dynamics of human interaction" (1999: 162).

What is outside biopower? Traversing worlds of risk and scarcity, constrained without being totally over-determined, people create small and fleeting spaces, through and beyond classifications and apparatuses of governance and control, in which to perform a kind of *life bricolage* with the limited choices and materials at hand (including being the subjects of rights and pharmaceutical treatments made available by state and non-state actors). Scholars and policy-makers are challenged to respect and to render publicly intelligible, without reduction, the angst, uncertainty, the passion for the possible and the travails that people like Evangivaldo, amid lifesaving interventions, are left to resolve by themselves and, too often, at the expense of others.

For over ten years, I have chronicled life in and out of Caasah. Repeatedly returning to the field, one begins to grasp what happens in the *meantime* – and I like to think of this work as a study of the meantime – the events and practices that enable wider social and political change, alongside those that debilitate societies and individuals, dooming them to stasis and intractability. In such returns, entanglements and intricacies are revealed. We witness how policies unfold over time – and the literalness of becoming, as AIDS survivors transition from patienthood back into personhood. I say *becoming*, for we have a responsibility to think of life in terms of both limits and crossroads, where technologies, interpersonal relations, and imagination can sometimes, against all odds, propel unexpected futures.

Evangivaldo, Torben, and I sat under a tree in the backyard, and we looked at the portraits Torben had made of the Caasah residents in 1997. With a simple chair and a black cloth against a brick wall, we had improvised a photo studio. Torben photographed each person and I recorded their life stories. "This work was important to me, it marked my history," Evangivaldo said. Celeste, the director, joined us: "You really captured the person," she told Torben, with a sigh. With a certain melancholy, Celeste admitted that "in the day-to-day work we really did not see this. . .we pretended that we knew who they were."

Image 5.5 Luis and Torben, 1997

As for Evangivaldo: "I know that this is a kind of scientific work for people to see what we go through in Brazil, but I also want to show it to my doctor and the nurses. I want them to see how I changed." Evangivaldo showed us the prescription for an anti-depressant that he also needed but couldn't purchase. "We already owe 75 reais at the pharmacy," he said – that was half of his disability pension. "I wake up at 4:00 am and ride my bike for two hours to get downtown. I go door to door asking for a job. There are days when I cannot get the money we need and I panic. I hide in a corner and cry. Then I don't know where I am." Yet, he found ways to transcend his sense of being choked. "I say 'focus, Evangivaldo, focus, you will find your bike and your way home.' And do you know why I manage to do this? Because my Juliana is waiting for me."

With antiretroviral therapies available, healthy residents like Evangivaldo, Fátima, and Rose had been asked to move out of Caasah. And, in the past year, Caasah itself had moved to a new state-funded building. It had been redesigned as a short-term recovery facility for patients sent by hospitals' AIDS wards and a shelter for orphans with HIV. Disturbingly, there was no systematic effort to track patients and their treatments once they left.

AIDS therapies are now embedded in landscapes of misery, and hundreds of grassroots services have helped to make AIDS a chronic disease also among the poorest in Brazil and beyond. This is not a top-down biopolitical form of control. The government is not using AIDS therapies and grassroots services as "techniques . . . to govern populations and manage individual bodies" (as anthropologist Vinh-Kim Nguyen has framed the politics of antiretroviral globalism – 2005: 126). As

I am arguing in the light of Caasah, the question of accountability has been displaced from government institutions, and poor AIDS populations take shape, if temporarily, through particular engagements with what is made pharmaceutically available. The political game here is one of self-identification. Proxy-communities, often temporary and fragile, and interpersonal dynamics and desires are fundamental to life chances, unfolding in tandem with a state that is pharmaceutically present (via markets) but by and large institutionally absent.

At the margins, both the institutional and pharmacological matters surrounding AIDS treatment undergo considerable flux. And poor AIDS survivors themselves live in a state of flux, simultaneously acknowledging and disguising their condition while they participate in *local economies of salvation* and articulate *public singularities*. Against the backdrop of a limited health care infrastructure and economic death and through multiple circuits of care, individual subjectivity is refigured as a *will to live*.

Philosopher Giorgio Agamben has also significantly informed contemporary biopolitical debates with his evocation of the *homo sacer* and the assertion that "life exposed to death" is the original element of western democracies (1998: 4). This "bare life" appears in Agamben as a kind of historical-ontological destiny – "something presupposed as nonrelational" and "desubjectified" (1999). A number of anthropologists have critiqued Agamben's apocalyptic take on the contemporary human condition and the dehumanization that accompanies such melancholic, if poignant, way of thinking (Das and Poole 2004; Rabinow and Rose 2006). Whether in social abandonment, addiction, or homelessness, life that no longer has any value for society is hardly synonymous with a life that no longer has any value for the person living it (Biehl 2005; Bourgois and Schonberg 2009; Garcia 2008). Language and desire meaningfully continue even in circumstances of profound abjection (Biehl and Moran-Thomas 2009). Such difficult and multifaceted realities and the fundamentally ambiguous nature of people living them give anthropologists the opportunity to develop a human, not abstractly philosophical, critique of the non-exceptional machines of social death and (self) consumption in which people are caught. Against all odds, people keep searching for social recognition and for ways to endure, at times reworking and sublimating afflictions and constraints.

Acknowledging the insights and alternative human capacities that grow out of abjection also forces us to inquire into how they can be part and parcel of the much needed efforts to redirect care. The need for subjective texture thus also raises broader anthropological questions about ethnography's unique potential to bring the private life of the mind into conversations about public health and politics. Rather than ethnographically illustrating the silhouettes of biopolitical theory, new ways of thinking about political belonging and subjectivity force us instead to reconsider this theoretical framework's very terms (Rancière 2004; Fischer 2009).

Gilles Deleuze (2006), who did not share Foucault's confidence in the determining force of power arrangements, is helpful here. According to Gilles Deleuze, desire, via the inventions, escapes, and sublimations it motivates, is constantly undoing – or at least opening up – forms of subjectivity and territorializations of power. Even the concept of *assemblage*, taken up not long ago by Aihwa Ong and Stephen Collier to name emergent global configurations – like "technoscience, circuits of licit and illicit exchange, systems of administration or governance, and regimes of ethics or values" (Ong and Collier 2005: 4) – has desire, in Deleuze and Guattari's definition in *Kafka: Toward a Minor Literature* (1986), at its core. For Deleuze and Guattari, assemblages are contingent and shifting interrelations among "segments" – institutions, powers, practices, desires – that constantly, simultaneously construct, entrench, and disaggregate their own constraints and oppressions (1986: 86). This emphasis on desire and the ways – humble, marginal, minor – that it cracks through apparently rigid social fields and serves as the engine of becoming figures centrally in Deleuze's divergences from Foucault, whose archaeology of the subject traces the ways in which he or she is constituted and confined by the categories of expert discourses, for example, in what, again, might be crudely sketched as a vertical or top-down movement.

Epistemological breakthroughs do not belong only to experts and analysts. The cumulative experiences of "the unpredictability of the political and social effects of technological inventions" – borne by people navigating contemporary entanglements of power and knowledge – are also epistemological breaks that demand anthropological recognition (Canguilheim 1998: 318). Long-term engagement with people is indeed a vital antidote to what economist Albert Hirschman identifies as "compulsive and mindless theorizing." The quick theoretical fix has taken its place in our culture alongside the quick technical fix. For Hirschman, as for the anthropologist, people come first. This respect for people, this attention to how policies are put together – how they take institutional hold and fit into unequal social relations – makes a great deal of difference in the kind of knowledge we produce. As Hirschman writes, "In all these matters I would suggest a little more reverence for life, a little less straitjacketing of the future, a little more allowance for the unexpected – and a little less wishful thinking" (1971: 338).

The anthropologist, upholding the rights of micro-analysis, brings into view the fields that people, in all their ambiguity, invent and live by. Such fields of action and significance – leaking out on all sides – are mediated by power and knowledge, and they are also animated by claims to basic rights *and* desires, as Evangivaldo affirms. It is not enough to simply observe that complicated new configurations of global, political, technical, biological (etc.) segments exist or are the temporary norm. We must attend to the ways these configurations are constantly constructed, un-done and re-done by the desires and becomings of actual people – caught up in the messiness, the desperation and aspiration, of life in idiosyncratic milieus. Nor is ours necessarily a choice between primarily "global assemblages" (Collier and Ong 2005) and principally local "splinters" of a "world in pieces" (Geertz 2000). At the horizon of local dramas, in the course of each event, in the ups, downs, and arounds of each individual life, we can see the reflection of larger systems in the making (or unmaking). And in making public these singular fields – always on the verge of disappearing – the anthropologist still allows for larger structural processes and institutional idiosyncrasies to become visible and their true impact known.

Persistent inequalities and the scientific aura of pretending not to know

By 2000, the Bahian health officials claimed that a plateau had been reached and that AIDS incidence was on decline, ostensibly in line with the latest statistics pointing to the success of the country's control policy. But the AIDS reality I saw in the streets of Salvador contradicted this profile, and a central concern of my ethnography has been to expose the limits of surveillance and to generate some form of visibility and accountability for the hidden AIDS epidemic experienced by the most vulnerable and marginalized.

While observing life literally in-the-making at Caasah, I also chronicled the work of Dona Conceição, a nurse, who provided meals and some form of care to one hundred homeless AIDS patients, involved in illicit economies and supporting their children. "Medical services never meet the demands, and civil society has abandoned them," Dona Conceição told me. "I try to alleviate things a bit. I am tied to them in spirit."

I met with officials at the state run epidemiological surveillance service and asked them to verify whether some of these street patients who reported being treated at the state's AIDS ward were registered in their database. These patients were nowhere to be found. Yet, from my vantage point, they were dying a very public death – a destiny that Evangivaldo other patient-citizens were trying to escape through extraordinary efforts.

Interestingly, Brazil's computerized registry of patients on antiretroviral therapy includes data on treatment combinations, dosages and CD4 counts, yet it does not include specific social indicators. Without knowing where these patients live, we

Image 5.6 Dona Conceição and her "street patients."

cannot assess the policy's national coverage; without knowing education or income levels, the class dynamics at work in treatment access remain unknown; without a sense of the social networks of this new medical population, we don't know about adherence patterns and how drug access translates into better health. The fact is that, on the ground, the AIDS treatment policy reproduces the fault lines of race and poverty and we see uneven outcomes for patients, as well continuous stigma and discrimination, even from health professionals. A recent survey on mortality in the state of São Paulo revealed that AIDS is two times more fatal among blacks than it is among whites.

These trends show the need for more in-depth program evaluations. Yet the field of global public health, AIDS notwithstanding, is dominated by econometric analysis with its powerful claims to statistical and epistemic superiority (Heckman and Vytlacil 2007) but skewed generalizability and short shelf-life. As economist Angus Deaton notes, a trial-based "randomization in the Tropics" is also unlikely to shed light on the keys to development because such endeavors do not offer insight into *why* specific programs do or do not work. Excluding observation and what deviates from ideal conditions, "the technical fixes fail and compromise our attempts to learn from the day" (2009:47). The all-too-human questions "why here, why me, why now?" – so crucial to anthropologists – are often elided.

To grasp how AIDS victims disappear from public accounting, I collaborated with local epidemiologists. We gathered the death certificates of all AIDS patients in the state's AIDS ward over six years. We found that over half died in their first hospitalization, suggesting that the majority of these people only gained access to hospital services at the point of death. We also discovered that only 26 percent of these AIDS cases were actually registered by the surveillance service. We were intrigued. What made some of these AIDS cases officially visible and the majority not? Compared to patients who died during their first hospitalization, patients who died during a later hospitalization were two thirds more likely to be registered. Moreover, men who self-identified as bisexual or homosexual were 50 percent less likely to be registered than those that reported heterosexuality.

These voids and biases were not just the result of precarious surveillance that could be addressed with a simple technical fix. The problem was rooted in three factors: first, the operating logic of a health care system that circumscribes service delivery to about 30 percent of the demand – those patients who autonomously search for continuous treatment, fighting for their place in the overcrowded and underfunded services; second, a powerful physician sovereignty that can neglect and deem some patients unworthy; and third, the problematic ways in which marginalized people living with HIV/AIDS respond to their disease given a fragmented system of care and the illicit economies they often engage. "They come in dying," the social worker told me. "They never heal. Without a home how can they adhere to treatment? There must be thousands like them." There are no records tracing these people's plights and the complex social and economic interactions that exacerbate infections and immune depressions remain unaccounted for.

My colleagues and I wrote a report to the Bahian Health Secretariat informing them of the existence of this hidden AIDS epidemic. I learned later that this report was shelved. Within these local force fields, the country's innovative AIDS treatment policy was coming alive and gaining international attention, islands of care and a triage-like state took form, and social death continued its course.

From collective epidemic to a highly privatized politics of survival

When I returned to Salvador in 2001, I worked with the roughly thirty homeless patients still under the care of Dona Conceição. They were now living on a concrete platform adjacent to the city's main soccer stadium. Many looked undernourished, had skin lesions, and complained of flu-like symptoms. But then, as Carisvaldo put it, "We push life forward anyway." Several said that they had begun picking up free antiretrovirals at the hospitals, but that they had stopped using them. According to Roberto, "Medication alone will not solve anything." His friend Luis said that he did not believe in the efficacy of the drugs: "My medicine is food, beans in my belly." A culture of "compliance" was far from here.

What I witnessed there speaks volumes to the tragic unfolding of the AIDS epidemic among the marginalized. One morning, two girls aged thirteen and fifteen, had joined the "street tribe." They had just escaped forced prostitution. Two of the homeless AIDS patients found them wandering and brought them to the soccer stadium, where they forced them again, into sex. "I want to go home," one girl told me, shivering. As I was talking to them, child welfare officers showed up and said that they would take the girls back to their hometown. As the police van was leaving, the street cleaners who had been there all along turned to the remaining people, and mockingly said: "Your fresh flesh is gone."

Image 5.7 Homeless

At the end of that week, I went to Brasília, the country's capital, where I met with Dr. Paulo Teixeira, then coordinator of the National AIDS Program. "The success of the Brazilian AIDS policy is a consequence of the activism of affected communities, health professionals, and government," he told me. Two years later, I would hear a similar explanation from Fernando Henrique Cardoso, Brazil's former president: "Brazil's response to AIDS is a microcosm of a new state-society partnership," he told me. Cardoso promoted the AIDS policy as evidence of the supposed success of his reform agenda – a state open to civil society, activist vis-à-vis the market, and fostering partnerships for the delivery of technology. "All the NGO work, treatment legislation, struggles over drug pricing are new forms of governmentality in action. . .engineering something else, producing a new world."

The AIDS policy emerged against the background of neoliberalization, and the politicians involved with it were consciously articulating a market concept of society. For Cardoso, citizens are consumers who have "interests" rather than "needs." Or, in the words of economist and former health minister José Serra, "The government ends up responding to society's pressure. If TB had a fifth of the kind of social mobilization AIDS has, the problem would be solved. So it is a problem of society itself." In this rendering, the government does not actively search out particular problems or areas of need to attend to – that is the work of mobilized interest groups. These public actions are seen as "wider and more efficacious than state action" (Cardoso's words). In practice, activism has enhanced the administrative capacity of the reforming state.

In my interview with Dr. Teixeira I mentioned the AIDS reality I had just observed in the streets of Salvador. "It is a portrait of Brazil," he said. "I am not happy with the work being done with AIDS and poor populations. We have to identify a working strategy." Dr. Teixeira made clear the state's position of basically deferring care to community organizations and added: "To work with these people is not the same as working with the elite in São Paulo, but effectiveness is also possible. If I get 20 to 30 percent of effectiveness with these people this is already a very important step." As I heard him, I was reminded of the 26 percent rate of AIDS registration in the Bahian surveillance service, of the ways in which the local state circumscribes populations for service. "We have to improve this," repeated Dr. Teixeira.

But the AIDS NGOs that were supposed to have taken over assistance "have long lost idealism and passion," activist Gerson Winkler bitterly told me. Winkler has lived with HIV/AIDS for over two decades and is one of the policy's most vocal critics: "Now it's all a game of make-believe," he said: "I select a clientele and pretend that I do a project, you pretend that you see it, the government pretends that it is monitoring it, and we all pretend that the results are true. It is a farce to think that NGOs can be the executor of state services." Asked about this criticism, Dr. Teixeira responded: "Did bad things happen in the process? Yes, but without outsourcing care there would not have been advancements either. Evolution is never unidirectional – it is forward and backward. We hope that it is two steps forward and one backward."

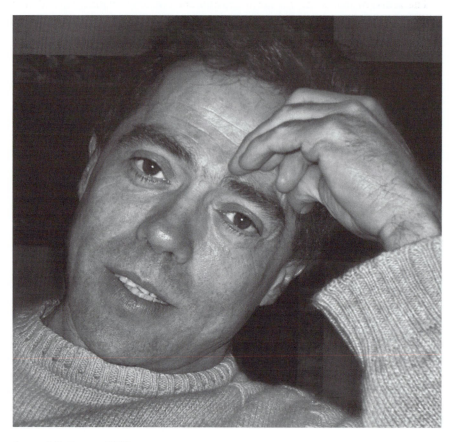

Image 5.8 Gerson, 2005

Winkler lamented that in this dance, AIDS politics had become increasingly atomized and that poor patients had been reduced to scavenging for assistance. "Stigma is constant," he continued. "Most poor patients have never been formally employed. It is very difficult to help them develop a work mentality and place them in the market." In other words, in the story I am telling you, we see a movement from collective epidemic to personalized disease; from public health to the pharmaceuticalization of health care; from governmental distance to the industrialization of the nongovernmental sector and to a highly privatized politics of survival.

Evangivaldo's trajectory, like those of many others, shows how empowering pharmaceutical access can be, but also how much additional effort is required to transform drugs that are "accessible" into drugs that are effective in the everyday lives of poverty-stricken patients. A vertical, top-down mass campaign against a disease, while valuable, leaves unaddressed the social realities that co-construct health outcomes (Easterly 2006, 2008). Health policies need to be directed at *people*, not simply disease.

The most prominent proponent of an approach that innovatively blends the vertical technological intervention with a horizontal focus on making health systems work is an anthropologist himself, Dr. Paul Farmer (2001, 2003). He and his colleagues at Partners in Health work with local communities in Haiti, Boston, Peru, and now Rwanda. Local clinics become a nexus of care, integrating HIV/ AIDS treatment and prevention activities, while also attempting to address co-infections and the new medical problems that AIDS patients face as they age. Accounting for individual trajectories and staying with patients through the progression of the disease and treatment (the work of paid *accompagnateurs*) is considered as important as tackling the economic and social factors that impact their families and mitigating the demise of clinical infrastructures. In rural Rwanda, for example, each patient receiving antiretroviral drugs also receives food for five people. Although this *biosocial* model may be rejected by public health orthodoxies on the basis that it is not "cost-effective" or "sustainable," it nonetheless expands the realms of feasibility and helps to shape new standards of care and intervention (Farmer 2008).

Real life has to be put back into the purview of AIDS policies. This requires going to where people are. Pauper patients, with no political voice, have been disregarded not due to the government's inability or ignorance necessarily. Where there has been an active HIV search, testing and care – in Brazilian maternity wards, for example – infection has been curtailed. Why not, then, reach out to other vulnerable groups and discuss interventions *with* them? We need to innovate and find ways to make testing, treatment and sustained assistance available to groups that escape categories yet suffer most from the epidemic.

Addressing the entanglement of people-disease-policy and market dynamics

Meanwhile, the magic bullet approach, with its focus on drug delivery and supply chain management, stretches far beyond the antiretroviral rollout. Many tropical diseases have also been subject to blanket treatment approaches, including child-hood malaria, river blindness, and parasitic infections. But as historians of the fight against syphilis and malaria remind us, the goal of eradication is an elusive target (Brandt 1985; Cueto 2007). Just as medical know-how, international political dynamics, and social realities change, so are biological systems in flux – bugs get resistant, new infections appear. A more complex model of this flux of people-disease-policy *and* market dynamics is required – and this, calls for innovative partnerships and methods.

In summer 2008, Amy Moran-Thomas shadowed health officials and medical NGO workers in northern Ghana as they worked on malaria prevention and deworming campaigns. And yet, during community visits she saw these efforts often went unheeded – treatments not taken, water filters for guinea worm and bed nets sitting in a corner, unused. Educational campaigns were trying to address this apparent negligence, but the problem ran deeper. When she asked parents what their primary health concerns were, they spoke of walking miles to find clean water during the dry season and struggling with diabetes in an environment where so much cheap imported food was the only alternative to hunger (Moran-Thomas 2009). In other words, interventions seemed to ignore the complex preventative and environmental health issues most central in their lives.

Consider the widely-cited study by Michael Kremer and Edward Miguel on curing worm infections in rural Kenya. Miguel and Kremer found that treating Kenyan schoolchildren with extremely cheap deworming medication increased their school attendance by some 10 percent. A *New York Times* op-ed heralded the study as "landmark" (Kristof 2007): with just a bit of cheap medication, poor countries could increase school attendance by leaps and bounds. Given the affordability and stunning success of the treatment, many commentators suspected that families who had not benefited from treatment during the study would very happily adopt this new technology.

But Kremer and Miguel then observed a puzzling turn of events when they followed a group of families outside the original study after the trial had ended (2007). Among these families, those who were friendly with families in the treatment group were *less* likely to treat their children than those who were friendly with families in the control group. They were also less likely to deem the medication effective at improving health. If deworming medicine is the panacea to anemia and school truancy, then why were better-informed families not treating their children?

Again, we have a case in which the interpersonal relations and needs of people on the ground elude controlled studies, and the question of how to learn to bring

Image 5.9 Guinea worm ad and market in Northern Ghana

local communities into the very design and implementation of feasible rather technology-enamored interventions is a continuous challenge. With international and national health policy's success largely re-framed in terms of providing and counting the best medicines and newest technology delivered, what space remains for the development of low-tech or non-tech solutions (such as the provision of clean water) that could prove more sustainable and ultimately more humanistic?

Back in Brazil, with patients taking advantage of new antiretroviral drugs, the annual HIV/AIDS budget increased to $562 million by 2006. In spite of the country's generic production capacity, about 80 percent of the drugs dispensed are patented. "We are moving toward absolute drug monopoly," Michel Lotrowska, an economist working for Doctors Without Borders in Rio de Janeiro, told me. "We have to find a new way to reduce prices; if not, doctors will soon have to tell patients 'I can only give you first-line treatment, and if you become drug resistant you will die'."

Recently, I asked a pharmaceutical executive how the private sector and the governments that his company was partnering with in Africa were thinking about this problem. For him, the crux of the matter was patient compliance. "If this is taken care of," he said, "then second-line treatments will not even be necessary." Moreover, he said "drug distribution systems should be improved to guarantee treatment consistency." Last, he told me "We need to invest in basic science and have better drugs to begin with." Patents and complex social and technical realities that might hinder best diagnostic practice or compliance do not enter into this global health conversation.

During recent research in Salvador, I learned that medical opinion-makers were urging doctors to prescribe T-20 as a first line treatment instead of using it as rescue drug. T-20 is a new drug that greatly helps patients with resistance to previous treatments at an annual cost of $20,000 per patient. I also heard of cases where doctors began prescribing the rescue drug Kaletra at the time of its launch in the United States, before its registration in Brazil. These doctors referred patients to a local AIDS NGO and to public-interest lawyers who pressured the state to provide drugs not yet approved by the country's FDA. In the face of pervasive pharmaceutical marketing enmeshed with social mobilization, regulatory incoherence thrives. Meanwhile, activist policy makers have to keep inventing strategies to keep the country's AIDS treatment rollout in place.

India has been a pivotal country in the last decade, taking advantage of the transitional period instituted by the World Trade Organization to allow member countries to enshrine strong patent protections into law. During this period, India specialized in the generic manufacturing of patented HIV/AIDS drugs, which played an integral role in driving down prices and ensuring treatment access in resource-poor countries. But since 2005, generic manufacturing of patented drugs is now strongly prohibited. This could not come at a worse time as patented drugs like Tenofovir and Efavirenz have replaced preexisting first line treatments and become the widely accepted standard of care.

As a last resort, governments might issue "compulsory licenses" which would allow them to manufacture or import generics in a time of crisis without consulting the patent holder. Although the license usually guarantees the patent owner a royalty fee of around 1 percent of generic sales, Thailand and Brazil still significantly lowered costs when they recently issued compulsory licenses for Efavirenz and imported an Indian generic. The drug is used by more by 75,000 Brazilians and activists praise this move as an important advance in the widening of access to the newest and most expensive therapies. But issuing compulsory licenses is not a long-term, sustainable solution. Due to recent restrictions on generic imports, the compulsory license requires countries to have internal pharmaceutical manu-facturing capacity, meaning that most resource-poor countries cannot utilize this tiny flexibility built into the reigning intellectual property regime.

The judicialization of the right to health

Across Brazil, patients are turning to courts to access prescribed drugs. The rights-based model of demand for access to AIDS therapies has "migrated" to other diseases and patient groups. Although Brazil has the developing world's most advanced HIV/AIDS program, many of its citizens still go to local pharmacies only to find that essential medicines are out of stock. Brazil is also one of the fastest-growing pharmaceutical markets in the world. Doctors increasingly prescribe and patients demand new drugs, some with questionable benefit. Faced with high cost or no availability, many individuals are suing the government to obtain drugs (see Biehl *et al.* 2009).

Although lawsuits secure access for thousands of people, this *judicialization of the right to health* generates enormous administrative and fiscal burdens and has the potential to widen inequalities in healthcare. Six thousand and eight hundred

Image 5.10 Patient filing a treatment lawsuit with a public defense lawyer

medical–judicial claims reached the Solicitor General's Office of the State of Rio Grande do Sul in 2006, for example, an increase from 1,126 in 2002. By 2008, an average of 1,200 new cases were reaching the Office per month. In 2008, US$30.2 million was spent by this state of 11 million people on court-attained drugs. This expense represents 22 percent of the total amount spent on pharmaceutical drugs that year and 4 percent of the state's annual projected health budget. About a third of current claims are for high-cost drugs not provided through the public health-care system. These claims surely account for a large proportion of state expenses.

Interestingly, a ruling by the Supreme Court in 2000 concerning an AIDS patient demanding access to a newer antiretroviral drug continues to inform the rulings of pharmaceutical provision in both state and federal courts. In his ruling, Minister Celso de Mello argued that the AIDS pharmaceutical assistance program was the actualization of the government's constitutional duty to implement policies securing the population's health. As the concrete embodiment of the need for "programmatic norms," the AIDS program acquired an inherent judicial value in Mello's ruling. As soon as the needy have medicines, according to Mello, the government's legal responsibility for implementing programmatic norms that secure health are fulfilled and cease to be "an inconsequential constitutional promise." In this rendering, the immediate assurance of the right to health through medicines circumvents questions about the limitations of policy, knowledge or resources.

Recent interviews reveal conflicting views. Many judges and public defenders working on the right-to-health cases feel they are responding to state failures to provide needed drugs, and some judges admit a lack of medical expertise to make informed decisions consistently. Administrators contend that the judiciary is overstepping its role, although some acknowledge that, because of these cases, distribution of several drugs has risen. Patient organizations have a highly contested role. Officials claim that at least some organizations are funded by drug companies eager to sell the government high-cost drugs. Patients encounter a bewildering and overburdened legal system in which injunctions granting access to life-saving drugs must be periodically renewed, typically resulting in interrupted treatment and medical complications.

The stakes are high and the debate is heated. What are the larger institutional and political implications of having the judiciary become the state executor? Are the courts a true alternative voice for those usually marginalized from the political process? Do we see a new form of "politicization" of the right to health that is making it more accessible, or an erosion of it, making it more privatized and more unequal?

Fragile islands of hospitality

Finally this brings me back to Caasah. "If you look carefully, nothing has changed," a tired Celeste told me during my last visit in June 2005. Caasah was still the only place in Salvador that provided systematic care to poor AIDS patients who had been discharged from public hospitals. "Some patients return to their families. Others go back to the streets. Disease keeps spreading and the government pretends not to know."

At the state's AIDS ward, Dr. Nanci lamented, "We are still full of wasting patients. The difference now is that they come from the interior, where no new services have been created. Access to therapies has been democratized, but health has not." Many doctors do not put drug addicts and the homeless on antiretroviral therapy. They say that there is no guarantee that they will continue the treatment and that they are concerned about the creation of viral resistance to ARVs. Thus, against an expanding discourse of human rights and pharmaceutical possibilities, we are here confronted with the on-the-ground limits of infrastructures wherein a new life with AIDS can be realized, but only on a limited way.

Out of the initial group of twenty-two Caasah patients with whom I had worked in 1997, seven were still alive in 2005 – among them Evangivaldo. Their added life was obviously a result of technological advancements, argued Celeste, "but it would not have happened if they had not learned to care for themselves." In the end, treatment adherence, she stated "is relative to each person. It requires a lot of will." Yet, far from representing a natural vitality, this *will to live* has to be fabricated and asserted in the marketplace and in local medical worlds by those with the means, as limited as they are, to do so. The AIDS survivors with whom I worked had all engineered fragile islands of hospitality in which they could inhabit their unexpected lives. They all had a place they called home, a small steady income, and a social network of sorts. In a pinch, they could still resort to Caasah. This institutional tie, as tenuous as it now was, remained vital to them.

To have someone to live for and to be desired by was also a constant thread in their accounts. "Fátima had a stroke," Evangivaldo told me the last time I saw him:

> She hurts inside because she cannot help. But I tell her that the important thing is that she is alive, that I do not mind being the man and the woman of the house. God knows everyone's gifts. The one who is strong now has to help the weaker. The important thing is to have a dignified life and to be healthy to see Juliana grow. That's what I have to say.

Without a doubt, Brazil has experienced a striking decrease in AIDS mortality. However, seen from the perspective of the urban poor, the AIDS treatment policy is not necessarily an inclusive form of care. Local AIDS services triage treatment, and social and economic rights for the poorest are sporadic at best. Brazil, which has innovated in access to treatment as a human right, must more fully define and implement a right to health that transcends medicines and individual demands, and ensure that primary health care and prevention are sufficiently robust to reduce vulnerability to disease. Likewise, at issue is a reconsideration of the systemic relation of pharmaceutical research, commercial interest, and public health care. We should think about a more sustainable solution to the obstacles posed by patentability and business control over medical science and care on the ground. Part of the solution may lie in comprehensive information and technology sharing among southern countries – a paradigm that would allow poorer countries to develop health technology assessment programs and to pool their manufacturing know-how and unite in the fight for fair pricing.

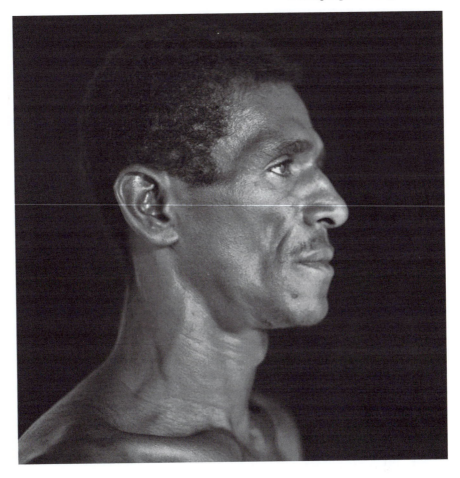

Image 5.11 Evangivaldo

Caasah's former residents are the new people of AIDS. They have by all standards exceeded their destinies. Now receiving treatment, Evangivaldo and many others refuse the condition of leftovers. And they face the daily challenge of translating medical investments into social capital and wage-earning potential. They live between-moments, between-spaces, scavenging for resources. At every turn, they must consider the next step to be taken to guarantee life. Theirs is the force of immanence. From these people, "you can learn that the human being is always the best measure, and that the measure of the human is not death but life."

Continually adjusting itself to the reality of contemporary lives and worlds, the anthropological venture has the potential of art: to invoke neglected human potentials and to expand the limits of understanding and imagination – a people yet to come. Thus at stake is also our formative power to generate a "we," an engaged audience and political community, that has not previously existed – our craft's potential to become a mobilizing force in this world.

Acknowledgments

I am deeply grateful to Torben Eskerod for his powerful photographic work and to Tom Vogl for his insightful editorial help and for making the figures presented in the chapter. I also want to thank Adriana Petryna, Joseph Amon, Amy Moran-Thomas, Alex Gertner, Ari Samsky, Peter Locke, and Mariana Socal for their comments and help. Research and writing has been supported by Princeton's Health Grand Challenges and the Program in Latin American Studies, the Center for Theological Inquiry, and the Ford Foundation.

References

Adams, Vincanne, T. Novotny, and H. Leslie. 2008. "Global Health Diplomacy" (editorial). *Medical Anthropology* 12(4): 315–323.

Agamben, Giorgio. 1998. *Homo Sacer: Sovereign Power and Bare Life.* Stanford, CA: Stanford University Press.

—— 1999. *Remnants of Auschwitz: The Witness and the Archive.* New York, NY: Zone Books

Banerjee, Abhijit V. 2005. " 'New Development Economics' and the Challenge to Theory." *Economic and Political Weekly*, October 1st, 4340–4344.

—— 2007. *Making Aid Work.* Cambridge, MA: MIT Press.

Biehl, João. 2005. *Vita: Life in a Zone of Social Abandonment.* Berkeley: University of California Press.

—— 2007. *Will to Live: AIDS Therapies and the Politics of Survival.* Princeton, NJ: Princeton University Press.

—— 2008. "Drugs for All: The Future of Global AIDS Treatment." *Medical Anthropology* 27(2): 1–7.

Biehl, João and Amy Moran-Thomas. 2009. "Symptom: Subjectivities, Social Ills, Technologies." *Annual Review of Anthropology* 38: 267–288.

Biehl, João and Peter Locke. 2010. "Deleuze and the Anthropology of Becoming." *Current Anthropology* 51: 317–351.

Biehl, João, Adriana Petryna, Alex Gertner, Joseph J. Amon, and Paulo D. Picon. 2009. "The Judicialisation of the Right to Health in Brazil." *Lancet* 373: 2182–2184.

Bleakley, Hoyt. 2009. "Economic Effects of Childhood Exposure to Tropical Disease." *American Economic Review*, Paper and Proceedings.

Bourgois, Philippe and Jeff Schonberg. 2009. *Righteous Dopefiend.* Berkeley: University of California Press

Brandt, Allan. 1985. *No Magic Bullet: A Social History of Venereal Disease in the United States since 1880.* New York: Oxford University Press.

Cabral Melo Neto, João. *Education by Stone.* New York: Archipelago Books.

Canguilheim, Georges. 1998. "The Decline of the Idea of Progress." *Economy and Society* 27 (2/3): 313–329.

Collier, Stephen J. and Aihwa Ong. 2005. "Global Assemblages, Anthropological Problems." In Aihwa Ong and Stephen J. Collier, eds., *Global Assemblages: Technology, Politics, and Ethics as Anthropological Problems.* Malden, MA: Blackwell Publishing.

Comaroff, Jean. 2007. "Beyond Bare Life: AIDS, (Bio)Politics, and Therapeutic Citizenship." *Public Culture* 19(1): 197–219.

Cueto, Marcos. 2007. *Cold War, Deadly Fevers: Malaria Eradication in Mexico, 1955–1975.*

Washington, D.C.: Woodrow Wilson Center Press; Baltimore, MD: Johns Hopkins University Press.

Das, Veena and Deborah Poole (eds). 2004. *Anthropology at the Margins of the State*. Santa Fe, NM: SAR Press.

Deaton, Angus. 2009. "Instruments of Development: Randomization in the Tropics, and the Search for the Elusive Keys to Economic Development." The Keynes Lecture, British Academy, unpublished paper.

Deleuze, Gilles. 2006. *Two Regimes of Madness: Texts and Interviews 1975–1995*. Los Angeles: Semiotext(e).

Deleuze, Gilles and Felix Guattari. 1986. *Kafka: Toward A Minor Literature*. Minneapolis: University of Minnesota Press.

Duflo, Esther, Rachel Glennerster, and Michael Kremer. 2008. "Using Randomization in Development Economics Research: A toolkit," Chapter 61 in T. Paul Schultz and John Strauss, eds., *Handbook of Development Economics*, Vol. 4, Amsterdam. Elsevier. pp. 3895–3962.

Easterly, William R. 2006. *The White Man's Burden: Why the West's Efforts to Aid the Rest Have Done So Much Ill and So Little Good*. Oxford: Oxford University Press.

——, ed., 2008. *Reinventing Foreign Aid*. Cambridge, MA: MIT Press.

Ecks, Stefan. 2005. "Pharmaceutical Citizenship: Antidepressant Marketing and the Promise of Demarginalization in India." *Anthropology & Medicine* 12(3): 239–254.

Epstein, Helen. 2007. *The Invisible Cure*: *Africa, the West, and the fight against AIDS*. New York: Farrar, Straus and Giroux.

Farmer, Paul. 2001. *Infections and Inequalities: The Modern Plagues*. Berkeley: University of California Press.

—— 2003. *Pathologies of Power: Health, Human Rights, and the New War on the Poor*. Berkeley: University of California Press.

—— 2008. "Challenging Orthodoxies: The Road Ahead for Health and Human Rights." *Health and Human Rights* 10(1): 5–19.

Fassin, Didier. 2007. *When Bodies Remember: Experiences and Politics of AIDS in South Africa*. Berkeley: University of California Press.

Ferguson, James. 2006. *Global Shadows: Africa in the Neoliberal World Order*. Durham, NC: Duke University Press.

Fidler, David. 2008. "Global Health Jurisprudence: A Time of Reckoning." *Georgetown Law Journal* 96(2): 393–412.

Fischer, Michael M.J. 2009. *Anthropological Futures*. Durham, NC: Duke University Press.

Foucault, Michel. 1990 [1976]. *The History of Sexuality*. New York: Vintage Books.

—— 2007. *Security, Territory, Population – Lectures at the Collège de France 1977–1978*. New York: Palgrave Macmillan.

Garcia, Angela. 2008. "The Elegiac Addict: History, Chronicity and the Melancholic Subject." *Cultural Anthropology* 23(4): 718–746.

Geertz, Clifford. 2000. "The World in Pieces: Culture and Politics at the End of the Century." In *Available Light: Anthropological Reflections on Philosophical Topics,* pp. 218–263. Princeton, NJ: Princeton University Press.

Hacking, Ian. 1999. "Making Up People." In M. Biagioli, ed., *The Science Studies Reader*. New York: Routledge.

Heckman, James J. and Edward J. Vytlacil. 2007. "Econometric Evaluation of Social Programs, Part 2: Using the Marginal Treatment Effect to Organize Alternative Econometric Estimators to Evaluate Social Programs, and to Forecast Their Effects in

New Environments" Chapter 71. in Heckman, James J. and Edward E. Leamer, eds., *Handbook of Econometrics, Volume 6B*, Amsterdam. Elsevier. pp. 4875–5143.

Hirschman, Albert O. 1971. *A Bias for Hope: Essays on Development and Latin America*. New Haven, CT: Yale University Press.

Kremer, Michael and Edward Miguel. 2007. "The Illusion of Sustainability." *The Quarterly Journal of Economics* (August 2007), pp.1007–1065.

Kristof, Nicholas D. 2007. "Attack of the Worms" (Op-Ed). *The New York Times*, July 2, 2007.

Moran-Thomas, Amy. 2009. "Disparity, Paradox, Ethnography: Observations on the Spread of Diabetes in the Developing World." *Anthropology News* 50(4).

Nguyen, Vinh-Kim. 2005. "Antiretroviral Globalism, Biopolitics, and Therapeutic Citizenship." In Aihwa Ong and Stephen J. Collier (eds), *Global Assemblages: Technology, Politics, and Ethics as Anthropological Problems*, pp. 124–144. Malden, MA: Blackwell Publishing.

Petryna, Adriana. 2009. *When Experiments Travel: Clinical Trials and the Global Search for Human Subjects*. Princeton, NJ: Princeton University Press.

Petryna, Adriana, Andrew Lakoff, and Arthur Kleinman, eds. 2006. *Global Pharmaceuticals: Ethics, Markets, Practices*. Durham, NC: Duke University Press.

Rabinow, Paul and Nikolas Rose. 2006. "Biopower Today." *BioSocieties* 1: 195–217.

Ramiah, Ilavenil and Michael R. Reich. 2005. "Public–Private Partnerships and Anti-retroviral Drugs for HIV/AIDS: Lessons from Botswana." *Health Affairs* 24(2): 545–551.

Rancière, J. 2004. "Who Is the Subject of the Rights of Man?" *The South Atlantic Quarterly*, 102(2/3): 297–310.

Reich, Michael R. 2002. *Public-Private Partnerships for Public Health*. Cambridge, MA: Harvard Center for Population and Development Studies.

Reynolds Whyte, S., M. Whyte, L. Meinert, and B. Kyaddondo. 2006. "Treating AIDS: Dilemmas of Unequal Access in Uganda." In A. Petryna, A. Lakoff, A. Kleinman eds., *Global Pharmaceuticals: Ethics, Markets, Practices,* pp. 240–287. Durham, NC: Duke University Press.

Robins, Steven. 2006. "From 'Rights' to 'Ritual': AIDS Activism in South Africa," *American Anthropologist* 108(2): 312–323.

Sachs, Jeffrey. 2005. *The End of Poverty: Economic Possibilities for Our Time*. New York: Penguin.

Samsky, Ari. 2009. *Pharmaceutical Philanthropy and Global Health: An Anthropological Study of Practices and Values Shaping Drug Donation Programs*. Princeton University, Ph.D. Thesis.

Scheper-Hughes, Nancy. 2008. "A Talent for Life: Reflections on Human Vulnerability and Resistance." *Ethnos* 73(1): 25–56

Singer, Peter. 2009. *The Life You Can Save: Acting Now to End World Poverty*. New York: Random House.

Todd, Petra E. and Kenneth I. Wolpin. 2006. "Assessing the Impact of a School Subsidy Program in Mexico: Using a Social Experiment to Validate a Dynamic Behavioral Model of Child Schooling and Fertility." *American Economic Review*, 96(5), 1384–1417.

Tsing, Anna. 2004. *Friction: An Ethnography of Global Connection*. Princeton, NJ: Princeton University Press.

Whitmarsh, Ian. 2008. *Biomedical Ambiguity: Race, Asthma, and the Contested Meaning of Genetic Research in the Caribbean*. Ithaca, NY: Cornell University Press.

Part II

Capital's margins: the political ecology of the slum world

6 Global garbage: waste, trash trading, and local garbage politics

Sarah A. Moore

Introduction

In September of 2009, Trafigura, a Dutch oil trading company with additional offices in Great Britain, settled a lawsuit brought against it by the people of Abidjan, the main city of the Ivory Coast. The suit alleged that hundreds of tons of waste dumped by the company around the city caused nausea, headaches, vomiting, violent rashes, and even death among thousands of people living near the dump sites. The company denied legal liability, and claimed that the waste was not toxic, but it also agreed to pay out 197 million US dollars.

About a year earlier, in October of 2008, Somali pirates demanded 8 million dollars in ransom for a Ukrainian ship that they had captured. They claimed that the money would be used to clean up toxic waste dumps along the coastal region of Somalia. The spokesperson for the pirates argued that the hijacking of the ship was, in part, a reaction to 20 years of illegal waste dumping on the Somali coast by European firms.

These two events, one a courtroom battle, the other a high seas drama, both point to one undeniable fact: as much as countries, people, and companies across the globe are connected (albeit unevenly) by the circulation of goods and services, they are also connected by the flows of waste. These flows of waste, moreover, are intimately connected to global flows of capital, processes of uneven development and marginalization. Because waste and its disposal are so closely tied to these other global processes, it is imperative to understand that where waste is produced and disposed of are not purely technical matters. Rather, they are inevitably political issues, deeply infused with power relationships, questions of justice and governance, and shaped by representational practices. In this chapter, I discuss several related political and economic aspects of garbage in order to highlight the importance of avoiding technocentric understandings of and solutions to environmental and public health problems. In doing so, I emphasize the roots of our garbage production and disposal practices and patterns in the global capitalist system.

This chapter is divided into the following sections. First, I discuss some basic issues in understanding the global garbage situation including definitions of garbage and waste, estimates of the quantities of waste produce across the globe, and how these definitions and quantities of waste vary spatially. Next, I turn to a larger discussion of the uneven production of waste and the evolution of the international trade in hazardous and municipal solid waste. In the following section, I discuss what happens at the end sites of these global flows of waste (i.e. dump and disposal sites) and issues of environmental justice as well as how garbage can be used as a local political tool by some groups. I conclude by highlighting how a political ecology approach, centered on the relations of capitalism, representation, and citizen/expert knowledge adds to the study of global garbage.

Global geographies of garbage

Before evaluating the political ecology of global garbage flows, some basic questions must be addressed. First, it is necessary to define the terms of the discussion. What, after all, is garbage? Second, some basic understanding of where and how garbage is produced and disposed of is essential to understanding the conditions under which garbage has become an important global commodity. These may at first appear to be straightforward questions, but they are not as easy to answer as one might think. In the first case, definitions of garbage differ. Is it simply something that has been thrown in the trashcan? What about items destined for recycling? Should they be considered part of the waste stream? Are garbage, waste, and trash all the same? In the second case, even if we agree on definitions of garbage, how do we collect information about how much of it there is and where it can be found?

There are no simple answers to these questions. For the purposes of this chapter, garbage, waste, and trash will be used interchangeably, but distinguished from hazardous waste. This might be a dubious distinction, given that most waste has some potential to be an environmental or public health risk, depending on how and where it is disposed of. On the other hand, though, most regulations deal differently with wastes identified as hazardous than with another large category of waste, municipal solid waste (MSW).The United Nations defines MSW as "waste originating from: households, commerce and trade, small businesses, office buildings and institutions (schools, hospitals, government buildings)." This, "includes bulky waste (e.g. white goods, old furniture, mattresses) and waste from selected municipal services, e.g. waste from park and garden maintenance, waste from street cleaning services (street sweepings, the content of litter containers, market cleansing waste), if *managed as waste*" (United Nations 2009a). This definition is broad and highlights the importance of the social context in deciding what is and what is not garbage: waste is what is "managed as waste." The management and treatment of MSW represents more than *one third* of the public sector's expenditures on pollution abatement and control (OECD 2008). This means that a large proportion of government money earmarked for all environmental needs goes to the management of MSW.

Hazardous waste, in contrast, can be broadly defined as "waste that, owing to its toxic, infectious, radioactive or flammable properties poses an actual or potential hazard to the health of humans, other living organisms, or the environment" (United Nations 2009b). This is also a broad definition, and one that leaves much room for debate over what should and should not be regulated as hazardous waste. For that reason, most waste must be listed in specific annexes according to national or international laws and agreements to be regulated as hazardous waste.

As is hinted at in the definitions of MSW and hazardous waste above, whether or not something is considered trash depends on time and place more than any inherent characteristics of the object itself. Most things bought (commodities) by consumers in wealthy countries end up at a dump, legal or illegal, far or near. But, this is not necessarily the whole story. Old clothes can be given away, handed-down, sold to a consignment store, or torn into rags and used for cleaning. A discarded computer could be smashed along with other items in the dumpster, or it could be carefully taken apart in a recycling center so that its various components can be shipped to Asia and reused. A banana skin could be thrown in a public garbage can, or tossed on the compost pile along with the grounds from your morning coffee. Then again, what if you throw your banana peel or your empty soda bottle in the trash and someone else picks it out of the trash to use in gardening or to return for a deposit. The point is that things can alternate, in the course of their social lives, from trash to treasure, useless to useful, valueless to invaluable, just as easily as they can go in the opposite direction (from treasure to trash, for example).

How garbage is defined and managed, then, are largely influenced by negotiated definitions of waste, like those of MSW and hazardous waste above. This has implications for how statistics on waste are compiled. At what point does something count as waste? Do we count everything disposed of by each household, or only the waste that goes to incinerators or dumps for final disposal? What about items that are disposed of locally, but then make their way overseas where they are recycled into other products? Such issues make it difficult to determine the amount of waste that is generated and disposed of in each household, city, or country. At the international level, efforts have been made over the last decade to keep better data on waste. The United Nations (UN) and the Organization for Economic Cooperation and Development (OECD) keep some of the most comprehensive statistics on garbage generation, disposal and trade.

There are many gaps in the data, and there are significant reporting differences between countries. Further, OECD data reflects only the situation of its 30 member countries. Nonetheless, there are some general trends that the data suggest. In general terms, municipal solid waste in OECD countries increased almost 23 percent between 1990 and 2006 from 530 to 650 million tons (OECD 2008). This is a per capita jump from 509–660 kg/year. This aggregate number, though, masks significant differences between OECD member countries. Some of the geography of global garbage production can be better understood by examining a few ways in which member countries differ, including total garbage production, per capita garbage production, and increases in per capita garbage production.

The overall largest producer of MSW in the world is the United States, with 222,863,000 tons/year. This by far exceeds the next largest producer, Japan, at 51,607,000 tons/year. Germany (49,563,000), Mexico (36,088,000) and the UK (35,077,000) round out the top five. These rankings change, however, if we consider the per capita, rather than total production of MSW in each country. The largest producers of MSW per capita are Ireland, Norway, and the United States. In each of these countries, per capita production of waste is near 800 kg/year. Denmark and Luxembourg also each produce more than 700 kg MSW per capita each year. Japan (400kg), Germany (600 kg), and the UK (580kg), some of the largest aggregate producers of waste, produce less waste per capita. The United States is the only country in the top five, both in total and per capita production of municipal solid waste. On the other end of the spectrum, four member countries have per capita numbers under 400 kgs/capita. These are Poland, Slovak Republic, the Czech Republic, and Mexico (one of the largest total producers).

The country with the largest increase in per capita production of waste since 1990 is Spain (over 70 percent). Other countries with a more than 50 percent increase in per capita waste production between 1990 and 2006 are Italy, Portugal and Greece. Of the countries reporting on this statistic, only Hungary and Poland saw a reduction in per capita waste production (about 10 percent in each case). The largest producers of waste vary in this category. While Ireland did not report on the percent change in per capita waste production, the US had little change (less than 5 percent), but Norway saw a near 50 percent increase in per capita solid waste production during the period. The OECD mean was just under 20 percent.

The country-level data cited above may help in part to indicate some of the factors creating the geography of waste production. While new data indicate that generation intensity per capita grew at a slower rate than the gross domestic product (GDP) and private final consumption expenditure (PFC) (OECD 2008), it is still generally agreed upon that economic growth, urbanization and the structure of consumption are all positively associated with increased garbage production (OECD 2008). These are all factors that many consider to be associated with *development* as popularly defined. That is, the high mass consumption culture often associated with places like the United States, much of Western Europe and Japan. In short, development, as the extension of capitalist relations (Wainwright 2008), produces garbage.

As described in the introduction to this volume, the success of global capitalism relies on growth. If, as happened in 2008 and 2009, growth is stopped, things can fall apart very quickly. An average 20 percent increase in per capita garbage production over the last 15 years reminds us that economic growth, while seemingly increasingly created through financial markets that appear detached from patterns of production and consumption, is still dependent on the creation and movement of goods and services across the globe. The creation and movement of goods is inevitably tied to the production of waste. Anything that is traded and consumed makes garbage both as a byproduct of production and as a remainder after the good has lost its utility for the consumer.

For people able to afford the high-consumption lifestyle of the "developed world" the most familiar form of waste is packaging. Food and beverages are a good example of the impact of packaging. While most of what is inside the package (meat, cheese, vegetables, fruits, soda, water) is consumed, the package (deli wrap, plastic bags, Styrofoam containers, aluminum cans, plastic or glass bottles) remains. These are necessary parts of the now global food system, but they present difficulties for waste managers, particularly in places where the consumption of packaged goods has increased quickly.

On the other hand, it is not just food that requires packaging. Most goods that are part of a globalized process of production come in some kind of container. All goods that are shipped across long distances require packaging. Many times, the bulk of the packaging is greater than that of the good itself. Think, for example, of a tiny halogen bulb in a big (nearly unopenable) plastic container. Packaging accounted for nearly one third of the MSW in the United States in 2005 (69,555,000 tons) (OECD 2008).

Clearly, though, if packaging is the first thing thrown out, it is not the last. The bulb, or the TV, or the Ipod or the sweater will eventually need to be disposed of. This too, presents a problem for waste management. The more goods that are produced and consumed, the higher the eventual waste stream. A good example of this is the growing amount of post-consumer electronic or e-waste. According to the United States Environmental Protection Agency, 41,100,000 computers (laptops and desktops) were disposed of in the US in 2007. Although 18 percent of e-waste in the United States was recycled that year, hundreds of millions of computers, televisions and cell phones were trashed, placing pressure on solid and hazardous waste disposal systems across the country (United States Environmental Protection Agency 2008).

Lack of sufficient disposal, though, has not kept companies from continuing to produce and market more and more goods, many of which face planned obsolescence. Even relatively "durable" goods like cars, refrigerators, computers and televisions, can also be replaced with items with more bells and whistles. Not to mention the stock of VCRs, tape players, and certain DVD players for which products are no longer made, due to investment by companies in other forms of technology. All of these end up somewhere, and chances increasingly are that they do not end up in your local dump. For many reasons, these and other items are often shipped to other countries for recycling and disposal. This has resulted in a multi-billion dollar waste trade industry.

In the next sections of this chapter, we will explore the imbrications of the global economy with waste through, first, the international flows of garbage and, second, the conditions associated with garbage's final resting spots, i.e. landfills.

Global flows of garbage

Capitalism makes garbage, and when local disposal systems reach capacity, the system is presented with a potential crisis situation. This crisis can be deferred through what many geographers, following David Harvey, refer to as a spatial fix (Harvey 2007). That is, garbage can be shipped to other places, often far removed from producers and consumers. In this way, waste, itself, has become a commodity. It is now bought and sold on a global market. Both municipal solid waste and hazardous waste are traded internationally. In this section, I focus on the international trade in hazardous waste, though, as discussed above, this can be a dubious distinction, given that all waste has the potential to cause environmental and public health problems.

Between 1976 and 1991, the cost of disposing of hazardous waste in industrial countries increased by a factor of 25 (Asante-Duah and Nagy 1998). This increased disposal cost, contributed to high levels of both legal and illegal transboundary shipments of waste. As transboundary shipments of hazardous waste increased in the 1980s and 1990s, it became increasingly clear that much of the waste was going from more developed countries with stricter environmental regulations to less developed countries with either less strict environmental regulations or without the capacity to enforce the laws in place (Asante-Duah and Nagy 1998; O'Neill 2000).

As one example, consider hazardous wastes shipped from the United States. The growth trends in notices to export hazardous wastes (required by the US EPA) between 1980 and 1990 are notable. In 1980, there were only 12 such notices filed. In 1986 there were 286. By 1988 that number had increased to 570. By the end of the decade, over 620 notices were filed each year. In global terms, by the early 2000s, the international trade in hazardous waste was a multi-billion dollar industry (OECD 2008).

International environmental economics suggests that in a "first-best" world of equal trade relationships, international trade in hazardous waste could be beneficial to all countries involved (Rauscher 2001). Indeed, such logic was echoed in the late 1990s by the US EPA, who argued that one reason to export hazardous waste was that in some cases, "hazardous wastes constitute 'raw' material inputs into industrial and manufacturing processes." The report continued: "This is the case in many developing countries where natural resources are scarce or non-existent" (US EPA 1998).

While it is true that hazardous wastes may sometimes be recycled, many are less sanguine about a cost-benefit approach to the waste trade (Asante-Duah and Nagy 1998; O'Neill 2000; Girdner and Smith 2002). In the "second best" world of environmental economics, it is argued that insufficient environmental policies in some countries could distort the market and make the international hazardous waste trade more harmful to the environment and to public health (Rauscher 2001). This is commonly known as the pollution haven hypothesis. The "pollution haven hypothesis" holds that some countries might voluntarily reduce environmental regulations in order to attract foreign direct investment. Some also argue that, even if producers are not inclined to relocate production, they might still decide to export the negative externalities of their production (like hazardous waste) to countries with less regulation and lower disposal costs (O'Neill 2000).

One way to prevent uneven, unjust, and potentially dangerous transboundary waste shipments is to force trading partners into international agreements such as the Basel Convention, which came into effect in 1992. The Basel Convention prevents richer countries from exporting their hazardous wastes to poorer countries. It has been ratified by 172 countries. The United States signed the agreement in the early 1990s but has yet to ratify it. Clearly, however, these regulations only apply to legal trade of hazardous waste. There is a significant illegal trade in haz-ardous waste, which for obvious reasons has been difficult to document. Lawsuits like the one filed against Trafigura by residents of Abidjan are one source of data on potentially illegal dumping.

In addition to directly banning certain waste transfers, parties to free trade agreements, which are proliferating in this period of global or regional economic integration, might also experience "harmonization" of regulation. This is generally seen as a way to bring regulatory standards of developing countries up to those of developed ones. If regulations are consistent across trading partners, this theoretically eliminates any incentive to ship hazardous waste across borders for treatment (O'Neill 2000).

There is little evidence, though, that harmonization necessarily promotes positive environmental outcomes (O'Neill 2000), despite the fact that it is much lauded as a way of greening environmental agreements. The North American Free Trade Agreement (NAFTA), for example, was originally considered a green free trade agreement because of its side agreements on the environment and steps toward harmonization of regulation across the United States, Mexico and Canada. But the environmental initiatives in the agreement are relatively weak. In fact, the environmental side agreement does not include specific provisions for the management of hazardous waste, but rather cedes these to pre-existing bilateral agreements between the US and Mexico (the La Paz agreement of 1986) and the US and Canada (the agreement was made in 1986 and amended in 1992).

It is not clear what impact these issues actually have on the environmental or public health. There are data to suggest that increased production associated with integration has led to higher levels of pollution along the Mexican border (Di Chiro 2004; Mumme 2007; Simpson 2008) and to an increase in illegal hazardous waste dumping and legal imports of hazardous waste (Slocum 2009).

There is evidence, though, that environmental initiatives created by NAFTA are less important to environmental quality than are some of its free trade articles (Sanchez 2002). One goal of economic integration is to reduce all barriers to trade, not just formal tariffs and quotas. Increasingly, free trade agreements focus on eliminating non-tariff barriers to trade (NTBs). Because environmental regulations have the potential to limit activities and profits of companies investing in foreign countries, such measures are considered NTBs. One important indication of this is the successful use of NAFTA Chapter 11 by corporations who would like to avoid more stringent environmental regulations.

Chapter 11 was written into NAFTA as insurance against expropriation of firm resources by states to protect foreign direct investment (FDI). In the case of NAFTA, it has been used by firms to argue against environmental regulations on the grounds that they represent expropriation because they diminish profits, particularly if they become more stringent over time. A number of prominent lawsuits, based on this interpretation, have been brought against the US, Mexico, and Canada by corporations under Article 1110. Many of these involve the treatment and disposal of hazardous waste. One early example of this is *Metalclad v. Mexico*. The US corporation Metalclad acquired a Mexican hazardous waste company and planned to construct a new waste facility in the city of Guadeleazar, San Luis Potosí in 1993. Metalclad believed that it had permission for the construction, but the municipal authorities shut down the construction after five months. Metalclad brought suit demanding compensation from the Mexican government under Article 1110 of NAFTA. The tribunal found in favor of the corporation, arguing that, since Metalclad was operating under the assumption that they had permission to build on the site, the denial of a permit by the Guadaleazar city council was "tantamount to expropriation." The tribunal also found that Guadaleazar did not have jurisdiction to deny a permit on the grounds of environmental hazards (Chiu 2003).

In these and other cases, it became clear that the interests of international corporations have been privileged over the rights of local people to decide whether or not to allow disposal of toxic substances in their communities. Such precedents ensure the continued growth of the transboundary trade in waste.

Living with trash: where the flows stop

The global nature of the garbage trade, though, must not be overstated. While the flows of garbage around the world may connect disparate places, the impacts of this trade are felt most deeply by specific people in the places where those flows stop. The garbage trade is truly international, but the effects are largely local. Contaminants leach into a specific community's aquifers, runoff soaks into soil in particular neighborhoods, roaches and rats congregate in locales with dumps. It is true that many of these problems have broader environmental impacts too. Waste incineration contributes to local air quality problems, but also global warming. But, much of the time the brunt of the consequences of the global garbage trade are borne by local communities. In this section, I discuss the politics of deciding who will live with and near waste.

The siting of waste disposal facilities is not primarily a technical issue at any level. Rather, it is always and everywhere a political one. This is well documented by scholars and activists in the area of environmental justice (Gottlieb 1993; Szasz 1994; Liu 2000; Westra and Lawson 2001; Kurtz 2005). Much of the original focus of environmental justice was on waste disposal facilities in the United States. Beginning with Bullard's path-breaking work on dumpsite locations in the Southern US, many scholars and activists brought attention to the fact that disposal sites were disproportionately located in minority and/or poor communities (Bullard 1994). Several potential explanations were put forth. Some argued that the siting was simply the result of land rents being lower in poor and minority neighborhoods. Others argued that such communities had a harder time being successful at NIMBY (not in my back yard) politics. Their marginal status meant that they had less political clout to prevent waste facilities from being located in their neighborhoods. Still others argued that this was a case of simple and blatant racism. In truth, all of these issues might be factors in some specific cases. What was missing in some of the early analyses of the problem, however, were ties to the related processes that created uneven development, marginalization of poor and minority communities, and the very problem of waste itself (Pulido 2000).

These related processes are tied to the way that capitalism works as a mode of production. Accumulation, as discussed in the introductory chapter, is necessary for the continued growth of the economy. But, it is also an uneven process, one that puts assets and wealth in the hands of the relative few, and forces countless others to meet their needs through wage labor. At the same time, those people fortunate enough to own the means of production are constantly at risk for declining profits, due to the crisis-prone nature of capitalism. In order to stave off under-consumption, firms must constantly create demand for new goods, thus the planned obsolescence of many commodities discussed above. As much as the processes of uneven development and increased production have differential impacts within a country like the United States, they have an analogous effect on an international scale. This is the problem that treaties like The Basel Convention are designed to solve. As noted above, however, such efforts have not been successful at amelio-rating the local effects of the transboundary trade in hazardous waste. Moreover, they say nothing about waste that, while not classified as hazardous (a constantly shifting category), could still have negative consequences for public and environ-mental health.

While there are obvious ethical problems with the fact that a small proportion of people (many of whom produce minimal quantities of waste themselves) are forced to live with wastes resulting from high-consumption lifestyles among a privileged few global citizens, there are other issues at stake when considering solutions to such problems. First, there are millions of people in places like Guatemala, Indonesia, India and Mexico who live on dumps. The plight of such people is well documented (Crocker 1988; Beall 1997). What is also evident, though, is that these dump communities have their own social structures and informal institutions that do provide livelihoods, however marginal, for many families. In many cases, dumps provide resource bases from which people glean materials to build housing, feed themselves and provide income through recycling (Castillo Berthier 1990; Castillo Berthier 2003).

It would be a mistake of course, to glamorize or romanticize life on a dump, but it would also be an error to assume that development institutions, philanthropic organizations and other well-meaning groups have an unquestionable right to interfere in such places, without seeking input from community members. Often, such groups are operating from within a set of discourses that identify waste with chaos, backwardness, and lack of purity. Because of this, they fail to recognize that, in some ways, these are well-ordered spaces complete with their own forms of governance and social practices that must be taken into account in finding just solutions to the ethical dilemma surrounding the unequal distribution of waste and disposal, locally, nationally, and internationally.

Protesting with trash

The production of waste is a necessary part of our global economy. But, even though millions of tons of waste flow across the globe into various facilities each year, it is a largely invisible one. In this section I discuss what happens when these flows are stopped short and garbage is left where it does not "belong." In these instances, garbage becomes a political tool.

As environmental justice teaches us, some people in particular places across the globe are forced to live their daily lives near dumps and other places with large quantities of trash. On the other hand, in many modern cities with well developed sanitation systems, the majority of residents are accustomed to relatively clean spaces where litter and debris, though still present, are minimal. In such places, garbage is mostly invisible and well-contained (in its place). This makes the unexpected sight of garbage in such places disturbing and gives a certain amount of power to people who can stop the flow of waste. There are a number of groups capable of employing this politics of manifestation – of making waste visible (Moore 2008). Two of the main groups are municipal sanitation workers and people who live near disposal sites. In the last decade, for example, numerous garbage collection strikes in Europe and North America have been effective in securing job benefits and better pay for workers.

This was the case in Philadelphia 1986, New York City 2006, Toronto, Ontario, 2002, Chicago 2003, Vancouver, British Columbia 2007, Athens, Greece 2006, Alumñécar, Spain 2007. The precise motivations behind these strikes differed, and they ranged in time from a few days to several months. They all, though, have in common the use of garbage as a political tool (Moore 2009). In each case municipal authorities, or private garbage haulers (as in the case of NYC), or legislators (as in Toronto) were forced to negotiate deals with employees and their union representatives to get trash off the streets and out of the sight of angry residents.

A similar way of using garbage as political leverage can be found in cases in which one group simply blocks access to disposal sites, thereby causing uncollected garbage to pile up across and urban area. One example of this comes from Oaxaca Mexico. Throughout the early 2000s residents of an informal settlement near the municipal dump regularly blocked the city trucks from dumping trash in the large open air dump used by the entire urban area. The blockades played out in a similar way each time. As the municipal authorities halted garbage collection, residents were told to keep their trash in their homes, rather than to pile it in the street. This advice, however, was not heeded by citizens who felt that the municipality was not complying with its obligation to keep the city clean. Hundreds of tons of waste piled up in parks, on street corners, and near market areas.

In addition to annoying residents with its visible presence and increasing smell, the garbage attracted rodents, insects and feral dogs. This, combined with the negative impression on tourists (a key source of the city's income) was enough to lead the municipal authorities to negotiate with the protestors to gain access to the dump. While some of the neighborhood's demands centered on garbage management issues: the inability of authorities to prevent fires on the dump, poor engineering of the landfill, and the fact that the dump was over capacity; the community also received electricity, a medical center and a meeting center as a result of the protests (Moore 2008). In this way, these marginalized citizens were able to use garbage to assert their rights to the city.

Whether wielded by municipal workers or neighborhood residents, the power of garbage as a political tool comes in part from the fact that it stinks and attracts pests. It comes equally, though, from the expectation of many modern urban citizens, particularly in developed countries, that garbage should be out of sight and thus off their minds. In this way, the unexpected presence of garbage reminds people, in a visceral way of the consequences of a high-consumption, disposable lifestyle.

Conclusion

This chapter has discussed numerous ways in which garbage is more than simply a technical issue. What garbage is, where garbage is, and how it gets disposed of are political issues. The very definitions of solid waste and hazardous waste are negotiated and differ according to context. Even when there is agreement on what constitutes garbage, questions remain about how to determine who has to live with it. In this way, state power and economic relationships are also important aspects of an international political ecology of garbage. This is as true of the two incidents that began this piece (dumping in the Ivory Coast and hijacking by Somali pirates) as it is of the hazardous waste flows influenced by NAFTA and other free trade agreements. On the other hand, municipal garbage workers and neighborhood protestors have demonstrated the ways that garbage can be used to challenge current power relationships and to improve the economic situation of workers and residents who live with it on a day to day basis.

These contradictory moments (garbage as threat vs. garbage as political tool) in the political aspects of garbage have commonalities that are revealed only by thinking beyond the familiar technological approaches that focus on responsible environmental *management* of garbage. A global political ecology of garbage instead points to how waste production, trade and disposal, just like global flows of (other) commodities, capital and services unfold across an always uneven terrain of development and power. While the former instances demonstrate the lack of local control over garbage flows and disposal and the disconnection between spaces of consumption and spaces of waste, the people who choose to protest with trash highlight the fact that global flows always have local nodes. Moreover, by stopping these flows and making waste visible in central areas, such protestors reconnect spaces of consumption and spaces of waste. This has radical implications for the continuation and extension of the high-consumption style of development associated with global capitalism, because it gives us insight into what might happen when there are no more spatial fixes available to resolve our garbage crisis. If continued production and consumption are enabled by the disposal of goods, what happens when trash remains in place?

References

Asante-Duah, D. K. and I. V. Nagy (1998). *International Trade in Hazardous Waste*. London, E. & F.N. Spon.

Beall, J. (1997). "Thoughts on Poverty from a South Asian Rubbish Dump." *IDS Bulletin* **28**(3): 73–90.

Bullard, R. D. (1994). *Dumping in Dixie: Race, Class and Environmental Quality*. Boulder, CO, Westview Press.

Castillo Berthier, H. (1990). *La Sociedad de la Basura: Caciquismo en la Ciudad de México*. México, D.F., Universidad Nacional Autónoma de México.

—— (2003). "Garbage, Work and Society." *Resources, Conservation and Recycling* **39**(3): 193–210.

Chiu, C. (2003). "NAFTA Chapter 11 and the Environment." *Environmental Policy and Law* **33**(2): 71–76.

Crocker, G. (1988). "Squatting on Garbage Dumps: Behind the Self-help Housing Debate: A Case Study of a Guatemala City Garbage Dump." *School of Planning*, Universidad Mariano Gálvez de Guatemala: 102.

Di Chiro, G. (2004). "'Living is for Everyone': Border Crossings for Community, Environment, and Health." *Osiris* 19: 112–129.

Girdner, E. J. and J. Smith (2002). *Killing Me Softly: Toxic Waste, Corporate Profit and the Struggle for Environmental Justice*. New York Monthly Review Press.

Gottlieb, R. (1993). *Forcing the Spring*. Washington, D.C., Island Press.

Harvey, D. (2007). *The Limits to Capital*. London, Verso.

Kurtz, H. E. (2005). "Reflections on the Iconography of Environmental Justice Activism." *Area* 37(1): 79–88.

Liu, F. (2000). *Environmental Justice Analysis: Theories, Methods, and Practice*. Boca Raton, LA, Lewis Publishers.

Moore, S. A. (2008). "The Politics of Garbage in Oaxaca, Mexico." *Society & Natural Resources: An International Journal* 21(7): 597–610.

—— (2009). "The Excess of Modernity: Garbage Politics in Oaxaca, Mexico." *The Professional Geographer* 61(4): 426–437.

Mumme, S. P. (2007). "Trade Integration, Neoliberal Reform, and Environmental Protection in Mexico – Lessons for the Americas." *Latin American Perspectives* 34(3): 91–107.

O'Neill, K. (2000). *Waste Trading among Rich Nations: Building a New Theory of Environmental Regulation*. Cambridge, MA, MIT Press.

OECD (2008). "OECD Environmental Date Compendium: 2006–8: Waste." Retrieved Nov 1, 2009.

Pulido, L. (2000). "Rethinking Environmental Racism: White Privilege and Urban Development in Southern California." *Annals of the Association of American Geographers* 90(1): 12–40.

Rauscher, M. (2001). "International Trade in Hazardous Waste." In G. Schulze and H. Ursprung (eds), *International Environmental Economics: A Survey of the Issues*. Oxford: Oxford University Press. pp. 148–65.

Sanchez, R. A. (2002). "Governance, Trade and the Environment in the Context of NAFTA." *American Behavioral Scientist* 45(9): 1369–1393.

Simpson, A. (2008). "NAFTA's Failure to Protect Public Health and the Environment: The Case of Metales v Derivados, Tijuana, Mexico." *Epidemiology* 19(6): S16-S16.

Slocum, R. (2009) "Rethinking Hazardous Waste under NAFTA." Americas Program Policy Report.

Szasz, A. (1994). *EcoPopulism: Toxic Waste and the Movement for Environmental Justice*. Minneapolis, University of Minnesota Press.

United Nations (2009a, August 2009). "Environmental Indicators: Waste." Available online at http://unstats.un.org/unsd/environment/wastetreatment.htm (accessed 15 November 2009).

—— (2009b, August 2009). "Hazardous Waste Generation." Available online at http://unstats.un.org/unsd/environment/wastetreatment.htm (accessed 15 November 2009).

United States Environmental Protection Agency (2008). "Electronics Waste Management in the United States: Approach 1." , Available online at http://www.epa.gov/osw/conserve/materials/ecycling/docs/app-1.pdf (accessed 15 November 2009).

Wainwright, J. (2008). *Decolonizaing development*. Malden, MA: Blackwell.

Westra, L. and B. E. Lawson (2001). "Introduction ." In *Faces of Environmental Racism: Confronting Issues of Global Justice*. Lanham, MD, Rowman and Littlefield Publishers, Inc.: xvii–xxvi.

7 Green evictions: environmental discourses of a "slum-free" Delhi

D. Asher Ghertner

Seeing the slum, seeing pollution

In 1994, an association of factory owners filed a petition in the High Court of Delhi to address the "great pollution problem resulting in dirt, filth and terrible sanitary conditions" in an industrial estate in South Delhi. Highlighting a litany of local environmental problems – including overflowing storm water drains, irregular electricity supply, congested streets, and open sewage – the petition made special mention of local slum settlements, asking "Whether it is not the statutory duty of the [Municipal] Corporation to destroy infectious huts and sheds in order to prevent the spread of any dangerous diseases."[1] With little mention of the other grievances raised in the original petition – namely, the Municipal Corporation's failure to provide adequate municipal services – the High Court responded in 2002 by placing the responsibility for poor environmental conditions squarely on the "unhygienic mushrooming of slums in the urban areas causing a lot of damage to the health environment [*sic*] of the city as a whole."[2] As a result of "the problem of the slum," the judgment went on to say, "citizens who have paid for the land and occupy adjacent areas [to the slums] are inconvenienced. An unhygienic condition is created causing pollution and ecological problems. It has resulted in almost collapse of Municipal services [*sic*]."[3] The Court concluded with the following order: "Encroachers and squatters on public land should be removed expeditiously without any pre-requisite requirement of providing them alternative sites."[4]

In the midst of the case's subsequent proceedings, the Court extended the logic equating slums with filth by taking arbitrary cognizance of the problem of pollution in the Yamuna River, an issue nowhere addressed in the original petition filed by the factory owners' association:

> What is required to be done in the present situation in this never ending drama of illegal encroachment in this capital city of our Republic? River Yamuna which is a major source of water has been polluted like never before. Yamuna Bed and both the sides of the river have been encroached by unscrupulous persons with the connivance of the authorities. Yamuna Bed as well as its embankment has to be cleared from such encroachment. Rivers are perennial source of life and throughout the civilized world, rivers, its water and its surroundings have not only been preserved, beautified but special efforts have

been made to see that the river flow is free from pollution and environmental degradation [*sic*].

The only connection between the original petition and the polluted Yamuna was a shared geographic imaginary of contagion, a territorialized fear of slums as contaminated spaces in a city in need of purification. The court thus stated,

> In view of the encroachment and construction of jhuggies [slum huts]. . .in the Yamuna Bed and its embankment with no drainage facility, sewerage water and other filth is discharged in Yamuna water [*sic*]. The citizens of Delhi are silent spectator [*sic*] to this state of affairs.[5]

While briefly acknowledging the multiplicity of sources of pollution in the Yamuna, the court took a complex hydrological and ecological problem – river degradation – and simplified it into the visible presence of a degraded population living on the banks of the river – a multi-generational group of slum settlements colloquially known as Yamuna Pushta. Despite a study by a non-governmental organization showing that Pushta contributed less than 0.5 percent of total effluent discharge into the river (D. Roy 2004), and despite the Delhi Water Board's open acknowledgement that the main cause of river pollution was the 22 open drains that carried untreated waste from mostly middle class residential colonies directly into the Yamuna, the court stood by its assertion that slums had destroyed the natural beauty and ecology of the river.[6] The most "scientific" evidence the court used to justify this claim was a set of photographs submitted by the Ministry of Tourism – part of its proposal to develop the site of Pushta into a riverside promenade and tourist attraction – ostensibly showing slum dwellers as "polluters" and carrying captions indicating the unsightliness of slums for foreign visitors and dignitaries.

In February and April of 2004, approximately 35,000 huts in Pushta were razed, displacing the more than 150,000 residents they housed (see Figure 7.1).[7] Months after the demolition, the court set up an expert committee – not to monitor the Yamuna's environmental quality, the basis for slum clearance in the first place, but rather to ensure that other "encroachments" on the river were removed apace. Between 2006 and 2008, an additional 10,000 huts were cleared under the watchful eye of the monitoring committee,[8] with 30,000 more identified for future removal.[9] After these demolitions – what the court labelled a "clean up" drive necessary to prevent the conversion of the Yamuna "into a huge sewage drain"[10] – evidence from the Central Pollution Control Board showed "no improvement in the quality of water. Instead, it had deteriorated over the years and several crore [1 crore = 10 million] rupees spent by the Government on the Yamuna Action Plan has virtually gone down the drain."[11] Despite this, it has become "common sense" that slums are the source of the Yamuna's pollution, with the Chief Minister publicly confirming this "truth" as recently as May 2009, even after almost all slums had been removed from the river banks.[12]

I begin by recounting these events because they highlight the metonymic association between slums and pollution that has become increasingly prominent

Figure 7.1 Site of the Yamuna Pushta settlements, before (top) and after (bottom) their demolition in April 2004. The upper-right hand corners of the images show the Yamuna River, the "clean up" of which provided the basis for the demolition. The lines on the left side of the images show Google's approximation of the location of major roads, which are a bit askew from the actual roads. © 2010 DigitalGlobe, © 2010 Google.

in state and judicial discourse in Delhi since the late 1990s. Indeed, over the past decade, approximately 1 million slum residents have been forcibly displaced from Delhi,[13] the vast majority thanks to court orders, like that in the Pushta case, depicting slum removal as a process of environmental improvement (Ramanathan 2006).

What stands out in these orders, and what I seek to explain below, is the scant empirical basis on which the environment is rendered legible as a category of knowledge and object of management in contemporary Delhi. If we take "legibility" to mean "an overall, aggregate, synoptic view of a selective reality" that makes "possible a high degree of schematic knowledge, control, and manipulation" (Scott 1998: 11), then it is clear from the Pushta case that the judiciary characterized river pollution – i.e. it made pollution legible – not in terms of effluent flow, hydrology, or other "scientific" measures of river ecology. Instead, it asked if the land uses in question accorded with what one would expect to find in a modern, orderly, and, in the language of the government, "world-class" city. Thus, the court compared the Yamuna with rivers "throughout the civilized world" and called upon the Ministry of Tourism's images of slum "unsightliness" to demonstrate that Pushta defied the aesthetic norms of a modern, "civilized" city. Based on this aesthetic appraisal – and *not* an assessment of Pushta's conformance with land use maps or pollution regulations – the court declared Pushta "polluting."

The environment, as a category of knowledge, operates here more as a culturally trained set of habits or aesthetic dispositions than as a calculated measure of ecological welfare or human health. Environmental expertise in Delhi, as I will elaborate below, thus derives its epistemological authority, or power to organize specific understandings of the world, not from those "mechanically objective" (Porter 1995) measures (e.g. statistics, maps, surveys) typically considered necessary to pare down entangled socio-natural environments into a governing legibility. Rather, this end is achieved from a strong aesthetic normativity – a hegemonic sense of how the city should look. In this chapter, I am broadly interested in showing how this *green aesthetic* – a distinct observational grid (or legibility) for making normative assessments of social space – came into being as a dominant "regime of truth" (Foucault 1976), how it is deployed, and with what material effects.

These questions are relevant not only in light of recent anti-poor environmental discourse in Delhi, what Baviskar (2003) calls "bourgeois environmentalism," but also because the government began to more broadly frame questions of urban development and political economy in environmental language at this time. It did so, in part, by branding infrastructure, public works, and real estate projects "green" that had never before been considered especially eco-friendly. Thus, a shopping mall or a flyover can now be deemed green on the basis of its contribution to a more ordered urban landscape – its "clean and green" look – regardless of its conformance with environmental laws or standards. For example, the Delhi Development Authority declared the construction for the 2010 Commonwealth Games – located on the banks of the Yamuna – "eco-friendly" based on "plans to ease traffic congestion and improve horticulture in and around stadia and important tourist destinations,"[14] despite the fact that it had commissioned "at least two scientific studies concluding that no permanent structures should come up in that area in order to protect the flood plain."[15] In the same period, the government launched its "Clean Delhi, Green Delhi" campaign, a city-wide public information drive aimed at instilling a sense of civic pride in the cleanliness and appearance of India's capital city, primarily through aesthetic projects (e.g. roadside landscaping, park rejuvenation) that do little to address underlying sources of environmental stress.[16] My aim in calling attention to this proliferation of "green speak" is not to say that the government has been straightforwardly "greenwashing" environmentally deleterious projects, although this is true in many cases. Rather, I want to suggest that the green aesthetic I am describing here acquired expanded epistemological authority in the context of new political-economic and governmental imperatives. A key aim of this chapter, then, is to ask why the environment arose as a newly significant legal and governance problem in Delhi in the late 1990s. That is, what were the administrative and legal compulsions that required the insertion of a new governmental object – the environment – into the field of politics?

What I will argue here is that green aesthetics arose as a governmental fix to the intractability of the slum problem. In the late 1990s, state officials and politicians began articulating the goal of turning Delhi into a "slum-free city," giving it a "world-class" look, promoting an efficient land market, and converting the "under-utilized" public land occupied by slum dwellers into commercially exploitable private property (DDA 1997). These were all part of the policies of economic liberalization initiated nationally in 1991 and concretely implemented in Delhi in the late 1990s (Ghertner 2005; Jain 2003). However, the planning and legal procedures for managing urban space and administering slum removal were based on a map- and survey-based rationality that faced numerous technical and political challenges at the time. The environment, as an aesthetic category, emerged out of this context as a new organizing logic upon which the question of urban governance could be framed. Obviating the need for maps, statutes, or statistics – those standard instruments of rational planning, modern law, and science – green aesthetics emerged as a regime of truth that allowed government to overcome the bureaucratic obstacles to Delhi's world-class ambitions and the violent displacements they entail. That is, the courts and state started using green aesthetics as a new governing legibility to exceed planning and statutory law, to invoke exceptional powers, and to project a new, bourgeois spatial imaginary of a "slum-free city." I make this argument in the following section, showing how challenges to map- and survey-based simplifications of slum space and the delays in slum removal they caused, compounded by the need to redevelop Delhi's visual landscape in preparation for the 2010 Commonwealth Games, led the courts to begin asserting general aesthetic norms for how the urban environment should appear. The environment as an aesthetic thus arose as a distinctly non-calculative political field, constructing a governing legibility without requiring the complicated and fraught calculative procedures of earlier governmental practice.

I follow in the third section by tracing how this green aesthetic concretely emerged through the law. I specifically show how the judiciary reinterpreted the meaning of public nuisance – an inherently aesthetic category defined in terms of codes of civility (Diwan and Rosencranz 2001: 97) – transforming the truism "slums are dirty" into the new truth claim that "slums are nuisances." Once defined as a nuisance, an actionable offense in environmental law, a new set of procedures was made available for remedying the problem of the slum – namely, demolition. Here and in a concluding discussion, I set out to establish the epistemological basis on which environmental expertise is established in contemporary Delhi, showing how the evaluation of nuisances, the foundation of environmental law in India (Jain 2005), consists of a necessarily aesthetic judgment. In asking what types of statements (and with what empirical backing) must be marshaled to define slums as nuisances – i.e. as illegal environments – I will suggest that aesthetic judgments do not differ so greatly from other, more strictly calculative forms of expertise. I will then conclude by showing how cultural construals of the environment – i.e. all discourses of the environment – necessarily engage in a type of aesthetic politics. In the words of Ranciere (2004), aesthetic discourses of the environment establish "a distribution of the sensible" that lays down universally recognizable boundaries between the beautiful/ugly, visible/invisible, legal/illegal, and green/polluting, enabling profoundly political decisions to appear as mere questions of taste and discernment: in Delhi today, slums are illegal because they appear so.

Calculative difficulties and the turn to aesthetics

In 2003, Delhi won the bid to host the 2010 Commonwealth Games, catapulting it into the international spotlight and strengthening the government's claim that a "world-class" future was within reach (DDA 2007). As the largest international sporting event ever scheduled to be held in India, and as part of India's broader ambition to host a future Olympic Games, the Commonwealth Games is a mega-event that has galvanized widespread support from the media and public and generated a general sense of the need for a physical and aesthetic upgrade to Delhi's cityscape. The government and media constantly narrate this ten-day event as an opportunity for Delhi, and by extension all of India, to showcase itself to the world. The celebration of and preparation for the Games, what Delhi's Chief Minister recently called the government's "top priority,"[17] is thus construed as an exercise in nation building that justifies broader changes in state policy and practice. For example, since the early 2000s, public finances were gradually shifted away from education, public housing, and food subsidies toward large, highly visible and "modern" infrastructure projects such as the Delhi Metro Rail (an underground subway system), more than 25 new flyovers, two new toll roads to Delhi's posh, satellite cities, and the Commonwealth Games Village – prestige projects built "to dispel most visitors' first impression that India is a country soaked in poverty" (Ramesh 2008).

The Commonwealth Games thus works as an impetus for deeper political-economic restructuring, defining the present as an exceptional moment that requires exceptional sacrifice on the part of the poor, exceptional investment from the state, and exceptional faith from the people and their representatives. It is in this context that the "world-class" city was explicitly defined as a "slum-free" city (DDA 2007), with slums emerging as the most visible obstacle to the city's global aspirations. But, the city's more than 1,000 slum settlements could not so easily be conjured away. Despite the clear mandate from above to remove slums, the practical means of doing so were limited. Through the 1990s, for example, various programs were launched to upgrade or relocate slums, but the slum population nonetheless increased from 260,000 to 480,000 families between 1990 and 1998.[18]

During this period, the decision to remove a slum lay primarily in the hands of the state agencies upon whose land slums were settled. Thus, if a slum on the land of the Delhi Development Authority (DDA) – Delhi's primary land administering agency – was to be removed, for example, the DDA was charged with notifying the slum residents, surveying the households to determine resettlement eligibility, collecting fees from those offered resettlement, purchasing and/or allocating the necessary land for establishing a resettlement colony, obtaining support from the police for protection during the demolition, hiring the demolition team for the appropriate day, and coordinating the resettlement exercise with the Slum Wing of the Municipal Corporation. Not only was each of these steps bureaucratically challenging, but the elaborate patronage relations extending from slums into the lower bureaucracy, what Benjamin (2004) calls India's "porous bureaucracy," made the assembly of accurate survey registers – a requirement before a demolition could be carried out at the time – nearly impossible. Surveys were tampered with, false names were appended, and between the time when the survey was completed and when the agency obtained the necessary clearances and land appropriations (typically multiple years), the number of people residing in the slum had changed, thus demanding a new survey and setting much of the same process in motion again (cf. Hull 2008). Furthermore, through the 1990s, the cost of obtaining and preparing land for resettlement colonies escalated (DDA 1997), creating a strong disincentive for land-administering agencies (like the DDA) to remove slums in the first place. In addition, the legal status of most slum settlements was ambiguous, with various forms of de facto regularization over the years (e.g. state-issued ration and voting cards, state-funded infrastructure improvements, the presence of government-run schools) making slum removal a charged political issue.[19] In short, the procedure for removing slums was costly, slow, and contentious.[20]

In the early 2000s, however, there was a drastic increase in public interest litigations (PILs) filed against slums by resident welfare associations (RWAs) (Chakrabarti 2008) – property owners' associations mobilized around neighborhood security and local environmental issues. Combined with the 2003 announcement of Delhi's successful bid to host the Commonwealth Games, this placed the municipal and state governments[21] under increasing pressure from both above and below to "clean up" the city. In the late 1990s, the courts had increasingly begun to take notice of "the dismal and gloomy picture of such jhuggi/jhopries [slum huts] coming up regularly"[22] and in 2002 observed that "it would require 272 years to resettle the slum dwellers" according to existing procedures and that the "acquisition cost. . .of land. . .and development . . . would be Rs. 4,20,00,00,000/-[\$US 100 million]."[23] This set of conditions was incompatible with Delhi's imagined world-class future, so the courts, in response to the PILs filed by RWAs, began intervening in slum matters and increasingly rebuked the DDA and other land-owning agencies for failing to address the "menace of illegal encroachment" and slums.[24] However, when the courts pushed these agencies to act more aggressively to clear slums, judges were befuddled by messy ground realities, missing government records, ambiguous tenure statuses, and outdated surveys. The courts found themselves in a position where they were unable to even assess the size of the problem, not to mention issuing informed action orders. For example, in a case against a slum in South Delhi, the High Court stated, "There are several controversies, claims and counter claims made by the learned counsel for the parties. The records are, however, scanty and the said claims and counter claims cannot be decided on the basis of existing material and documents on records."[25] In the case of the clearance of slums along the Yamuna River discussed in the introduction, the High Court observed that

> . . . in spite of repeated directions no progress has been made by the DDA as the DDA has not submitted area-wise sketch plans showing clusters of jhuggis [huts] and other structures on various parts of Western embankment of the river Yamuna. It seems that the DDA itself does not have [a] plan.[26]

Such an absence of cadastral precision and accurate plans is widespread in slum-related cases, which by the late 1990s led to the absence of a synoptic vision, or governing legibility, by which upper-level bureaucrats and the courts could "survey a large territory at a glance" (Scott 1998: 45) and "govern from a distance" (Rose 1999). For Latour (1987), such "action at a distance" relies on a "cascade" or relay of measurements and inscriptions (e.g. survey registers) that can be combined and simplified into more generalizable and thus legible re-presentations of the territory (e.g. maps and statistical tables) as they move up the chain of administrative command to "centers of calculation," like courtrooms and centralized government offices. The absence of reliable baseline surveys in Delhi, however, broke this cascade, rendering knowledge of slum space highly localized rather than abstractly knowable and manipulable from above. As a result, land-administering agencies could easily delay slum-related court decisions for years by postponing court hearings in order to survey and reassess the ground situation. Until accurate visual simplifications of slum space were secured (i.e. until the "cascade" of inscriptions was complete), bureaucrats sitting in state offices and judges in courtrooms had their hands tied, or so it seemed.

But, this was not merely a matter of conducting better or more accurate surveys. As Blomley (2008) reminds us, "simplification is complicated," meaning it both requires a great deal of inscriptive effort and generates often conflicting, incoherent, or non-aligning representations of reality. He notes that processes of simplification necessary to make legal decisions require the production of sharp categorical distinctions between, for example, private/public, planned/unplanned, and legal/illegal. But such a binary logic encounters diverse and entangled socio-natural environments that do not in any simple way conform to the categories of state legibility. Blomley specifically examines a property dispute along the boundaries of the shifting Missouri River, noting how different simplification practices presented in the court created different "rivers," each with distinct meanings and linked property claims.[27] In Delhi too, slum cases inevitably involve a proliferation of historical claims to property, with slum residents organizing a diverse array of legal documents, written affidavits and other verbal/textual evidence of historical residency to substantiate their right to occupancy. The neat legal categories employed by surveyors hence butt against competing property claims, porous boundaries, and incompatible evidences. Surveyors entering slums usually rely on some degree of local knowledge in order to grasp the lay of the land, but this knowledge often conflicts with the bureaucratic grid of the survey (uniform plots, single owners, stable boundaries, etc.), calling the accuracy/truthfulness of the survey itself into question.[28]

In many instances in Delhi, the ownership of the land occupied by slums is ambiguous in the property records themselves, putting the court in the strange position of being prepared to order a slum demolition, but not knowing which agency is obligated to carry out the order. In a case that ultimately resulted in more than 2,800 homes being razed in 2006, one party claimed that the land in question belonged to the Municipal Corporation, but "Thereafter it was difficult to find out as to who was [*sic*] the owner of the land as all the land owning agencies abdicated their responsibilities and none was prepared to own the land."[29] In response to such ambiguity, the court in the early 2000s began appointing its own monitoring committees and court commissioners to do ground level field assessments in place of the state bureaucracy. The court thus viewed a lack of legibility as a technical failure – the product of an incompetent or corrupt state – that could be overcome by more efficiently implementing the existing survey-based calculative practices. But, producing calculations capable of administering the law and accurately reflecting local conditions required extensive field knowledge of not only the current ground reality, but also the history of such spaces. These court-appointed surveyors ended up producing equally (or more) flawed simplifications of the ground reality, as was pointed out by a civil writ petition contesting a court committee's recommendation to demolish a slum in north Delhi:

> . . . it is apparent that the inspection and scrutiny performed by the Learned Court Commissioner appears, at best, perfunctory . . . [and contains] marked discrepancies about the area and size of the basti [slum] . . . [The Committee's report] is also incomplete, cursory and factually inaccurate. [The letter by the Court Commissioner] requests the Court to give directions for removal of encroachments without clarifying what are considered encroachments. . . . further the Monitoring Committee also differs from the Learned Court Commissioner in its assessment of the size of the basti . . . the authorities appear to be unclear even to the extent and demarcation of the land area in question – the land of two Khasras [plots] (110, 111) are shown in the Revenue record as merely Government land, without designating a specific land owning agency.[30]

In this particular case, the legal boundary between "planned" developments and "encroachments" did not accurately reflect the mixed land uses found on the ground – the state had itself placed the so-called "encroachers" on vacant lots as part of a temporary housing policy in the 1970s. The effort to parse dynamic and amorphous tenure arrangements[31] into this clear binary produced "factually inaccurate" simplifications.

By the 2000s, through a combination of an increasingly complex and unruly ground situation and the inability of existing calculative practices to render that ground sufficiently legible to the courts and upper-level bureaucrats, the epistemological foundation of slum surveys was called into question, both through legal challenge and community-led counter-surveys (see Ghertner 2010). Despite regulators', bureaucrats', and jurists' best efforts to devise categories of knowledge suitable to the domain to be governed, simplification is by definition partial, forcing processes and patterns that arose independently of state plans into a state rationality (cf. Scott 1998: 124). Compounding these calculative challenges is the fact that, according to the Municipal Corporation, 70 percent of Delhi is "unauthorized," meaning it violates land use codes or building bye-laws in some way or another.[32] If the court were to begin removing all unauthorized land uses, most of Delhi would have to be razed, including many developments central to Delhi's "world-class" ambitions. Thus, strict enforcement of the Master Plan or development codes, which had been avoided for almost fifty years, would lead not just to a "slum-free" city, but also a business-, mall-, and industry-free city. Recognizing this dilemma, the Municipal Corporation submitted in the High Court that the problem of unauthorized constructions and slums is "mammoth in nature – and cannot be controlled by simply dealing under the existing laws or under the provisions of [Delhi's] master plan" (Biswas 2006).[33] That is, it called upon the judiciary to exceed existing law to remake the city.

The courts did so by abandoning the previous bureaucratic and statutory requirement that land-owning agencies create calculative, map- and survey-based simplifications of slum space. Through the 1990s, government surveys were conducted to summarize slums according to the duration of the slum population's occupation of the land in question, residents' eligibility for resettlement, the land use category of the occupied land, and the density and size of the population settled thereupon. Only then would summary statistical tables and maps that simplified messy ground realities into compact "planes of reality" (Rose 1991: 676) be relayed up the bureaucratic chain so that state decision makers and judges could assess their legality. But, as shown above, assembling such inscriptive and "mechanically objective" simplifications was slow, contentious, and sometimes impossible given the ambiguity in property records. So, instead of requiring these complex calculative procedures, the courts started using a surrogate indicator to identify illegality: the "look" or visual appearance of space. In lieu of accurately assessing (i.e. creating paper re-presentations that correspond to) physical space, a set of visual determinants began to be used to render slums legible and locatable within the new, predominantly aesthetic "grid of norms" (Rose 1991).[34] As I will now show, the environment, as framed through the legal category "nuisance," would become the mechanism for carrying out this transition from a calculative to a more aesthetic regime for evaluating physical space.

Green aesthetics and the nuisance of slums

In the early 2000s, the courts began making widespread mention of Delhi as a "showpiece," "world-class," and "heritage" city. In a landmark judgment from 2000, the Supreme Court stated,

> In Delhi, which is the capital of the country and which should be its showpiece, no effective initiative of any kind has been taken by the numerous governmental agencies operating there in cleaning up the city. . . . Instead of "slum clearance" there is "slum creation" in Delhi. This in turn gives rise to domestic waste being strewn on open land in and around the slums. This can best be controlled . . . by preventing the growth of slums.[35]

The court thus established the presence of slums as the clearest obstacle to Delhi becoming a world-class city, a link made even clearer when the Delhi High Court noted that at the current pace, it would "require 1,263 years to demolish the illegal constructions carried out over the last 50 years, and convert Delhi into a world-class city."[36]

Court documents from this period, like in the Pushta case discussed in the introduction, show that the growing concern for the city's world-class appearance increasingly came to be expressed in the early 2000s through an environmental discourse of cleanliness and pollution. Popularized through the Delhi Government's "Clean Delhi, Green Delhi" slogan, this discourse tied deficiencies in environmental well-being and appearance to the presence of slums, largely through the legal category of nuisance – the statutes for which provide the underlying basis for environmental law in India (Jain 2005). For example, in 2001, the Delhi High Court stated: "Delhi being the capital city of the country, is a show window to the world of our culture, heritage, traditions and way of life. A city like Delhi must act as a catalyst for building modern India. It cannot be allowed to degenerate and decay. Defecation and urination cannot be allowed to take place in the open at places which are not meant for these purposes."[37] Before 2000, nuisance-causing activities like open defecation or unhygienic living conditions was not a sufficient justification for demolishing a slum. Unsanitary conditions in slums and general slum-related public nuisances were legally considered the responsibility and fault of the Municipal Corporation through the 1980s and 1990s: slums were dirty because the state did not provide them with basic services.[38]

However, as I have argued elsewhere (see Ghertner 2008), the early 2000s introduced a new legal discourse of nuisance that reconfigured the parameters and mechanisms by which slum-related nuisances were to be remedied. The juridical category of nuisance is broadly considered any "offense to the sense of sight, smell, or hearing" (Jain 2005: 97) and is as such directly linked with aesthetic norms and codes of civility.[39] In Indian law, nuisances are of two types, public and private, where the former is an "unreasonable interference with a right common to the general public" and the latter is a "substantial and unreasonable interference with the use or enjoyment of land" (Ibid.). Because slums are almost entirely settled on public land, slum-related nuisances have always been addressed through public nuisance procedures. The definition of public nuisance, according to statute and precedent, had until this time included only particular *objects possessed* or *actions performed* by individuals or groups that interfered with a public right. Aesthetically displeasing, annoying, or dangerous actions or objects could only be addressed by improving municipal services or fining individuals for their violation.[40]

The state's inability to improve, clean up, or remove slums, as well as the court's failure to efficiently provide order to the city by removing slums through existing statutes, led to two gradual shifts in how public nuisance was interpreted in the early 2000s, which together produced "the environment" as a discursive platform for removing slums according to their aesthetic impropriety. First, the courts increasingly began accepting petitions under *public* interest litigation from *private* parties (mostly RWAs, but also hotel and business owners) claiming that neighboring slums were interfering with their quality of life and security. That is, concerns of a distinctly *private* nature were granted legal standing as matters of *public* purpose, or, as Anderson (1992: 15–17) noted of colonial jurisprudence in India: "Propertied groups were able in many instances to invoke public nuisance provisions against anyone threatening the value of their property," making nuisance "the coercive arm of property rights." This elevation of the concerns of propertied residents, or blurring of public and private nuisance, was based on the High Court's 2002 distinction laid out in the Pushta case between "those who have scant respect for law and unauthorisedly squat on public land" and "citizens who have paid for the land."[41] This ruling established land ownership as the basis of citizenship as such, rendering the preservation and security of private property a public priority and setting the conditions for a broader reworking of nuisance law based on bourgeois/private understandings of the environment.

The second shift in the interpretation of public nuisance made the *appearance* of filth or unruliness in and of itself a legitimate basis for demolishing a slum. This change took place by redefining the categories of nuisance such that not only *objects* or *actions*, but also *individuals* and *groups* themselves could be declared nuisances, a shift carried out by equating slum-related nuisances with slums themselves – that is, slums do not just improperly dispose of "matter" (e.g. trash, sewage), but are themselves "matter out of place" (Douglas 1966). For example, in 2002, building on the distinction between propertied citizens and unpropertied "encroachers," the High Court ruled: "The welfare of the residents of these [property-owning RWAs'] colonies is also in the realm of public interest[,] which cannot be overlooked. After all, these residential colonies were developed first. *The slums have been created afterwards which is the cause of nuisance* and brooding [*sic*] ground of so many ills."[42] This judgment, widely cited in subsequent slum-related cases (including the Pushta case), vastly expanded the range of procedures that could be administered to remove nuisances: no longer simply through imposing fines and penalties, but by displacing entire populations.

Once the interpretation of nuisance was expanded to include categories of people or entire population groups, the legal (and calculative) basis for slum demolition was simplified. Demolition orders no longer require complex mapping and survey exercises to determine the nature of land use or demand even the confirmation of land ownership in slum cases. Today courts ask for little more than the demonstration by a petitioner (who is usually a neighboring RWA) that the slum in question is (i) on public land (which is the definition of "slum" and has never been a sufficient condition for demolition orders in the past), and (ii) a nuisance. Evidentially, this is most commonly and effectively done by furnishing photographs that show the slum's "dirty" look and poor environmental conditions: open defecation, overcrowded living conditions, children playing in and "taking over" the street, stagnant water, municipal waste, etc.

Since approximately 2002, the courts have considered such photographs sufficient evidence to confirm that the slum in question does not conform to the desired "clean and green" look and have, in the majority of such cases, issued demolition orders. For example, in a case in South Delhi, an RWA prayed to the High Court "for better civic amenities and for nuisance caused by open wide drain [*sic*]" without making a single mention of the neighboring slum in its petition. Only in the petition's annexures, containing photos with such captions as "Jhuggi [slum] dwellers defecate in nallah [drain]," was it revealed that a slum existed beside the drain. Nonetheless, the court observed that "Photographs were filed of the area showing the filth at site and encroachments in and around the nallah" and ordered that "The area should also be cleaned and the encroachments removed."[43] Without initiating an inquiry into the settlement's size, legal basis, or effluent discharge, the court ordered the slum's demolition.

Official statistics, which are notoriously inaccurate and underreported, suggest at least a tripling in the pace of slum demolitions since 2000 (Dupont 2008), when the judiciary reinterpreted nuisance law and made the violation of bourgeois codes of civility and appearance a legitimate basis for slum removal. This new aesthetic ordering of the city, in which the legality and essential features of space can be determined entirely from a distance and without requiring accurate survey or assessment, marks a clear shift away from the previous approach to carefully surveying, monitoring and assessing the land use status of areas under question. In this new, more aesthetic framework, the law crafts a governing legibility by disseminating standardized aesthetic norms. Spaces are then known to be illegal or legal, deficient or normal, based on their outer characteristics and adherence to these norms. The ability to look at a building, plot of land, or population and immediately locate it within a "grid of norms" (Rose 1991) is an entirely different way of knowing and evaluating urban space than the calculative, inscriptive approach typified in much of the literature on legibility and state simplification, which I discuss further below. This more aesthetic approach allows government to overcome the (political and bureaucratic) difficulty of translating messy "reality out there" (e.g. population densities, land use designations, territorial area, pollution levels, settlement history, etc.) into a numerical or cartographic legibility. Instead of having to inscribe the population and its complex relation with things into quantifiable forms that can be aggregated, compiled, and then calculated, this aesthetic normativity works to ascribe an aesthetic sense of what ought to be improved and what ends achieved. Nuisance has thus become not just the coercive arm of property rights, but also the legitimizing basis for extending a bourgeois vision of a "slum-free" Delhi.

Environmental knowledge beyond maps and numbers

In this chapter, I have shown how the environment, as a category of knowledge, operates in Delhi more as an aesthetic than a "scientifically" calibrated proxy for human health or ecological welfare. In this arrangement, spaces that violate settled codes of appearance and urban order are labeled "polluting," regardless of their contribution to pollution levels or resource degradation. The flip side of this, which I have discussed in less detail here, is that spaces that adhere to a certain environmental sensibility or green aesthetic, even if they violate environmental law and are ecologically harmful, acquire the label "green." Indeed, this green aesthetic explains the underlying basis of Delhi's major land use decisions over the past ten years: almost one million slum residents displaced for being a "nuisance," shopping malls constructed on protected green belts, and the world's largest Hindu monument (the Akshardam Temple) appearing on the Yamuna floodplain. Green aesthetics has thus become the overarching means by which a governing legibility is secured in contemporary Delhi. This has two broad implications that I will briefly explore in this concluding discussion.

First, the Delhi case demonstrates how the environment does not exist as a stable domain to be measured and analyzed, but is rather constituted and acquires meaning only through the techniques experts use to define it. As a result, it operates as a very different biopolitical object depending on how it is rendered visible. Had pollution been quantified and tracked in Delhi using the accepted tools of modern science and ecology instead of as a criterion of appearance and order, environmental knowledge would have produced vastly different material and political-economic effects. For example, had the Yamuna case discussed in the introduction been evaluated according to scientific standards, it is quite likely that the Pushta settlement would never have been demolished and the Commonwealth Games Village never approved. I make this point not to say that there is some "true" environment out there that was concealed in this more aesthetic rendering, but rather to emphasize the epistemologically distinct techniques by which the environment can be discursively produced, a point which should come as no surprise to political ecologists (see Braun and Castree 1998; Peet and Watts 1996).

This leads me to my second point, which is to expand the epistemological treatment of expertise and state simplification beyond the domain of techno-science. In the simplest sense, all simplification is necessarily visual. As Scott (1998: 184) says, "Any substantial state intervention in society . . . requires the invention of units that are visible." But, the question then becomes how such visibility is rendered – a core concern of political ecological studies of environmental knowledge. For example, a rich array of empirical studies drawing from Foucault's (2007) insights on governmentality examines how the environment is constructed as a governmental object of analysis and management used to structure both political-economic possibilities and environmental subjectivities. Braun (2000) thus traces how the science of geology, with its prospecting and mapping techniques, brought the Canadian landscape into view as terrain of calculable resources. Murdoch and Ward (1997) show how statistical surveys of British agriculture in the 1940s manufactured a shared vision of the "national farm," enabling policy interventions into the everyday practices of farming. Agrawal (2005), Demerritt (2001), Sivaramakrishnan (1999) each describe the statistical production of the forest – either as a national resource to be managed or as a "scientific" unit to be measured, exploited and maintained. A common theme throughout these works, and one shared more broadly with literature on environmental expertise,[44] is the role of mechanically objective techniques in securing a governing legibility, where "mechanical objectivity" is understood as the repetition of standardized procedures of measurement, demarcation, quantification, and reportage (Porter 1995).[45] In Scott's words: "Whatever the units being manipulated, they must be organized in a manner that permits them to be identified, observed, recorded, counted, aggregated, and monitored" (Scott 1998: 124). Foucault (2007: 348), too, insists that effective governmental simplification must follow the "rule of evidence" and "be scientific in its procedures" (350): "A government that did not take into account this kind of analysis . . . would be bound to fail." And Rose, in what has become a foundational text on governmentality, argues that "To govern a problem requires that it be counted" (Rose 1999: 221).

Numbers and maps are indeed powerful objectifying techniques, as these authors effectively show, and my intention here is not to argue that these studies have missed the point or been overly narrow in focus. Indeed, censuses, statistics and maps have a long history in the operation of colonial and postcolonial government in India, effectively defining social categories like caste, religion and language group on terms amenable to governmental intervention.[46] Without denying the power of techno-scientific rationality, then, I want to suggest that environmental knowledge be considered beyond the ambit of science and its discursive mobilizations. For, the case of Delhi shows how the environment is known as much through aesthetic criteria as through rigorous statistical and cartographic techniques. This confirms that the dissemination of a strong normative sense of how a place should look, even without the backing of mechanical objectivity, can lead to a codification of aesthetic norms and a real ability to structure physical space (cf. Duncan and Duncan 2004).

Hannah (2000), in perhaps the most systematic treatment of governmental calculation, hints at this fuzzy boundary between aesthetics and science in identifying what he considers the two processes necessary to form governmental objects: "abstraction" and "assortment." Abstraction involves creating an "observational field" across which "agents of the governmental gaze can travel without significant impediment, and throughout which they can expect to be provided with complete and accurate information" (124). In Hannah's case, abstraction is accomplished through mapping, which allows objects to be easily locatable within a broader terrain, or "grid of specification" (56). Once objects are individually locatable and rendered discrete, they can be organized, contrasted, and evaluated – what Hannah calls "assortment." Hannah's empirical case is the late nineteenth-century U.S. Census, where abstraction consisted of the formation of enumeration districts, each a discrete unit with a set of properties (e.g. population growth, resource availability, income) that, once quantified and assorted, became governmental objects to be managed.

Abstraction and assortment, according to Hannah, provide a means for discerning and evaluating qualitative differences and conveying those differences to a broader audience (decision makers, a scientific community, the public at large). Science, on this reading, is a means to establish easily discernible divisions in the world and to mobilize representational difference – to parse interrelated processes, establish hierarchies, and make objects visible and therefore governable. But, from the Delhi case we see how an aesthetic normativity can achieve this same end, effectively abstracting a complex ground scenario into units that differ in appearance and allowing those units to be assorted according to codes of order, visuality, and desirability. Hannah hints at this possibility in arguing that abstraction and assortment are processes by which a "field of vision" is oriented in such a way that viewers perceive it on terms provided by experts. This reading of governmental practice closely aligns with what Bourdieu describes as "social categories of perception," or "principles of vision and division": "schemes of action which orient the perception of the situation and the appropriate response" (Bourdieu 1998: 25).

Both Hannah and Bourdieu's formulations, then, show that governmental objects are produced through the establishment of visible boundaries between what is/is not in the designated class of objects (e.g. what counts as pollution or a forest), thus suggesting that expertise as discernment – the ability to judge consistently and convincingly – is not so different from aesthetic expressions of taste or preference. The principles of classification upon which expert judgments are passed must be more durable and appear natural in order to gain authority. They must be "invested with all the force of universally binding proposition" (Eagleton 1990: 95) in order to appear to be more than mere subjective fancies, but "mechanically objective" methods are but one way of establishing socially agreed upon classificatory principles.

In Delhi, the law of nuisances became the basis upon which a particular bourgeois structuring of the "field of vision" was codified, giving an aesthetic preference for how the urban environment should be ordered the force of law and reason. The ability to disseminate aesthetic norms, then, must be considered a key component of environmental politics – something that shapes not only popular opinion, but expert legal and state evaluations as well. Without attention to this more aesthetic/sensory basis of environmental decision-making, I argue, studies of environmental politics risk being overly formalistic and privileging more empiricist epistemologies over a broader consideration of affect and "structures of feeling" (Williams 1977). Giddens (1990), in describing what he calls "reflexive modernity," thus argues that the inability to fully understand the techno-scientific make-up of contemporary societies leads to a necessary mixing of affect and reason in our modern knowledge systems. Debates in political ecology and science studies have taken this proposition a step further, showing such a mixing of reason and non-reason to be not the product of an inability to understand techno-science, but rather a premise of techno-science itself (e.g. MacKenzie 2008). This chapter extends this insight by showing how the law and governmental reason intersect with and, in the case of Delhi at least, are built upon a distinctly aesthetic domain.

Notes

1 Okhla Factory Owners' Association vs. Govt. of NCT of Delhi, CWP No. 2112 of 2002 (Delhi High Court).
2 108(2002) DLT 517, paragraph 10.
3 Ibid., paragraph 44.
4 Ibid., paragraph 49.
5 Ibid., order dated March 3, 2003.
6 "Yamuna pollution issue: Delhi High Court summons top officials", *The Hindustan Times*, New Delhi, February 17, 2006.
7 For details on the history of Pushta and its residents' experience of displacement, see Menon-Sen and Bhan (2008).
8 Okhla case, order dated August 11, 2006
9 See note 6.
10 Okhla case, order dated June 1, 2006.
11 See note 8.
12 "CM concern for green lung, seeks expert panel", *The Times of India*, New Delhi, May 14, 2009.
13 Combined demolitions (notoriously under-)reported by the DDA and Slum and JJ Wing of the Municipal Corporation from 1997–2007 led to the conservative estimate of 710,000 displaced residents. *The City Development Plan of Delhi*, prepared by private consultants, on the other hand, estimates that 1.8 million residents were displaced in 1997–2001 alone.
14 "Delhi is gearing up for a new green revolution", *The Hindu*, New Delhi, July 6, 2008. See also Sridharan, E., "Restrict Yamuna with walls and develop low-lying areas", *Times of India*, New Dehli, May 20, 2009.
15 "How Yamuna-bed plans got green light", *Mint*, New Delhi, March 24, 2008. India's largest shopping mall complex being built in Delhi was also approved by the Supreme Court for its role in attracting top international retailers, despite the failure of all the developers involved to submit the mandatory environmental impact assessments.
16 For example, see the primarily aesthetic function of the 34 million rupee Green Delhi Action Plan, "Delhi is gearing up for a new green revolution", *The Hindu*, New Delhi, July 6, 2008.
17 "C'wealth Games top priority of Govt.", *The Hindu*, New Delhi, February 6, 2008.
18 Municipal Corporation of Delhi. 2002. *Annual Report of the Slum and JJ Wing, 2001–2002*.
19 Furthermore, the Delhi Master Plan entitles the poorest segments of the population to 25 percent of total residential land. However, the 25 percent of the population living in slums today occupies less than 2 percent of Delhi land. Planner Gita Dewan Verma therefore calls slum residents "Master Plan implementation backlog", rejecting the label "encroachers" in favor of a term that signals their legal entitlement to land in the city and the DDA's failure to build the mandated low-income housing (Verma 2002).
20 For further details on the slum removal process, see Ghertner (2010).
21 Delhi is a city-state with both a state legislature and an elected municipal government, serving under the Municipal Corporation of Delhi. As India's capital, land management and the police remain under the control of the central government. Thus, the DDA is not directly accountable to elected representatives of the municipal or state governments.
22 Pitampura Sudhar Samiti versus Government of India, CWP 4215/1995, order dated May 26, 1997.
23 Okhla judgement (see note 1), paragraph 22.

24 Affidavit filed by Mr. Satish Kumar, Under Secretary, Ministry of Urban Development and Poverty Alleviation (Delhi High Court), CWP 2253/2001.

25 Resident Welfare Association vs. DDA and Ors. (Delhi High Court), CWP 6324/2003, order dated August 29, 2007.

26 Okhla case, order dated March 29, 2006.

27 For other examples of the slippages that emerge in creating classificatory schemes suitable to the needs of capital accumulation, regulation, or governance, see Bowker (1988), Hayden (2003), McAfee (2003), and Robertson (2004). Robbins (2001) shows how differently situated viewers interpret the same representations in sharply divergent ways. What is remarkable in Delhi is the convergent interpretation of slums as aesthetically and legally deviant across interest groups. For more on both how this hegemonic reading is constructed and the symbolic violence it entails, see Ghertner (2010).

28 For a further discussion of the negotiability of the survey as a mode of knowledge assembly and the subjectivity implicit in other processes of objectification, including map-making, see Sivaramakrishnan (1999: 122) and Appadurai (1993).

29 Hem Raj vs. Commissioner of Police (Delhi High Court), CWP 3419/1999, order dated March 1, 2006. Compare with Roy's (2004) discussion of the "unmapping" of Calcutta and the territorialized flexibility to which it gave rise.

30 Civil Misc. Petition 6982/2007 (Dayavanti and Ors.) in CWP 4582/2003 (Delhi High Court).

31 Such a diversity of tenure arrangements is common throughout India, providing the basis for economic clustering and dynamic informal growth economies (Benjamin 2005). In Banglore, Benjamin (2005: 30), identified more than 10 forms of tenure, and in Madikiri, a town in Karnataka, 24.

32 See Municipal Corporation of Delhi affidavit filed in 2006 in Kalyan Sansthan vs. GNCTD (Delhi High Court), CWP 4582/2003.

33 In fact, the Municipal Corporation confronted this dilemma after the Supreme Court had ordered it to close and seal all commercial establishments operating in residential zones of the city in late 2005. This led to the sealing of thousands of businesses, with tens of thousands more threatened, citywide protests by traders leading to the death of three young men, the demolition or partial demolition of hundreds of private residences not conforming to building codes as well as a shopping mall under construction in South Delhi, and a political nightmare for the ruling Congress Party. In 2006, the Lower House (Lok Sabha) of the Indian Parliament passed a legislative act postponing all demolitions and sealing drives in Delhi for one year. While this act also included slums, the courts did not acknowledge their protected status and continued with slum clearance apace. The DDA finally modified the Master Plan ex post facto to regularize Delhi's commercial land use violations in 2007 (DDA 2007).

34 Slum surveys did not stop, but are now conducted almost entirely for the purposes of establishing resettlement eligibility after a settlement has been found "illegal" – i.e. not to adjudicate on the legality of the slum in the first place.

35 Almrita Patel vs. Union of India (2000 SCC (2): 679).

36 "'So, it'll take you 263 years to wash sins!'", *The Hindustan Times*, New Delhi, August 19, 2006.

37 CWP 6553/2000 (Delhi High Court), order dated February 16, 2001.

38 See, for example, Ratlam Municipal Council vs. Vardichan (AIR 1980 SC 1622) and Dr. K.C. Malhotra vs. State of M.P. (M.P High Court), CA 1019/1992.

39 The Oxford English Dictionary defines "civility" as "conformity to the principles of social order, behaviour befitting a citizen; good citizenship".

40 See The Indian Code of Criminal Procedure (1973), Section 133, the primary statute dealing with public nuisance and a key component of environmental law.

41 Okhla judgment, paragraph 44.

42 Pitampura Suhdar Samiti vs. Government of the National Capital Territory of Delhi (Delhi High Court), CWP No. 4215/1995, paragraph 19, emphasis added.
43 CWP 1869/2003 (Delhi High Court), order dated November 14, 2003.
44 Other approaches that we might consider include ecological modernization, which, according to Scott and Barnett (2009: 373), "is the dominant approach to environmental governance and adopts a science-based policy approach"; studies informed by Beck's (1992) work on "risk society"; or studies of political ecology drawing on science studies that see "science as the underlying basis through which environmental change is understood" (Forsyth 2003: 9). Blomley's (2008) above-mentioned discussion of legal simplification also focuses exclusively on such calculative practices.
45 Mitchell, too, describes how it is through a well-defined set of mechanical procedures that objectivity is established: 'The performance of the law will gain its authority from following this particular sequence of acts [granting land, survey of boundaries, placing of boundary stones, recording of measurements. . .]' (2002: 58).
46 On the role of the census as a key "investigative modality" organizing the imperial capacity to govern, see Cohn (1987). Also see Appadurai (1993) and Dirks (2001). For similar studies related to land use and environmental classification in India, see Agrawal (2005), Chatterjee (2004) and Sivaramakrishanan (1999).

References

Agrawal, A. (2005). *Environmentality: Technologies of Government and the Making of Subjects*. Durham, NC: Duke University Press.

Anderson, M. R. (1992). Public nuisance and private purpose: Policed environments in British India, 1860–1947. *SOAS Law Department Working Paper* 1, London: School of Oriental and African Studies.

Appadurai, A. (1993). Number and the colonial imagination. In C. Breckenridge and P. van der Veer (eds.), *Orientalism and the Postcolonial Predicament: Perspectives on South Asia* (pp. 314–338). Philadelphia: University of Pennsylvania Press.

Baviskar, A. (2003). Between violence and desire: Space, power, and identity in the making of metropolitan Delhi. *International Social Science Journal, 55*(1), 89–98.

Beck, U. (1992). *Risk Society: Towards a New Modernity*. London: Sage.

Benjamin, S. (2004). Urban land transformation for pro-poor economies. *Geoforum, 35*(2), 177–187.

—— (2005). *'Productive Slums': The Centrality of Urban Land in Shaping Employment and City Politics*. Cambridge, MA: Lincoln Institute of Land Policy.

Biswas, S. (2006). Why so much of Delhi is illegal [Electronic Version]. *BBC News*. Retrieved February 8, 2006 from http://news.bbc.co.uk/go/pr/fr/-/2/hi/south_asia/4665330.stm.

Blomley, N. (2008). Simplification is complicated: property, nature, and the rivers of law. *Environment and Planning A, 40*(8), 1825–1842.

Bourdieu, P. (1998). *Practical Reason: On the Theory of Action*. Stanford, CA.: Stanford University Press.

Bowker, G. (1988). Not hung – drawn and quartered; Pictures from the subsoil, 1939. In G. Fyfe and J. Law (Eds.), *Picturing Power: Visual Depiction and Social Relations* (pp. 221–254). London: Routledge.

Braun, B. (2000). Producing vertical territory: Geology and governmentality in late Victorian Canada. *Ecumene, 7*(11), 7–46.

Braun, B. and Castree, N. (eds.). (1998). *Remaking Reality: Nature at the Millennium*. New York: Routledge.

Chakrabarti, P. D. (2008). Inclusion or exclusion? Emerging effects of middle-class citizen participation on Delhi's urban poor. *Institute for Development Studies Bulletin, 38*(6), 96–104.

Chatterjee, P. (2004). *The Politics of the Governed: Reflections on Popular Politics in Most of the World*. New York: Columbia University Press.

Cohn, B. (1987). *An Anthropologist Among Historians*. London: Oxford University Press.

DDA. (1997). *Delhi Development Authority Annual Report, 1996–1997*. New Delhi: Delhi Development Authority.

—— (2007). *Master Plan for Delhi 2021*. New Delhi: Delhi Development Authority.

Demeritt, D. (2001). Scientific forest conservation and the statistical picturing of nature's limits in the Progressive-era United States. *Environment and Planning D: Society and Space, 19*(4), 431–559.

Dirks, N. (2001). *Castes of Mind: Colonialism and the Making of Modern India*. Princeton, NJ: Princeton University Press.

Diwan, S. and Rosencranz, A. (2001). *Environmental Law and Policy in India: Cases, Materials and Statues*. New Delhi: Oxford University Press.

Douglas, M. (1966). *Purity and Danger: An Analysis of Concepts of Pollution and Taboo*. New York: Praeger.

Duncan, J. S., and Duncan, N. G. (2004). *Landscapes of Privilege: The Politics of the Aesthetic in an American Suburb*. New York: Routledge.

Dupont, V. (2008). Slum demolitions in Delhi since the 1990s: An appraisal. *Economic and Political Weekly, 43*(29), 79–87.

Eagleton, T. (1990). *The Ideology of the Aesthetic*. Oxford: Basil Blackwell.

Forsyth, T. (2003). *Critical Political Ecology*. London: Routledge.

Foucault, M. (1976). *Knowledge/Power*. New York: Pantheon.

—— (2007). *Security, Territory, Population* (G. Burchell, trans.). New York: Palgrave Macmillan.

Ghertner, D. A. (2005). Purani yojana ki kabr par, nayi yojana ki buniyad: *Dilli Master Plan 2021* ki chunauti aur sambhavnae [Building the new plan on the grave of the old: The politics of the *Delhi Master Plan 2021*]. *Yojana, 24*(7), 14–20.

—— (2008). An analysis of new legal discourse behind Delhi's slum demolitions. *Economic and Political Weekly, 43*(20), 57–66.

—— (2010). Calculating without numbers: Aesthetic governmentality in Delhi's slums. *Economy and Society, 39*(2), 185–217.

Giddens, A. (1990). *Consequences of Modernity*. Stanford, CA: Stanford University Press.

Hannah, M. G. (2000). *Governmentality and the Mastery of Territory in Nineteenth-Century America*. Cambridge: Cambridge University Press.

Hayden, C. (2003). *When Nature Goes Public: The Making and Unmaking of Bioprospecting in Mexico*. Princeton, NJ: Princeton University Press.

Hull, M. S. (2008). Ruled by records: The expropriation of land and the misappropriation of lists in Islamabad. *American Ethnologist, 35*(4), 501–518.

Jain, A. K. (2003). Making planning responsive to, and compatible with, reforms. *Cities, 20*(2), 143–145.

—— (2005). *Law and Environment*. Delhi: Ascent.

Latour, B. (1987). *Science in Action: How to Follow Scientists and Engineers through Society*. Cambridge: Harvard University Press.

McAfee, K. (2003). Neoliberalism on the Molecular Scale: Economic and genetic reductionism in biotechnology battles. *Geoforum, 34*(2), 203–219.

MacKenzie, D. (2008). *An Engine, Not a Camera: How Financial Models Shape Markets.* Cambridge, MA: MIT Press.

Menon-Sen, K. and Bhan, G. (2008). *Swept off the Map: Surviving Eviction and Resettlement in Delhi.* New Delhi: Yoda Press.

Mitchell, T. (2002). *Rule of Experts: Egypt, Techno-Politics, Modernity.* Berkeley: University of California Press.

Murdoch, J. and Ward, N. (1997). Governmentality and territoriality: The statistical manufacture of Britain's 'national farm'. *Political Geography, 16*(4), 307–324.

Peet, R. and Watts, M. J. (1996). Liberation ecology: Development, sustainability, and environment in an age of market triumphalism. In R. Peet and M. J. Watts (eds.), *Liberation Ecologies: Environment, Development, Social Movements.* New York: Routledge.

Porter, T. M. (1995). *Trust in Numbers: The Pursuit of Objectivity in Science and Public Life.* Princeton, NJ: Princeton University Press.

Ramanathan, U. (2006). Illegality and the urban poor. *Economic and Political Weekly, 41*(29), 3193–3197.

Ramesh, R. (2008). Delhi cleans up for Commonwealth games but leaves locals without sporting chance. *Guardian*, January 8.

Ranciere, J. (2004). *The Politics of Aesthetics* (G. Rockhill, trans.). New York: Continuum.

Robbins, P. (2001) Fixed categories in a portable landscape: The causes and consequences of land cover categorization. *Environment and Planning A, 33* 161–179.

Robertson, M. M. (2004). The neoliberalization of ecosystem services: Wetland mitigation banking and problems in environmental governance. *Geoforum, 35*(3), 361–373.

Rose, N. (1991). Governing by numbers: Figuring out democracy. *Accounting, Organizations and Society, 16*(7), 673–692.

—— (1999). *Powers of Freedom: Reframing Political Thought.* Cambridge: Cambridge University Press.

Roy, A. (2004). The gentleman's city: Urban informality in the Calcutta of new communism. In N. AlSayyad and A. Roy (eds.), *Urban Informality* (pp. 147–170): Lanham, MD: Lexington Books.

Roy, D. (2004). *Pollution, Pushta, and Prejudices.* Delhi: Hazards Centre.

Scott, D. and Barnett, C. (2009). Something in the Air: Civic science and contentious environmental politics in post-apartheid South Africa. *Geoforum, 40*(3), 373–382.

Scott, J. C. (1998). *Seeing Like a State: How Certain Schemes to Improve the Human Condition Have Failed.* New Haven: Yale University Press.

Sivaramakrishnan, K. (1999). *Modern Forests: Statemaking and Environmental Change in Colonial Eastern India.* London: Oxford University Press.

Verma, G. D. (2002). *Slumming India: A Chronicle of Slums and their Saviours.* Delhi: Penguin Books.

Williams, R. (1977). *Marxism and Literature.* Oxford: Oxford University Press.

Part III

Risk, certification, and the audit economy: political ecology of environmental governance

8 The politics of certification: consumer knowledge, power, and global governance in ecolabeling

Sally Eden

Introduction

Political ecology originated in studies of developing countries and that continues to be its main focus today. However, to be truly global, political ecology needs to consider the connections between developing and developed countries and particularly between producers and consumers in the North and South. This chapter is about those connections and about the politics of traceability that surround attempts to make consumers in rich countries care about and support issues like environmental protection and the health and quality of life of people in other countries, especially agricultural producers and industrial workers. It is specifically about "ecolabels" that stamp products with guarantees of sustainability, of fair trade and of worker welfare, stamps that carry not only political, but also ecological and economic complexities.

Such ecolabels are symbols of attempts at global green governance by harnessing consumer power to address dysfunctional elements of the economy. This applies especially where the search for profits and for cost reduction leads to environmental and human damage, through pollution, resource depletion, dangerous working conditions, poor rates of pay, employment insecurity and other forms of exploitation. But because ecolabels seek to do this within the existing capitalist system, these attempts at global green governance have been heavily criticized by those who would rather see a stronger challenge to that system itself. In this chapter, we will consider both the attempts and the criticisms, to evaluate the politics of certification and traceability through the lens of political ecology.

The politics of traceability

We need to begin by relating ecolabels more clearly to theories of political ecology. But what do we mean by political ecology? Although he seems reluctant to provide a single definition, Robbins (2004, pages 11–12) does offer some key criteria for political ecology: that it considers environmental processes to be the product of political processes, that it therefore requires researchers especially to analyze politics and power, and that it does not seek merely to analyze these processes and power relationships, but to try to change them for the better, especially by exposing their flaws and offering better, more sustainable, more equitable alternatives.

The intentions of many of the ecolabeling approaches reflect these criteria well, especially those set up by nongovernmental organizations (NGOs). This is because these approaches assume that power in global trade rests with developed countries, but that their consumers are in some ways misusing that power by buying (perhaps unknowingly) products that damage environments, human health and community integrity elsewhere. Providing traceability through certifying and labeling "good" products thus tries to harness this consumer power for positive effect, through persuading consumers to buy products defined in some way as "better" for the environment or people because of how they were produced. Hence, ecolabels are suitable for analyzing through political ecology because they are both prompted by politics and unequal power relationships, yet also seek to harness those power relationships and turn them from detrimental to beneficial purposes.

Because of what it owes to Marxist theory, political ecology has tended to concentrate on production and products, rather than on consumption and commodities, and especially on agro-food supply chains. Geographical research into certification expanded in the 2000s, covering a range of commodities such as organic food, coffee, forest products and ethical clothing (Giovannucci and Ponte 2005; Guthman 2004; Hatanaka *et al.* 2005; Hughes 2006; Mutersbaugh 2002; Mutersbaugh *et al.* 2005; Gulbrandsen 2005, 2004; Klooster 2006; Morris and Dunne 2004). The standard to which products are certified may be negative, in that it bans undesirable practices or contents, or positive, in that it requires desirable ones or, more usually, a mix of the two. Products are checked against the standard and, where they pass, they are licensed to use the certifying organization's logo on their promotional material, to make supply chains traceable for consumers (Guthman 2004).

This process raises two important theoretical issues: knowledge and power. I will consider these in more detail before turning to some specific examples of certification.

The problem of knowledge in certification

> The production knowledge that is read into a commodity is quite different from the consumption knowledge that is read from the commodity. Of course, these two readings will diverge proportionately as the social, spatial and temporal distance between producers and consumers increases.
>
> (Appadurai 1986: 41)

Knowledge is at the core of certification. First, the assumption is that without certification, modern consumers have little or no knowledge about many products that they buy, because they are distanced from production systems, as Appadurai implies. Hence, although they can tell from a physical product what it tastes like and how much it costs, they cannot tell the circumstances of its production. An egg salad sandwich cannot tell a consumer whether it came from free range chickens or from a battery farm; a cotton tee-shirt cannot tell a consumer whether it was made by workers paid a living wage and protected by good health and safety systems or made under exploitative and dangerous working conditions.

Such "distancing" from production can be geographical but it is also social, not least because the numbers of people working in agriculture and food manufacture in developed countries has decreased so greatly in the last century or so. In other words, most of us do not visit farms and factories very often, even if we live close by. This means that, instead of firsthand information, consumers increasingly rely on secondhand information about products, such as from commercial advertising (e.g. Goodman 2004), media stories in newspapers or the internet, popular books about the nasty underbelly of production and retail processes (Schlosser 2002; Lawrence 2004) or less formal means, such as folklore and gossip. Or they simply do without information at all.

But it is problematic to assume that distancing is bad and "close up" is good in terms of consumption (Barnett *et al.* 2005), because geographers know that space does not simply determine human relationships in this way. This distancing can instead be understood theoretically by drawing on the concept of alienation, where workers are alienated from the products of their own labour through automated, fragmented, unskilled industrial manufacturing, a process that deprives their work of meaning and of connection. The argument is that, in a similar way, consumers are alienated from the producers, so that the damage that production causes is hidden from the consumer when they buy the product – the product thus seems innocent, detached (distanced) from its legacy of pollution and exploitation (e.g. Vos 2000: 246). Hartwick (1998) takes gold as an example and argues that advertising and cultural norms associate buying gold jewelry with showing love, pleasure and wealth, whereas its production is associated with agricultural poverty, destitution and family breakdown under the pressures of working in goldmines in South Africa.

This is sometimes referred to as "consumer fetishism," because a product, for example a sports shoe, becomes not merely an object made of rubber and cloth, but a cultural symbol of much greater fantasy: of style and exclusivity, for example. The consumer can buy the symbol in order to buy into that fantasy – but the consumer does not want their fantasy to be spoiled by thoughts of exploited workers and tortured animals. And marketing and advertising perpetuates this selective and stylized presentation of commodities, especially through association with celebrities. Hence, the cultural politics of consumption enable environmental damage and human exploitation to persist in support of industrial capitalism.

Researchers have therefore argued that it is necessary to expose this innocence, to turn around this distancing and "educate" or otherwise force consumers to think about production when they are indulging in consumption, to defetishize the commodities produced by modern capitalism (Hudson and Hudson 2003) and to "unravel the magic of the commodity, rather than reveling in its seductive delights" (Hartwick 2000, page 1178). In other words, the aim is to change consumption and make consumers see the effects of the production system that the shiny new product (and its advertising campaign) hides from us.

> We are more than what we eat. We are our relations with distant others. These relations are hidden by the sign of the commodity. Deconstruction, understood in a geomaterialistic sense, uncovers these hidden dimensions of the sign.
>
> (Hartwick 2000: 1183)

To do this and to correct this distancing, a vast range of labels have been developed to provide information to consumers about the hidden effects of production. This gives a product what have been termed "credence," "proxy" or "secondary" qualities (e.g. Jahn *et al.* 2005), because they rely not on direct consumer perception but on belief and confidence in this secondhand information. In this way, it is argued, labels can enable consumers to exert their political power through their purchasing.

The result has been a vast array of labels, run by different kinds of groups for different kinds of purposes (see Table 8.1). For example, the European Union set up its own ecolabel scheme in 1993, to harmonize various different national initiatives that had developed. This covers a wide range of product groups, from detergents, soaps and shampoos, to light bulbs, computers, televisions, tissue paper, bed mattresses and footwear, and certification is managed by each national government. But the complexity of the scheme has meant that few products have used its daisy-like label on packaging or in other promotional ways, and it still has little consumer visibility. In a separate development, the UK government's Food Standards Agency set up the Assured Food Standards (AFS) scheme to promote food produced or processed in Britain in compliance with national regulation, using part of the UK's national flag to symbolize its "Britishness" and appeal to consumers in a very different way from the ethics of other ecolabels. Other labels do not involve any certification, especially where the ecolabel is a generic product claim, such as "GM-free," "free-range" and "not tested on animals."

In this chapter, I will focus on ecolabels that use certification – which are often run by NGOs but sometimes with government input – because these are specifically intended to contribute to global environmental governance through guaranteeing traceability for consumers to choose "better" products, rather than merely supporting commercial promotion.

Certified ecolabels also illustrate well this problem of knowledge because they offer information as part of what I call a "knowledge fix" (Eden *et al.* 2008) to the problem of distancing. This "fix" assumes that consumers will understand the information and act on it to change production and retailing systems for the better. In the sociology of science literature, this is referred to as the "deficit model" of public understanding, a model that assumes that first, public distrust is based on a deficit of information (for example, about how food production is regulated) and that second, distrust can be corrected if (good) information is provided by "experts," especially scientists. But research (e.g. Wynne 1995; Irwin 1995; Irwin and Michael 2003; Gregory and Miller 1998) shows that providing information does not necessarily change attitudes or counteract public distrust, because consumer knowledge is much more complicated than that and is produced not simply by receiving information passively, but through various and highly interactive sociocultural processes.

This means that this "knowledge fix" is simplistic as a solution to "bad" consumption (e.g. Jackson 2002; Goss 2004), because consumers do not necessarily process information in the way that such labels assume. Few consumers fit the perfect profile (taken from classical economics) of "rational" decision making when buying food, clothing, cars or other products. We rarely research our decisions thoroughly, especially for small purchases, and we often buy things that we do not

Table 8.1 Examples of certified ecolabels, organised by theme.

Ecolabel theme	Examples of certifying organisations operating in the UK	Examples of logos
environmental	European Union governments	
organic	Organic Farmers and Growers Ltd Scottish Organic Producers Association Organic Food Federation Soil Association Certification Ltd Bio-Dynamic Agricultural Association Irish Organic Farmers and Growers Association Organic Trust Limited Quality Welsh Food Certification Ltd Ascisco Ltd	
sustainable fish	Marine Stewardship Council (MSC)	
sustainable forestry	Forest Stewardship Council (FSC)	© 251658240
animal welfare	Royal Society for the Prevention of Cruelty to Animals (RSPCA)	
fairtrade	Fairtrade Labelling Organizations International via FLO-CERT	251658240

Note: the list of organic certification bodies includes all those approved by the UK's governmental department, Defra, as of June 2009.

really need or even end up throwing away. We may buy things on impulse, not because of what we know about them, but because of the way that they make us feel or because of celebrity endorsement. Consumers therefore do not respond to information about products in a straightforward way, so the knowledge fix does not necessarily change the system so easily. Moreover, even where more information is provided by an ecolabel and that ecolabel is backed by a reputable organization, consumers may not make sense of it in the way that is intended – they may not automatically trust it nor act on it when buying products.

The knowledge fix in the case of consumption also relies on the modern mantra of "choice" as the basis for the global economy, a choice that is assumed to be a) free and b) informed. But is there such a thing? Most consumers would like to choose cheap, good quality products that have not harmed people or environments in their production – but this choice may not even be offered to us where we shop. Sometimes, also, consumers have to choose between two options – such as fairtrade or organic, organic or local – when they would rather a product fulfilled both conditions and avoid that difficult "trade-off" choice completely. So, in that sense, even certifications may compete against each other, depending on the consumer's own priorities. And purchase itself is constrained by cost and our own incomes, rather than being "free."

Overall, therefore, we can say that knowledge is a problem in certification because, first, there is not enough of it amongst consumers, second, even where there is enough information, consumers may not believe it or act on it when purchasing and, third, changing either of these situations is very difficult within existing capitalist systems.

The problem of power in certification

The second major theoretical issue is power. Researchers have become concerned that certification does not merely verify and circulate information on labels, but can be used by powerful corporations, especially food manufacturers and retailers, as yet another tool to exploit powerless producers, especially small agricultural businesses and cooperatives. For example, large retailers may seek to gain competitive advantage in ethical markets by requiring suppliers to gain independent certification for their products as a condition of trading with them (Klooster 2006; Morris and Dunne 2004). Certification involves costs – of applying to use the eco-label and of implementing any "corrective actions" to production and management, as required by the certifiers in the process – and retailers often expect producers to absorb these costs (Klooster 2006; Mutersbaugh 2002). Although some products certified as organic or fairtrade may be able to charge the consumer a higher price to cover these costs (for the "premium" on coffee, see Giovannucci and Ponte 2005 and Mutersbaugh 2002), others cannot (especially certified timber and paper, Morris and Dunne 2004; Eden 2009). And even where they do, large retailers may pocket these premia as profits, rather than pass them on to suppliers.

Although the power of large retailers/manufacturers is sometimes uncritically accepted in political ecology and political economy, the power of the consumer is often, in turn, derided. The consumer is seen as weak, deluded by commodity fetishism, fantastical marketing and disinformation and at the mercy of large retailers, who encourage them to desire unnecessary and luxurious commodities. Hence, large retailers and producers are theorized as not only powerful in designing and operating supply chains to their own advantage, but also powerful in making consumers think in certain ways, ways that perpetuate the search for commodities as the answers to life's problems.

There is, however, another way to look at the question of consumer power, by seeing consumption as an important way in which people build their identities and engage with the world – what Miller (1995: 41) calls "a relatively autonomous and plural process of cultural self-construction." Approaches like Miller's, from anthropology, sociology and human geography, consider consumption as an active sociocultural process, by which status is defined, relationships are made and broken and meanings conveyed.

But how are ethics drawn into consumption? Barnett *et al.* (2005) discuss ethics and responsibility through the idea of "action at a distance" in ordinary consumption. They argue that, as well as being politically influential, ethics can be used also to build consumer identity and effectively to promote oneself (and possibly denigrate others) through social distinction.[1] Miller (1998: 127) writes about shopping as love and sacrifice, especially through household provision for family and friends (to please "the objects of our devotion") and through a wider outreaching based on values such as thrift and sacrifice. So such approaches have a more positive view of consumers as active appropriators of consumption for their own purposes, rather than merely being dupes of consumption.

But one problem in consumption ethics is that using consumption as a political vehicle also prioritizes the individual over the collective (Barnett *et al.* 2005) and economic channels over more traditional political channels, such as voting or activist campaigning. Because of this negative perception, the power of big retailers and producers is often assumed to be huge, even insurmountable, but the power of consumers is assumed to be negligible, even illusory. An opposing and more positive view is that the power of consumers is latent, but capable of being roused and used for good, even within existing capitalist systems, because of the collective power of consumer demand to hit commercial producers through buying less from them – or buying more from their competitors.

To draw attention to these ironies of power in the global economy, the anthropologist Daniel Miller (1995: 34) suggests that we consider "the housewife as global dictator," especially the figure of the housewife in a developed country, who is traditionally constructed as small and insignificant. He also points out that shifting power from producers to consumers does not mean the overthrow of capitalism, because there may well be profits to be made in new fields, such as ethical consumerism. So even allowing for consumer power still leaves us within the capitalist system that political ecology frequently seeks to challenge. But that power is not immanent or static; rather, power is produced through ongoing economic but also cultural, social and environmental relationships, emphasizing that we cannot simply take power differentials or effects for granted, especially within the global economy.

Some examples of certification

Let us consider some examples of ecolabel certification schemes in more detail, to explore these ideas.

Example 1, the Forest Stewardship Council

The first example is the "tick-tree" ecolabel of the Forest Stewardship Council (FSC). The FSC is a nongovernmental organization (NGO) that wants to protect forests and the environment, thus promoting the three pillars of sustainable development through "environmentally responsible, socially beneficial and economically viable management of the world's forests" (FSC 2000, p.2) and especially to prevent illegal logging and biodiversity loss through poor forestry practices.

In 1993, FSC set up a system of global certification to mark products from forests that meet its standards with a trademarked symbol – a "tick-tree" (see Table 8.1) – that companies and organizations must apply for a license to use. This system was prompted by the failure of the 1992 UN Conference on Environment and Development (UNCED or the Rio "Earth Summit") to produce a binding convention to prevent deforestation in countries across the world. In effect, because the national governments failed to act through regulation, the FSC stepped into the gap and sought to act through the economy and specifically through linking production and consumption of forest products globally. This fits the idea of environmental governance well – as "governance without government" (Pattberg 2005, page 187; also Gulbrandsen 2005). FSC seeks to include diverse nongovernmental stakeholders (especially business and NGOs) in more open, proactive and socially relevant decision making and to use mechanisms such as market incentives, voluntary schemes and partnerships, rather than legislation and prosecution, to achieve its goals.

How does this certification system operate? First, FSC has ten International Principles of sustainable forest management and 57 indicators or criteria against which the principles can be measured. Together, these define the standard that FSC expects forest managers to meet in order to qualify to use its tick-tree. However, these are also adapted to suit different environmental conditions, often on a national or subnational basis. For example, there is only one standard for the UK, but at least nine for the USA, due to the USA's much more diverse ecological and climatological conditions.

Second, FSC needs to verify that applicants for its tick-tree meet the standard and can be given a forestry management certificate. Verification is done by independent "certification bodies," who are accredited by the central FSC office and act as quasi-regulatory agencies or a certification police force. Such "third-party certification" is commonly argued to be the most credible and the least susceptible to conflicts of interest (e.g. Jahn *et al.* 2005; Hatanaka *et al.* 2005), compared, for example, to companies certifying themselves.

Third, the tick-tree must reach the customer, on consumer products (such as furniture and printing paper) and off-product information (such as business letterheads and websites). The FSC's "chain of custody" standard stipulates that the certified product must be kept separate from non-certified products throughout the (often global) supply chain, as it travels from the forest to factories and warehouses, and eventually to retail outlets. This is the key to traceability and this process is again verified by certification bodies, to protect the FSC's reputation against false claims on uncertified products.

So, the tick-tree becomes a symbol of the FSC's certification, with the idea that anyone seeing it on a product who is concerned about the environment will buy that product in preference to non-certified product. The tick-tree is a trademark, a set of knowledge packaged as part of the product and a brand that commodifies a complex set of certification processes to sell products. In this way, FSC works within existing capitalist systems, aiming to lever consumer power not to radically overhaul them, but to gradually reform them and to push the forestry industry towards more sustainable practices.

It is worth remembering also that the FSC was originally sponsored by WWF and B&Q, bringing together a large environmental NGO and a successful company in the home improvement sector in the UK. FSC is funded primarily through donations from a range of organizations, including charitable foundations, governments and companies like IKEA and Home Depot, supplemented by membership subscriptions and accreditation fees from certification bodies, producing a total revenue in 2004 of US$ 3,729,625 (FSC 2004). This is very small for a global network, especially compared to the multinational companies that it seeks to influence.

By May 2007, forests had been FSC-certified in 78 countries, covering 125 million hectares globally (up from 10 million in 1998) and representing perhaps 5% of forests used for primary production and total sales of $20 billion globally (FSC 2010; UNECE/FAO 2006) and £1 billion in the UK (Cooperative Bank 2008). But the geography of FSC-certified forests is skewed towards temperate and boreal softwoods, reflecting the pattern of industrial forestry. So, although FSC's original concerns were about deforestation and illegal logging of tropical hardwoods in the global South, in fact more of the economically productive forests in the North have sought certification. Some would see this as FSC's failure to change the forestry system, especially where it is ignored by companies involved in illegal logging of biodiversity-rich rainforests.

Moreover, despite what is often seen as a successful initiative in industrial terms, awareness of FSC is still low amongst consumers. FSC estimates that its label is recognized by 21 percent of consumers in Switzerland and the Netherlands, 23 percent in the UK and 33 percent in Denmark (FSC 2007). This is not helped by some retailers of timber products not promoting FSC even where they sell FSC-certified products. For example, IKEA argues that labeling products in its stores with the FSC mark would dilute its own brand, and other small timber merchants frequently find it too expensive to pay for "chain of custody" certification to be able to use the FSC logo themselves, even where they sell FSC-certified timber. So FSC-certified products may not be marked as such for the ordinary consumer, producing "leakage" of FSC-certified products out of the certification system.

Even where there is press coverage, this may have little effect on consumers. The final Harry Potter book, *Harry Potter and the Deathly Hallows*, was published by Scholastic in the USA and Bloomsbury in the UK on paper certified by FSC as from "mixed sources" (which means 30–65 percent is certified, in this case), but it is doubtful whether many of the millions of readers were aware of this – the (small) FSC symbol only appeared with the copyright declaration on the inside front page. Without consumer knowledge, the FSC certification process is limited in how far it can harness consumer power as the NGO would wish.

Despite this, the FSC is often seen as a model for other schemes, particularly in the way that it set up its certification procedures globally. For example, the Marine Stewardship Council (MSC, see Table 8.1) was set up in 1997 to promote sustainably managed marine fisheries, based on the FSC's certification model. There are also competing ecolabels for sustainable timber, such as PEFC, often set up by commercial interests and trade associations, but which again emulate the structures and processes of FSC. However, these are often regarded as less credible because of their closeness to commercial interests and less stringent because of their lower or more flexible forestry standards. It is therefore not solely how certification is managed that affects how it is interpreted, but also the perceived independence of the ecolabel itself from existing power relationships in global trade.

Example 2, Fairtrade products

Fairtrade products have a more complex history, but share similarities with the FSC example in terms of how certification is managed today. Labeling specifically to promote products produced in socially responsible ways arose in the 1980s in different countries, particularly for coffee. But national initiatives adopted different names, which became a problem for global supply chains exporting and importing between countries, as the different labels did not carry their meanings across national borders. To harmonize across this confusion, an umbrella NGO was set up in 1997 in Germany, called Fairtrade Labelling Organizations International (FLO), to operate transnationally and to develop global standards for fairtrade labeling across 21 countries, especially in northern consumer markets (Renard 2005). In 2002, FLO launched an International Fairtrade Certification Mark (now run by its certification arm, FLO-CERT), and applicants for this are verified by approved certification bodies in a similar way to the FSC verification process.

Today, Fairtrade specifically seeks to support small producers and plantation workers and, like FSC, they cite the three pillars of sustainable development (environmental, social and economic development) as important to their work. The key objectives of Fairtrade standards focus on guaranteeing a fair minimum price to producers, supporting long term (more secure) trading relationships and ensuring that conditions of both production and trade for fairtrade products are fair and responsible in social, economic and environmental terms. So, unlike FSC, Fairtrade guarantees a premium above normal world market prices to the producer, thus emphasizing more strongly the economic pillar of sustainable development, to try to correct the unequal balance of world trade.

In 2008, retail sales of Fairtrade-certified products globally reached 2.9 billion euros. In the UK, sales reached £712.6 million, a strong trend upward from £290 million in 2006, and the most important sectors were bananas (£184.6 million) and coffee (£137.3 million), which together made up nearly half the total (Fairtrade Foundation 2009). But this is still a tiny proportion of the billions of pounds spent on food every year.

Consumer awareness is higher for Fairtrade than for FSC: FLO estimated that 57 percent of UK adults recognized their Fairtrade label in 2007. With the rising awareness of Fairtrade, large retailers are increasingly interested in Fairtrade-certified products. For example, Starbucks has long been a target for campaigners because of its global sales and seeks to counter this bad press in various ways. It used to promote Fairtrade-certified coffee only once a month, but in November 2008 it announced that all coffee sold in its 700 outlets in the UK and Ireland would use only Fairtrade sources by the end of 2009. This would be a huge increase from the existing 6 percent of its sales and would probably make Starbucks the largest buyer of Fairtrade coffee globally (Hickman 2008). The notion of such a commercially successful corporation getting involved in a supposedly alternative form of production raises concerns about capture of this agenda (e.g. Renard 2005). It reflects Goodman's (2004; also Freidberg 2003) concerns about how the marketing of fairtrade coffee draws on ethics and values as part of commoditizing marginalized producers and ends up producing a new consumer fetish – a fetish that perhaps emphasizes rather than hides the conditions of production, but still a fetish that capitalism can also benefit from.

Example 3, organic products

Another key example is that of organic products. Unlike the other two examples, organic production is not certified through an NGO that applies the same procedures globally; instead, it is partly regulated in many countries, including regulation of exporting and importing. This means that different countries may not only have different definitions of "organic," but also that the organizations performing the certification will be managed differently in each country. An international supply chain will therefore have to deal with more than one certification system, unlike the FSC's label with its global "chain of custody" system.

In EU countries, the label "organic" is defined through European Regulation 2092/91 in terms of allowable production practices (from the size of hen houses to permitted veterinary medicines) and permissible inputs and treatments (from farmyard manure to feed additives). At least 95 percent of the agricultural ingredients of a product must come from such practices for it to be labeled as "organic" in the EU, and it must not have been produced using genetically modified organisms. The Regulation is then implemented by national governments in each EU country. In the UK, it is implemented through the Organic Products Regulations 2004, managed through the government's Department for Environment, Food and Rural Affairs (Defra). Producers applying to use the organic label are checked through a complex set of audit practices by one of nine "certification bodies" that verify organic production under Defra's supervision (Table 8.1). To provide trace-ability for consumers, all foods certified as organic are labeled with the relevant certification body's code, e.g. UK5 for the Soil Association, UK2 for Organic Farmers and Growers. In Sweden, by comparison, the Regulation is managed by the national KRAV label – a label managed not by a government department, but by the eponymous KRAV, an umbrella NGO since 1985, which seems to command a great deal of respect nationally (Boström and Klintman 2006).

In the USA, organic labeling developed from the 1960s but was often unregulated or regulated by different states in different ways. The federal government sought to standardize this diversity, to promote commerce between states and to provide clarity for consumers. In 1990, the Organic Food Protection Act was incorporated into the Farm Bill (Boström and Klintman 2006; Vos 2000) and National Organic Standards were later developed. Internationally, the not-for-profit International Federation of Organic Agriculture Movements (IFOAM) oversees a global certification process through independent certification bodies similar to FSC's and Fairtrade's, with the difference that products certified as organic also have to negotiate national organic regulations on entering some markets (such as the UK's). So, organic certification is a more complex system, with multiple interacting and partly regulated elements.

Organic-certified food and drink sales in the UK were estimated at £1,911 million in 2007 (Cooperative Bank 2008), making this the biggest of the three examples in terms of consumer purchase, but still a tiny proportion of total sales. But consumers also make sense of the label "organic" in different ways. It is defined officially by the consequences of its production for the environment and animal welfare, not because of product qualities or consequences for human health. But a survey (Mintel 2007) of people who bought organic food reported that they were motivated to do so because of health, taste and environmental effects (45 percent, 42 percent and 40 percent respectively), two out of three of which are not included in the "official" definition of organic. So, to refer back to Appadurai's comment earlier, the knowledge put into a commodity by producers can be very different from the knowledge read off from that commodity by consumers, even when ecolabels are provided to supposedly guarantee clarity and traceability.

Mainstreaming these examples

These three examples show similarities and contrasts. For example, unlike FSC-certified timber (which rarely gains a price premium), many organic products gain a price premium over non-organic products in retail outlets. This has generated both concern about uncertified products falsely claiming organic qualities and also concern about big business expanding organic production in order to increase profits without subscribing to the original ideologies of the organic movement (Vos 2000; Guthman 2004), especially in its holistic relationship with the environment, animal welfare and human health and its support for small-scale, often family-run or community-run farms. Organic and fairtrade are often referred to as "alternatives" to mainstream production and consumption, but growth and potential capture by commercial (especially multinational) corporations is considered to threaten this status and indeed the integrity of organic farming as a whole (Vos 2000; Mansfield

2004; Guthman 2004) or even, in the case of production in developing countries for export, to represent "ecological neocolonialism" (Mutersbaugh 2002, page 1181; also Freidberg 2003). Indeed, the 1997 proposals by USDA for National Organic Standards generated a storm of protest within the organic movement (Vos 2000; Mansfield 2004; Boström and Klintman 2006) because they were seen as lowering the standard and perverting the real meaning of organic for many who had long practised organic production and consumption, not least through defining organic to include the use of genetic modification, irradiation and sewage sludge. (These three processes were later excluded.)

However, Giovannucci and Ponte (2005: 298) argue that the division between alternative and mainstream markets in the case of sustainable coffee is "increasingly blurred," and this applies to many other products also, not least because large proportions of sustainable, organic and fairtrade certified products are bought in "conventional" outlets, such as big supermarket chains in the UK (Soil Association 2006).

This is the mainstreaming dilemma at the heart of much ecolabeling: is it better to be purist, to implement the highest standards of production in environmental and social terms, but as a consequence to remain an expensive niche, or is it better to modify the standards (some would say, dilute or downgrade the standards) and expand the mass market? The first option produces a "deep green" niche, where those involved make a substantial change individually, but a small change collectively because they are too few in number. The second option produces a "light green" mass market, where those involved make a less substantial change individually (leading to accusations of pandering to the yuppie market – Talbot 2004), but the collective impact of many more people buying the better products may be more substantial on a global scale. This is clearly a political decision, but one couched not in traditional realms of politics, regulation and law, but in terms of the everyday consumer decisions that parlay knowledge and care into the wider arena of global environmental governance.

Conclusions

Certification schemes have expanded and diversified in the last ten years or so, producing and verifying ecolabels in attempts at global green governance using traceability and consumer power. But they remain problematic. From a theoretical point of view, they fit with political ecology's emphasis upon the problems of world trade and the power relationships and inequalities associated with it, especially in terms of addressing the problem of knowledge in global consumption. But the solution that such schemes promote often fails to find favor with political ecologists, because of the burdens of cost and time that they put on small producers and sometimes because researchers simply are suspicious of success shown when certification schemes seem to validate – to "greenstamp" or "greenwash" or "bluewash"[2] – big business (e.g. Goodman 2004: 910).

From a practical point of view, it is also very difficult to measure their impact on the global economy in terms of trade balance or environmental protection, not least because so many other factors are involved in the related production and consumption processes. But certainly the markets for organically, sustainably and/or fairly produced goods remain a small minority of the whole. And ecolabels themselves are not necessarily interpreted in the way that their promoters would wish – consumers often continue to be suspicious of the claims that such labels make about sustainability or fair trading conditions and, like brands, some ecolabels have far greater consumer awareness and credibility than others, so it can be unhelpful to generalize.

Because of this, some have argued that such ecolabels are not the answer and instead we need more state-backed regulation to protect vulnerable people and environments (e.g. Klooster 2006; Talbot 2004). That we need stronger regulation is certainly true. But it is worth remembering that FSC, for example, was set up precisely because the UN failed to get nation-states to agree on how to globally regulate against deforestation. And many ecolabeling schemes would happily acknowledge that their greatest achievement would be to be replaced by regulation, because that would mean that they had changed production norms sufficiently for regulation to become politically feasible. But the bald fact is that this is unlikely at present.

Given this, it is better to see certification as one of many tools for environmental global governance, rather than the only solution. Indeed, many NGOs belong to schemes like FSC at the same time as lobbying forestry companies and national governments about improving standards for forestry management. And there are numerous other campaigns that do not specifically stamp ecolabels on products, but use other mechanisms to seek to improve the sustainability and equity of global production processes within existing capitalist systems. The UN's Global Compact and the not-for-profit Ethical Trading Initiative are two examples of schemes that operate largely within business networks to improve corporate responsibility training and practices, rather than through seeking strong consumer visibility.

With this in mind, we can see that ecolabels will not change the world on their own, but they are one weapon in a much wider arsenal and one that at least usefully addresses the complexities of consumption and consumer power within a political ecology of the global economy.

Notes

1   This also relates to the common (but wrong) assumption that ethical consumers are all "liberal, middle-class, environmentalist[s]" (Hartwick 2000, page 1186).
2   To "bluewash" means to make superficial claims of corporate social responsibility. The term refers to the blue of the UN flag and specifically to companies pledging to the UN's 1999 Global Compact.

References

Appadurai, Arjun (1986) Introduction: commodities and the politics of value. Pages 3–63 in Arjun Appadurai (ed.) *The Social Life of Things: Commmodities in Cultural Perspective.* (Cambridge University Press, Cambridge)

Barnett, Clive, Paul Cloke, Nick Clarke and Alice Malpass (2005) Consuming ethics: articulating the subjects and spaces of ethical consumption. *Antipode* 37, 1, 23–45

Boström, Magnus and Mikael Klintman (2006) State-centred *versus* nonstate-driven organic food standardization: a comparison of the US and Sweden. *Agriculture and Human Values* 23, 163–180

Cooperative Bank (2008) *The Ethical Consumerism Report 2008*. http://www.goodwith money.co.uk/ ethical-consumerism-report-08/

Eden, Sally (2009) The work of environmental governance networks: traceability, credibility and certification by the Forest Stewardship Council. *Geoforum* 40, 383–394

Eden, Sally, Christopher Bear and Gordon Walker (2008) Mucky carrots and other proxies: problematising the knowledge-fix for sustainable and ethical consumption. *Geoforum* 39, 1044–1057

Fairtrade Foundation (2009) Facts and figures on fairtrade. http://www.fairtrade.org.uk/ what_is_fairtrade/facts_and_figures.aspx

Freidberg, Susanne (2003) Cleaning up down South: supermarkets, ethical trade and African horticulture. *Social and Cultural Geography* 4, 1, 27–43

FSC (2000) *Principles and Criteria for Forest Stewardship.* (FSC, Washington DC)

—— (2010) FSC Facts & Figures, available online at http://www.fsc.org/facts-figures.html (accessed 25 May 2010)

Giovannucci, Daniele and Stefano Ponte (2005) Standards as a new form of social contract? Sustainability initiatives in the coffee industry. *Food Policy* 30, 284–301

Goodman, Michael K. (2004) Reading fair trade: political ecological imaginary and the moral economy of fair trade foods. *Political Geography* 23, 891–915

Goss, J. (2004) Geography of consumption I. *Progress in Human Geography* 28, 3, 369–380

Gregory, Jane and Steve Miller (1998) *Science in Public.* (Plenum Trade, New York/London)

Gulbrandsen, Lars H. (2004) Overlapping public and private governance: can forest certification fill the gaps in the global forest regime? *Global Environmental Politics* 4, 2, 75–99

—— (2005) Explaining different approaches to voluntary standards: a study of forest certification choices in Norway and Sweden. *Journal of Environmental Policy and Planning* 7, 1, 43–59

Guthman, Julie (2004) The trouble with "organic lite" in California: a rejoinder to the "conventionalisation" debate. *Sociologia Ruralis* 44, 3, 301–315

Hartwick, Elaine (1998) Geographies of consumption: a commodity-chain approach. *Environment and Planning D: Society and Space* 16, 423–437

—— (2000) Towards a geographical politics of consumption. *Environment and Planning A* 32, 1177–1192

Hatanaka, Maki, Carmen Bain and Lawrence Busch (2005) Third-party certification in the global agrifood system. *Food Policy* 30, 354–369

Hickman, Martin (2008) All Starbucks' coffee to be Fairtrade. *The Independent* 26 November.

Hughes, Alex (2006) Learning to trade ethically: knowledgeable capitalism, retailers and contested commodity chains. *Geoforum* 37, 1008–1020

Hudson, Ian and Mark Hudson (2003) Removing the veil? Commodity fetishism, fair trade, and the environment. *Organization and Environment* 16 (4), 413–430

Irwin, Alan (1995) *Citizen Science.* (Routledge, London)

Irwin, Alan and Mike Michael (2003) *Science, Social Theory and Public Knowledge.* (Open University Press, Maidenhead).

Jackson, Peter (2002) Commercial cultures: transcending the cultural and the economic. *Progress in Human Geography* 26 (1), 3–18

Jahn, Gabriele, Matthias Schramm and Achim Spiller (2005) The reliability of certification: quality labels as a consumer policy tool. *Journal of Consumer Policy* 28, 53–73

Klooster, Daniel (2006) Environmental certification of forests in Mexico: the political ecology of a nongovernmental market intervention. *Annals of the Association of American Geographers* 96, 3, 541–565

Lawrence, Felicity (2004) *Not on the Label*. (Penguin, London)

Mansfield, Becky (2004) Organic views of nature: the debate over organic certification for aquatic animals. *Sociologia Ruralis* 44, 2, 216–232

Miller, Daniel (1995) Consumption as the vanguard of history: a polemic by way of an introduction. Pages 1–57 in Daniel Miller (ed.) *Acknowledging Consumption* (Routledge, London)

Mintel (2007) *Organics*. London: Mintel.

—— (1998) *A Theory of Shopping*. (Polity Press, Cambridge)

Morris, Mike and Nikki Dunne (2004) Driving environmental certification: its impact on the furniture and timber products value chain in South Africa. *Geoforum* 35, 251–266

Mutersbaugh, Tad (2002) The number is the beast: a political economy of organic-coffee certification and producer unionism. *Environment and Planning A* 34, 1165–1184

Mutersbaugh, Tad, Daniel Klooster, Marie-Christine Renard and Peter Taylor (2005) Certifying rural spaces: quality-certified products and rural governance. *Journal of Rural Studies* 21, 4, 381–388

Pattberg, Philipp (2005) What role for private rule-making in global environmental governance? Analysing the Forest Stewardship Council (FSC). *International Environmental Agreements* 5, 175–189

Renard, Marie-Christine (2005) Quality certification, regulation and power in fair trade. *Journal of Rural Studies* 21, 419–431

Robbins, Paul (2004) *Political Ecology: A Critical Introduction* (Blackwell, Oxford)

Schlosser, Eric (2002) *Fast Food Nation*. (Penguin, London)

Soil Association (2006) *Organic Market Report*. (Soil Association, Bristol)

Talbot, John M. (2004) *Grounds for Agreement: The political economy of the coffee commodity chain*. (Rowman & Littlefield, Lanham MD)

UNECE/FAO (2006) *Forest Products Annual Market Review 2005–6*. (UNECE/FAO, Geneva)

Vos, Timothy (2000) Visions of the middle landscape: organic farming and the politics of nature. *Agriculture and Human Values* 17, 245–256

Wynne, Brian (1995) 'Public understanding of science,' in Sheila Jasanoff, Gerald E Markle, James C. Petersen and Trevor Pinch (eds.) *Handbook of Science and Technology Studies,* pages 361–388. (Sage, London)

9 Climate change and the risk industry: the multiplication of fear and value

Leigh Johnson

Without fear, and without value to protect, there is never any insurance purchased.
[. . .]
We would say that more or less everything is insurable.

(Director of Emerging Risks division in a
major international reinsurance company, 2009)

Inundated port cities bringing trade and shipping to its knees: Gulf of Mexico oil production crippled by a series of major hurricanes: lower Manhattan under water, or an overtopping Thames paralyzing global financial markets. These images no longer read like scenes from the wildly apocalyptic film *The Day After Tomorrow*. They are narratives regularly reported in the press and are commonplace in scenario planning conducted by the phalanxes of risk consultants, policy wonks, and academics concerned with the challenges of adapting to the medium and long term implications of global climate change. Future climate projections coupled with disastrous economic scenarios are not hard to come by, thanks to a forest of reports turned out by multilateral organizations, environmental agencies, global insurers, industry associations, policy think tanks, university research centers, environmental non-governmental organizations, and so on. As the economic risks posed by global warming have become increasingly apparent and publicized, the planetary scale of both financial interdependence and climate change impacts have made it impossible for the institutions managing global capital to ignore them. Potential threats often cited include damage to property and infrastructure from sea level rise and coastal subsidence, declines in agricultural and marine productivity, water shortages, increasing health care and health insurance costs and work days lost, and economic losses due to more frequent or severe weather-related disasters such as heat waves, droughts, floods, winter storms, and tropical cyclones (cf. Nicholls *et al.* 2007, Epstein and Mills 2005, Downing *et al.* 1999).[1] These are the sorts of impacts that prompted sociologist Ulrich Beck to place climate change in a new category of risks generated by industrial society that are fundamentally "uninsurable" (Beck 1999). But is this in fact the case? In what follows I show that the insurance industry is plowing ahead with its attempts to make climate change impacts insurable, and therefore profitable, lines of business. Moreover, I argue that the ways in which the industry is going about this project reveal the constellations of science, value, and fear – a political ecology of risk – through which the modern insurance industry broadly construed (the risk industry) reproduces itself.[2]

Given the total values potentially at stake in a warming world – some 5 to 20 percent of annual global GDP by one estimate (Stern 2006) – adherence to basic corporate risk management practices dictates that boardrooms and the business elite cannot dismiss the possibility that dire future climate projections could become a reality. The reinsurance industry[3] – which holds hundreds of billions of dollars in exposures to insurers' losses – has been warning business and government about the potentially devastating consequences of global warming since at least the mid-1990s (cf. Swiss Re 1994). These admonitions have come with increasing urgency as both the scientific consensus around the anthropogenic causes of present climate change and our ability to make projections about impacts have grown. A number of companies and industry associations have thrown their weight behind several government-run loss mitigation programs and continue actively advocating for adaptation policies as well as emissions reductions. Within the (re)insurance industry itself, increasing concern about the quantification of climate risks has magnified an already-existing trend towards quantitative "catastrophe modeling" of geophysical hazards for the purposes of pricing and risk management (Grossi and Kunreuther 2005). The adoption of these modeling practices has also entailed a turn towards an increasingly technical workforce; major property insurers and reinsurers are likely to employ numerous PhD-credentialed scientists with hazard-specific expertise. Likewise, many companies are involved in sponsoring a proliferating number of academic climate studies, commissioning research projects, and funding symposiums, university chairs, and PhD students. In short, (re)insurance as an industry has become a hub for climate change risk assessment.

The simplistic explanation for this phenomenon is that (re)insurers are motivated by the bottom line; that is to say, by their need to fully gauge both the degree of their exposures and the potentialities for profit. But while this explanation is probably true, it is also superficial. In this chapter I argue that the industry's encounter with global warming and scientific research is more than a contingent result of contemporary circumstances and profit pressures. Rather, it logically follows from the position that (re)insurance occupies within modern capitalism. The financial mechanism of insurance has long been a central pillar in the global economy's organization of nature and production of value. The guarantee of property and liability insurance[4] – that potential losses will be covered in exchange for a premium paid in advance by the insured – is fundamental to securing private property and credit and establishing faith between counterparties. This security is a prerequisite for the vast number of transactions that collectively comprise the "global economy": global trade, shipping, construction, manufacturing, transportation, energy production, and so on.

The risk industry, then, denotes an enormous constellation of institutions and individuals whose activities all hinge upon the analysis, placement, financing, and management of others' property risks. Although the term "risk industry" is not technically used in the business or academic literature, I have adopted it here for several reasons. First, this field exceeds the boundaries of the (re)insurance industry as typically conceived (limited to direct insurers, reinsurers, and brokers). It includes institutions whose access to capital is relatively limited, but who are nevertheless powerful due to their control over information and analytical techniques: private catastrophe modeling firms, science and engineering consultancies, and university research partnerships. It also includes players from the capital markets – most notably multi-strategy hedge funds and dedicated catastrophe bond funds – who do not themselves buy or sell insurance, but whose investment activities as buyers (and secondary market sellers) of insurance-linked securities contribute to the total underwriting capacity of the (re)insurance industry.

Most importantly, by using the moniker of "the risk industry" to refer to all of these players despite their often oppositional positions as buyers, sellers, analysts, brokers, investors, and speculators in risk, I aim to emphasize their shared interests and roles in the *reproduction of risk*. That is to say that the proliferation of new kinds of quantifiable risks and the data with which to measure them, as well as their trading and bundling into new investment products, are not reactive phenomena. Rather, the normal functioning of the risk industry involves active product creation and management, in which scientific research plays a significant part. Other scholars of insurance have taken note of how the process of risk identification lends itself to infinite replication. As Defert (1991) has argued with regard to the development of life insurance technologies, "each new risk identified has a new cost . . . each new protection . . . makes visible a new form of insurable insecurity . . . security becomes an inexhaustible market" (215). But whereas Defert associates this multiplication of insecurities with the expansion of governmentality and statistical science in the nineteenth century, I want to suggest that, at least for property insurance, the reproduction of "insurable insecurities" is a byproduct of the competitive and expansionary nature of the capitalist economy. This chapter takes the case of climate change to trace how such insecurities emerge, how they are made insurable, and how fears about the destruction of asset values ultimately make the purchase of the new insurance product necessary.

Attention to this process seems especially important given the extreme visibility and rhetorical power of global warming impacts (Demeritt 2006, Luke 2008, Beck 2009). And as both climate risks *and* our knowledge of them grow, so will the market for new insurance and risk-management products, thus reproducing the industry's own conditions of existence. Beyond its traditional market, the risk industry will gain influence in the larger economy as the pace of environmental change quickens and some regions are faced with potentially massive devaluations of fixed capital and real estate through both catastrophic losses and steady systemic changes.[5] The industry's analysis and pricing of climate-linked risks may in some cases determine which impacts are deemed too costly or too dangerous to ignore, and thus which are deserving of dedicated adaptation policies. It is also likely to influence which sorts of risks companies and governments feel the need to plan for and insure against.

I pursue several convergent paths to investigate the dynamics of knowledge production and value creation emerging as the risk industry seeks ways to measure and manage the impacts of climate change. To provide a basis for discussion, the first section briefly enumerates the principles of modern insurance, its history, and the contemporary organization of the market. The second section discusses the pivotal and growing role that natural science and scient*ists* play within the risk industry by identifying and quantifying environmental risks in order to make them insurable. The chapter concludes with an inquiry into the relationship between scientific research, climate anxieties, and economic value.

Rather than beginning this chapter with any "principles" of insurability or insurance practice, I want to emphasize the historically evolving nature of insurance forms. As philosopher and scholar of insurance François Ewald points out, the principles on which this technology of risk is based did not "fall from the sky"; they were only developed in retrospect, following centuries of diverse insurance practice. Thus, insured risks have varied in the degree to which they meet all the technical standards for insurability enumerated in theoretical literature and textbooks (cf. Denenberg 1964, 615). Insurers may be willing to cover risks not typically considered "insurable" in the face of heavy competition in traditional markets, or if high investment returns have buoyed their balance sheets. But given the extent to which these standards of insurability are nevertheless invoked in discussions of the industry's ability to cope with emerging risks like climate change, they bear repeating (cf. Swiss Re 2005; Mills *et al.* 2005). For the private insurance mechanism to function, the probability and severity of an event should be calculable, although the precise location and time of loss must be unforeseeable. This implies that: (1) the likelihood of the peril is known; (2) loss events are frequent enough that the "Law of Large Numbers" (convergence toward the mean value of the population) applies; (3) the correlation between separate loss events in an insurer's book of business is limited; (4) the potential and probable maximum losses for an event are calculable and financially manageable; and (5) that the pool of insured does not pose a greater risk of loss than the general population (Berliner 1982). By now it should be obvious that climate change could generate a number of environmental scenarios that violate most, if not all, of these criteria. Indeed, in hewing to the letter of these criteria, Ulrich Beck argues that industrial society's "manufactured uncertainties" such as genetic engineering, nuclear power, and climate change are simply "too risky" to be insured (1999, 4; 2009). Leaving aside the internal problems in Beck's vague and teleological formulation – since climate change is a long-term trend in weather patterns and not a single event, by definition it is not in itself an "insurable risk" – I interpret his argument to be that environmental hazards that become more extreme or more frequent due to climate change are uninsurable (156).[6] But, as we shall see, circumstances seem to indicate otherwise.

Insurance, capitalism, and catastrophe

In one form or another, insurance has attempted to provide security against the loss of assets since the birth of the capitalist world economy. First some clarification of terminology is in order, given the tremendous and diverging scholarship on risk and institutions of insurance. By "insurance," I do not mean to refer to the abstract technology of pooling the social or economic risks of a population, nor the multitude of forms this technology has taken in history (including social welfare systems, mutualist societies, private companies, social security, annuity schemes, etc) (Ewald 1991). My focus is on the institution of private insurance, through which a firm promises financial compensation to a policyholder in the event of loss, in exchange for an advance premium payment from the policyholder.

The proliferation of various types of these private arrangements over the centuries has perhaps made the private insurance transaction seem like a natural matter of course. But the widespread exchange of money for a guarantee of future financial security developed in response to a set of logistical problems posed by a very specific organization of economy and society, at a particular time and location; namely, problems raised by merchant trade in early modern Europe. The centrality of the private insurance mechanism to the development of early capitalism[7] is worth mentioning insofar as it lays the groundwork for this chapter's discussion of the risk industry's role in the maintenance of value.

Formal marine insurance contracts appeared in the thirteenth century and achieved significant scale by the fourteenth, as traders in the hubs of merchant capitalism developing around the Mediterranean sought guarantees of financial security for their investments and transactions (Lopez 1976, Lane 1973). Vessels and their cargo were commonly lost to storms, shipwreck, and piracy; poor ship-building and inexperienced crews made matters worse. Since sea trading already relied upon credit and bills of exchange, bankers were well positioned to enter the market providing maritime insurance for voyages. By the fifteenth century marine risks were shared broadly between the banking classes; Venetian brokers could obtain dozens to hundreds of underwriters for a single voyage (Lane 1973, 382). This general system of marine insurance was later replicated by the Dutch and then the British, and organized around the financial centers of Amsterdam and London. By around 1700, the Dutch had for the first time developed a specialized insurance sector with full-time brokers, underwriters, and dedicated companies (Braudel 1992).

As the scale of European imperial ventures expanded, so did the necessity of insurance. The length of voyages in the open ocean grew, exposing ships to new hazards like cyclones and leaving them less defensible against piracy. The value of the cargo itself was also mushrooming. Insuring the value of ships and cargo – including slaves, bullion, finished goods, and raw materials – became ever more critical to the maintenance of trade and financial capital flows (Baucom 2005). The extension of credit, the purchase of insurance, and the expansion of Europe's imperial footprint were inseparably bound.

In his searing account of eighteenth-century finance and the Atlantic slave trade, Ian Baucom points out the peculiar ontological "magic" performed by insurance that made it so critical to the expansion of finance capital:

> Insurance thus does not confer a monetary value upon lost things, it sets the money form of value free from the life of things . . . Absent the security

insurance provides, finance capitalism could not exist. The world of things would stage its revenge on value each time some object or another was destroyed, would refasten value to embodied things and make one as mortal as the other.

(Baucom 2005, 96)

In less elegant terms, we might say that the insurance mechanism ensures that capital is not tethered to the final fate of the particular earthly goods in which it has been invested, and is instead connected to them only through the universal equivalent of the money form – which is to say, through their exchange value. Insofar as this is true, an investor's loss can always be compensated by monetary remuneration. Thus insurance's promise of abstract security removed much of the typical reluctance that accompanied any proposal to transact business in an unfamiliar and potentially dangerous environment.

The "magic" of property insurance extends beyond the world maritime trade to fixed capital investments in the built environment. In the nascent urbanizing centers of the eighteenth century, stores, factories, warehouses, and residential structures clustered tremendous property values in close proximity. And "as the century progressed, new sources of combustion hazard appeared and there were novel, larger and more complex risks in the shape of new machinery, processes and materials to insure" – docks, mills, sugar refineries, and breweries among them (Pearson 2004: 4–8). Fire insurance thus provided some assurance to industrialists and merchants that large investments in new machinery, technology, and commercial goods would be compensated in the event of a conflagration (Kopf 1929). But insurance companies could do little to change the fundamental vulnerability of urban built environments to fire,[8] and the total accumulated value of claims from major city fires routinely exceeded the financial capacity of direct insurers, forcing them into bankruptcy. A new institutional form – reinsurance – was deliberately developed in response to city fires, which were financially crippling for direct insurers due to the *simultaneous* and *extreme* nature of losses incurred across a large number of properties in one location. Reinsurance companies had larger capital bases and more geographically dispersed risk pools; they were explicitly concerned with developing international books of business from their inception in 1846 (Kopf 1929).

"Modern" catastrophe reinsurance coverage, including that for tropical storms, winter storms, and earthquakes, arguably has its roots in the fallout from the 1906 San Francisco earthquake and fire. Earthquake coverage had typically been excluded from property (re)insurance because – like other large natural catastrophe perils – it was considered uninsurable.[9] But after many (re)insurers survived the disaster despite paying out enormous sums, some companies reconsidered its business potential and began writing California earthquake coverage as a new line of business (Freeman 1932; Guatteri *et al.* 2005). The fact that insurers began actively pursuing the market for its particular hazard profile in the decades following 1906 exemplifies a familiar pattern in which risks once thought to be unmanageable become tractable in light of new technologies and assignations of financial responsibility. It is significant that these reorganizations tend to take place at times in which both hazard-related fears and agglomerations of value are growing. I revisit this idea and its implications below.

Today, reinsurance coverage is still the primary mechanism by which property insurers absorb losses to natural catastrophes.[10] Changes in the frequency of major loss events would reverberate throughout the industry on a global scale. Insurers typically purchase a number of layers of coverage from different reinsurers, each with different "attachment" and "exhaustion" points. In the example reinsurance program diagramed in Figure 9.1, four separate reinsurers are contracted to cover different "layers" of the direct insurer's losses. The figure demonstrates how all five companies involved are financially exposed to the catastrophe risks that the direct insurer takes on in its book of business. Similar reinsurance programs are brokered for many thousands of direct insurers globally, each with varying geographic concentrations of exposure to different perils. Reinsurers also sell reinsurance coverage *among themselves*; that is to say, Reinsurer B may have also purchased $50 million in "retrocession" cover from Reinsurer D that triggers once B's ultimate net losses exceeds $500 million. Thus, the very characteristics that spread losses throughout the broadest capital base possible also expose the entire reinsurance industry to the systemic changes in event frequencies and severity that global warming may bring.

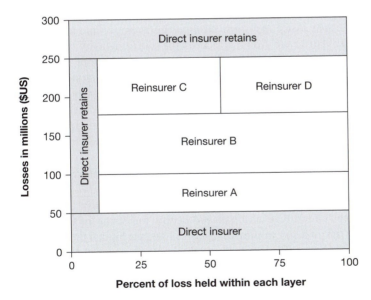

Figure 9.1 Example of a hypothetical standard "CAT XL" (extreme natural catastrophe) reinsurance program for a direct insurer. In this case, the direct insurer would be liable for all of the first $50 million in losses, 10 percent of the next $200 million, and all losses above $250 million. Figure adapted from Murnane 2004.

Although by definition no individual weather *event* (a Hurricane Katrina, for example) can be directly attributed to climate change, recent event trends have already raised concerns at an industry-wide level. A series of European winter storms, floods, and heat waves throughout the last decade have caused significant economic losses and fatalities in a region that is not typically a major source of insurance industry losses. Although there is no global trend in overall tropical cyclone frequency and there is a great deal of natural variability on decadal timescales, the number of *intense* hurricanes (Saffir-Simpson Category 4 and 5 storms) around the globe has nearly doubled in the past 35 years (Webster *et al.* 2005). The total power of tropical cyclones – an integral of maximum windspeeds over the lifetime of the storm – has more than doubled in the North Atlantic over the last 30 years (Emanuel 2005). And regardless of the ongoing scientific disagreement about the extent to which these increases in North Atlantic cyclone intensity and frequency can be attributed to global warming, the abnormally active hurricane seasons of 2004 and 2005 demonstrated how the industry's exposure to losses in the coastal built environment has ballooned thanks to population growth and unprecedented real estate development.

A number of game-changing events occurred in these seasons that raised panic about the potential link between climate change and tropical cyclones. By its end in December, the 2004 North Atlantic hurricane season was the costliest on record. It also logged the first hurricane ever recorded in the South Atlantic, demonstrating the potential for unexpected changes in extreme event patterns and locations (Pezza and Simmonds 2005). But these records were summarily surpassed by 2005's series of storms, which formed faster and grew stronger on average than any before (NOAA 2005a). The year broke Atlantic records for storm intensity (Hurricane Wilma), number of named storms (27), financial destruction (more than $100 billion), and fatalities (over 2,250) (NOAA 2005a, 2005b). It generated a paradigmatic shift in the way the risk industry imagined the damage potential of successive "super catastrophes." Industry insiders still categorize many market conditions, underwriting strategies, and risk perceptions as belonging to "pre-" or "post-KRW" times, underlining the epochal transformation affected by Katrina, Rita, and Wilma.

It is hardly surprising, then, that panic about global warming and extreme events reached a fever pitch in the aftermath of the 2005 season. *Time Magazine* published an October 2005 cover story asking "Are we making hurricanes worse?" and environmental groups called Katrina-displaced people the first climate change refugees. Within the risk industry, some (re)insurers raised alarms that their balance sheets were already suffering from the effects of climate change (cf. Munich Re 2006, Association of British Insurers 2004). A post-2005 publication by the world's largest reinsurer, Munich Re, warned that the industry's continued underwriting ability depends "on the development of adequate insurance solutions for catastrophe scenarios that have hitherto been considered inconceivable – we have to *think the unthinkable*" (2006, 1, italics mine).

But despite this language of emergency, insurance and reinsurance business was generally good. Eleven new reinsurance start-ups appeared in Bermuda (known as "the Class of 2005") to capitalize on high premiums, industry-wide profits soared, and loss ratios remained low in 2006 and 2007 (Mallon 2006). Billions of dollars of underwriting capacity was generated through increasingly popular catastrophe bonds, sponsored by reinsurers in efforts to tap into the broader capital markets. Meanwhile, catastrophe modeling firms dramatically reworked their tropical cyclone models, adopting hypothetical event sets based on 5-year forward-looking projections of climate conditions rather than using the standard long-term historical record of landfalling hurricanes (1900–2005) as the basis for event sets. These "near term" models resulted in suggested risk prices that were typically between 20–40 percent higher than those generated by previous models. Reinsurance prices increased due to these new model results, as did the capital reserves deemed necessary to underwrite these coastal risks. Already facing capital shortages due to hurricane losses, insurers responded by raising rates dramatically in many hurricane-prone U.S. states. This was more easily achieved in the industrial and commercial property insurance markets, whereas residential rate increases were often limited by state regulators in the interests of consumer protection. In Florida, a number of companies ceased writing new residential policies and refused to renew existing ones, or insisted on "wind exclusion" clauses. Companies faulted Florida's populist Governor and Insurance Commissioner for expanding the state-run insurance fund and suppressing private companies' premium increases to such a degree that it became financially dangerous for them to do business in the state. Regardless of whether these claims were exaggerated, the Florida case is illustrative of a larger point I pursue below.

Despite the numerous hurricane landfalls of 2004–2005, insurers did not exit the market because of any inherent characteristics of the state's geographic vulnerability that made risks absolutely incalculable or uninsurable. They abandoned Florida because political regulation prevented them from securing the premium increases necessary to earn their desired rate of return in the post-KRW insurance world. The fact that insurers continued writing Florida commercial and industrial property coverage (for which they charged, and received, greatly increased premiums) demonstrates that suppressed rates of return on residential lines, as opposed to fundamental problems of epistemological incalculability, were at issue. I point this out in order to demonstrate, *contra* Beck, the extent to which the risk price on offer determines the insurability of a peril. Or, to return to the quote with which this chapter began: given the right conditions, "more or less everything is insurable." And if it is not, it must be *made to be so*; as a reinsurance executive explained to me: "you must be constantly assessing and expanding the number of risks you are willing to write, because if you don't, insurers end up retaining them and your business shrinks and eventually, you cease to exist." The rest of the chapter turns its attention to what these conditions of insurability might be, and how they are being actively produced by the risk industry as it confronts climate risks in particular.

Making the insurable: science and risk price

If extreme weather risks such as North Atlantic tropical cyclones and European winter storms will likely increase due to climate change, then they also provide a test case for the insurability of these perils. Scientific and technical developments in computing power, numerical modeling, and climate risk assessment have trans-formed the market, making these perils insurable despite the uncertainty involved. Additionally, major players have secured access to highly technical expertise by establishing significant relationships with academic climatologists. Much as Ericson and Doyle (2004) have found in the case of terrorism insurance, it seems that Beck may have underestimated the risk industry's capacity to reconfigure circuits of research and knowledge and produce devices of calculation to suit its purposes even in the face of "incalculable uncertainties" (1999, 156: see also Collier 2008).

Since the late 1980s, private catastrophe modeling firms have developed and modified a set of tools to simulate hundreds of thousands of loss events and generate probability curves to calculate the chances of insurance losses exceeding a parti-cular dollar value. Models for any given peril combine thousands of assumptions and estimations of climatology, meteorology, features of the built environment, structural engineering, and financial behavior into hundreds of thousands of lines of computer code. This is not to say that catastrophe models generate perfect repre-sentations of risk; in my experience many modelers are quick to point out their limitations. But somewhat irrespective of these limitations, there is a sense in which the shared nomenclatures and repeated applications of catastrophe models as calculative devices has standardized the market and distilled diverse geographical risks into a common language. The fact that the models are created by third party firms who have no perceived attachment to any particular segment of the industry gives them added credence. Their strength lies in their ability to create and stabilize shared sets of meanings and quantifications of risk that are used by insurers, reinsurers, brokers, ratings agencies, and bond investors alike (cf. NAIC 2007a, 2007b).

New "emerging" perils are also the subjects of modeling endeavors. Recent examples include models for coastal flooding, terrorism, pandemic flu, longevity risk (for life insurers and life insurance-based bonds), catastrophic mortality, and "litigation epidemics." Like earthquakes at the turn of the twentieth century, all of these perils were at one point considered "uninsurable." All have since been subjected to statistical manipulation and modeling in order to estimate their return periods (the frequency with which an event of a certain magnitude can be expected to recur) and damage curves (proprietary algorithms that model the types and extent of damages resulting from a simulated peril). This is not to say that such events are now well understood. For a number of perils, modelers have yet to completely achieve what historian of science Ian Hacking (1990) calls "the taming of chance," by which individual events are explained in terms of their belonging to a large, statistically regular event set (or population). By definition, catastrophic events fall in the long tail of any such distribution and are so infrequent as to make statistical calculations difficult. But because these calculations are in such high demand by the industrial and financial sectors, modeling firms are engaged in constant competitive processes to produce more models for new regions and new risks.

The industry-wide adoption of these proprietary third party vendor models is a rather obvious market vulnerability that illustrates another way in which the risk industry produces and multiplies new risks. Here I concur with Beck (2009) that "the separation between production and management of risks becomes untenable when risk management can itself become a source of risks." (136). Catastrophe model source code is strictly proprietary and unavailable for scrutiny even by clients, who pay hundreds of thousands to millions of dollars to license the software. And although the vendors dedicate considerable human resources to interfacing with clients to explain the models, ultimately no outside scientific assessment is possible.[11] The risk of depending on a black box is, of course, that it is impossible to know whether the fundamental assumptions governing parts of the model's output are good ones.[12] This is particularly worrying since virtually 100 percent of the market for catastrophe models is controlled by just three major firms, two of which (Risk Management Solutions [RMS] and Applied Insurance Research [AIR]) are overwhelmingly dominant. Thus, if the models dramatically underestimate probable losses, the entire industry may be overexposed and take an uncomfortably large financial hit (as it did with the 2005 hurricanes). On the other hand, if the models overestimate probable losses, insurers and reinsurers may be adopting overly conservative underwriting standards, reserving more capital than is necessary to pay losses, and missing competitive opportunities to write new profitable business. The technically oriented reinsurers (typically based in Zurich, Bermuda, and London) are especially cognizant of this potential problem and have begun funneling tremendous resources to in-house efforts to conduct "sensitivity testing" and essentially reverse engineer model components.[13] Upon hiring a climatology PhD to evaluate the vendor models' strengths and weaknesses for his reinsurance company, one research director reportedly explained: "the biggest risk we face is, *what if RMS is wrong?*" This awareness is quite common; of the reinsurance companies studied for this research, most have at least a handful of young employees with recent Masters or PhD degrees in geophysical sciences working to critically assess how their company can better evaluate third party models' views on risk.

In general terms, reinsurers are using geophysical scientific expertise in order to conduct a sort of arbitrage on catastrophe risk prices. In finance practice, "arbitrage" is the general strategy of simultaneously entering into a set of opposing transactions to exploit pricing discrepancies between economically equivalent assets in two different markets. This is commonly practiced in commodity, futures, and currency markets; if a commodity is trading at $65.50 in the London market and $65.35 in Tokyo, an arbitrageur purchases a large volume on the Tokyo exchange and simultaneously sells the same amount on the London exchange, pocketing the difference as profit. In the case of catastrophe risk price arbitrage, reinsurers hire climatological and meteorological experts whose analysis of models allow the company to identify instances in which a vendor-modeled risk price is higher or lower than what the reinsurer's "best science" indicates it should be for their particular book of business, allowing them to strategically buy or sell coverage accordingly.

In addition to the credentialed professionals they hire to oversee the analysis of modeled risk price, a number of the major reinsurers and brokers have also involved outside scientists in larger applied research projects, which typically function through long-term partnerships or specific arrangements between the private companies and university research centers or departments. In some cases, they are specific and goal-oriented; for example, Swiss Re worked with a group at the Swiss Federal Institute of Technology (ETH) and the Federal Office of Meteorology to quantify climate change's effect on European winter storm damages projected through the year 2085 (Schwierz *et al.* 2006). In others, the relationship might be described loosely as one of academic patronage in which a reinsurer or broker partially funds a university research group, but without direct commercial intentions. For example, Munich Re has sponsored £3 million worth of research at the London School of Economics to quantify the business impacts of climate change, and Swiss Re has donated 5 million Swiss francs for a Chair in Integrative Risk Management at ETH. The reinsurance broker Aon Benfield sponsors the Benfield Hazard Research Center at University College London, which organizes applied physical and social research efforts on hazards including climate change, floods, earthquakes, landslides, windstorms, and volcanoes, and offers a masters degree in geophysical hazards. The broker Guy Carpenter has more directly targeted climate research by funding the Guy Carpenter Asia-Pacific Climate Impact Centre at the City University of Hong Kong. Willis reinsurance brokerage runs its own "Willis Research Network" which funds and commissions extreme event research from scientists from at least twenty academic institutions worldwide (including the U.S. National Center for Atmospheric Research) in what it calls the "world's largest partnership between academia and the insurance industry" (Willis 2008). There are similar examples of member-based industry organizations (the Risk Prediction Initiative in Bermuda and the Lighthill Risk Network in London among them) that also commission research or solicit consulting expertise from academic scientists.

But, despite this significant enrollment of academic expertise, reinsurers and brokers who I interviewed could think of very few academic collaborations *or* in-house applied research projects that had resulted in changes to reinsurance underwriting practices or rates. While reinsurance giants such as Munich Re funnel a significant amount of money into applied research, and their public relations divisions generate dramatic reports calling for higher rates in response to global warming, there is rarely any institutionalized connection between a company's public risk face and its internal risk pricing structures with regard to climate change risks. This tremendous disconnection seems to be a result of both the timescales on which reinsurance contracts operate and the competitive pressures on pricing within the reinsurance industry. This reveals a fundamental irony of reinsurers' engagement with climate change research: because reinsurance contracts are renewed and repriced on an annual basis, *in the short term, climate change risk is largely irrelevant.* Currently natural annual modes of climate variability such as El Niño and La Niña have a much greater (and still poorly understood) impact on a given year's catastrophic weather risk. One broker summarized his clients' perspective as such:

> The insurance industry is remarkably short-termist . . . Anybody that has a three-year view is actually an oracle. Most people are concerned with this year and next. So people saying, oh god mate, it's all going to go wrong in twenty years time, temperatures are going to be up by two degrees, and sea surfaces will be up by a meter. Couldn't give a damn, frankly. It's not important, right. Doesn't matter.

If climate change accelerates and more data become available to quantify impacts, reinsurers will likely raise rates to match perceived catastrophe risks. But thanks to the yearly renewal structure, this can be done over a number of years in the future, which is in any case far beyond the typical two year business perspective of underwriting departments. In the meantime, any first movers are at an immediate competitive disadvantage. If a reinsurer were to significantly increase its rates today based on its own internal analysis of climate change risks, it would immediately lose underwriting business to other firms who did not take the same view, and whose risk prices were consequently lower.

If applied climate research does not result in changes to day-to-day operation, why do reinsurance companies continue to fund it? On the one hand, this can be understood as a corporate risk management practice, intended to prevent companies from being caught off guard and help them more fully understand and manage their current weather-related risks. This might also include using up-to-the-minute meteorological data to hedge these risks in the so-called "live CAT" bond market. It is in essence an attempt to avoid becoming the post-hoc referent of Warren Buffet's famous aphorism, penned following the insurer bankruptcies from Hurricane Andrew in 1992: "it's only when the tide goes out that you find out who's been swimming naked." Establishing lasting relationships with research institutions in the present also cements connections that can be relied on into the future as research demands increase.

On the other hand, funding climate research is also a strategic public relations and investor relations strategy. Regardless of the practical realities of annual renewals and price competition, reinsurers' business also depends on being seen as forward-looking, technically savvy, and attentive to potential future dangers. So "their public persona [usually] says one thing while their business perspective does something else," according to a number of brokers. "Reinsurers must appear to understand the issue" in the eyes of clients and investors, and they fund research and produce publications on global warming "because it's the right thing to be seen to be doing." In response, the major brokers fund their own research collaborations; if they are to win contracts, they must appear to their clients (the primary insurers) to be "ahead of the game" and in possession of equal or greater information and technical expertise than the reinsurers and other brokers. A sort of one-upmanship prevails, resulting in an avalanche of glossy publications, press releases, and sponsored projects. Niklas Luhmann's (2005) perceptive comments on risk research suggest the paradoxical results:

[We] must abandon the hope that more research and more knowledge will permit a shift from risk to security. Practical experience tends to teach us the opposite: the more we know, the better we know what we do not know, and the more elaborate our risk awareness becomes. The more rationally we calculate and the more complex the calculations become, the more aspects come into view involving uncertainty about the future . . .".

(Luhmann 2005, 28)

This finally brings us full circle to Defert's (1991) argument about the multiplication of "insurable insecurities." Through the expansion of risk identification and calculation, here exemplified by the proliferation of research initiatives and publications organized by the risk industry, the provisioning of climate security becomes an inexhaustible market.

Making markets: the dialectics of fear and value

"The underlying business model of banks is greediness; the underlying business model of insurance is fear." This was how a director of emerging risk research at a large reinsurer candidly explained the industry's general business strategy to me in 2009. He continued straightforwardly: "without fear and without value to protect, there's no insurance purchased." These statements may be rather intuitively obvious, but they seem to me remarkable nonetheless because of their source. His conscious identification of fear as an organizing principle of both insurance sales and purchases provides an extremely useful avenue for exploring the proliferation of risks as a function of insurance institutions themselves.

In the *Grundrisse*, Marx insists that production under capitalism creates not only the commodity, but also its consumer – "not only. . .an object for the subject, but also a subject for the object" (1993, 92). I turn to this dialectic to show how it might elucidate both the multiplication of insurable insecurities and their consuming subjects. As the quotes above indicate, markets for weather-related catastrophe coverage are made as the political economy of the risk industry articulates with economic anxieties about climate change, producing both the *objects* and *subjects* of the insurance transaction.

Besides fear, economic value is the other component motivating the transaction. As private insurance has long been inextricably bound up with modern capitalism's expansion and its production of surplus value, it is no surprise that insurance purchases and the total insured value exposed both rise alongside economic growth. And as values rise, so do the fears about their potential loss. Florida is a commonly cited example: due to massive population growth and a real estate market booming on the heels of cheap credit, one industry trade association estimates that 1992's Hurricane Andrew would have cost double its original $15 billion in insured losses if it had hit Florida a decade later (ABI 2004). So, as anxieties grow about climate change's impacts on fixed capital and assets, in some senses (re)insurers find themselves with an "inexhaustible market" for products.

Some companies publicly contribute to a discourse of climate emergency, despite underwriters' closed-door claims that global warming is essentially a non-issue in daily practice. For example, a former CEO of Swiss Re has called climate change "the number one risk in the world ahead of terrorism, demographic change and other global risk scenarios" (Coomber 2006). Around the same time, a Munich Re corporate publication called for radically recalibrating hurricane risk estimates in response to anthropogenic climate change. Its portrayal of a perilous world of hazards spiraling utterly out of control was conjured in one chapter's title, "Peak meteorological values and never-ending loss records" (2006: 17).[14]

While I by no means intend to dispute the extreme seriousness of climate change or the necessity of reducing carbon emissions, it seems to me that these sorts of invocations of immediate emergency – and the ways in which they are marshaled to make certain conclusions about insurance solutions self-evident – are problematic. In this chapter I have avoided making the facile claim that insurers and reinsurers intentionally exaggerate risks and adopt scare tactics about global warming out of profit-seeking. There are probably instances of this occurring, but, as a general explanation, I believe such a claim misses the point (and in any case, it generates evidentiary requirements that would be next to impossible to fulfill). In fact, it is not necessary to suppose the existence of any sort of industry collusion or exaggeration. Rather, the risk industry – through its definitions of risk and the institutions for its management – reproduces the conditions for its own existence through its everyday operations. This is true in both a discursive and material sense. That is, the industry identifies new risks about which investors and/or the consuming public should be alarmed, and also reinforces the structures of compensation and the organization of property that make particular populations relatively more vulnerable – and other populations more able to pay for security. Such an understanding of risk production makes it possible to both believe in the dire threats posed by global warming and maintain a critical perspective on how these risks are measured, priced, packaged, and distributed. For as Francois Ewald (1991) reminds us, no particular form of insurance is an inevitable response to a certain set of problems, it is always just one of the possible ways of applying a technology of risk. This is to say, it is a question of politics.

Acknowledgements

Many thanks to the brokers, reinsurers, and modelers who so generously shared their time and experiences with me. This material is based upon work supported by the National Science Foundation under Grant No. 0928711.

Notes

1   There is a large and constantly mushrooming subdiscipline attempting to quantify the economic impacts of climate change at various levels of uncertainty and using a range of future scenarios for CO_2 emissions. See for example, "The Stern Review on the Economics of Climate Change," commissioned by the Treasury of the UK (Stern 2006).
2   The research on which this chapter is based was carried out between 2006–2009 in the U.S., U.K., Germany, and Switzerland. I employed a diverse set of methods including non-participant and participant observation at over a dozen academic and industry conferences; in-depth interviews with nearly 50 climatologists, meteorologists,

catastrophe modelers, reinsurers, brokers, insurers, and catastrophe bond fund investors; and ongoing review of industry literature including corporate reports, white papers, newsletters, press releases, and regulatory transcripts.

3   The world's major reinsurers underwrite more than a trillion dollars in coverage for thousands of individual insurance companies across a multitude of policy types, regions, and scales. Insurers seek reinsurance in order to expand their own underwriting capacity and prevent massive claims payments from sending them into bankruptcy. A large part of the risks that insurers attempt to reinsure fall in the property-casualty (P/C) and business interruption sectors, which may also be the most vulnerable to climate change. Hereafter this chapter adopts the shorthand of "(re)insurance" to denote both the reinsurance and insurance industries. There are quite significant differences between the two industries; however, climate change confronts both with similar epistemological, evidentiary, and financial challenges.

4   Including related lines of insurance such as marine (i.e. shipping), aviation, and energy.

5   There are faint rumblings of such recognition emerging with regard to places subject to the most visible (and most easily visualized) risks of sea level rise and flood. See Stycos 2009, Nicholls *et al.* 2007.

6   Since "climate" is the average of weather patterns over many years, "climate change" describes a systematic and directional change in these same weather patterns over a period of years (of which "global warming" is only one example). It is therefore logically and scientifically false to claim that any one event was specifically "caused" by global warming. We can only hypothesize about the mechanisms through which global warming may have increased the *average* likelihood or severity of these events.

7   Lopez (1976) considers the development of marine insurance and marine law as one of the most significant drivers of the "commercial revolution" culminating in the fourteenth century, alongside other "inventions" such as credit, navigational aids, and the improvement of shipbuilding.

8   Although there were efforts to establish rudimentary building codes and fire prevention practices after the 1666 fire (Pearson 2004).

9   Claimants and insurers clashed over what portion of damages had been caused by the earthquake (in which case they were excluded) versus the conflagration that followed. After a great deal of confusion, most insurers adopted a common policy of paying all claims for buildings consumed by the fire unless there was clear evidence that the property had been completely destroyed by the initial quake. Most reinsurers (the majority of which were foreign firms) eventually agreed to pay out and "follow the fortunes of their cedents" (Guatteri *et al.* 2005). Today, residential property insurance coverage in the U.S. generally includes losses due to fire and wind/snow storm damage, but excludes flooding (covered by the National Flood Insurance Program) and earthquakes (policies administered separately, often by states in conjunction with the private sector). In some especially hurricane-prone states such as Florida, wind coverage is sometimes also excluded or sold separately. Commercial property coverage typically includes flooding.

10 For tropical cyclones and earthquakes alone, reinsurance underwrites roughly US $300 billion of insured values (highly concentrated in the United States and Japan).

11 Although my interviewees identified AIR as making more efforts to flesh out the scientific assumptions and components of its models to its clients.

12 This leaves the system open to a self-amplifying cycle of risk taking, much in the same way that trading of collateralized debt obligations spiraled based on the analysis of a few third party ratings agencies in the mid-2000s.

13 Such attempts to "reverse engineer" the models are usually explicitly forbidden in the licensing contracts between modeling firms and their clients.

14 I asked several brokers about whether Munich Re's public position on the immediate threat of global warming was carried through to its business pricing practices. One said simply, "They are full of shit. But they talk a good game!"

References

Association of British Insurers. 2004. "A changing climate for insurance: A summary report for chief Executives and policy makers." London: ABI.

Baucom, I. 2005. *Specters of the Atlantic: Finance Capital, Slavery and the Philosophy of History*. Durham, NC: Duke University Press.

Beck, U. 1999. *World Risk Society*. London: Wiley-Blackwell.

—— 2009. *World at Risk*. London: Polity.

Berliner, B. 1982. *Limits of Insurability of Risks*. Englewood Cliffs, NJ: Prentice-Hall.

Braudel, F. 1992. *Civilization and Capitalism, 15th–18th Century: The Wheels of Commerce*. Translated by S. Reynolds. Berkeley: University of California Press.

Buffet, W. 1992. Chairman's letter to shareholders. Berkshire Hathaway Inc. http://www.berkshirehathaway.com/letters/1992.html. Last accessed July 2 2009.

Collier, Stephen. 2008. Enacting catastrophe: Preparedness, insurance, budgetary rationalization." *Economy and Society* 37(2): 224–250.

Coomber, J. 2006. Interview with the climate group, 7 March. Available online at http://www.theclimategroup.org/news_and_events/john_coomber/ (accessed 2 July 2009).

Defert, D. 1991. "'Popular life and insurance technology'" In: G. Burchell, C. Gordon and P. Miller, eds. *The Foucault Effect: Studies in Governmentality*, 211–233. Chicago: University of Chicago Press.

Demeritt, D. 2006. "Science studies, climate change and the prospects for constructivist critique." *Economy and Society* 35(3): 453–479.

Denenberg, H.S., 1964. *Risk and Insurance*, Englewood Cliffs, NJ: Prentice-Hall.

Downing, T., Olsthoorn, A. and Tol, R. eds. 1999. *Climate, Change and Risk*. London: Routledge.

Emanuel, K. 2005. "Increasing destructiveness of tropical cyclones over the past 30 years." *Nature* 436: 686–688.

Epstein, P., and Mills, E. 2005. *Climate Change Futures: Health, Ecological, and Economic Dimensions*. Cambridge, MA: Harvard Medical School Center for Health and the Global Environment.

Ericson, Richard V. and Doyle, Aaron 2004. "Catastrophe risk, insurance and terrorism," *Economy and Society* 33:2, 135–173.

Ewald, F. 1991. Insurance and risk. In C. Gordon, G. Burchill and P. Miller, eds. *The Foucault Effect: Studies in Governmentality*, 197–210. Chicago: University of Chicago Press.

Freeman, J. 1932. *Earthquake Damage and Earthquake Insurance*. New York: McGraw-Hill.

Grossi, P. and Kunreuther, H., 2005. *Catastrophe Modeling: A New Approach to Managing Risk*. New York: Springer.

Guatteri, M., Bertogg, M., and Castaldi, A. 2005. "A shake in insurance history: The 1906 San Francisco Earthquake," Swiss Re Natural Hazards Series. Zurich: Swiss Reinsurance.

Hacking, I. 1990. *The Taming of Chance*. Cambridge: Cambridge University Press.

Kopf, E. 1929. "Notes on the origin and development of reinsurance." *Proceedings of the Casualty Actuarial Society* 16, 22–91.

Lane, F. 1973. *Venice: A Maritime Republic*. Baltimore, MD: Johns Hopkins University Press.

Lopez, R. 1976. *The Commercial Revolution of the Middle Ages, 950–1350*. Cambridge: Cambridge University Press.

Luhmann, N. 2005 [1993]. *Risk: A Sociological Theory*. London: Aldine Transaction.

Luke, T. 2008. "The politics of true convenience or inconvenient truth: Struggles over how to sustain capitalism, democracy, and ecology in the 21st century." *Environment and Planning A* 40: 1811–1824.

Mallon M, 2006. "Bermuda: The sky's the limit." *Reinsurance*, 1 June.

Marx, K. 1993 (1857). *Grundrisse*. New York: Penguin.

Mills, E., Roth, R., and Lecomte, E. 2005. "Availability and affordability of insurance under climate change: A growing challenge for the U.S." CERES Report.

Munich Reinsurance 2006. "Hurricanes – More intense, more frequent, more expensive: Insurance in a time of changing risks." Munich: Münchener Rückversicherungs-Gesellschaft.

Murnane, R. 2004. "Climate research and reinsurance." *Bulletin of the American Meteorological Society* 85:55, 697–707.

National Association of Insurance Commissioners 2007a. *Property and Casualty Insurance Committee Public Hearing on the Regulation of Catastrophe Modelers.* Washington, D.C., 28 September.

National Association of Insurance Commissioners 2007b. *Property and Casualty Insurance Committee Public Hearing on the Use of Catastrophe Modeling by Ratings Agencies.* Houston, TX, 1 December.

National Oceanic and Atmospheric Agency 2005a. "Noteworthy records of the 2005 Atlantic hurricane season." *NOAA Magazine*. Nov. 29.

National Oceanic and Atmospheric Agency 2005b. "Monthly tropical weather summary." National Hurricane Center, Miama, FL. Dec 1.

Nicholls, R., Hanson, S., Herweijer, C., Patmore, N., Hallegatte, S., Corfee-Morlot, J., Chateau, J. and Muir-Wood, R. 2007. *Ranking Port Cities with High Exposure and Vulnerability to Climate Extremes: Exposure Estimates.* Paris: Organisation for Economic Cooperation and Development.

Pearson, R. 2004. *Insuring the Industrial Revolution: Fire Insurance in Great Britain, 1700–1850.* London: Ashgate.

Pezza, A. and Simmonds, I. 2005. "The first South Atlantic hurricane: Unprecedented blocking, low shear and climate change." *Geophysical Research Letters* 32, L15712.

Risk Management Solutions. 2007. *Testimony to NAIC Property and Casualty Insurance Committee Public Hearing on the Regulation of Catastrophe Modelers.* Washington, D.C., 28 September.

Schwierz, C., *et al.* 2010. "Modelling European winter wind storm losses in current and future climate." *Climatic Change* 101(3): 485–514.

Stern, N. 2006. "Review on the economics of climate change." London: HM Treasury, Government of the United Kingdom.

Stycos, S. 2009. Value of coastal property is overrated, global warming expert says. *The Jamestown Press*, June 18.

Swiss Re 1994. *Global Warming: Element of Risk.* Zurich: Swiss Reinsurance Company.

—— 2005. "Innovating to insure the uninsurable." *sigma* 4. Zurich: Swiss Reinsurance Company.

Webster, P., Holland, G., Curry, J. and Chang, H.R. 2005. "Changes in tropical cyclone number, duration, and intensity in a warming environment." *Science* 309: 1844–1846.

Willis Research Network. 2008. "Confronting the normality of extremes." www.willis researchnetwork.com. Last accessed July 2 2009.

10 Carbon colonialism? Offsets, greenhouse gas reductions, and sustainable development

A. G. Bumpus and D. M. Liverman

The agreement to establish the UN Framework Convention on Climate Change (UNFCCC) at the 1992 Rio Earth Summit saw the beginning of formal negotiations to reduce emissions of greenhouse gases in order to avoid the risks of dangerous anthropogenic climate change. In 1997, the Kyoto Protocol committed industrialized countries to binding greenhouse gas reductions based on their 1990 emissions, but eased the task through creating a market that would allow countries to trade emission reductions or to purchase emission reductions from projects in Eastern Europe (Joint Implementation – JI) and the developing world (Clean Development Mechanism – CDM) rather than make them domestically (Liverman 2009). Around the same time private sector companies and NGOs were creating a parallel voluntary market that would allow firms and individuals to compensate for their emissions by purchasing credits from emission reduction projects in the developing world. The emission reduction credits from the CDM and the voluntary market became known as "carbon offsets." The first official carbon offset was a voluntary agreement between an American Electric Utility and a forestry project in Guatemala in 1989 (Bayon *et al.* 2007). Because deforestation is a major source of greenhouse gas emissions and new forests absorb carbon dioxide, reforestation was covered under the CDM. However, forest protection could not generate credits under Kyoto, and it was not until the UNFCCC negotiations in Bali in 2007 that proposals were made to allow credits for Reducing Emissions from Deforestation and forest Degradation (REDD) within the international climate regime (Neeff and Ascui 2009).

Climate change can be seen as a threat to capital accumulation from climate impacts and expensive mitigation. However the carbon market creates new opportunities for profit in the development and marketing of carbon offsets (Bakker 2005; Buck 2007; Bumpus and Liverman 2008). The business rationale for reducing emissions includes internalizing external costs, seeking competitive advantage via innovation, and as a response to the concerns of environmental groups, consumers, and investors (Newell and Paterson 1998; Stern *et al.* 2006). Spatially differentiated emission abatement costs mean that Kyoto's suite of flexible mechanisms allows for cost-effective final allocation of climate change mitigation (Barrett 1998) that will minimize and harmonize "marginal abatement costs across space through the use of market-based instruments" (Copeland and Taylor 2005). Carbon reductions under the Kyoto Protocol can be seen as a material and discursive response to scientific and public pressure for regulation on emissions, and the inclusion of carbon trading mechanisms, including offsets, represent an example of market environmentalism which assumes that the way to protect the environment is to price nature's services, assign property rights, and trade these services within a global market (Liverman 2004).

The CDM, the voluntary carbon offset market, and REDD are innovations in environmental governance that are intellectually fascinating and highly controversial. The market has grown dramatically over the last decade, dominated by carbon offsets generated by the CDM and with a value of almost $10 billion (Figure 10.1). Critical perspectives from political ecology provide rich and revealing opportunities to analyze carbon offsets as a new commodity that links north and south through a complex set of technologies, institutions and discourses. This chapter explores some of these perspectives drawing on our own work and those of colleagues who have used political economy, governmentality, and science studies to think about climate governance and carbon offsets.

Carbon offsets have emerged as a new strategy to manage greenhouse gases (GHGs) and promote sustainable development. By compensating for emissions in one area by reductions in another, governments, companies, and individuals are using offsets to address their impact on climate change, claim "carbon neutrality" and invest in projects in the developing world. The fundamental rationale conveyed by offset advocates is that paying for greenhouse reductions elsewhere is easier, cheaper, and faster than domestic reductions, providing greater benefits to the atmosphere as well as to sustainable development, especially when offsets involve

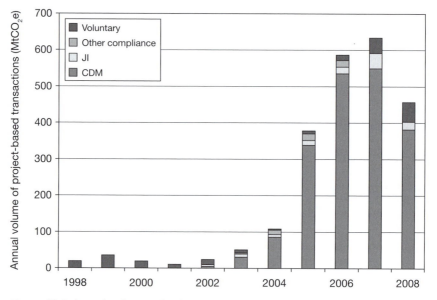

Figure 10.1 Annual volumes of carbon transactions associated with the CDM, the voluntary offset market and Joint Implementation from 1998 to 2008. Source: (Capoor and Ambrosi 2009).

projects in the developing world. However, the dubious effectiveness of some offsets in reducing emissions, coupled with controversy surrounding their local impacts and issues concerning unequal terms of trade, have led to a rush of criticism from NGOs, activists and academics (Bachram 2004; Lohmann 2005; Smith 2007; Wara 2007). Offset advocates include business, multilateral and market environmentalist actors supporting offsets as part of broader neoliberal approaches to environmental governance and climate change (Bayon *et al.* 2007; Bumpus and Liverman 2008; Newell 2008; Capoor and Ambrosi 2009).

Offsets are complex in their nature, involve multiple stakeholders at multiple scales, and include a wide variety of project types and governance forms. As a result, and in part because carbon offsets are created by various individuals within companies and communities for reasons that range from pure profit motives and leadership aspirations to care for the planet and elimination of poverty, there are variable definitions of the technical and sustainable development components of projects, and even of the overall definition of a carbon offset.

The creation of a carbon offset

A carbon offset is created when a project is approved that reduces or sequesters greenhouse gases and when those reductions are converted into a measurable and marketable commodity. Many different sorts of projects and technologies can be used to create carbon offsets including renewable energy, energy efficiency, industrial gas capture, methane capture from waste, forestry, soil management, and switching from coal to less carbon intensive fuels (Table 10.1). The politics of technologies mean that as yet, nuclear power and carbon capture and storage are not seen as acceptable offset projects. Although offset technologies are often lumped together by both advocates and critics our research suggests that the governance, material effectiveness and sustainable development benefits of offsets vary considerably between different types of technology (Bumpus 2009; Lovell and Liverman 2010). For example, some offsets, such as HFC destruction, allow carbon accounting principles to be easily applied and provide a better assertion of carbon reductions, but have limited development benefits, whilst others, such as decentralized improved cookstoves, are harder to prove emissions reductions, but may have much wider development and livelihood benefits.

Offsets rely on "baseline-and-credit" trading systems that create "assets" in the form of carbon credits which are measured in tons of carbon dioxide equivalent (tCO_2e). Reductions in different types of greenhouse gases (e.g. carbon dioxide, methane) are made equivalent through a conversion based on the global warming potential of different gases. These credits are supposed to represent carbon reductions compared to what would have happened without the project, starting from a baseline (Yamin 2005). Thus, investment in energy efficiency could reduce emissions from a power station compared to what might have happened. This is the fundamental notion that offsets should demonstrate "environmental additionality" where the offset has contributed to a net reduction in atmospheric CO_2, which is measurably different to a business as usual trajectory of emissions.

Table 10.1 Examples of offset project types, gases reduced and their relation to additionality, methodologies, and ability to be included in markets

Offset type	Project type	Gas reduced and technology	Financial additionality	Example of technology baseline methodologies and monitoring	Carbon saving / sustainable development implications
Industrial/ waste gas destruction	Hydrofluoro-carbon reduction (HFC)	HFC23, incinerators and scrubbers at plant-level	No environmental law or economic incentive to reduce HFC23 therefore proof of additionality easy	Created from existing factory emissions and after technology implementation, flow meters in emissions stacks; on-site monitoring in control rooms	Easily passed in CDM process; extremely profitable because of high GWP; technology for reductions and monitoring clear, and additionality easy to determine, but low inherent sustainable development benefits, although some projects are exposed to a 'sustainability tax' from host governments
	Landfill capture	Methane (CH4), capturing emissions from open landfill sites (flaring or using methane for energy production)	Implementing flaring or energy technology is not economically viable without carbon finance. Problem if energy provision is profitable and carbon finance difference is marginal	Created from methane emissions from landfill before and after technology implementation, monitored and recorded at site through flow meters. Some problems with methane leaking, lower emissions reductions than anticipated	Initially clear technology and reductions, however technically difficult problems with technology implementation and actual reductions. Some studies find landfill capture to have high development benefits (Olsen and Fenhann 2008), whilst others note little 'development dividend' (Cosbey et al. 2005).
Fossil fuel substitution/ reduced use	On-grid wind farm On-grid Biomass energy	Substitution of fossil fuel electricity generation by wind farms; Biomass burners. Replaces grid-based emissions	Project should not be economically attractive without carbon finance. More difficult: renewable energy can be	Baselines methodology justified on prevailing technologies and projected investment in the national grid. Monitoring created through recording amount	Technology well understood and can be incorporated into both compliance and voluntary carbon markets relatively easily. Renewables (especially wind and hydro) have been shown to have

		(national emissions factor) or decentralized emissions	profitable depending on local context.	of energy created and fed into the grid	high development benefits, with principal benefits accruing as employment, welfare, growth and access to energy
	Energy efficiency	Reduction in use of fossil fuel based electricity through energy efficiency	Additionality can be difficult: Energy efficiency can be economically advantageous without carbon finance.	Baselines determined on before and after technology implementation. Difficult if energy efficiency is decentralized or based on behavioral factors.	More difficult to include due to problems with decentralized energy efficiency measures (such as people not using efficient light bulbs).
Biological Sequestration	Afforestation and Reforestation	Increase in carbon sinks due to increasing sequestration through planting/restoring forests	Has to show that forests were not regenerating naturally, that it was not possible to (re)plant locally without carbon finance	Baselines difficult because of complexity in understanding above/below ground biomass in local contexts, and extrapolating this methodologically to wider areas. High transaction costs in monitoring forestry	Difficult methodologies for understanding amount of carbon sequestered, understanding leakage effects from forests and providing adequate cost-effective monitoring mean entering compliance markets has been difficult to include some of the poorest communities (Smith and Scherr 2003) in poorer countries (e.g. Sub-Saharan Africa) in carbon finance
	Reduced Emissions from Deforestation and Degradation (REDD)	Reduced deforestation and degradation	Has to show that forest was at risk of deforestation or degradation without carbon finance support	National baselines of deforestation used for understanding net reduction in deforestation or more individual project-basis for specific areas.	Difficult methodologies for understanding national baselines vs. local effects, especially with relation to opportunity cost and compensation for local forest dwellers and users. But large potential to assist in biodiversity conservation and local human development (Ebeling and Yasué 2008)

Offset projects must also show that they would not have happened without carbon finance. This is the requirement of "financial additionality" which asserts that the project would not have been financially viable without carbon finance or that is has overcome barriers to implementation as a direct result of carbon finance input (Greiner and Michaelowa 2003).

Both environmental and financial additionality are socially constructed through technical definition and argument, and critics claim that both the CDM rules and voluntary offset projects have allowed credits from projects that would have happened anyway (Bumpus 2009).

Despite these problems, offsets must rely on an accurate determination of whether emissions *are* materially different because of the impact of carbon finance. This is created through techniques and rules that require justification of financial additionality, an analysis of the existing and potential future baseline scenarios and by ex-post analysis of emissions post-project implementation through the monitoring of the project.

These complex processes of creating and legitimating offsets are clear examples of governmentalities of the new carbon economy (Oels 2005; Bäckstrand and Lövbrand 2006) and have produced a new form of expertise and consultancy in the development of carbon projects.

In order to create the ton of carbon dioxide equivalent (tCO_2e) that is to be traded as a carbon reduction in carbon markets, or sold to consumers or companies for their "carbon neutrality," carbon offset developers use a series of documents and tests to establish, and justify to potential investors, buyers and carbon standards, that the carbon reductions are verifiable, true and permanent. Documents such as the Project Design Document (PDD) specify the project, its methodologies and monitoring procedures. These documents are checked by third party organizations, and submitted to standards bodies or, in the case of the CDM, the CDM Executive Board (CDM EB), for validation, registration and ultimately issuance of credits. This process aims to ensure that a ton of carbon dioxide equivalent on paper is the same as a ton reduced from the atmosphere (see Figure 10.2). This is important in providing stability and worthiness to the credit as a commodity, but, moreover, that the papers traded represent a real additional benefit to the atmosphere.

Offsets rely on these processes in order to measure and assert the carbon reduction, so that governing bodies, such as the formal regulatory CDM EB, or self-regulatory voluntary standards boards such as the Gold Standard or the Voluntary Carbon Standard, can issue a commensurate carbon credit. The fundamental rationale for implementing most carbon offset projects is the ability to sell or use carbon credits for compensating for emissions in one area over another. If the credit is not able to be commensurate with other carbon emissions (i.e. to create an effective balance in CO_2 emissions and reduction), and ultimately with money (Castree 2003), the underlying rationale for offsetting under market environmentalism is funda-mentally eroded. These processes, which have rules specific to different projects and technologies (Gillenwater *et al.* 2007), are crucial to "hem-in" the carbon in order to assert a ton of carbon reduced on paper equals a ton reduced in the atmosphere and therefore is commensurable and able to be traded in a market (see Figure 10.3).

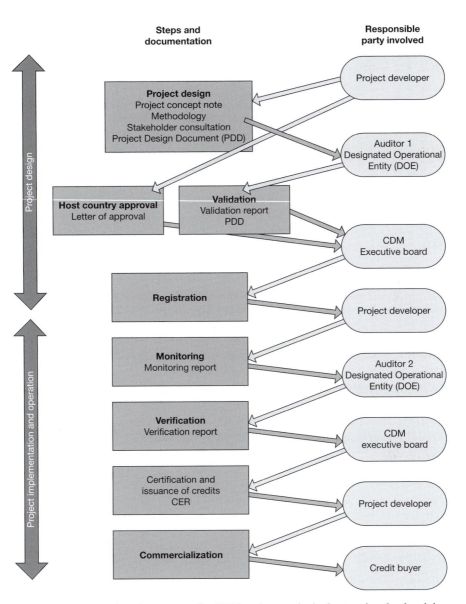

Figure 10.2 The registration process for CDM projects, principal actors involved and the role of documents in showing the project design, validation of the project to a mechanism, such as the CDM, conveying information on monitoring, verification of information, and creation of new carbon credits. Voluntary offset projects that are registered to other standards, such as the Gold Standard or the Voluntary Carbon Standard, go through similar processes of checks and measures. Source: Kollmuss *et al.* (2008).

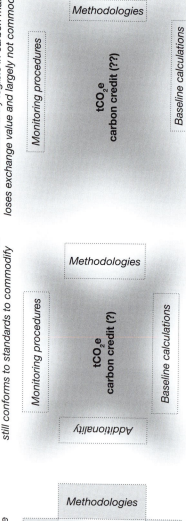

A:
– Strong determination of additionality, methodologies, baselines
– Accurate and measured monitoring data (e.g. flow meters in smoke stack)
– Credit highly legible in carbon markets, higher exchange value

B:
– Components to determine integrity of credit are weaker and less interconnected
– Monitoring data is not measured (e.g. statistical samples of cook stoves)
– Credit less legible in carbon markets, but still conforms to standards to commodify

C:
– Components to determine integrity of credit are weak or not included (e.g. no additionality)
– Monitoring data guessed and not measured (e.g. assumed and not measured forest biomass)
– Credit becomes hardly legible in carbon markets, loses exchange value and largely not commodified

Increase in material complexity to reduce carbon, commodity and make legible in carbon markets; reduction in carbon market exchange value

Figure 10.3 A diagram illustrating the "hemming-in" of a ton of carbon dioxide equivalent (tCO$_2$e) in order to help guarantee its carbon reduction effectiveness. These practices determine the material reduction of a ton of carbon from the atmosphere. Projects that have a material basis that allows these components to be easily defined create a more certain commodity (*A*). As projects become more complex and technologies less well-understood, accurate accounts of these four components can become weaker (moving from *A* to *C*). The commodification of the ton of carbon dioxide equivalent becomes more difficult and finally the integrity of the carbon credit as a saleable commodity on the market becomes difficult to assure (*C*). Source: Bumpus (2009)

Carbon offsets and sustainable development

When the Clean Development Mechanism was included in the climate regime in 1997 it was presented as a way for the south to benefit from northern obligations to reduce emissions through projects that would help decarbonize economies and provide side benefits for people and ecosystems in the developing world. This promise was governed through the an approval process for CDM projects where host country "designated national authorities" must sign off that projects meet their criteria for sustainable development. However, countries vary in the rigor and definitions of what constitutes sustainable development, and have an incentive to sign off on projects that bring carbon finance to their business or conservation sector.

Voluntary carbon offsets are not confined to the rules of the CDM system, leading some to note that the VCO markets are able to assist in development where the CDM cannot, especially given their more networked governance and links to development NGOs (Taiyab 2006; Lovell *et al.* 2009). For example, improved cookstoves, which can have multiple development and livelihood benefits, in addition to GHG reductions (Mann 2007), were excluded from the CDM until 2008, whilst they were included in VCO markets for some time. In addition, the reduced costs for implementing a project in the voluntary sector may have allowed smaller, community-based projects funded by companies specifically targeting local development benefits.

Environment-development implications of offsets vary according to the technical components of offsets. The environment-development contradictions in offsets have been well documented (Brown and Corbera 2003; Olsen 2007; Sutter and Parreño 2007; Boyd *et al.* 2009), but specific empirical work on this area is still thin (Bozmoski *et al.* 2008; Bumpus and Liverman 2008) and poses a double-edged sword for the climate and development community (Boyd *et al.* 2009).

Political ecology provides an excellent framework for assessing the effects of carbon offset projects in the developing world because it allows for local agency to react to institutional rules and structures, engages the material nature of on-the-ground carbon reductions, and integrates multi-level and networked[1] environment-development interests at its core (Zimmerer and Bassett 2003; Bryant and Bailey 1997; Robbins 2004). Understanding the local social and environmental implications of multiscalar projects and policies has been an important contribution of political ecology (Blaikie and Brookfield 1987; Zimmerer and Bassett 2003; Peet and Watts 2004). Offsets can also therefore be approached through an analysis of the relationship between transnational (carbon) capital and its effects in specific communities in the global South.

A number of recent studies, as well as several dissertations currently underway, have visited carbon offset projects in the field to ask questions about the existence and distribution of development benefits, the amount and legitimacy of the carbon reductions, and local perceptions of the projects (Brown *et al.* 2000; Corbera 2005; Bumpus 2009).

Bebbington and Batterbury (2001) identify livelihood as one of four key components, alongside place, scale, and network, in understanding development implications in political ecology. In addition, political ecology seeks to explain how local-level cultural and ecological communities form part of (and are influenced by) a much wider set of political and economic structures that often have national and global linkages (Neumann 2009). Carbon offsets can therefore be theorized through an understanding of the frontier of local livelihoods and globalization (de Haan and Zoomers 2005): as they expand carbon markets into new locations in the South, Northern offset companies often (cl)aim to assist local communities in which they operate, provide co-benefits, or actively engage local people as governors of the carbon reductions themselves. Some claim that there can be a "social carbon" that better integrates local needs and broader sustainable development, and that carbon finance is well placed to engender positive change (Schlup 2005; Reis 2009).

Others, however, note that carbon offsets are controversial in their local impacts. "Carbon colonialism," increased local inequity, and restrictions of access to resources crucial to some of the poorest local people, such as landfill sites and community land, have all been accusations leveled at different forms of carbon offsets in the developing world (Lohmann 2000; Bachram 2004; Smith 2007). An important component that development theory brings to analysis of carbon offsets is the importance of process in understanding offsets' local effects (rather than just tick boxes) and the creation of uneven development through carbon finance (Kiely 2007). Following this, unequal power relations among stakeholders at multiple scales enable resource-strong stakeholders to define the terms of the carbon trade and marginalization of others from potential benefits.

A critical political ecology analysis that reflects the ways in which carbon finance is resisted or "reworked" into more positive local impacts is an important practical and theoretical endeavor. An analysis of "external influences" (Bridge 2002), such as carbon finance flows and structures, policies and governmentalities, in combination with an actor-oriented livelihoods analysis, allows an exploration of the "carbon-development interface" in offsets. Carbon offsets must be conceived as relational: only by analyzing the relationships that exist between the carbon emitter in the North and the carbon reducer in the South can the environment-development implications for offsets are understood.

Environmental governance and a political economy of carbon offsets

Offsets represent a form of governance of the environment that is "multi-sited, marketised and increasingly transnational" (Newell 2008: 528). State and non-state actors all play important parts in the negotiation, creation and daily governance of offsets. Environmental governance associated with offsets is therefore conceived as occurring through a multitude of actors at multiple levels (Swyngedouw 2000; Betsill and Bulkeley 2004; Bulkeley 2005) from international companies, to UN processes and intergovernmental agencies, through to multinational project verifiers, non-governmental agencies, and local organizations, households and individuals associated with projects. Following Okereke *et al.* (2009) we propose that insights into the governance of offsets can be gained from both neo-Gramscian political economy and Foucauldian inspired ideas about governmentality, with the former focusing attention on issues of power, the state and the privatization of governance and the latter on the processes of governing through rationalities and technologies that include scientific research, audit, monitoring, regulation and the design of markets (Bäckstrand and Lövbrand 2007; Pattberg and Stripple 2008; Lovell and Liverman 2010).

The political economy of how offsets are structured – from international mechanisms, to national policies and local dynamics of implementation – plays a significant role in who defines offsets, how they are enacted and their associated distributions of climate and development benefits (Brown and Corbera 2003; Corbera *et al.* 2009). A political ecology of offsets therefore engages the structures that underpin their existence and daily functioning, and is concerned with under-standing "the broad logics of production and the distributive consequences of these logics" (Castree 2002: 361). This approach holds in tension offsets' international components with their local implications and their linkages that span multiple scales, sites and networks.

By enlisting the help of the developing world, international offsets not only provided a spatial fix for capital entities that were mandated to make emissions reductions, but also opened new channels of finance that allowed capital to create cheap carbon credits in the South and sell them into Northern markets where emissions reduction activities were more expensive. The use of this spatial fix to find cheap emissions reductions parallels other ways that capital avoids economic crises under neoliberalism and enlists the developing world in the pursuit of further accumulation, as locally specific nature is incorporated as new revenue streams (Jessop 1998; Katz 1998; Harvey 2005). Offsets represent new forms of "green capitalism," which has the potential to leverage the vast resources and innovation of the private sector to invest in clean technologies. A thoughtful account of the emergence of offsets, their governance and multiscalar implications, however, must understand the relative political economic dimensions, and the variable use of power, by different actors with different interests, for different outcomes. This can lead to unwanted effects, such as the initial flooding of the carbon markets with cheap credits from hydrofluorocarbon projects, difficulties in financing additional renewable energy projects, and the possibility of creating profitable, but atmos-pherically ineffective, carbon reduction credits (Wara 2007).

Analyses of the current carbon offset markets show an unequal spatial distribution of projects and a bias towards certain technologies. As of mid-2009 the CDM had a total of 1,732 projects registered, and it was expected to generate over 1.6 billion tons of certified emissions reductions by 2012.[2]

The majority of carbon credits issued under the CDM have been for the reduction of high-global-warming potential gases such as HFCs, PFCs and N_2O in countries such as China, Korea and India (Figure 10.4). Although there are more and more projects associated with renewables – perhaps of greater benefit to local com-munities – these are smaller scale. The CDM is a project-based system, meaning that emissions credits are generated and issued on a project-by-project basis, and not as a result of more general policy, sectoral or programmatic approaches to managing GHGs. One of the strongest criticisms of the CDM is that is has created perverse incentives for developing country governments to curb domestic policies that promote energy efficiency, renewables or other activities that reduce emissions for fear of pre-empting additionality of future CDM investments (Figueres 2006). This lack of transformation has serious environment-development implications considering the growth of emissions in developing countries and the consequent need for a shift in developing country investment patterns to mitigate climate change.

Although projects claim development benefits for host countries and communities, development in the CDM is often similar to development associated with business-as-usual capitalism: uneven, variable and ephemeral (Olsen and Fenhann 2008; Bumpus 2009). A more fundamental reason why the CDM is not

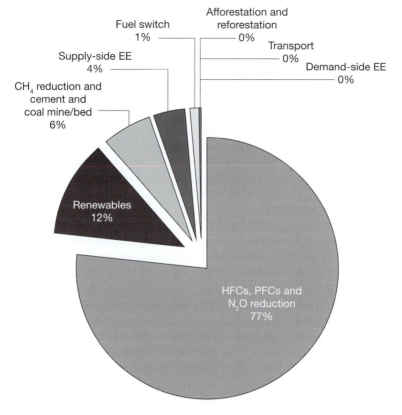

% CERs issued in each sector by July 2009

Figure 10.4 Amount of Certified Emissions Reductions (CERs) generated by the CDM for specific sectors. Although renewables make up the largest percentage of projects, carbon reductions have principally come from industrial gas destruction (B) leading critics to note that the CDM has not contributed to wider transformation of energy infrastructure and "clean" development in the global South. Source: UNEP Risoe CDM project pipeline (July, 2009).

"clean," as Newell *et al.* (2009) note, is because, despite the CDM's rapid growth, it still only represents a fraction of the total capital flows to the developing world focused on fossil fuel energy. In this way "clean" development from the mechanism is vastly overshadowed by business-as-usual "dirty" development, when, for example, less that 50 percent of the World Bank's US$1.8bn energy portfolio contains climate change considerations (WRI 2008).

The voluntary market has largely started to follow the CDM because of the linking of the markets through the possibility of selling credits pending registration as a CDM project. The top three project types by volume in the VCO markets are hydropower (32 percent), landfill gas (17 percent) and wind (15 percent), with the remaining one-third of transactions coming from fourteen other technologies. Thus, both in terms of innovative technology and geographical distribution, principal VCO credits in 2008 largely mirror CDM project types, and only 1 percent are sourced from Africa, compared to Asia, the Middle East and the US (at 45 percent, 15 percent and 28 percent), leading us to more fundamentally question the role of the *both* the CDM and VCO offset markets in materially achieving the rhetorical sustainable development in South they promote (Bozmoski *et al.* 2008; Bumpus 2009).

Commodity and value chain analysis (Gereffi and Korzeniewicz 1994; Hartwick 1998; Hughes 2001) offers another interesting possibility for understanding who benefits in the production and consumption of a carbon offset, by drawing attention to the multiple steps from the local project to the international carbon market and from a variety of brokers to the consumer of the offset.

Governmentalities and disciplining the material nature of carbon reductions

Political ecology seeks to understand human processes and their relation to the environment through an analysis of society–nature interactions. The "nature" of carbon is therefore an important theoretical, and practical, component of under-standing carbon offsets as a political ecology problem. We therefore need to attend to the "matter of nature" (FitzSimmons 1989; Bakker and Bridge 2006) in order to understand "what" carbon reductions we are dealing with, how it is reduced, and with what socio-natural consequences. A governmentalities approach under-stands material elements as simultaneously discursive, allowing room to explore a post-structural political ecology in offsets.

As we have noted, carbon reductions in offsets are governed over space by multiple actors, at multiple levels. Most of the finance for carbon offsets is chan-neled from the North, whilst most offset projects take place in the South (Capoor and Ambrosi 2009). Verifying the carbon reductions in these projects are inter-national consultants, verification companies (known as Designated Operational Entities, DOEs, in the CDM) and carbon experts whose knowledge is used to justify the creation of carbon reductions. Such knowledge is also used to turn reductions into credits and commodify them through the complex technical procedures of defining a ton of carbon dioxide equivalent, as described above. The CDM

Executive Board (CDM EB), which is administered by the United Nations Framework Convention on Climate Change (UNFCCC), tightly governs emissions reductions under the CDM in order to guarantee the reductions it has created, and to provide a legal basis for substituting one ton emitted with one ton reduced. The legal, and inter-governmentally negotiated-basis of the CDM is perceived to be the most rigorous carbon standard, with strict rules to maintain its "environmental integrity."

As a result of the voluntary market's lack of formal regulation, serious critique has also been leveled at VCO projects and companies for unjustified profiteering and inaccurate carbon accounting. As a result, the VCO industry has begun to self-regulate through the creation and promotion of carbon standards, such as the Voluntary Carbon Standard (VCS), and governments have stood in to provide optional codes of conduct (ENDS 2009). To counteract criticism, proponents of VCOs have maintained the need for flexibility in regulation in order to maintain innovation, avoid overly bureaucratic processes (such as in the CDM) and promote the need for easier, but legitimate, ways to combat climate change. Despite this, public skepticism, and the proposed need for standards and government oversight to ensure VCOs' environmental value rather than just profiteering (Gillenwater *et al.* 2007) has meant that the VCO market has started to converge with the CDM in forms of standardization. This has meant that the environmental worth of VCO credits is taken more seriously, even to the point where VCO credits may become compliant with emerging systems, such as in the US, but it necessarily means that the innovation for development in poor communities, seen as a benefit of the market (Taiyab 2006), is lost in the prioritization of carbon reductions over local development (Olsen 2007).

A governmentalities approach therefore understands carbon reductions as a product of discursive structuring of reductions, through documents and procedures stipulated by some actors over others, and ultimately the disciplining of carbon reductions and local social relations in offset projects by the exercise of power through the production of knowledge. Documents such as the PDD, its stipulations for national and local consultations, and specific methodologies that require carbon accounting (see Figure 10.1), discipline and control local processes from "far-flung" places in what can be seen as a broad "global environmental governmentality" (Lövbrand *et al.* 2009). Such discursive acts are inextricably linked to material effects, thus discursive approaches, and the use of documents to confine and "hem-in" carbon (Bumpus 2009) 104), are inherently related to the extent to which actual reductions of carbon are created and different material options for sustainability are discursively envisioned and materially created (Redclift 2009).

A second way in which post-structural political ecology aspects of offsets are enacted is through environment and development imaginaries (Peet and Watts 2004) that are used to convey images and stories about offsets that are consumed in Northern markets (Bryant and Goodman 2004; Lovell *et al.* 2009). High local sustainable development projects, such as those under the Gold Standard or those promoted with large community development benefits, are described by carbon offset retailers to consumers in order to tap into the emerging "moral economies" of "fair trade" or ethical consumer markets. High value development carbon credits may be harder to source, and more expensive, however this leads certain buyers to be attracted to their "boutique" status, and buy them for public relations and marketing reasons *beyond* just carbon reductions per se (Taiyab 2006). The imaginaries of local stories of offsets are clearly reproduced on retailer's websites and communication(Lovell *et al.* 2009), however livelihoods and the local development reality of projects is less clearly verified and asserted.

Carbon offsets provide an interesting, and understudied, issue through which regulation theory to manage a newly created environmental commodity (Bumpus 2009), narratives on consumer choices for certain credits over others (Lovell *et al.* 2009), and the material linkages between the North and South through commodity chains (cf. Goodman 2004), can be brought to bear within political ecology to better understand the mechanisms, and multi-scalar implications, of what could become the world's largest commodity market.

The political ecology of reduced emissions from deforestation and forest degradation (REDD)

REDD has become a prominent issue in the climate change and carbon offset debate. Carbon credits for avoiding deforestation were largely left out of the formal carbon mechanisms until 2007, when the Bali UNFCCC negotiations (re)opened up possibilities to discuss the inclusion of reducing emissions through the conservation of forests, either through transnational funds or through international carbon offsetting and trading mechanisms.

Emissions from deforestation and degradation constitute approximately 17 percent of total global annual emissions (Global Carbon Project 2007), and groups such as the Coalition of Rainforest Nations see channels of carbon finance for offsetting as a mechanism for paying for a reduction in deforestation and degradation (Miles and Kapos 2008). Reducing deforestation through the use of carbon finance has been seen by some as the key to a North–South bargain on climate change (Reuters 2009), given the agreement on most sides that reducing deforestation is a laudable goal in and of itself.

Proposals for REDD policies suggest that carbon credits could be issued for the conservation of forests that would have ordinarily (i.e. under business-as-usual scenarios) been deforested or degraded, although the particulars of how this is to be funded are still controversial (Global Carbon Project 2007). Various alternatives have been proposed for the design of the REDD market, including making it part of the CDM, creating a separate market, financing through direct funds from the North (rather than the carbon market), and governing it through national rather than project-based accounting (Ebeling and Yasué 2008; Streck *et al.* 2008; Neeff and Ascui 2009).

Theoretically, REDD promises the transfer of funds from Northern countries to the South in order to reward developing countries for reducing deforestation and improving conservation. Operationalizing REDD, however, is technically and ethically difficult. One key disagreement is over how to define a forest. The UN system allows plantations to be considered as forests, but NGOs such as the World Rainforest Movement (WRM) and the Global Forest Coalition (GFC) have claimed that this definition endangers indigenous peoples, peasants, small farmers, biodiversity, and even exacerbates climate change (WRM 2008). Others note that prioritizing carbon stocks over other non-commodified ecosystem services, such as biodiversity and watersheds, will endanger them by opening them up to what O'Connor (1998) calls the second contradiction of capitalism: carbon is monetarily valued, but associated cultural and biological processes are not, and therefore continue to be market "externalities" and are thus left open to unrestricted destruction (Wilson *et al.* 2006). Others note that these externalities can be brought into the mechanism. For example, the Climate, Community and Biodiversity Standards (CCBS), argue that REDD projects can "generate multiple benefits to community and biodiversity while reducing GHGs" (CCBA 2009). The inclusion of REDD into the climate regime is highly controversial, with debates on the effectiveness of understanding levels of carbon sequestration, national- or project-level accounting, the rights of indigenous peoples and forest communities and the incorporation of other ecosystem services into a carbon commodity.

Developing countries are also mixed in their support for REDD: whilst countries like Papua New Guinea support the idea because of possibilities for revenue generation from conservation, others like Brazil do not support tradable commodity generation from forest conservation because it allows for continued pollution by developed countries, since they can meet their commitments through offsets. They argue instead for donated funds from developed countries that pay for avoided deforestation. Concerns about property rights are also emerging. Does the carbon consumer in Europe "own" the carbon sequestered into the wood in Brazil? What if they would want to "call in" that ownership on the forest (i.e. deforest it) and balance the carbon credit with a different project? A problem with the perceived commensurability of carbon credits is that different credits reflect very different local, socio-environmental relations in specific places across the world, and that buying one type of credit does not necessarily have the same local effects as buying another. A key concern, therefore, is who defines ownership over carbon, over what scales and with what outcomes.

The interaction between human livelihoods and international political negotiations on trading mechanisms, and the variable definitions of forests and local implementation of conservation through REDD, make it a key issue for a political ecology analysis to understand the translocal impacts of REDD policy, the increasing mobilization, and differentiated knowledge politics of indigenous groups resisting *and* supporting REDD initiatives. Pilot projects are just beginning to be created for REDD, and therefore will provide participatory research case studies.

Political ecology's long concern with questions of forest governance suggests that it can offer important insights into the debates over REDD including the governmentality of forest definition, measurement, and certification, the struggles over property and indigenous rights, the causes of deforestation and most effective solutions, and the institutional roles of the World Bank, environmental NGOs, and other actors (Bäckstrand and Lövbrand 2006; Coomes *et al.* 2008; Turner and

Robbins 2008; Peluso 1995; Fairhead and Leach 1996; Klooster and Masera 2000; Robbins 2001; Klooster 2005).

Conclusions

Political ecology has much to offer the analysis of the international climate regime including the political economies of responsibility for emissions, the distribution of vulnerability to climate changes, the decisions to use market solutions, the agency of non-nation state actors, the governmentalities of climate science and monitoring, and the interactions of climate policy and development. In this chapter we have focused on a particular element of the international response to climate change – that of carbon offsets – describing their production, consumption, and governance. We have shown how, due to their complex technical and material nature, offsets provide challenges in converting them into commensurate commodities and rely on standards in order to manage public perception and ensure that an environmental benefit has actually occurred. Offsets rely on private market functions, but are controlled by strict rules of governance in order to achieve socially desired emissions reductions.

The different offset forms (CDM, VCOs and REDD) open up important specific questions for political ecology. The possible transformational effects and multilevel implications of CDM reform, the changing governance of standardisation in VCOs, and understandings of the multi-level environmental and knowledge politics in REDD, are just a few of the potential research agendas. A political ecology approach also creates space for analysis of more fundamental problems: for example, the extent to which a *carbon* market can realistically attend to certain forms of development, the limitations of market mechanisms in governing emissions reductions more generally, and the crucial questions of power relations in North–South (and South–South) environments and development outcomes as a result of offsetting and wider flows of capital for clean or dirty development.

Drawing on both Gramsci and Foucault we have provided an introduction to work that has explored the political economy and governance of carbon offsets as well as the new governmentalities that are governing the carbon economy. As market environmentalism has become the *de rigueur* way to manage the environment (Castree 2008), and carbon markets expand to incorporate more space, countries and communities, we must attend to understanding emissions reductions in offsets, their resultant broader ecological outcomes, and the socio-political effects that new channels of carbon capital create. A "global political ecology of carbon offsets," created through careful, comparative, in-depth, critical analysis of offset policies, projects and actors at multiple linked levels and networks, will help inform approaches to the issue and provide politically-nuanced, socio-ecologically-centred, interpretations of future offset options.

Acknowledgments

We wish to acknowledge the support of the UK Research Councils in the form of an ESRC/NERC fellowship for Adam Bumpus's doctoral research and funding from the Tyndall Center for Diana Liverman's research program. We are also grateful for inspiration from colleagues in ECI at Oxford University and the Tyndall consortium especially Heather Lovell and Harriet Bulkeley.

Notes

1   "Level," "scale" (and the politics thereof) and "network" are contentious and hotly debated theoretical topics in political ecology. This chapter is not the place to examine this issue in depth (for review on this issue, see Neumann 2009), however offsets are an interesting political ecology case study because of their inherently multi-scalar *and* horizontally, spatially linked and networked nature. The role of scale, networks and scalar politics is therefore a continuing and interesting avenue of enquiry in a political ecology of carbon offsets.

2   This is according to UNFCCC data, July 2009.

References

Bachram, H. (2004). "Climate fraud and carbon colonialism: The new trade in greenhouse gases." *Capitalism Nature Socialism* 15(4): 5–20.

Bäckstrand, K. and E. Lövbrand (2006). "Planting trees to mitigate climate change: Contested discourses of ecological modernization, green governmentality and civic environmentalism." *Global Environmental Politics* 6(1): 50–75.

—— (2007). Climate governance beyond 2012: Competing discourses of green governmentality, ecological modernization and civic environmentalism. In Mary E. Pettenger, ed., *The Social Construction of Climate Change: Power, Knowledge, Norms, Discourses.* Aldershot, Ashgate, 123–148.

Bakker, K. (2005). "Neoliberalizing nature? Market environmentalism in water supply in England and Wales." *Annals of the Association of American Geographers* 95(3): 542–565.

Bakker, K. and G. Bridge (2006). "Material worlds? Resource geographies and the 'matter of nature'." *Progress in Human Geography* 30(1): 5–27.

Barrett, S. (1998). "Political economy of the Kyoto protocol." *Oxford Review of Economic Policy* 14(4): 20–39.

Bayon, R., A. Hawn, and K. Hamilton (2007). *Voluntary Carbon Markets: An International Business Guide to What They Are and How They Work.* London, Earthscan.

Bebbington, A. and S. P. J. Batterbury (2001). "2001. Transnational livelihoods and landscapes: Political ecologies of globalization." *Ecumene* 8(4): 369–380.

Betsill, M. and H. Bulkeley (2004). "Transnational networks and global environmental governance: The cities for climate protection program." *International Studies Quarterly* 48(2): 471–493.

Blaikie, P. M. and H. C. Brookfield (1987). *Land Degradation and Society.* London, Methuen.

Boyd, E., N. Hultman, J. T. Roberts, E. Corbera, J. Cole, A. Bozmoski, J. Ebeling, R. Tippman, P. Mann, K. Brown, and D. M. Liverman (2009). "Reforming the CDM for sustainable development: Lessons learned and policy futures." *Environmental Science and Policy.*

Bozmoski, A., M. Lemos, and E. Boyd (2008). "Prosperous negligence: Governing the clean development mechanism for markets and development." *Environment: Science and Policy for Sustainable Development* 50(3): 18–30.

Bridge, G. (2002). "Grounding globalization: The prospects and perils of linking economic processes of globalization to environmental outcomes." *Economic Geography* 78(3): 361–386.

Brown, K. and E. Corbera (2003). "Exploring equity and sustainable development in the new carbon economy." *Climate Policy* 3 1469–3062: S41-S56-S41-S56.

Brown, S., M. Burnham, M., Delaney, M., Powell, R., Vaca, and A. Moreno (2000). "Issues and challenges for forest-based carbon-offset projects: A case study of the Noel Kempff Climate Action Project in Bolivia." *Mitigation and Adaptation Strategies for Global Change* 5(1): 99–121.

Bryant, R. L. and S. Bailey (1997). *Third World Political Ecology*, London, Routledge.

Bryant, R. L. and M. K. Goodman (2004). "Consuming narratives: The political ecology of 'alternative' consumption." *Transactions of the Institute of British Geographers* 29(3): 344–366.

Buck, D. (2007). "The ecological question: Can capitalism survive?" *Coming to Terms with Nature: Socialist Register*: 660–71.

Bulkeley, H. (2005). "Reconfiguring environmental governance: Towards a politics of scales and networks." *Political Geography* 24(8): 875–902.

Bumpus, A. (2009). "Carbon development: A political ecology analysis of carbon offset projects for local development and global climate benefit." School of Geography and Environment, Oxford University.

Bumpus, A. and D. Liverman (2008). "Accumulation by decarbonization and the governance of carbon offsets." *Economic Geography* 84(2): 127.

Capoor, K. and P. Ambrosi (2009). *State and Trends of the Carbon Market 2009.* Washington, DC, The World Bank.

Castree, N. (2002). "Environmental issues: from policy to political economy." *Progress in Human Geography* 26(3): 357–357.

—— (2003). "Commodifying what nature?" *Progress in Human Geography* 27(3): 273–297.

—— (2008). "Neoliberalising nature: the logics of deregulation and reregulation." *Environment and Planning A* 40(1): 131–152.

CCBA (2009). "Launch of forest carbon standards in new languages reaches key audiences." Available online at http://www.climate-standards.org/news/news_languages_apr 2009.html.

Coomes, O.T., F. Grimard, C. Potvin, and P. Sima (2008). "The fate of the tropical forest: Carbon or cattle?" *Ecological Economics*, 65(2) 207-212.

Copeland, B. R. and M. S. Taylor (2005). "Free trade and global warming: A trade theory view of the Kyoto protocol." *Journal of Environmental Economics and Management* 49(2): 205–234.

Corbera, E. (2005). "Interrogating development in carbon forestry activities: A case study from Mexico." School of Development Studies, University of East Anglia.

Corbera, E., M. Estrada, and K. Brown (2009). "How do regulated and voluntary carbon-offset schemes compare?" *Journal of Integrative Environmental Sciences* 6(1): 25–50.

Cosbey, A., J.E. Parry, J. Browne, Y.D. Babu, P. Bhandari, J. Drexhage, and D. Murphy (2005). *Realizing the development dividend: making the CDM work for developing countries.* Ottawa: IISD.

de Haan, L. and A. Zoomers (2005). "Exploring the frontier of livelihoods research." *Development and Change* 36(1): 27–47.

Ebeling, J. and M. Yasué (2008). "Generating carbon finance through avoided deforestation and its potential to create climatic, conservation and human development benefits." *Philosophical Transactions B* 363(1498): 1917.

ENDS (2009). "Carbon offset providers to reject government code." *Environmental Data Services (ENDS)* 409: 13–14.

Fairhead, J. and M. Leach (1996). *Misreading the African Landscape: Society and Ecology in a Forest-savanna Mosaic*, Cambridge, Cambridge University Press.

Figueres, C. (2006). "Sectoral CDM: Opening the CDM to the yet unrealized goal of sustainable development." *McGill International Journal of Sustainable Development Law and Policy* 2: 5–5.

FitzSimmons, M. (1989). "The matter of nature." *Antipode* 21(2): 106–120.

Gereffi, G. and M. Korzeniewicz (1994). *Commodity Chains and Global Capitalism*, New York, Greenwood Pub Group.

Gillenwater, M., D. Broekhoff, M., Trexler, J., Hyman, & R. Fowler (2007). Policing the voluntary carbon market. *Nature Reports Climate Change*. Available online at http://www.nature.com/climate/2007/0711/full/climate.2007.58.html.

Global Carbon Project (2007). "GCP – Carbon Budget." Available online at http://www.globalcarbonproject.org/carbonbudget/07/index.htm.

Goodman, M. K. (2004). "Reading fair trade: Political ecological imaginary and the moral economy of fair trade foods." *Political Geography* 23(7), 891–915.

Greiner, S. and A. Michaelowa (2003). "Defining investment additionality for CDM projects – practical approaches." *Energy Policy* 31(10): 1007–1015.

Hartwick, E. (1998). "Geographies of consumption: A commodity-chain approach." *Environment and Planning D* 16: 423–438.

Harvey, D. (2005). *A Brief History of Neoliberalism*. Oxford, Oxford University Press.

Hughes, A. (2001). "Global commodity networks, ethical trade and governmentality: Organizing business responsibility in the Kenyan cut flower industry." *Transactions of the Institute of British Geographers* 26(4): 390–406.

Jessop, B. (1998). "The rise of governance and the risks of failure: The case of economic development." *International Social Science Journal* 50(1): 29–45.

Katz, C. (1998). "Whose nature, whose culture? Private producations of space and 'preservation' of nature." In B. Braun, ed. *Remaking Nature at the Millenium*. London, Routledge, 46–63.

Kiely, R. (2007). *The New Political Economy of Development: Globalization, Imperialism, Hegemony*, Basingstoke, Palgrave Macmillan.

Klooster, D. (2005). "Environmental certification of forests: The evolution of environmental governance in a commodity network." *Journal of Rural Studies* 21(4): 403–417.

Klooster, D. and O. Masera (2000). "Community forest management in Mexico: Carbon mitigation and biodiversity conservation through rural development." *Global Environmental Change* 10(4): 259–272.

Kollmuss, A., H. Zink, and C. Polycarp 2008. *Making Sense of the Voluntary Carbon Market: A Comparison of Carbon Offset Standards*. WWF Germany.

Liverman, D. M. (2004). "Who governs, at what scale and at what price? Geography, environmental governance, and the commodification of nature." *Annals of the Association of American Geographers* 94(4): 734–738.

—— (2009). "Conventions of climate change: constructions of danger and the dispossession of the atmosphere." *Journal of Historical Geography* 35(2): 279–296.

Lohmann, L. (2000). "Shopping for carbon: A new plantation economy." From http://www.thecornerhouse.org.uk/item.shtml?x=52186.

—— (2005). "Making and marketing carbon dumps: Commodification, calculation and counterfactuals in climate change mitigation." *Science as Culutre* 14(3).

Lövbrand, E., J. Stripple and B. Wiman (2009). Earth System governmentality: Reflections on science in the Anthropocene. *Global Environmental Change*, 19(1): 7–13.

Lovell, H., H. Bulkeley and D. Liverman (2009). "Carbon offsetting: Sustaining consumption?" *Environment and Planning A* 41(10): 2357–2379.

Lovell, H. and D. Liverman (2010). "Understanding carbon offset technologies." *New Political Economy* 15(2): 255–273.

Mann, P. (2007). "Carbon finance for clean cooking – time to grasp the opportunity." *Boiling Point* 54: 1–2.

Miles, L. and V. Kapos (2008). "Reducing greenhouse gas emissions from deforestation and forest degradation: Global land-use implications." *Science* 320(5882): 1454.

Neeff, T. and F. Ascui (2009). "Lessons from carbon markets for designing an effective REDD architecture." *Climate Policy* 9(3): 306–315.

Neumann, R. P. (2009). "Political ecology: Theorizing scale." *Progress in Human Geography* 33(3): 398–398.

Newell, P. (2008). "The political economy of global environmental governance." *Review of International Studies* 34: 507–529.

Newell, P. and M. Paterson (1998). "A climate for business: Global warming, the state and capital." *Review of International Political Economy* 5(4): 679–703.

Newell, P., N. Jenner and L. Baker (2009). "Governing clean development: A framework for analysis." *Development Policy Review*, 27(6):717–739.

O'Connor, J. (1998). *Natural Causes: Essays in Ecological Marxism*, New York, London, Guilford Press.

Oels, A. (2005). "Rendering climate change governable: From biopower to advanced liberal government?" *Journal of Environmental Policy & Planning* 7(3): 185–207.

Okereke, C., H. Bulkeley and H. Schroeder (2009). "Conceptualizing climate governance beyond the international regime." *Global Environmental Politics* 9(1): 58–78.

Olsen, K. (2007). "The clean development mechanism's contribution to sustainable development: A review of the literature." *Climatic Change* 84(1): 59–73.

Olsen, K.H. and J. Fenhann (2008). "Sustainable development benefits of clean development mechanism projects: A new methodology for sustainability assessment based on text analysis of the project design documents submitted for validation." *Energy Policy* 36(8): 2773–2784.

Pattberg, P. and J. Stripple (2008). "Beyond the public and private divide: remapping transnational climate governance in the 21st century." *International Environmental Agreements: Politics, Law and Economics* 8(4): 367–388.

Peet, R. and M. Watts (2004). *Liberation Ecologies: Environment, Development, Social Movements*, London, Routledge.

Peluso, N. (1995). "Whose woods are these? Counter-mapping forest territories in Kalimantan, Indonesia." *Antipode* 27(4): 383–406.

Redclift, M. (2009). "The environment and carbon dependence: Landscapes of sustainability and materiality." *Current Sociology* 57(3): 369–369.

Reis, R. (2009). "Brazilian NGO creates innovative social carbon methodology." *Environmental Communication: A Journal of Nature and Culture* 3(2): 270–275.

Reuters (2009). "REDD presents best chance for new climate pact." Available online at http://communities.thomsonreuters.com/Carbon/364752.

Robbins, P. (2001). "Fixed categories in a portable landscape: The causes and consequences of land-cover categorization." *Environment and Planning A* 33(1): 161–180.

—— (2004). *Political Ecology: A Critical Introduction*, Oxford, Wiley-Blackwell.

Schlup, M. (2005). "The Gold Standard: Linking the CDM to development and poverty reduction." *Conference on Climate or Development?* Hamburg, Hamburg Institute of International Economics (HWWI), pp. 28–29.

Smith, J. and S.J. Scherr (2003). "Capturing the value of forest carbon for local livelihoods." *World Development*, 31(12): 2143–2160.

Smith, K. (2007). *The Carbon Neutral Myth: Offset Indulgences for your Climate Sins*, Amsterdam, Carbon Trade Watch, Transnational Institute.

Stern, N., S. Peters, V. Bakhshi, A. Bowen, C. Cameron, S. Catovsky, D. Crane, S. Cruickshank, S. Dietz and N. Edmonson (2006). *Stern Review: The Economics of Climate Change*, London, HM Treasury.

Streck, C. *et al.*, eds. (2008). *Forests, Climate Change and the Carbon Market: Risks and Emerging Opportunities*, London, Washington, Earthscan.

Sutter, C. and J. C. Parreño (2007). "Does the current Clean Development Mechanism (CDM) deliver its sustainable development claim? An analysis of officially registered CDM projects." *Climatic Change* 84(1): 75–90.

Swyngedouw, E. (2000). "Authoritarian governance, power, and the politics of rescaling." *Environment and Planning D: Society & Space* 18(1): 63–76.

Taiyab, N. (2006). *Exploring the Market for "Development Carbon" Through the Voluntary and Retail Markets*, London, International Institute for Environment and Development.

Turner, B. and P. Robbins (2008). "Land-change science and political ecology: Similarities, differences, and implications for sustainability science." *Annual Review of Environment and Resources* 33: 295–316.

UNFCCC (2009). United Nations Framework Convention on Climate Change. From http://cdm.unfccc.int/index.html (accessed July 21, 2009).

Wara, M. (2007). "Is the global carbon market working?" *Nature* 445(8): 595–596.

Wilson, K. A. *et al.* (2006). "Prioritizing global conservation efforts." *Nature* 440(7082): 337–340.

WRI (2008). Correcting the world's greatest market failure: Climate change and the multilateral development banks, World Resources Institute, Washington DC. Available online at http://www.wri.org/publication/correcting-the-worlds-greatest-market-failure (accessed July 24, 2009).

WRM (2008). "Groups unite to challenge the definition of forests under UNFCCC/REDD." From http://www.wrm.org.uy/actors/CCC/cop14/Goups_REDD.html.

Yamin, F. (2005). "The international rules on the Kyoto mechanisms." In F. Yamin, ed. *Climate Change and Carbon Markets: A Handbook of Emissions Reductions Mechanisms*, London, Earthscan, 71–74.

Zimmerer, K. S. and T. J. Bassett (2003). *Political Ecology: An Integrative Approach to Geography and Environment-Development Studies*, New York, London, Guilford Press.

Part IV

War, militarism, and insurgency: political ecology of security

11 The natures of the beast: on the new uses of the honeybee

Jake Kosek

The use of the honeybee

The current state of the honeybee is undeniably dismal, the consequences serious. Aside from honey and beeswax, over one-third of current global agriculture production depends on the honeybee for pollination (Cox-Foster and van Engelsdorp 2009). A considerable decline in honeybee populations began even before the latest reports of "colony collapse disorder." In 2006 the number of hives in the U.S. was approximately 2.4 million, less than half of what it was in 1950 (Cox-Foster and van Engelsdorp 2009). Global environmental changes have been devastating, whether the intensification of industrial agriculture, toxic pollution, climate change, loss of habitat, or the spread of disease and parasites. But the most recent trouble came in 2006 and 2007, when almost 40 percent of honeybees in the U.S. disappeared and millions of hives around the world were lost (van Engelsdorp, *et al.* 2009; Cox-Foster and van Engelsdorp 2009). The decline in honeybee populations was so dramatic it eclipsed all previous mass mortality in the bee world, making it the worst recorded crisis in the multi-millennial history of beekeeping. There is still no consensus about the cause of this devastation.

In response to the crisis, geneticists are combing through the newly mapped bee genome, insect pathologists are trying to isolate a viral culprit, toxicologists are tracing chemical residues, and bacterial entomologists are scouring the intestines of sick bees. Few researchers, however, are systematically situating the crisis, whatever its cause, within the broader historical, political, and economic relationships between bees and humans.[1] To ask only what has happened to the bee to cause this crisis is to miss the more fundamental question. How has the changing relationship between bees and humans brought the modern bee into existence in a way that has made it vulnerable to new threats?[2]

Answering this question demands that we pay attention both to the quotidian and co-constitutive histories of humans and bees and to current re-makings of the bee. Moreover, it requires an epidemiology of the crisis based on the understanding that society has not only influenced the making of the modern honeybee, but that human interests, fears, and desires have become part of its material form. This remaking is not just symbolic, but rather it is about the bee's exoskeleton; its nervous system; its digestive tract; its collective social behavior. There are, of course, many places (from federal laboratories to backyard beekeepers), as well as many new pressures (from industrial agriculture to global climate changes), involved in the remaking of the bee. However, there are less visible yet significant pressures on the honeybee. In fact, the largest funding for bee research and bio-engineering during the Bush administration was by military intelligence and weapons research agencies who hope to harness and develop bees' abilities as part of the "war on terror." This chapter explores how the honeybee was remade during the Bush Administration – both symbolically and materially – as a military technology and strategic resource for battlefield tactics.

The natures that are present in the rhetoric and practices of empire-building become manifest in the natures of modern ecologies. Honeybee biopolitics has become part of a shifting terrain of the politics of nature and culture, or more specifically, the human and the non-human, and is at the heart of many con-temporary debates about the "war on terror."[3] It is particularly surprising that those of us committed to political ecology and the cultural politics of nature have not weighed in more directly and forcefully in debates about the war on terror. This chapter is an initial step towards a critical politics of the nature of this war during the Bush administration.[4] To understand the making of the nature of the bee in the current moment, we will have to explore the links between contemporary understandings of nature, modern bee physiology, and human sociality. But this is going to require that we first think about the politics of nature and the human/non-human divide in the war on terror before we return to the honeybee.

The nature of evil

On September 11, in his first major address to the nation just a few hours after the collapse of the World Trade Center, President George Bush declared that the attacks were "evil, despicable acts of terror," adding that "today, our nation saw evil, the very worst of human nature. . . ." (Barron 2001). The theme of the evil nature of individuals became a central component of the administration's struggle to make the bombing of the World Trade Center intelligible in a particular way. A few days later, Condoleezza Rice echoed these sentiments about the nature of evil, stating that the newly declared war on terrorism was a "war against evil people and a war against the evil of terrorism (Jackson 2005: 67)." Donald Rumsfeld added to the growing analysis of the essential nature of terrorists, rhetorically asking, "Are we ever going to be able to stop people from wanting to terrorize others? No, I suspect not. . .no one around the Pentagon is going to change the nature of human beings (Rumsfeld 2001a)."[5] Bush boldly added, "Our responsibility to history . . . is already clear: to answer these attacks and rid the world of evil (Barron 2001)."[6]

There are a lot of things going on in these quotes – arrogance and ignorance, among others – but I want to point to this particular formulation of nature in its relationship to evil. Of course the nature of Evil, writ large, is beyond the scope of this chapter, but these remarks point to a strange and potent conjuncture of a particular formation of nature and evil. There is nothing new about the invocation of evil to legitimate war and conquest, from the Crusades of Europe to colonization and genocide in the Americas. But what is strange in this current form is the bold and rapacious return of a particular Divine Nature, combined with Linnaean taxonomies, that seem to fix the essence of individual behavior in particularly universal ways. Perhaps most egregiously, the turn to this dangerous essentialism has been taking place in the name of democratic liberal secular humanism.

Raymond Williams has famously pointed out the extraordinary amount of human history that goes unnoticed in the formulation of nature.[7] Here there seems to be a Divine Natural Order that returns with a vengeance in the language and logic of the Bush administration. This interpretation of Judeo-Christian nature holds that individuals have God in them as an inherent component of their living being – part of the human essence that is eternal. When, for example, in the fifteenth century, De Las Casas and Sepúlveda had their famous debate within the Catholic Church about colonial slavery, it hinged on whether Indians and blacks in the colonies had souls; that is, whether they had the essence of God inside them. If they did not have a soul then they were in fact not human and there was no theological, moral, or political reason not to treat them as animals and force them into slavery. Here the distinction between human and non-human and human and animal is being made in early racialized terms, where race is understood not primarily in terms of skin tone or biological qualities, but in terms of the humanness that was defined in relationship to the Christian soul. Those who are designated human can falter and be redeemed, but God is always in essence who they are. The colonial subject as animal Other is at once a threat to the stability of colonial rule and the rationale and possibility for that very rule. When Condoleeza Rice says, "terrorists are evil" by their very "nature," or Bush says the terrorists have declared "war on humanity,"[8] they are arguing that neither God nor reason drives them, and they are threatening the category of what it means to be human.

In effect, the evil of terrorism is rooted in the essence of certain people, in their immutable nature, which of course implies that rooting out terrorism actually requires the elimination of certain types of individuals or groups of people. They cannot be redeemed and they cannot be negotiated with because their evil lies beyond reason; it is in their very nature.[9] Again Bush stated, "By their inhuman cruelty, the terrorists lie on the hunted margins of mankind. By their hatred, they have divorced themselves from the values that define civilization itself."[10] This is a strange brew of medieval natures in an era of Darwinian evolutionary discourse, genetic engineering, biotechnologies and Lamarckian social engineering, but these resurrected natures form potent assemblages that have powerful material consequences.[11]

But there is something else going on here. It's not just nature, but rather nature in relationship to the human, that is at the heart of the war on terror and more directly relevant for this discussion. Attorney General John Ashcroft's testimony to Congress just after the attack is instructive:

> Ladies and gentlemen of the judiciary committee, the attack of September 11 drew a bright line of demarcation between the civil and the savage, and our nation will never be the same. On the one side of this line are freedom's

enemies, murderers of innocents in the name of a barbarous cause. On the other side are friends of freedom. Today I call upon congress to act to strengthen our ability to fight this evil wherever it exists, and to ensure that the line between the civil and the savage so brightly drawn on September 11 is never crossed again.

(Ashcroft 2001)[12]

Consider, too, the words of former Senate minority leader and Ambassador Howard Baker, who asserted that "the attacks [of September 11th] were attacks on not just the United States but on enlightened, civilized societies everywhere. It was a strike against those values that separate us from animals – compassion, tolerance, mercy (Jackson 2005: 48)."[13] It is important to note that the animal itself does not have to be directly declared, as its propensities are well known: it "lives in caves," it "hides in the shadows," it "burrows," it "scurries," and "crawls." It warrants responses such as being "hunted down" or "smoked out" and "totally destroyed (Jackson 2005: 12)." To understand the workings of evil nature in the war on terror necessarily returns us to the questions of the animal and the human.[14] Animal products of this colonial crucible return to animate contemporary understandings of the nature of terrorists and necessitate certain interventions in the name not just of civilization, but of humanity. Here Schmitt's friend/enemy distinction is delineated and preserved on the backs of animals.

To take seriously the question of the animal in relationship to war means more than just a discussion of how animal symbolism is tied to humans as part of lived atrocities between humans. As Agamben (1998) points out in his analysis of the violence of the camps in Nazi Germany:

> The correct question to pose . . ., therefore, [is] not the hypocritical one of how crimes of such atrocity could be committed against human beings. It would be more honest and above all more useful to investigate carefully the deployments of power by which human beings could be so completely deprived of their rights and prerogatives that no act committed against them could appear as a crime.

He goes on to say that "Jews were exterminated not in a mad and giant holocaust but exactly as Hitler had announced, 'As lice', which is to say as bare life." The ban, the wolf, and bare life are all about transgression of the human/non-human divide in relationship to violence.

Judith Butler's arguments in *Precarious Life* (2006) about the politics of indefinite detention dovetail with and also hinge in part on the boundaries of the human and non-human. She states that "if they [detainees] are [portrayed] as killing machines, they are not humans with cognitive functions entitled to trials, due process, to knowing and understanding the charges against them. They are something less than human." She observes "the reduction of these human beings to animal status where the animal is figured as out of control, in need of total restraint" (Butler 2006).[15] In these and many other cases, the boundaries of the human and the non-human, and often the human and the animal, are the crux of a politics of bare life, a precarious life in indefinite detention and the horror of human bodies becoming killable.[16]

It would be too simplistic to claim that violence simply implements what has already happened in discourse, that dehumanization somehow produces violence necessarily or directly. That's not what I'm arguing. What I do want to argue is that dehumanizing discourses matter, and those of us interested in the cultural politics of nature could be more attentive to the making of the non-human, against which the human is partially defined and into which individuals and populations violently cross in modern biopolitics, both in military contexts and in human rights discourse. This analysis of the animalization of the human is an important part of a cultural politics of nature and militarism, but it is only a start.[17]

That said, I think it is a profound mistake to stop our analysis of politics and the nature of modern militarism with humans becoming simply symbolic animals, so that they can be extra-juridical, immoral, and killable. These resurrected natures are made material not only in the rhetorics of war, in making subjects killable, but in the materialization of modern ecologies.[18] This matters not simply because other animals and ecologies are being remade through war and empire, but because human sociality is forged through the production of knowledge of other creatures. The intimate relationships between material creatures who are involved in other relationships, part of other stories and histories, comprise the unstable, shifting ground on which human sociality is formed. These material relationships between beasts are not necessarily pleasant imbrications, as much parasitic formations as companion species.[19] So let's turn to the honeybee as both a site for the making and remaking of human natures of war and a site through which war – in this case the "war on terror" – is remaking the bee.

Homeland security detective devices

There is, of course, a long history of writing on insects, both as models and as metaphors for human sociality, morality and politics. From fighting ants to racialized lice to industrious bees, their size, sociality, and ubiquitous presence has made them the source and site of creative and scholarly writing. More recently, there has been a renewed interest by scholars in the role insects have played in politics and human sociality and in the intimate relationship between these six-legged creatures and people. But insects are more than metaphors. Edmund Russell's (1996) treatment of the connections between insect and human annihilation in World War I and World War II; Anna Tsing's "Gleanings in Bee Culture" (1995), which explores how the traffic of nature between bees and human sociality works as a means of naturalizing forms of social difference; Timothy Mitchell's (2002) now classic article "Can the Mosquito Speak?"; Joseph Masco's (2004) giant fighting ants in cold war "Mutant Ecologies"; Hugh Raffles' (2010) explorations of human insect relations in his *Insectopedia*; these works have all made explicit the material relationships between humans, insects, and the politics of nature. For if animals are human Others, insects are the Others of animals, intimately involved our lives, much maligned, but powerful sites and sources for the production of human nature.

Insects and their environs are an intimate part of broader changing ecologies of empire. Much has been written about Green Imperialism (Cosgrove 1984, Crosby 1986; Grove 1995, Sparke 2004, Weizman 2007), tracing the contours of power and territorial expansion and its radical transformation of landscapes, both as sources and limitations of the products and profits of colonial endeavors. Plants and gardens serve as spatial and taxonomic representations of race, hierarchy, and territorial ambitions (Mukerji 1997) and form the basis of Nature Governance (Drayton 2000; Matless 1998) through imperial practices of the science of "improvement" of the world. Schiebinger (1993) has explored these histories and rhetorics of gender, race, and empire through the science of botany, while McClintock (1995) and Stoler (2001, 2008) make clear the centrality of Nature to the violence and geography of imperial projects. Anthropologists have explored how the science and practice of ecology become intertwined in broader questions of cultural politics of nature and difference (Moore *et al.* 2003; Comaroff and Comaroff 2001). While these and many other works call attention to the connection between ecologies and empire, there is little treatment of the political economy behind the production of the organism itself (Haraway 1989; Haraway 1991; Schrepfer and Scranton 2003; Vivanco 2001). Moreover, the focus on ecologies of empire has been almost entirely based in the eighteenth and nineteenth centuries.

Following the openings of science and technology studies (see for example Haraway 1989; 2008; Franklin 2007; Zylinska 2009) and political geography (Carney 2002; Schrepfer and Scranton), my work here attends to a more intimate remaking of the modern ecology of the honeybee. It is only in the last century that honeybees have been hived in a space made for easy observation and manipulation by the beekeeper and transported thousands of miles on the back of semi-trucks to serve as pollinators. Colonies' social organization has been transformed, with fewer guard bees, a shortened or non-existent hibernation season, and a different sized and prefabbed comb. The bodies of individual workers have changed color from black to yellow, become almost one-third larger in size, and sport more hair. Bees now have a different digestive tract and an exoskeleton almost twice as thick as a hundred years ago. Workers are more docile than they once were and have a life span shortened by 15 percent (Berenbaum 2009, Michener 1974, Preston 2006, Stephen 1969, Winston 1987). Moreover, the honeybee has served as an archetype for understanding human collective society, and as such, more has been written about bees than almost any other single non-human species – not just by apiarists and scientists but also by philosophers, kings, sociologists, criminologists, physicists, and poets (Crane 1999, Preston 2006). These cultural texts of bees help make human collective behavior intelligible, and in turn, these understandings influence our relationship with the honeybee. The political, economic, and cultural histories through which we have come to know and understand the bee are part of how we breed, select, and relate to the bee. The cultural frameworks we have mobilized to understand the races of bees, the organization of bee labor, gender in bee society, or the character of hierarchy in bee worlds, have now become physically part of the bees' biology. To treat the bee as a wild and instinctual object of a bucolic nature is to erase the political human history of the honeybees' contemporary biology and its long history as a militarized species.[20]

The bee is not alone among insects in the service of militarized campaigns and torture. The Emir of Bukhara used beetles to eat the flesh of his prisoners. Massive research projects took place in Germany, Japan, Russia, and the U.S., during which hundreds of millions of insects were cultivated and tens of millions of beetles and mosquitoes were deployed to infest crops, soldiers, and civilians. General Ishii Shiro released hundreds of millions of infected insects across China during WW II, which caused the deaths of tens of thousands of people. In the Korean War, U.S. airplanes dropped plague-infested fleas on North Korea and later used mosquitoes, wasps, and bees as part of torture techniques against the Vietcong in Vietnam. The Cold War also saw crop-eating beetles dropped on Vietnam, North Korea, and Cuba and fostered research that transformed modern entomology (Lockwood 2008; Tucker and Russell 2004). In the "war on terror" the Bush administration approved the practice of placing bees and spiders in confinement boxes as part of the torture of US detainee Abu Zubaydah (Scherer 2009).[21]

Bees have been widely used in warfare since antiquity: hives were dropped on invading armies or launched into fortified tunnels, caves, forts, and bases. The well-documented decline in the honeybee population during the late Roman Empire is now believed by many to be due to their extensive use in warfare. In the sixteenth century there was even a multi-armed catapult, like a windmill, that launched hives at enemy fortresses. In fact, the entomology and etymology of the bee is intertwined in war. The word *bombard* comes from *bombos,* which in Greek means bee, making an association of the threatening hum of an angry swarm and incoming projectiles (Lockwood 2008). Later, the bee became central to the war machine not as a projectile but as a source of beeswax that was used to coat almost all ammunition, small and large, during WW II. As a 1944 *Popular Science* article, "How Science Made a Better Bee," explains, "Amazing new discoveries [new breeding technologies] bring improvement to nature's masterpiece, enabling the busy little insect to do a better job for war" (Sinks 1944).

But my interest departs from this long history of violent inter-species relations to explore how the honeybee was enlisted in the war on terror. War always produces particular ways of seeing and knowing, but an amorphous war like the recently declared "war on terror" – a war not oriented towards a single state or clearly defined theatre of battle, declared against a diverse foe who is hard to identify – most dramatically illustrates the centrality of practices of knowledge-production in the work of war.

As General Colin Powell made clear in his presentation to the United Nations in 2003: the war on terror is a war of intelligence. The enemy's lack of coherence – institutionally, ideologically, and territorially – makes the search for the enemy central to a politics of the war on terror, both in maintaining that there is an enemy and in demonstrating the connections, coherence, and intention of the terrorists. The incoherence of the enemy opens up the possibility of terrorists anywhere, making anyone a potential target or suspect. Objects themselves take on the possibility of being implicated in this network: a lost piece of luggage; an oddly parked van; a suspicious-looking individual.

How then are we to discern the intent of individuals, animals, and objects? We must know them, see beyond them, look inside them, and listen past what they claim for something inside, something more deeply hidden. As Secretary of Defense Donald Rumsfeld stated, "The war on terror requires new technologies of warfare, but even more importantly, new technologies of surveillance" (Rumsfeld 2001a). U.S. intelligence agencies had to make the human and non-human speak their intentions (cf. Latour 2004), to hear not just what the tortured said, but to know their truth, to discern it, and when necessary make this truth intelligible. "We still talk about the end product of intelligence being a cursor on a target," said Michael Hayden, ex-director of the CIA, but in fact, "the war on terrorism requires a fundamental shift in how the military and intelligence communities do their jobs. Instead of focusing on the so-called 'find/fix/finish' strategy used in World War II and the Cold War, we have [a strategy] in which the enemy is hard to find but relatively easy to finish off, it's all about 'find.'" This "finding" requires ways of seeing and knowing, techniques of warfare that are not just aimed at what the enemy is doing, not just "actionable intelligence" (Rumsfeld 2001b), but what it is *intending*, and what is possible.[22] Intelligence-gathering is not just limited to sociologists, lawyers, and military planners, but includes biologists, anthropologists, epidemiologists, and even etymologists.

Here is where the bee and other animals enter the story. Rather than being used just as weapons of war, bees have become involved in the search for what was beyond our reach, what was beyond our senses. The behavior and physiology of bees has become instrumentalized to extend the capacity of the human senses. The deployment of bees, or what military scientists call "six-legged soldiers," has resulted in new and intimate relationships. Humans are not only experts in bee behaviors but also, by developing bees to serve our interests, we have become quite literally part of the honeybee's nervous systems, migration patterns, and community relations.

More money is spent on bee research for intelligence and military purposes than all other forms of federally funded bee research combined (Haarmann 2009). This research has supported a new regime of managerialism with bees, involving the harnessing of innate capacities of bees for detection and intelligence-gathering – or, as the Office of Homeland Security has stated, "deploying bees as efficient and effective homeland security detective devices."[23] Apiary entomologist Jerry Bromenshenk traces the use of bees as "micro sensor technologies" to fears about the health effects of pollution on honeybees, which in turn led to the use of these insects as "bio-monitors" for all kinds of toxic materials.[24] Bromenshenk realized that the sensitivity, social behaviors, and ecology of the honey bee could – as he explained to me with the passion of a preacher – be a "apiary revolution. . .an incalculable boon for eco-toxicologists" (Bromenshenk 2009). Others at Los Alamos National Laboratory and Sandia National Laboratory picked up on his enthusiasm and began to use the honeybee to monitor contaminated sites in and around Los Alamos, where the radioactive legacy of the Cold War will emanate for millennia to come.

As Paul Fresquez, director of the environmental sciences monitoring group at Los Alamos, explained to me as we watched bees busily flying back and forth over the 16 foot-high, barbed-wire-laced security fences of Los Alamos' top secret areas, "You can simply place a hive in an area that you are worried is contaminated and the bees, thousands of them, will do field samples, literally hundreds a day, of almost any pollinating plant within two miles of the hive without disturbing anything" (Fresquez 2004). He explains that traces of radionuclides, many of which are structurally similar to the calcium that plants take from the soil, are detectible in flower pollen and nectar near contaminated sites. Honey made by bees from these contaminated flowers can be tested for the presence and concentration of tritium and strontium-90. Honeybee bodies also have small-branched hairs with a static charge, causing them to attract chemical and biological particles, including a diversity of pollutants, biological warfare agents, and diverse explosives (Fresquez 2004). They also inhale air and water for evaporative cooling of the hive. Bees thus sample air, soil, water, and vegetation, as well as diverse chemical forms of gaseous, liquid, and particulate matter. If a hive is well placed, very accurate gradient maps of the distribution of radioactive materials and other toxic contaminants can be produced (see Bromenshenk *et al.* 2003).

Bees were used as environmental monitors for almost a decade before their applications in espionage were seriously considered. After years of failing to develop mechanistic means for detecting chemical explosives, many researchers turned to animals to do this work. Part of the program was funded by the Defense Advanced Research Project Agency's (DARPA) Controlled Biological and Biomimetic Systems Program for work at Los Alamos, Sandia National Laboratory, and other research sites. Hives eventually were deployed around the world to test areas suspected to contain nuclear material, according to one anonymous source in the stealth insect sensor project team whom I interviewed in 2006. Bees were next developed with the goal of mapping the large number of mines in northern Afghanistan (Hanson 2006).

DARPA-funded research also trained free-flying bees to detect certain scents – say that of a landmine – by placing traces of the explosive chemicals near food sources (Bromenshenk *et al.* 2003). The bees associate the scent of the mine with food, and when placed in a minefield, the bees will fly more intensive patterns around the mines. The bees are tagged using infrared tagging technology, and their flight patterns are recorded in order to create a map of the areas they have traveled (Figure 11.1). The bees' foraging behavior is not completely changed, but their purpose is transformed to forage for landmines rather than food (German 2002: 1–3).

Bees have a sense of smell more sensitive than a dog's. With upwards of 50,000 individuals per hive, they have an ability to cover a greater area than canine companions. They need less attention than a dog and only need a fraction of the time in training (McCabe and Wingo 2008). Like dogs, bees have a large number of chemo-receptors that recognize signals identifying kin, as well as pheromones that enable social communication within the hive. The receptors also detect external food sources and other chemical agents. Each antenna is covered with thousands

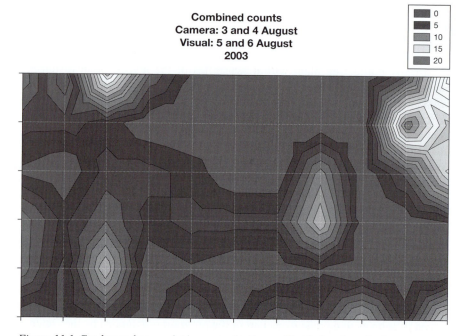

Combined counts
Camera: 3 and 4 August
Visual: 5 and 6 August
2003

■	0
■	5
■	10
□	15
■	20

Figure 11.1 Can honey bees assist in area reduction and landmine detection?

of separate individual receptors, and, with paired antennae, bees can very quickly determine the direction and intensity of an odor. Moreover, their ability to detect suites of chemicals, including those most common in various sorts of land mines (such as 2.4-DNT, TNT, 2.6 DNT and RDX) in concentrations as low as 50–70 parts per trillion, has made them, in Bromenshenk's words, "indispensable agents for future chemical and biological warfare detection teams" (Bromenshenk 2009) (Figure 11.2).

The technology of bee deployment, however, presented some problems: as one member of the team pointed out, "It turns out bees have minds of their own, and that they can be delinquent from their training, for while they are easily reined in some respects, they do not always do as they are told.. . .We would like to be able to get bees to fly right past an apple bloom to the explosive or human target every time, but this would require more intensive training or more intensive intervention into the bio-physiology and genetics of the bee than we have yet been able to do" (McCabe 2008). Training bees to fly past flowers would involve feeding them entirely in the lab, never bringing them into contact with living flowers outside. Even then bees don't always behave as they are taught, and only some bees are consistently trainable. It also became clear that in complicated conditions, where there are a lot of other "distractions" such as the "instinctive behaviors for feeding and mating as well as responses to temperature changes" (Wingo 2008), it is even harder for the bees to do detection work. It seems that the collective bee was less controllable and less reliable than researchers would like. In some cases, laboratories kept hives in small tent-like structures and never let the bees out; in other cases, greenhouses an acre in size were set up to control non-experimental variables of the bees' habitat.

Bromenshenk, along with collaborators from intelligence agencies, then tacked in a slightly different direction. The research team focused intensive training efforts on a specific response of individual bees. Bees were placed in individuated Styrofoam cells, taped in place, and then over the period of a few days or even a few hours were given the scent of whatever chemical the researcher wanted them to identify with food. They learned, in a way that would make Pavlov proud, to stick their tongues out when they smelled the scent of the chemical. The bees that did this reliably were placed in a cartridge and inserted into a machine. This gave these researchers a computer readout – both magnifying and graphing the bees' response (Figure 11.3). When bees stick out their tongues in this cyborg assemblage, it is an interspecies signal translated by computers into an alarm or flashing message on a screen, identifying a chemical, a bomb, or a biological agent. With military grade TNT, this tongue response is 99 percent accurate. The trained bees last a few days to a few weeks. Then a new replacement cartridge is shipped, and "like a razor, you simply slip out one cartridge and replace it with another" (Anonymous 2006) (Figures 11.3, 11.4, 11.5).

Biomechanical engineers are developing still more intimate relationships with bees, inserting new technologies into larva. This DARPA project is aimed at developing tightly coupled machine-insect interfaces by placing micro-mechanical systems inside the insects during early stages of metamorphosis, allowing greater control over insect locomotion (Lal 2007).[25] In theory, if these bio-electromechanical interfaces are placed early enough in insect larvae, they will be able to heal and incorporate the technology. This interface would allow humans to control insect behaviors and motion trajectories via specialized GPS units along with optical or ultrasonic signals. The control can happen through direct electrical muscle excitation, electrical stimulus of neurons, and the projection of pheromones (Johnson 2007).[26] Many of these insects, whose nerves have grown into internal silicon chips, are slotted to tote cameras into new shadowy domains. DARPA researchers are also raising cyborg beetles with power for various electronic devices harvested from the insect itself (Zerner *et al.* in press).

I found myself skeptical of the likelihood these projects would fully come to fruition. My interviews with DARPA-funded scientists, such as Wingo, Bromenshenk, Haarmann, McCabe, and others, revealed complex relationships between technology and biological physiology – more complex than DARPA's published material would have you believe. It is easy to fall into a kind of techno-conspiracy theory formulation, which overstates efforts to totally control insect natures through these intimate reworkings of technology, bee behaviors, and physiology. But it is also true that a great deal of research funding is meant to do just that, and most of this research is classified. Moreover, some of the successes that Charles Zerner *et al.* and Joe Masco have documented elsewhere make clear that even if insect biology is less mechanical than it is often portrayed, such trans-formations and manipulations of insects' physical and social architecture should be taken seriously (Zerner *et al.* in press; Masco 2006). Potentially even more significant are the new breeds of bees are being created, albeit slowly and cautiously in light of what happened when a Brazilian crossbreeding experiment resulted in "killer bees." However, now that the bee genome has been mapped (it was one of the first insects to be mapped), there are new efforts in military research labs to restart breeding in order to make a more useful militarized bee (McCabe 2008).[27]

The new uses of the honeybee blur the line between the human and this insect species. The modern bee is already a product of a history of breeding and selection and behavior modification related to agriculture, economic interests, race and immigration relations, but this is a different engagement, one that uses these animals not as weapons but as sources and technologies of intelligence. Honeybees form part of a growing militarized ecology in which new relationships and new forms of both insects and humans are being made. Through the bee, humans develop the ability to extend their own senses beyond the capacity to see, such that the bee's sense becomes part of human intelligence. At the same time, the bee's nature is remade, as its very genetics, biology, and physiology are transformed by military interests and desires. Bees become more human, and humans come to know the world in part through the bee, but in a particularly militarized form.

Figure 11.2 Bees' response I

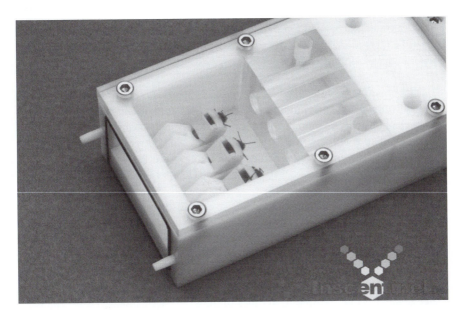

Figure 11.3 Bees' response II

Figure 11.4 Bees' response IIII

Swarms: the animalization of military strategy

I have outlined some of the remaking of the bee related to intelligence, but the human/bee has more work to do, particularly in this time of war. For as Deleuze and Guattari astutely point out,

> War contains zoological consequences. . . . It is in war, famine, and epidemics that werewolves and vampires proliferate. Any animal can be swept up in these packs and the corresponding becomings. . . . That is why the distinction we must make is less between kinds of animals than between the different states according to which they are integrated into . . . war machines.
>
> (Deleuze and Guattari 1980: 243)[28]

Here the animal takes different forms; it is itself transformed through its integration into battlefields and tactics, becoming part human, part animal as both are remade.[29]

What I find most useful here is that Deleuze and Guattari and others make explicit that Human Nature is largely forged in the domain of the non-human; or more accurately, as the editors of this volume make clear, through interspecies relationships (cf. Haraway 1989, 2008, Wolfe 2003). This may be particularly so, as Deleuze and Guattari point out, in times of imperial expansion and war. The Bush administration's war on terror enacts this relationship perhaps with particular intensity, tracing the evil of modern terrorism to an assumption about the essential character and immutable nature of threatening outsiders. This sense of innate maliciousness, which is derived from the link between terrorists and animals, implies that rooting out terrorism actually requires the elimination of certain types of individuals or groups of people. They cannot be redeemed and they cannot be negotiated with because their evil lies beyond reason; it is in their very nature.[30]

So as we learn to combat this savage adversary, learn to see the war on terror as a very different type of war, the language segues into defining a new type of enemy and an appropriate response. As Bush states, it is a war without "frontlines," without a "definable territory," without a singular ideological definable group, and without a "nation state." What is clear is that the nature of the enemy has crossed the boundaries of civilized engagement and, as such, either necessitates a new type of surveillance to discern the true nature of its subjects and objects or warrants a certain type of violent response. As Donald Rumsfeld asserts, "The nature of our response needs to be directly related to the nature of the terrorist threat" (Rumsfeld 2001a). There is a parallel analysis made on the battlefield related to terrorist strategy: they will not fight by the rules of ethical warfare, which further confirms their uncivilized status and requires different tactics.

Here the animal returns – not in the same way it did before, but as a different beast. So how are bees being integrated into war machines? One of the central new approaches to operational, strategic, and tactical approaches to the war on terror is through the swarm. There are many different forms of the swarm, but the most often cited in military strategy is that of the ant and the bee. Take, for example, the work of John Arquilla, an early proponent of swarming in the Department of

Defense Analysis, an advisor to many generals, and a chief military advisor to Donald Rumsfeld. He says in the opening of his famous RAND Corporation study, *Swarming and The Future of Conflict,* that swarming needs to replace the AirLand Battle doctrine that has been the conceptual framework for the U.S. Army's European war fighting policy from 1982 up to the Shock and Awe techniques of the Iraq war.

AirLand Battle emphasized close coordination between land forces acting as an aggressively maneuvering defense and air forces attacking frontline enemy forces. Swarming, as Arquilla and others have defined it, is about decentralizing force operations in a manner that values mobility, unit autonomy, and continuous and synchronized real-time communication. It entails the "systematic pulsing of force/or fire by dispersed, interknitted units, so as to strike the adversary from all directions simultaneously" (Arquilla and Ronsfeldt 2002). Sean Edwards, another RAND Corporation researcher, explains that "swarms are complex adaptive systems, but have no central planning, simple individual rules, and non deterministic behaviors that evolve with the specific situation" (Arquilla and Ronsfeldt 2002). Arquilla told a Congressional hearing that the war on terror is driven by an "organizational race" to build networks and swarms. Flexible, adaptive, collective responses, according to Arquilla, are at the heart of future military struggles (Arquilla 2008). Swarm strategies were outlined by the U.S. Joint Forces Command in 2003 and are expected to be fully operational in the war on terror by 2012.

These strategies are explicit in the use of bees and ants as models. Arquilla told congressional representatives:

> Swarming appears in the animal kingdom long before it did in human affairs. [. . .] As the name suggests, the concept of swarming comes from the nature of insect behavior, and many of these behaviors are directly applicable to military strategic and tactical operations. [. . .] The form of swarming that goes on in the hive, bees and ants, employ blanketing tactics when foraging outside the hive – striking their adversaries or prey from all directions. The goal is to overwhelm any cohesive defenses that might be mustered. Although these insects often move in linear formations, they are quite adept at shifting into a swarming mode at any point of engagement.
>
> (Arquilla and Ronsfeldt 2002)

Here biological descriptions of social and collective behavior of bees and ants serves as the foundational logic and model for human strategies of war; socio-biology meets military planning. Arquilla and other military planners, strategists, and modelers draw directly from the behaviors of insects, as well as from entomologists and animal behaviorists such as E.O. Wilson, to make sense of and make possible new ways of organizing human behavior. Some researchers are mapping the patterns of swarm movement mathematically into algorithms, others more conceptually, but the insect is part of the constitution of this strategy of war not simply as metaphor but as model (Arquilla and Ronsfeldt 2002; Edwards, 2000; Booker 2005).[31]

French entomologist Pierre-Paul Grassé's 1950s work on bees and wasps has also been resurrected. Now Grassé is commonly cited in military strategy, particularly his notion of lattice swarm behavior, or what he called "stigmergy," in which bees and wasps are enabled to build complex structures by taking their cues from the structure and behavior of their neighbors. As the hive is built, bees observe its current state, changing their behavior accordingly to build the next piece:

> An individual agent has a repertoire of actions it can use to move through this space and modify the environment. An agent's sensors detect information derived from local properties of the agent's current position in the lattice and the positions directly adjacent to it. Since each agent has only a local view of the overall activity of the swarm, some additional mechanisms of communication are available to coordinate the collective behavior of the swarm.
>
> (Booker 2005; Lambert 2003)

These mimetic relationships, as Taussig and Caillois among others make clear, are not simply about imitation or representations of the real, but simultaneously a means for the production of alterity (Taussig 1993) and distinction (Caillois 1984). Here mimetic practices at once create the distinction of the civil and the savage by dehumanizing the terrorist and making them intelligible as nonhuman (cf. Butler 2006) while at the same time the animal (the bee) is used as the natural non-human rational strategic answer to barbaric aggression. Military strategists have returned to the bee as a model for human soldier strategic behavior. The bee's technique of communication and de-centered coordination allows for advantage in fighting an enemy that has itself been made non-human. Here the natural history of the swarm returns to animate militarized battlefield strategy, but not as the same animal as the terrorist and not in the same way. This time the nature of the animal becomes a model for efficient and effective warfare. Nature is no longer a threat. Civilization's relationship to the nature of the bee and the swarm is one of imitation and, as one military strategist put it, "of deep respect for a complex system" (Edwards, 2000). These are the new zoological consequences of the war on terror's remaking of animal/human natures and apiary ecologies.

Ecologies of empire

There are some basic questions at stake here. Under what conditions and through what means are some human lives eligible for human rights, ethical treatment, due process, etc., while others become killable as "cowardly," "burrowing," sub-terranean animals? What is the legacy for bees and humans in their work as technological instruments of espionage and architects of the military strategies of U.S. empire? How might we better understand and remake these militarized ecologies? These questions are part of a larger natural history of modern warfare, one that seems to be woefully absent from our work on the cultural politics of nature and the animal.

Making subjects killable is partly, as Butler (2006), Agamben (1998), and others have pointed out, about making subjects non-human and, as such, placing them beyond the domain of the rights, norms, and ethics that have been defined within the realm of the human. This has been a deep philosophical problem for a long time. However, it is rare, as Derrida points to (and others, such as Donna Haraway, demonstrate), that the specificity of animals (the actual animal) and their cultural histories are taken seriously in relationship to how people are made killable. As I have discussed, both terrorists and Marines are made into animals, but the type of animal they are, and more specifically the particular cultural history and form the animal takes, matters. Moreover, the animal can be both a means of making people killable by making people sub-human or it can be a means of positioning sovereign power above the rule of law. Rethinking the animal would necessitate a non-essentialist taxonomy and biology that would undermine, or at least complicate, the ways in which people and animals are made into spies or into killable sub-humans – in either case, beyond the rule of law. This attention to hierarchy, plurality and specificity also enables openings or crossings, where animals are already partially human, allowing us to rethink the animal through the specific and changing qualities both of and within species.

In this article I have also sought to bring the relationship of animals and militarization into view because it's a critical site through which the bee is being remade, both materially and symbolically. I believe that at the current moment, when we are facing the most serious crisis of the honeybee in its/our millennia-long relationship, we need a close accounting of the many places and forms through which our relationships to bees are being made and remade. This accounting would be part of a critical natural history of the honeybee, one that is attentive to the political economy of industrial agriculture, to the chemistry and molecular biology of international chemical corporations, as well as to the genetic laboratories in which we are searching for the bee's "social gene." It also requires, as I have begun to explore here, attention to the instrumentation of the bee as a means of tracking and tracing the boundaries of dangerous subjects and suspect objects. These new uses of the honeybee are part of a remaking of its material body, as well as the new ecological contours of empire. These ecologies of empire matter, for they constitute the materials out of which future relationships and future bodies, human and apiary, will be forged.[32]

Notes

1 Moreover, a focus on the broader political economy of beekeeping does not get us to the making of the bee itself and often remakes the divide between nature and culture – a divide that leaves too much of the politics of the contemporary crisis out of bounds of the politics of beekeeping.

2 As Raymond Williams pointed out a few decades ago, nature itself has a past that requires an exploration into the material histories and politics of its making. In fact, the honeybee in modern history is so bound to industrialism, modern capitalist agricultural production,

contemporary forms of breeding, and genetic manipulation that to call the bee fully non-human is to miss the intimacy of the relationships that have made not just the environment but the bee itself – its nerves, digestive tract, skeleton, flesh, size, behavior (individual and social), and its molecular and genetic structure. As Hackenberg told me during an interview, "the bee that I work with today is not the same creature that my dad worked with and is not the bee that God made. He did not make the bee to travel 15,000 miles in a year on the back of a semi, or subsist on pesticide-laced, pollen-enriched corn paddies imported from China, and to pollinate one crop and one crop alone for weeks at a time. But what can we do? The crops need pollinating."

I believe we need a political geography of this modern creature, both as a means of understanding how the current crisis came about and also to understand the intimate transformations wrought by modern science and capitalism on social and ecological relationship, human and non-human. This requires that the analytical tools of ecology, geography, history, culture, and political economy be used to explore and illuminate the tiny spaces and intimate behavior of the body of the bee.

3   I use the terms nature/culture and human/non-human, acknowledging that they are false binaries, as Canguilhem (1976) Haraway (1989) Butler (1990), Latour (1991) and many others have pointed out. As the authors of this volume make clear, the non-human is a relational category that implies the absence of its constitutive Other. I use these binaries here because it is precisely these fictitious delineations of human and non-human that are that are at work in the making human subject animals and in the process making certain individuals killable or beyond the law. I trust that the larger argument – of the inseparability of nature and culture, the human and non-human – is clear from the way I demonstrate the making of the honeybee: in fact this inseparability is the foundation on which this article rests.

4   http://www.whitehouse.gov/news/releases/2001/09/20010911–16.html. White house press release.

5   http://www.usatoday.com/news/sept11/2001/10/25/rumsfeld-transcript.htm

6   http://www.washingtonpost.com/ac2/wp-dyn?pagename = articleandnode = andcontent Id = A30485–2001Sep14andnotFound = true

7   Raymond Williams, 1980, *Ideas of Nature*. New York, Verso.

8   BBC news "Bush warns of a 'War on Humanity'" October 6th 2005.

9   Bush said that negotiating with terrorists is a "foolish delusion." See NYT May 22 2008, item by Helene Cooper http://www.nytimes.com/2008/05/22/washington/22assess.html?scp=1andsq=foolish+delusionandst=nyt.

10 Bush October 20th 2001, as cited in Jackson 2005: 49. Not even Ronald Reagan's famous speech about the Evil Empire makes these naturalizing moves. For Reagan and others who have invoked evil in relation to communists, this evil was an ideology that could be redeemed, not an uncivilized essence residing in some humans. The problem lies not inherently in communists' nature but in their evil reasoning, their ideology. In his famous Evil Empire speech, Reagan quotes C.S. Lewis:

> The greatest evil is not done now in those sordid 'dens of crime' that Dickens loved to paint. It is not even done in concentration camps and labor camps. In those we see its final result. But it is conceived and ordered in clean, carpeted, warmed, and well-lighted offices, by quiet men with white collars and cut fingernails and smooth-shaven cheeks who do no need to raise their voice.

Evil is not in the dark natures of communist individuals, but in the dangerous communist ideology of communist governments. (It's particularly sad, I must say, that I have to invoke Ronald Reagan as a more progressive alternative in my argument.)

At the same time, there seems to be a modern form of a Linnaean taxonomy of humans in which terrorists are defined as a unified, identifiable group. The political process through which people come to enact terror on others becomes irrelevant, or even treasonous to talk of, when a collective group is defined by Bush as simply a "species of evildoers." As such, terrorists are no longer defined by the *act* of creating terror, (i.e. a terrorist is one who practices an act that causes terror in others) but instead become a group or coherent population that is defined by the nature of its cause. As such, the torture at Abu Ghraib does not become a terrorist act, but the act of terror by Al Qaeda always confirms the evil nature of terrorists. This unified population, the making of a collective equivalence, requires the same action everywhere a violent assemblage of an essential collective identity, whether they be jihadists, communist rebels, drug lords, animal rights activists, or ex-Weather Underground members. But the list certainly doesn't include U.S. soldiers, prison guards, police officers, CIA operatives, or U.S. mercenaries who use the torture techniques of water boarding, electric shock, etc. So a global war on terror makes sense because it identifies a group of people whose inner essence is evil and immutable and classifies these people as part of a population that shares a collective form of evil tendencies.

11 Anne McClintock, from a talk given at University of California Berkeley 10/09/08 entitled Paranoid Empire. In fact, a very unconventional type of person had to be constructed – one who was exempt from both the laws of peacetime (due process) and the accepted laws of war (Geneva Conventions), so their human rights could be suspended in the kind of war the Bush administration envisaged. Moreover, as Anne McClintock (2008) has pointed out, the animal essences of these liminal beings, people who are not quite human, need to be confirmed, they need to be made to speak and be spoken for through interrogation and the most extreme type of confessions to confirm the their evil nature.

12 Attorney General John Ashcroft's testimony to Congress. Ashcroft September 24, 2001.

13 http://www.globalsecurity.org/military/library/news/2001/09/mil-010924-usia12.htm

14 Here the civilized stands against the savage, the division of nature and culture, the human and the animal. This division follows tracks set down over hundreds of years in the relationship between colonialism and empire, in the natural history of race and the religious doctrine of Divine nature and Divine right, in the tradition of defining the savage against the saved, the lawful against the unruly, the animal against the civilized.

15 Talal Asad, in his new work on suicide bombing, makes the point that one of the most horrifying aspects of death is not the extinguishing of life but that "when no signs of the living body can be relied on, the ground that sustains the sense of being human – and therefore what it means to be humane – collapses. What seems to horrify is the ease with which the boundary between what is alive and what is not – between the sanctity of the human corpse and the profanity of an animal carcass – can be crossed . . . of life transformed into meat."

16 See also Talal Asad. 2007. *On Suicide Bombing*, New York, Columbia University Press. Especially pages 80–81.

17 Jackson, Richard, 2005, *Writing The War on Terrorism: Language, Politics and Counter-Terrorism*, Manchester, Manchester University Press.

18 We need to make sense of those forms of interaction, association, and intimacy that arise between animals and humans and transform the material composition of both human and animal, in which the human arises from the animal and the animal from the human. This, I believe, requires attention to the material of the animal itself, for not all animals are the same. Cockroaches are not bees, wolves are not horses, so there is a need for those of us interested in militarized ecologies to talk of *animals*, not *the animal*, both symbolically and materially. It is interesting to note that Butler actually discounts this entirely, stating that "it is important to remember that the bestialization of the human in this way has little, if anything, to do with actual animals, since it is a figure of the animal against which the human is defined" (p. 78). I think this statement too quickly discounts the relationship between the "figure of the animals" and the "actual animals," against and through which the human is defined. At a moment when so much rests on the lines between the human and the non-human, it's worth looking materially and

historically at the making of the human and the non-human. We need an account of the way that intimate relationships across human/non-human boundaries are themselves central to the the making and remaking of these boundaries between. What is the violent transgression through which some humans become animals, beyond reason and, to quote Ambassador Baker again, without "compassion, tolerance [or] mercy"? By what process do we produce the vulnerabilities and abject violence of being made outside the human and outside human ethics, human rights, and morality? That is, how do people become killable? Declarations of essence yoke social behavior to inherent propensities in ways that mark the animal from the human.

19 Haraway, Donna 2008. *When Species Meet*. Minneapolis: University of Minnesota Press and Serres, Michel 2007 [1980]. The Parasite. Minneapolis: University of Minnesota Press

20 I want to highlight the intimacy of the relationship between the human and the non-human. As a human geographer interested in questions of environment and society, I want to argue that we too often fail to acknowledge that the human is bound up in the material of the natural; to simply explore how humans have impacted natural systems is to miss much of the history and politics of the making of those systems and species. This question necessitates attention to the intimate relationship that bees and humans have had for centuries. During that time both species have considerably influenced each other, both behaviorally and biologically. But it is over the last hundred and fifty years that changes to the bee, particularly in industrialized societies, have been both dramatic and almost entirely unexplored. These changes have radically altered the structure and behavior of the hive, from logs and skips to a fully industrialized hive modeled on the modern factory. The bee's range has also been radically altered, from a radius of two miles to the migratory geography of the modern bee, who travels thousands of miles of on the back of semi-trucks and is fed on corn syrup and soy protein supplements in order to pollinate single crops for eight weeks at a time. Through modern breeding and genetic manipulation, the very biology of the modern honey bee has been transformed, whether to enable bees to be shipped long distances or to make them hairier for more efficient pollination.

But probably the most important change for contemporary beekeeping was the unprecedented portability and management of the hive in ways that had not previously been possible. This mobility in turn allowed for the rise of the industrial geography of beekeeping, in which 80 percent of the hives in the US are now trucked around the country, serving the mono-crop blooms of large scale industrial agriculture. Without this service, a large portion of contemporary agriculture would simply not be biologically or economically possible. In a attempt to make the bee even more suited for industrial agriculture, bees thus must move not only to be profitable but to survive; the geography of bees and beekeepers has thus been radically remade in the last 100 years in ways that both correspond to and enable the geography of industrial agriculture. In turn, modern industrial agriculture has been enabled by and transformed the honeybee: they work 2–4 more months than they used to, they are nomadic, they are treated by more chemicals for more diseases and given large quantities of supplemental high fructose corn syrup and cheap soy protein to boost their pollen production. Even more radical interventions have been made to the honeybee in the US through breeding and genetic engineering programs which I explore elsewhere.

21 The legal memorandum for the CIA, prepared by Assistant Attorney General Jay Bybee, reviewed 10 "enhanced interrogation techniques" for interrogating Abu Zubaydah and determined that none of them constituted torture under U.S. criminal law (Scherer 2009).

22 Again, as Rumsfeld put it, "we need intelligence that tells us what is coming before it is even planned." This has brought new resources into intelligence gathering dedicated to what Donald Rumsfeld famously called the "known unknowns" in a defense department news briefing on February 2nd 2002, when he stated:

> As we know, there are known knowns. There are things we know we know. We also know there are known unknowns. That is to say we know there are some things we do not know. But there are also unknown unknowns, the ones we don't know we don't know.

23 Interview with the stealthy insect sensor project team at Los Alamos National Laboratory, Los Alamos, NM, May 2006. There is, of course, a deep irony here, for philosophers from Aristotle, to Marx, to Heidegger, to Geertz, as well many others, have turned explicitly to the bee as a social being with a complex society to explore the similarities between humans and bees. But all have ultimately delineated the human from the bee by the human ability to think, to not act from natural instinct or essential qualities but to possess *intelligence*. After centuries of philosophical work that differentiates the animal from the human based on the bee's inability to possess intelligence, the bee becomes employed as an agent of intelligence gathering. In this case the bee becomes a technological apparatus for extending the capacities of the human, behaviorally and biologically possessing the desires and interests of a militant state in the gathering of military *intelligence*.

24 These original observations were tested in a much larger way after the Chernobyl disaster. See J.J. Bromenshenk, S.R. Carlson, J.C. Simpson and J.M. Thomas, 1985, "Pollution Monitoring of Puget Sound with Honey Bees," *Science*, 277: 632–634.

25 From interview with Dr. Amit Lal. Also see DARPA Micro Systems Technology Office program descriptions.

26 Johnson, Colin. 2007. DAPRA Hatches Plan for Insect Cyborgs to Fly Reconnaissance. EE Times, October 3rd,http://www.eetimes.com/news/semi/showArticle.jhtml;jsession id = W5RSN0BIVNGNYQSNDLPSKH0CJUNN2JVN?articleID = 202200707and pgno = 3. The Hi-mem efforts funded by DARPA are funding both military and U.S. Universities to carry out this work. This research falls under what DARPA calls "Bio-Revolution," which is a program designed to reengineer living organisms to improve Department of Defense (DOD) capabilities. DARPA's Bio-Revolution programs are focused on four thrust areas: Protecting Human Assets, Maintaining Human Combat Performance, Biology to Enhance Military Systems, and Restoring Combat Capabilities after Severe Injury. All of DARPA's Bio-Revolution programs have one mission in mind: to use the life sciences to benefit the U.S. military.

27 Kirsten McCabe, LANL insect sensor project team, Los Alamos, NM, May 2008. Gene Robinson, one of the leaders of the Honeybee Genome Sequencing Project team, told me that the mapping of the bee genome "marks a new chapter in the relationship between the bee and man. No longer are we going to have to accept the bee as is, in its natural form."

28 For a thoughtful and critical take on Deleuze and Guattari's treatment of the animal human see Haraway 2008:27–35. As the previous section of this essay should demonstrate, I agree with Haraway's critique of Deleuze and Guattari's "distain for the daily, the ordinary, the affectional . . . [and the] profound absence of curiosity about and respect for and with actual animals" (Haraway 2008:29).

29 Here that Vampire and the Werewolf are the part human, part non-human becomings that result from the contagion of the battlefields. This is not simply a process of imitating animals, as Massumi (1992: 93) makes clear, but a "contamination" that combines affects from abstract bodies and incarnates them as human matter. These reincarnations are incomplete, partial formations – part human, part animal, werewolves and vampires. The "war machine" is a form of social subjection where animals, in this case bees, become constitutive pieces or working parts of a human animal form.

30 Bush said that negotiating with terrorists is a "foolish delusion." See Helene Cooper NYT May 22 2008, http://www.nytimes.com/2008/05/22/washington/22assess.html.

31 There has been detailed research of ant and bee uses of pheromones to coordinate, forage, swarm, and attack. Then the pheromone-based algorithms derived from the studies are

mapped and applied to simulation-based experiments, which in turn are used to design troop communication strategies or used to control the swarm patterns of unmanned aerial vehicles conducting attacks against mobile targets (Booker 2005).

32 Finally, for those of us interested in the cultural politics of nature, it is critical that we attend to the ways that nature returns through the animals in different forms here – as divine essence, as machine or technology, as an architectural drawing or model of human sociality, as part of the war-machine.

References

Agamben, Giorgio 1998. *Homo Sacer: Sovereign Power and Bare Life*. Stanford, CA: Standford University Press.

Anonymous 2006. Interview with Author, Los Alamos, NM, June 13.

Arquilla, John 2008 Testimony of Dr. John Arquilla before the House Armed Services Subcommittee on Terrorism, Unconventional Threats and Capabilities, presented September 18. See http://armedservices.house.gov/pdfs/TUTC091808/Arquilla_Test

Arquilla, John and David Ronfeldt 2002 *Swarming and the Future of Conflict*. Santa Monica, CA: RAND Corporation.

Ashcroft, John 2001. Testimony to Congress. Ashcroft September 24. http://www.global security.org/military/library/news/2001/09/mil-010924-usia12.htm.

Barron, James 2001. Thousands Feared Dead as World Trade Center is Toppled. *New York Times*, September 11, A1.

Berenbaum, May 2009. Interview with Author, Reno June 12.

Bishop, Holley 2005. *Robbing the Bees*. New York: Free Press.

Booker, Lashon 2005. Learning from Nature: Applying Biometric Approaches to Military Tactics and Concepts. *The Edge* 9(1).

Bromenshenk, Jerry J. 2009 Interview with the Author, Reno Nevada, January.

Bromenshenk, Jerry J., S.R. Carlson, J.C. Simpson, and J.M. Thomas. 1985. Pollution Monitoring of Puget Sound with Honey Bees. *Science,* 277: 632–634.

Bromenshenk, Jerry J., Colin B. Henderson, Robert A. Seccomb, Steven D. Rice, Robert T. Etter, Susan F.A. Bender, Philip J. Rodacy, Joseph A. Shaw, Nathan L. Seldomridge, Lee H. Spangler, and James J. Wilson 2003. Can Honey Bees Assist in Area Reduction and Landmine Detection? *Research, Development and Technology in Mine Action*, 7.3, December. See http://maic.jmu.edu/journal/7.3/focus/bromenshenk/bromenshenk.htm

Butler, Judith 1990. *Gender Trouble: Feminism and the Subversion of Identity*. New York: Routledge.

—— 2006 *Precarious Life: The Power of Mourning and Violence*. New York: Verso.

Caillois, Roger 1984. Mimicry and Legendary Psychasthenia, trans. John Shepley. *October*, 31 (Winter): 17–32.

Canguilhem, Georges 1976 Nature Denature et Nature Naturante. In *Savoir Faire, Espérer, Les Limites de Raison*, Brussels: Publications des Facultés Universitaires Saint-Louis.

Carney, Judith 1999. *Black Rice,* Boston, MA: Harvard University Press.

Carney, Judith 2002. *Black Rice: The African Origins of Rice Production in the Americas*. Cambridge: Harvard University Press.

Comaroff, J., and J.L. Comaroff 2001 Naturing the Nation: Aliens, Apocalypse and the Postcolonial State. *Journal of Southern African Studies* 27(3): 627–651.

Cooper, Helene 2008. New York Times, May 22. http://www.nytimes.com/2008/05/22/ washington/22assess.html?scp = 1&sq = foolish+ delusion&st = nyt.

Cosgrove, D. 1984. *Social Formation and Symbolic Landscape*. London: Croom-Helm.

Cox-Foster, Diana and Dennis van Engelsdorp 2009. Solving the Mystery of the Vanishing Bees. *Scientific American*, April 2009.

Crosby, Alfred 1986 *Ecological Imperialism*, Cambridge: Cambridge University Press.

Crane, Eva. 1999. *The World History of Beekeeping and Honey Hunting*. New York: Routledge.

Deleuze, G., and F. Guattari 1980. *A Thousand Plateaus. Becoming-Intense, Becoming-Animal, Becoming-Imperceptible. . .,* trans. Brian Massumi. London and New York: Continuum, 2004. Vol. 2 of *Capitalism and Schizophrenia*. 2 vols. 1972–80. Translation of *Mille Plateaux*. Paris: Les Editions de Minuit.

Drayton, Richard H. 2000. *Nature's Government: Science, Imperial Britain, and the "Improvement" of the World*. New Haven, CT: Yale University Press.

Edwards, Sean J.A 2000. *Swarming on the Battlefield: Past, Present, and Future*. Rand Monograph MR-1100. Santa Monica, CA: Rand Corporation.

Franklin, Sarah 2007. *Dolly* Mixtures: The Remaking of Genealogy. Durham, NC: Duke University Press.

Fresquez, P.R. 2004. Interview with the Author, Los Alamos, NM, October 3.

German, John 2002. Can Bees Detect Landmines? Sandia Helps Montana Researcher Try Training Bees To Find Buried Landmines. News Release Sandia National Lab, April 23, 51(8):1–3.

Grove, Richard 1995. *Green Imperialism: Colonial Expansion, Tropical Island Edens and the Origins of Environmentalism, 1600–1860* (Studies in Environment and History). Cambridge: Cambridge University Press.

Haarmann, Tim 2009. Interview with Author, Los Alamos, NM.

Hanson, Todd 2006. Bringing in the "Bee Team." *Los Alamos National Laboratory NewsLetter*, November 20 7(24):1–3.

Haraway, Donna 1989. *Primate Visions: Gender, Race, and Nature in the World of Modern Science*. London: Routledge.

—— 1991. *Simians, Cyborgs and Women: The Reinvention of Nature*. New York: Routledge.

—— 2008. *When Species Meet*. Minneapolis: University of Minnesota Press.

Horn, Tammy 2005. *Bees in America*, Lexington: University of Kentucky Press.

Jackson, Richard 2005. *Writing the War On Terrorism: Language, Politics and Counter-Terrorism*. Manchester: Manchester University Press.

Johnson, Colin 2007. DARPA Hatches Plan for Insect Cyborgs to Fly Reconnaissance. *EE Times*, October 3, http://www.eetimes.com/news/semi/showArticle.jhtml;jsessionid =

Lal, Amit 2004. Phone interview with the author, 2006.

Lambert, Captain John D. 2003. Unmanned Undersea Vehicles (UUV) Program and Potential Swarming Applications (PPT) in Conference on Swarming and Network Enabled Command, Control, Communications, Computers, Intelligence, Surveillance, and Reconnaissance (C4ISR). *Complexity Digest*. Retrieved on 2007-12-11.

Latour, Bruno 1991. *We Have Never Been Modern*, trans. Catherine Porter. Cambridge, MA: Harvard University Press.

—— 2004. *Politics of Nature: How to Bring the Sciences into Democracy*. Cambridge, MA: Harvard University Press.

Lockwood, Jeffery 2008. *Six-Legged Soldiers: Using Insects as Weapons of War*. Oxford: Oxford University Press.

McCabe, Kirsten 2008. Interview with Author, Los Alamos, NM.

McCabe, Kirsten and Robert Wingo, 2008 Interview with the Los Alamos National Laboratory Insect Sensor Project Team, Los Alamos, NM, May.

McClintock, Anne, 1995. *Imperial Leather: Race, Gender and Sexuality in the Colonial Contest*. New York: Routledge.

Marx, Karl 1990. [1867] *Das Kapital*. Vol.1. New York: Penguin Classics.

Masco, Joseph 2004. Mutant Ecologies: Radioactive Life in Post-Cold War New Mexico. *Cultural Anthropology*. 19(4): 517–550.

——— 2006. *The Nuclear Borderlands: The Manhattan Project in Post-Cold War New Mexico*. Princeton, NJ: Princeton University Press.

——— 2008. Paranoid Empire: Spectors from Guantanamo and Abu Ghraib. BBRG Keynote Lecture with the Townsend Center for the Humanities, UC Berkeley, October 9.

Massumi, Brian 1992. A User's Guide to Capitalism and Schizophrenia: Deviations from Deleuze and Guattari. Boston: MIT Press.

Matless, David 1998. *Landscape and Englishness*. London: Reaktion.

Michener, Charles 1974. *The Social Behavior of the Bees*. Cambridge, MA: Harvard University Press.

Mitchell, Timothy 2002. "Can the Mosquito Speak?" in *Rule of Experts: Egypt, Techno-politics, and Modernity*. Berkeley: University of California Press.

Moore, Donald, Jake Kosek, and Anand Pandian 2003. *Race, Nature, and the Politics of Difference*. Durham, NC and London: Duke University Press.

Mukerji, Chandra 1997. *Territorial Ambitions and the Gardens of Versailles*. Cambridge: Cambridge University Press.

Preston, Claire 2006. *Bee*. London, Reaktion.

Raffles, Hugh 2010. *Insectopedia*. New, York: Pantheon.

Rumsfeld, Donald 2001a. Stakeout at NBC October 1st. See http://www.defenselink.mil/transcripts/transcript.aspx?transcriptid = 1964.

——— 2001b Interview with NBC "Meet the Press" with host Tim Russert, September 30.

Russell, Edmund 1996. Speaking of Annihilation: Mobilizing for War Against Human and Insect Enemies, 1914–45. *Journal of American History*, Mar 96, Vol. 82 Issue 4, pp.1505–1529,

Scherer, Michael 2009. U.S. Bush Torture Memo Approved Use of Insects. *Time Magazine*, April 16.

Schrepfer, Susan, and Philip Scranton 2003 *Industrializing Organisms: Introducing Evolutionary History*. New York: Routledge.

Seeley, Thomas 1995. *The Wisdom of The Hive: The Social Physiology of Honeybee Colonies*, Cambridge, MA: Harvard University Press.

Sinks, Alfred 1944. How Science Made a Better Bee. *Popular Science*, September: 98–102.

Sparke, Matthew 2004. In the Space of Theory: Postfoundational Geographies of the Nation State. Minneapolis: University of Minneapolis Press.

Stealth, Don 2006. Interview with the LANL Insect Sensor Project Team, Los Alamos National Laboratory, Los Alamos, NM, May.

Stephen, William 1969. *The Biology and External Morphology of Bees*. Corvallis, OR: Oregon State University.

Stoler, Ann Laura 2001. Tense and Tender Ties: The Politics of Comparison in North American History and (Post)Colonial Studies. *Journal of American History* 88(3): 829–865.

——— 2008. Imperial Debris: Reflections on Ruins and Ruination. *Cultural Anthroplogy* 23(2): 191–219.

Taussig, Michael 1993. *Memesis and Alterity: A Particular History of the Senses*. New York: Routledge.

Tsing, Anna L. 1995. Empowering Nature, or: Some Gleanings in Bee Culture. In S. Yanagisako and C. Delaney, eds., *Naturalizing Power: Essays in Feminist Cultural Analysis*, New York: Routledge, pp. 113–143.

—— in press. Unruly Edges: Mushrooms as Companion Species. In S. Ghamari-Tabrizi, ed., *Thinking With Donna Haraway*. Cambridge: MIT Press.

Tucker, Richard and Edmund Russell 2004. *Natural Enemy, Natural Ally: Towards an Environmental History of War*. Corvallis, OR: Oregon State University Press.

van Engelsdorp, Dennis, Jerry Hayes, and Jeff Pettis 2010. A Survey of Honey Bee Colony Losses in the U.S. Fall 2008 to Spring 2009. *Journal of Apiculture* 49(1): 7–14.

Vivanco, Luis Antonio 2001. Spectacular Quetzals, Ecotourism, and Environmental Futures in Monte Verde, Costa Rica. *Ethnology* 40(2): 79–92.

Weizman, Eyal 2007. *Hollow Land: Israel's Architecture of Occupation*. New York: Verso.

Wingo, Robert 2008. Interview with Author, Los Alamos, NM.

Winston, Mark 1987. *The Biology of the Honeybee*. Cambridge, MA: Harvard University Press.

Wolfe, Cary 2003. *Zoontologies: The Question of the Animal*. Minneapolis: University of Minnesota Press.

Zerner, Charles, Miriam Ticktin, and Ilana Feldman in press. In Ilana Feldman and Miriam Ticktin, eds., *In the Name of Humanity: The Government of Threat and Care*. Durham, NC: Duke University Press.

Zylinska, Joanna 2009. *Bioethics in the Age of New Media*. Cambridge, MA: MIT Press.

12 Taking the jungle out of the forest: counter-insurgency and the making of national natures

Nancy Lee Peluso and Peter Vandergeest

Introduction

Forests versus jungles – what is the difference? Are all jungles tropical? Are they always violent? How do forests or jungles relate to expressions of state power? And what do violent jungle landscapes of the past tell us about "scientifically" managed forests, forested nature reserves, and "conflict timbers" of the present?

We explore these questions using examples from the Southeast Asian countries of Indonesia, Malaysia, and Thailand, although the relevance of our arguments extends globally, into the forests/jungles of South and Central America and Africa. The objective of this chapter is to put together lessons and ideas from the literatures on the political ecologies of war and of forestry, thereby expanding and deepening recent interventions on "resource wars" (e.g. Klare 2002; Watts 2004; Le Billon 2000, 2001, 2004, 2008).

In critical geography, political ecology, and political science literatures, the notion of resource wars takes various forms, depending on the "nature" of the resource in its geographic and commodity form, and its relative value and accessibility (Peluso and Watts 2001). Much of this literature has focused on oil, diamonds and other gems, and what is known as "conflict timber." In some civil and international wars of the last decade or so, these valuable resources have been harvested and sold on black markets and allegedly legal channels alike, their profits used to buy guns and other arms and to fund revolutionary or other sorts of political violence. In Nigeria, Nicaragua, the Congo, Burma, Cambodia, Angola, Sierra Leone, Bolivia, and other countries experiencing conflict or civil war, the capture and sale of these conflict resources enable and perpetuate violence, a situation that has not escaped the notice of academics, human rights and environmental activists, and government officials (e.g. Reno 1999; Watts 2001; Le Billon 2000, 2001, 2004, 2008; Jarvie *et al.* 2003).

Less well-documented, however, are the ways resources such as forests played different sorts of roles in wars of earlier decades, when access to the heavy equipment of contemporary forest exploitation and multiple market outlets for forest products were neither easy nor common. We argue here that to understand the terms as well as the claims and counter-claims around resource wars, legal and illegal logging, and conservation's discontents it is important to understand the changes in the roles forests played in earlier resource wars.

In West Kalimantan, Indonesia, for example, 1967 was a critical year in the region's and the nation's histories of both political violence and forestry. For the first time, national forests were established in that province and in the three other provinces of Indonesian Borneo. The establishment of national forests not only disregarded the land and forest claims of most resident peoples, but were intertwined with the violent evictions of over 100,000 rural Chinese and Chinese-Indonesians from their homes and farms and a racialized redefinition of the "native" citizenry in these rural areas. These evictions were carried out as part of an intensive and extensive militarization of the province, for two different but related national projects related to the establishment of national power, boundaries, and territorial control (Peluso and Harwell 2001; Davidson and Kammen 2002). Within a few short years, national forest territories had been established and production forest concessions allocated across nearly 60 percent of the province. The biggest concessions went directly to various branches of the army and other military services, and to corporations and companies in joint ventures with military leaders (Robison 1986). Leading to the most extensive deforestation of all time on the island of Borneo – in the aftermath of most of the brutal political violence – these wars over control of the people and spaces of the interior were fought guerrilla style in "jungles" known as "Southeast Asia's Second Front" (Brackman 1966) long before this latter term came to be attached to contemporary discourses of Islamist terrorism.

Guerrilla warfare and the physical enclosure – and dispossession – of rural Chinese also went hand-in-hand in forested areas known as "jungles" of peninsular Malaysia in the preceding decade. During the Malayan "Emergency" (1948–60), when British colonial forces fought an insurgency run by the Malayan Communist Party, Chinese were rounded up and put into villages surrounded by barbed wire, both to facilitate their surveillance and to "protect" the non-communists among them. Malays in border areas as well as Orang Asli were also consolidated into settlements within which they were given new land to grow crops and build better homes, while reserve forests were demarcated in the areas that were cleared of people. Even before the Emergency was declared officially over, foresters talked about things being back to "normal" as they returned to work in the upland and lowland forests occupied by "jungle insurgents" from about 1948 to 1959/60 to implement new forest management models that allowed for accelerated rates of logging. Similar barbed wire enclosures and village consolidations were enforced during a later "Emergency" in Sarawak, one of the Borneo states (along with Sabah) that joined the Federation of Malaysia in 1963 (Porritt 2004).

In Thailand, insurgency rendered jungles out of forests in the 1960s and 1970s when the Communist Party of Thailand (CPT) and the student movement left the cities for the "countryside" of the forested border mountains. These jungles, like the others mentioned above, were re-appropriated by the state through counter-insurgency actions, which lasted until the end of that insurgency in the early 1980s. Some became production forests and protected areas, others industrial agriculture, and others were allocated as smallholdings. All were territorial solutions meant to quell insurgent violence.

We argue here that the nature of forests and their representation – as jungles or national or local territories – matters to understandings of political violence, nation-state building, and forestry alike. This is important to contemporary debates about resource wars, which need to better examine the ways tropical forests as theatres of insurgency have been key in shaping them as *political* entities (political forests) in the first place. Le Billon (2004, 2008) and Watts (2004) have argued this as well, emphasizing the importance of historical, geographic, and spatial specificities of resource-related conflicts, and their embeddedness within conflicts broader than the spaces of the resources themselves (see also Peluso and Watts 2001). Where the previously mentioned authors have focused primarily on gems (especially diamonds) and oil, we focus here on tropical forests. Current work on "conflict timber" literally does not see the forests for the trees – except as sites of "deforestation," articulating a different sort of argument than the one we are forwarding here.

This blind spot thus obscures two important lessons about nation-building and the construction of national natures. First, forests are not simply ecological configurations, but political and politicized zones. When they are depicted as "jungles," a particular set of geographic and political imaginaries used to justify state violence comes into play and generates particular kinds of territorial controls. Second, whether or not trees remain in these contested spaces, the very sites of contestation – often borders, often mountainous or swampy, all dubbed "jungles" – help in realizing nation-building projects *through* violence, militarization, resettlement, and other territorial practices of counter-insurgency.

The articulations of war and forestry thus help make both territorial nation-states and political forests. Yet, the effects of war as a specific form of political violence on the making of forests, on the practice of forestry, and on the consolidation of national states' territorial control through forestry are barely documented and narrowly theorized.[1] We argue here for differentiating forms of and motivations for political violence in order to better understand how insurgency and counter-insurgency can bring national forests into being or strengthen existing state forestry institutions (Haraway 1991; Neumann 2004; Sundberg 2009). The cases from Southeast Asia in the 1950s–1970s serve to demonstrate the ways both insurgencies and counter-insurgencies have enabled the extension, establishment, and normalization of political forests. The insurgencies at this time, in the immediate aftermath of nationhood or during the creation of post-colonial states, constituted what we call "alternative civilizing projects." They took place in historical moments and geographic sites where the territorial influence of new nation-states on everyday practices and loyalties was still tentative. Counter-insurgency helped normalize political forests as components of the modern nation-state and spatially and institutionally differentiated forests and agricultural areas. In the process, counter-insurgency operations laid the groundwork for newly racialized and nationalized forests and citizen-subjects. They also enabled the transfer of military technologies to state forestry departments, which both benefited and suffered from the militarization of forests during and after violence stopped.

We therefore demonstrate that state forestry and contemporary ideas of political forests emerged not just from the dissemination and local transformations of scientific models from Europe (during the colonial period) and from the FAO and other international forestry agencies (during the early postcolonial period),[2] but also out of the violent politics through which new national states and their ideologies were made after World War II. The Cold War period established or secured many national identities, but the critical importance of forests – and the jungles from which they were carved – are rarely considered in political analyses of the Cold War (for recent examples see, Mamdani 2004; Westad 2006). Yet, forests were configured as quintessentially national or state natural resources during the Cold War, when the "jungles" in Asia, Africa, and Latin America became synonymous with political violence.

We focus here on three Southeast Asian nation-states, Indonesia, Malaysia, and Thailand. None of these became communist or Islamic states in the Cold War era or after, but communist or Islamist insurgencies or fears of them were important elements in national politics of the time. In all of these nation-states, extensive insurgent activities were based in forested areas called "jungles," "mountains," or "hills" – terms with specific political valences in those times. In all three nation-states, government forces prevailed through counter-insurgent activity, unlike some Southeast Asian countries where insurgent forces won and took over the national government. As insurgent violence was repressed, the shifting cultural politics of states' discursive and spatial practices around forests and forest-based subjects became fundamental to understanding the making of the nation-states and national natures.[3]

Our argument consists of four connected parts: first, the alternative civilizing projects (political opposition) and the violence characterizing them produced effects on forests and forestry that cannot be understood only through the lenses of either colonial-era forest-making, or of late twentieth-century notions of "resource wars." By constraining some types of land use practices and enabling others, insurgency and counter-insurgency brought political forests into being and vastly extended national forest territories. Concurrently, the particular materialities of tropical forests – their biological and ecological properties and their spatialities in terms of locations and extents, enabled the kind of violent engagement – guerrilla warfare – associated with Cold War era "jungle wars," insurgencies, civil wars, or revolution.

Second, counter-insurgency strategies included transforming the "jungles" of wartime discursively, practically, and institutionally to "forests" (Slater 1995; Peluso 2003a; Sioh 2004). This process involved constructing, undoing, and reconstituting the spatialized society–nature relations inherent in the term "jungle," in particular the practice of shifting cultivation as a form of agro-forestry. Shifting from jungles to forests distinguished *forests* from *agriculture* politically, spatially, and territorially in terms of both agency jurisdiction and legitimate land users, building on earlier legal-institutional efforts to do so (Dove 1993; Sivaramakrishnan 1999; Agrawal and Sivaramakrishnan 2001; Peluso and Vandergeest 2001; Potter 2003; Vandergeest and Peluso 2006a, 2006b; Forsythe and Walker 2008).

Third, State responses to jungle-based insurgencies involved massive spatial reorganizations of populations through resettlement, colonization, and the territorial re-zoning of property rights. These reorganizations of space and of peoples' relations to national space were racialized – i.e. differentiated along ethnic and racial lines, producing citizen-subject categories of "minorities" and "majorities" in relation to the *national* identities emergent in this time period (Anderson 1991). Racialization created political subjectivities as state authorities represented some racial/ethnic groups as more loyal to the new nation-state than others. These representations were the bases for differentiated state territorial practices toward subjects labeled minorities or majorities, particularly peoples who were "tribalized" by their associations with jungles, and rendered "violent" through their territorial associations with insurgent groups.

Fourth, Insurgency and nation-state building processes stimulated both a militarization of jungles and the production of expensive technologies for accessing and surveilling them. Both militarization (the mobilization of troops for fighting, patrolling, or other security activities) and the deployment of military technologies articulated with the intentions and needs of forest managers.[4] The expense of such technologies had previously precluded their extensive development and application for *forest* surveillance, particularly before the timber industry became an important part of these new national resource-extracting economies and before the rise of big conservation (Leigh 1998). After insurgencies were quelled, forestry and conservation benefited from the technologies developed for jungle warfare and counter-insurgency.

In sum, we bring the political ecologies of war and forests together through our analysis of a specific moment in global conflict. We contend that the making of national political forests (legislated, zoned, mapped, classified, and managed by professional, "scientific" government agencies) was interwined with the violent making of nation-state territories and political subjects through common repertoires of violent state practices. We argue that certain forms of political violence are more likely than others to bring forests into being as political entities/institutions, and enable their continued "recognition" as territorial subjects.[5] Indeed, the formation of both territories and people as subjects of state governance and surveillance have much in common, as usefully demonstrated by Sioh (1998) for forests in Malaya. State territories are thus embodied, and bodies are territorialized through racialization.[6]

Positioning the practices of insurgency and counter-insurgency in relation to the making of national forests makes several contributions to both the geographical studies of war and to political ecology scholarship. Until now, the main area of intersection has been through the lens provided by the notion of "resource wars." The resource wars argument – that valuable resources motivate, fund, and fuel many contemporary civil wars – does not account for all the ways forest-based violence has affected forest-making and state-making. The slippages between insurgent places – such as "jungles" and "mountains" – and insurgent bodies, often caused the people living in these places to be targeted as disloyal subjects and recalcitrant citizens long past the time when political differences were resolved. Recognizing the importance of forest-based insurgent activity in the development of political zones such as "forests" or "agriculture" thus helps us understand better larger questions of state territoriality, sovereignty, and governance.

Antecedents of "resource wars": violence making forests, and forests making nations

> Ironically, twenty years of war saved Cambodia's forests from the destruction associated with economic growth in the ASEAN region.
>
> (Le Billon 2000:786).

The territorial dimensions of war have been discussed in various ways in geographical treatments of interstate wars, guerrilla warfare and other forms of violent insurrection, and repression. Geographical scholars[7] have examined how political territories and borders are made through war and violence, and how wars have in turn been shaped by these territories and borders (Flint 2005:5; Murphy 2005; Thongchai 1997). Political territories and borders that are created through wars are fluid, and not limited to those associated with national states. They can include, for example, a barricaded neighborhood in an urban insurrection, transnational spaces partially controlled by guerillas during the night (Flint 2005:6; Feldman 1991; McColl 1967), or a sacred space (Stump 2005). Among the diverse ways that political violence and territoriality are bound up with each other, the processes of state territorialization in the pursuit of sovereignty – understood as control over the deployment of "legitimate" violence in a territory – looms large. State territorialization works at multiple scales and through diverse encounters with those who continue to contest state territorialization "from within" (Wainwright and Robertson 2003). These encounters can range from battles over the building of a highway through land claimed by indigenous people in the US, to struggles over forest access in South Asia (Sivaramakrishnan 1997) and Southeast Asia (Vandergeest and Peluso 1995b). State territorialization, moreover, is always unfinished and contested by people who remember the violence by which colonial and postcolonial states fought challenges to their territorialization practices (Vandergeest and Peluso 1995a; Wainwright and Robertson 2003:213).

Geographers have also demonstrated that territories and spaces of nature are not fixed but fluid, and produced through human and non-human activities/agency.[8] Forests have drawn attention in several distinct geographies of war. Recent quantitatively-oriented research that seeks to understand the spatial distribution of conflict and violence includes attention to the question of whether the cover or resources provided by forests increases the likelihood of political violence in these areas (Rustad *et al.* 2007; Buhaug and Lujala 2005; O'Loughlin and Witman forthcoming). More important for our purposes is the scholarship on resource wars that has focused on the relationship between warfare and access to valuable natural resources, although the ways war helps construct forests as a category of state power and jurisdiction has not generally been a part of that discussion.

How do we understand the connections between forests, territoriality, and war? In what is usually a distinct literature from that on the geography of war, authors writing on the political ecology of forests have spent a great deal of effort analyzing forests' constitution as territories or territorialized property rights, whether these are claimed or held by states, corporations, kinship groups, customary institutions of various sorts, or individuals.[9] Yet, where forest-based violence is concerned, most political ecologies of forestry do not deal with the kind of political violence associated with war *per se* (as opposed to violence between foresters and villagers, for example), nor do they attempt to understand war's role in the discursive, institutional, and material making of the forest (cf. Hecht and Cockburn 1989). Indeed, it is quite a strange lacuna in this literature, because of the important roles that forests played – albeit dressed up as "jungles" – in the revolutionary wars that led to the establishment of many post WWII nation-states. In addition, an extensive, virtually stand-alone literature on the conflictual, often violent social relations around conservation areas (only sometimes constituting forests) and nature reserves in Africa has called up the question of "what difference nature makes" in conservation wars or wildlife wars (Neumann 2001, 2004; Anderson and Grove 1989; Brockington 2002). Neumann (2004) in particular has theorized the normalization of conservation areas and practices in nation and state building and how conservation has changed notions of sovereignty and state territory.

Philippe Le Billon is one of the few authors who explicitly uses a political ecology approach to frame war and post-war social relations in forests. In a path-breaking article in 2001, he showed how shifting political conditions in and around a troubled but re-emergent nation-state – Cambodia – created a "new frontier of capitalism" (p. 791) in the nation's as-yet-unexploited, timber-producing, forest areas. Like other forest frontiers constructed through the conjunctural convergences of newly commoditized timber resources, access to markets and global consumers, and the activities of various types of entrepreneurs (from corporate to freelance pirates), Cambodia's forest frontier was soon subject to massive resource extraction.

The actors in this 1990s forest drama, unfolding in the aftermath of the Khmer Rouge's previous regime of extreme violence and social displacement, were both connected through and competitive for the as-yet-untapped timber wealth of the democratizing country's forests. The democratizing state and its panoply of supporting international and national institutions was one set of contenders. Incorporating into the global political economy at a time when national state power was decentralizing and weakening across the globe, Cambodia was dependent on both international aid and international advice on how to restructure and benefit from its debut as a capitalist economy. Almost forgotten in international representations were its previous characterizations on the world stage as a rogue state, whose very name evoked horrific images of "killing fields." This first set of players had gained a foothold on the domain of the law, setting the terms of the "legal" in their own favor. This brought them into immediate contention with another set of players – insurgents – hoping to similarly benefit from the concurrent rise of democracy and capitalism. The insurgents' claims and practices, however, had been rendered "illegal," by the nascent nation-state. What constitutes "illegal logging," including some carried out by associates of high government officials, is one of Le Billon's key concerns in that landmark paper.

Various iterations of this and other resource wars arguments have traveled widely; geographers and political scientists have applied differently defined notions of "resource wars" and "conflict timber" to forests beyond Cambodia (see, e.g. Ross 2003; Bryant 2007) or critiqued overly quantified and generalized versions of it (e.g. Watts 2004, Le Billon 2004, 2007). The argument also does not extend well to forests as strategic territories, whether during war or peacetime, nor to forests before the times that occupying insurgents were able to easily access international markets to sell "conflict timber" to fund their revolutionary activities. For example, the violent rush of colonial states and other actors to claim teak (in Southeast Asia and India) during the eighteenth and nineteenth centuries must be seen as an initial step in the making of political forests and fundamental to the extension of these colonial states' territorial power (e.g. Bryant 1997; Peluso 1992; Barber 1989, Vandergeest 1996).

Thus, we argue that the pre-transition phase in Cambodia was not an irony at all, as suggested in this section's epigraph. Rather, those twenty years of war can be seen as exemplary of a common pattern. In Indonesia, Malaysia, and Thailand, the insurgencies fought between 1950 and 1980 helped construct the political forests in those spaces – or led to their state-sanctioned agricultural conversion. In all those conflicts, forests served the purposes of warfare not for the commodity value and wealth of their marketable timbers, but in their roles as both cover and strategic territory. Both ruling national states and alternative civilizing powers fought over these spaces and resources in order to govern or control access to them. As such, the relationship between war and forests can be understood in relation to a longer term history of insurgency in which forests have been important primarily as cover for insurgent guerrillas, producing in turn a systematization of military counter-insurgency practices aimed at controlling both forest territories and resident subjects. These strategies of war and rule have been documented in manuals of war as well as in military books and journals.[10] It was only after the consolidation of national forests through state territorialization and enclosure in Malaysia, Indonesia, and Thailand that the massive commercial logging and conservation efforts that characterized the 1970s through the present moment (2010) could take place. Counter-insurgency contributed to how these spaces were produced as political forests, to widely accepted definitions of legal and illegal logging, and to notions of national sovereignty and territory.

Jungle insurgencies and counter-insurgency in Southeast Asia

The Japanese occupation of much of Southeast Asia (1942–1945) generated forest-based (among other kinds of) political resistance, first to the Japanese troops and military government occupation officials and subsequently to colonial powers returning to the region after WWII. Both periods of political violence involved occupation, war, and revolution, resulting in forest destruction and major population movements (Soepardi 1974; Kathirithamby-Wells 2005). During these wars, state forestry continued in those parts of Southeast Asia that had been organized by colonial-era forestry departments, but the mandate for state agencies was generally to contribute to the war effort. In Java, Indonesia, for example, the Forestry Department was put under the Japanese Department of War. The legacies of war included extensive timber cutting and the production of other crops (such as castor oil plants) for strategic purposes. The Japanese forced people to colonize certain forests to cut timber for industrial fuel or to grow castor oil; in some places villagers hid in forests to escape Japanese occupiers. Even colonial-era foresters destroyed forests and forestry infrastructure as part of the allies' scorched earth policies (Soepardi 1974).

After World War II, the British returned to Malaya, and the Dutch to Indonesia. Thailand had not been formally colonized, and as a Japanese ally, was not occupied formally during the war. In Indonesia, the returning Dutch faced immediate resistance. The Indonesian Revolution (1945–1949) still affected forestry in Java primarily, the main site of colonial forestry control – especially in the island's teak forests, from which the republican revolutionary government ordered fuelwood to be harvested. Teak fuelwood was used to power trains and teak timbers were used to build railroads and roads. In late colonial Malaya, the British declared an Emergency in 1947, as the Malayan Communist Party (MCP), which had conducted an armed, jungle-based resistance to the Japanese occupation, became increasingly militant against the returning British. The planned decolonization of the peninsula was delayed as the British responded with counter-insurgency tactics that became a model for subsequent counter-insurgencies around the world, as described below (Osborne 1968). Malaya became independent in 1957 after the insurgency had been effectively defeated. In 1963, Singapore and the Borneo states of Sarawak and Sabah joined Malaya to form the Federation of Malaysia. Singapore soon withdrew, establishing itself as a nation-state. In Sarawak, armed resistance to this Federation by communist forces was supported internationally by, among others, neighboring Indonesia's President Sukarno. This produced another jungle-based international border conflict, an international one, the so-called *Konfrontasi* (Confrontation) (1963–1966).

The Malayan and Sarawak Emergencies and Confrontation were early among these Cold War-era, jungle-based insurgencies, which affected every Southeast Asian country with significant non-urban territories, as well as many other countries around the world.[11] Anti-colonial revolutions of the post-WWII period had been viewed as the models for these Cold War period insurgencies (see, e.g. McColl 1967). Such insurgencies are increasingly dubbed "civil wars," but by any name, characterized the situations in many new nation-states during the post-WWII decades, as competing groups fought for control of national regimes' ideological and practical structure. We base our arguments here on the following cases: the 1948–1957 Malayan Emergency in Peninsular Malaya; the Communist Party of Thailand's insurgency from the mid-1960s to the early 1980s; the violence between Indonesia and Malaysia (Sarawak) in Borneo as part of Confrontation (1963–1966), and the complex communist-led insurgencies in Indonesia and Sarawak in the 1960s and early 1970s.[12] The Indonesian national government carried out some eight years (1966–1974) of counter-insurgency operations in Sarawak and West Kalimantan.[13]

Most of these insurgencies were inspired by Maoist revolutionary ideas, encapsulated in the phrase, "Let the countryside surround the cities." This slogan was meant to mobilize insurgents to influence, organize, and inspire peasants and other rural subjects (jungle-based or not) to rise up and take over the cities where newly emergent nation-state governments were based. It was a call to arms explicitly in search of territory (McColl 1967). Not all forest-based insurgencies of this period were communist, however. Islamic militants desiring an Islamic state in newly independent Indonesia launched rebellions in certain regions of Indonesia (1950–1957), particularly in mountainous areas of western Java, Sumatra, and Sulawesi. We include them in our considerations here, as they also represented alternative civilizing projects, and used the tactics of guerrilla warfare employed by Maoist-inspired insurgents. They were part of the *Darul Islam* (Islamic State) movement and involved *Tentara Islam Indonesia* (the Islamic Army of Indonesia), referred to in Indonesia and among scholars as "DI/TII." Unlike the Maoist rebellions, the US, Britain, and other "western" powers supported these Islamist conflicts, as at that time they worried more about the geopolitical loyalties of then-president Sukarno – who they saw as too accommodating toward communism.

We differentiate the political nature of insurgent forest-based violence from the local resistance and violent contestations that resulted from the imposition of colonial or post-colonial forestry controls.[14] What was contested in the insurgencies and counterinsurgencies discussed here was not access to forest resources or land *per se*, but the ideologies, territorial forms, and hegemony of the emergent nation-state itself – a point aptly demonstrated by E.P. Thompson in his 1975 classic text on eighteenthcentury England, *Whigs and Hunters*. "Power and property rights" were at issue then as well as in these late twentieth-century jungles. Indeed, many jungle-based insurgencies took place in what were at the time heavily forested border or difficult-access mountainous areas, where political forests had yet to be created or where both state forest management and state power was ineffective and weak because it was either not economic or practical (see Table 12.1 and Figure 12.1). Further, insurgents did not reject the nation-state form but opposed the guiding ideologies, and in some cases, the territorial composition, of the nation-states taking shape in the wake of the Japanese occupation of World War II.

Some insurgent groups enlisted, attracted, or forced local people already living in and around these forested areas to engage in anti-state violence or to provide shelter and provisions to their guerrillas. However, the political violence *generally* was not started by resident forest villagers. Rather, students, organizers, party members, combatants, and other participants "went down" to the countryside or "into" the jungles and mountains intending to carry out (to launch or continue previously started) insurgencies from there. Their strategies included training or convincing villagers of the advantages of resisting the nation-state and winning their "hearts and minds." This latter strategy became a standard counter-insurgency strategy as well.

The degree to which local people were actually engaged in these alternative state projects varied, which is also important to understanding the site-specific ways in which political ecologies of war and forestry come together. In some insurgencies, for example, a strict differentiation of the actors as "external" insurgents and "local" people is misleading. But these representations are crucial to understanding the responses and legitimating narratives of ruling states in the course of violence and the subsequent imposition of state forestry, and also to comprehend the continuing divisions of the forested and cultivated components of these agrarian/agroforestry environments into "forests" and "agriculture." When insurgent forces took control of these jungle areas, they helped create the conditions for more intense and violent nation-state activities in these localities, solidifying the incorporation of remote/ border territories and subjects within the political geobodies of these still contested nation-states (Thongchai 1997; Trouillot 1991; Li 1999).

Table 12.1 shows the difference in areas of political forests that were gazetted or reserved by states in these three countries during the 1930s, at the peak of the region's pre-WWII colonial power, and again during the mid-1980s, when insurgencies were effectively over and most forest reservation completed. The table indicates that the regions most associated with the international trade boom in tropical hardwoods of the 1950s through 1980s were NOT the same as those reserved as forest in the colonial era. Java and the Federated Malay States (now part of peninsular Malaysia) were the most successful sites of colonial forest practice, and were the sites where the most permanent, political forests were formally created under colonial rule. It was only after World War II that Sarawak (Malaysian Borneo), Thailand, and Kalimantan (Indonesian Borneo) had significant percentages of their landed area set aside as political forests.

These upland and border "forests" were rarely if ever pristine, untouched forests at the onset of the insurgencies. These areas have very deep histories of human occupation. Most had been long occupied and farmed by swidden cultivators, settled agriculturalists, and hunter-gatherers; thus the term "jungles" actually better described the mélange of conditions on the ground. Everyday access to these areas

Table 12.1 Political forest areas in Malaysia, Indonesia, and Thailand

Site	% Land Reserved by Government as Forest 1930 (approx.)	% Land Reserved by Government as Forest, mid-1980s	Increase in Percent of Land Reserved as Forest 1930s– 1980s (approx.)
Java (Indonesia)	17 (1929)	19.9	2.9
FMS (Malaysia)	27.6 (1939)	24 (1976)[a]	−3.6
Kedah (Malaysia)	27 (1939)	32.6[b]	5.6
Siam (Thailand)	0	42	42
Sarawak (Malaysia)	0.8 (1929)	37.6	36.8
Dutch Borneo/Kalimantan (Indonesia)	0.007 (1927)	82[c]	82

Note: the post-war nations that these colonial territories became part of are in parentheses

Source: Vandergeest and Peluso 2006a:36.

a Mahmud 1979:90.
b "Logging Industry in Peninsular Malaysia: A case Study in Perak."
c This amount includes about 15% of the land cover of all four provinces of Kalimantan. That land, according to the Tata Guna Hutan Kesepakatan (TGHK; Forest Map Governance Agreement), could be converted to other uses, but at the time was under the jurisdiction of the Ministry of Forestry. Calculated from Departemen Kehutanan 1986: Vol. III: 87; the number seems high as the amount estimated for national forest territory varies in published accounts from 70–74%. For West Kalimantan alone, the amount was 59%.

for the purpose of farming, hunting, and other livelihood activities was still controlled largely by the people who lived there when insurgencies broke out (Bowie 1992; Jonsson 2005, Li 1999, Peluso 1992). Even in cases where colonial state territories had encompassed them on paper, state control of them through effective practice on the ground was still elusive or tentative. In each region, the specific ways that governments asserted control over these peopled, violent spaces they pejoratively called jungles or mountains or hills shaped the future practices of state forestry, the forests themselves, and the relationships between the forest-based subjects and the national states.

A clash of civilizations

Before going into detail on the processes of "taking the jungle out of the forest," we should clarify the point that many of the groups involved in forest-based political violence during the Cold War were no longer resisting territorial incursions by returning colonizing states, but aimed to build alternatively-oriented states with strong rural bases, based on Marxist, Maoist, or Islamist ideas. A mixing of ideas at this time was common across these various "people's" movements. For example, a famous Indonesian general wrote about the strategic importance of jungles and mountains in his book on guerilla warfare, drawing freely on Maoist strategy (Nasution 1953). He had been positioned in two different ways in relation to guerrilla warfare: first as a republican guerrilla fighter during the Indonesian revolution against Dutch colonial power and later as a general in the national Indonesian army. In the latter role, he used his own experience as a guerrilla to strategize against Islamist insurgents fighting Sukarno's syncretic nationalist vision. At the same time, these insurgents saw themselves as alternative nationalists, *not* as outsiders living in areas that might be construed as "non-state spaces" (Scott 1998). In the case of DI/TII, they were former members of Indonesia's national army and seeking to control the state, not secede from it.

Territorial control was a central feature of the Maoist model that emphasized the need for a "base area" from which insurgents could operate and eventually surround the cities. According to various theories of revolution, base areas were believed most effective if they had access to major political targets such as cities or transportation infrastructure, were sites of previous political violence indicating rural populations alienated from urban states, had potential for gaining logistic and provisioning support from local populations, and had terrain with cover (McColl 1967).

Except in the Borneo territories, the communist parties in Southeast Asia were initially largely urban-based, and moved to the "countryside" after they were criminalized or violently attacked by national forces. Indonesia differed from Thailand and Malaysia in the sense that the Communist Party (PKI) and other leftwing parties and affiliate organizations were legal; they constituted key players in Indonesian politics until March 1966, when Indonesia's second president, Suharto, criminalized them. Communism was perhaps best organized in urban areas of Indonesia, but various pushes to engage workers in the forestry and plantation industries in Java and Sumatra, as well as peasants all over the country, had also generated sizable rural organizations.[15]

Sarawak was something of an exception here, having only one population center, Kuching, that could be called a "city," and its population was small. Sarawak was also a state of smallholder agriculturalists, with few landless peasants or urban proletarians of which to speak (Porritt 2004). However, nation-state-related territorial questions *did* concern some of its citizens, who opposed an alliance of Peninsular Malaya with Sarawak and Sabah in the form the British proposed as The Federation of Malaysia. Communists and some other organizations were more interested in either a "North Kalimantan" state comprised of Brunei, Sabah, and Sarawak, or some kind of territorial connection with the Indonesian parts of Borneo; imaginings that constituted what we have been calling an alternative civilizational project. Insurgents operated out of border-area jungles on either side of the inter-national border between Malaysia and Indonesia.

While jungle and mountain-based insurgents may have understood themselves as part of alternative civilizing projects, state powers in the region depicted them as "uncivilized" and opposed to the modernizing goals of the urban-based states. Similar "wild" associations were made with the jungle spaces insurgents occupied. Among other things, national state leaders recognized that these competing civil-izational projects had potential to dispose and replace them. This was demonstrated, not least, by the ultimate successes of communism in neighboring nations such as China and the nations formed of the former Indochina.

Articulations of counter-insurgency and political forestry

The making of the forest was articulated with state-making by differentiating forest territories – political forests – from other kinds of land use zones, and the making of racialized political subjects. In the remainder of this chapter we draw on some specific cases in today's Malaysia, Indonesia, and Thailand. We examine some of the discursive and material practices through which nation-state territorialities were constructed through insurgency and counterinsurgency, and how counter-insurgency contributed to the making of political forests and the racialization of bodies and territories. We focus on three key processes by state institutions and actors:

1. "Taking the jungle out of the forest": making clear boundaries between forests and agricultural areas, in large part by criminalizing what was deemed to be agriculture in forested areas. This involved discursive strategies as well as material practices such as reservation of forests and the designation of certain areas for settlement and permanent agriculture (industrial or smallholder private).

2. The relocation of people into and out of these forest areas through resettlement, evictions, and consolidations of settlements, with specific practices frequently based on racialized understandings of loyalty to the nation-state.

3. The militarization of forest areas through the deployment of troops, establishment of military bases, and transferring personnel and technologies of counter-insurgency and surveillance from the military to forest management agencies and timber companies.

Reorganizing space and reconstituting the nation: taking the jungle out of the forest

Taking the jungle out of the forest involved first depicting the jungle as a wild place occupied by wild people. The rhetoric of jungles was important to both insurgents and counter-insurgency; jungles, like mountains, hills, uplands, and mangrove swamps were marginal politically – at the edges of state power, which was one reason insurgents chose them for their base areas. They also provided cover – physically and politically – both in the nature of their terrain and by winning over or terrorizing the inhabitants of these areas. Insurgents saw themselves as alternatively oriented nationalists, but states characterized them as recalcitrant subjects, wild people, not quite full citizens or even fully human. When they went into the jungles or the mountains, they could be further represented as living in wild, uncontrolled places. These representations lent legitimacy to state projects to control these regions and the people associated with them – both long-settled peoples and the insurgents who moved in.

A key aspect of counter-insurgency practice was thwarting insurgents' access to local people by cutting the links between insurgents and forest residents (McColl 1967). This meant cutting off food and supply lines, as well as preventing physical access between the two sets of jungle dwellers as much as possible, to prevent prior residents' political re-education, recruitment, and logistic or empathetic support. In this way, counter-insurgency goals clearly coincided with those of state forestry (and later, conservation areas and "reserves"): to transform "jungles" from peopled, untamed, dangerous mixtures of people and allegedly wild and separate natures, into more orderly, state managed (or at least administered), and integrated through differentiated forests and agricultural areas, with people settled neatly and securely in villages next to them (Peluso 2003b; Slater 1995).[16]

Counter-insurgency operations involved moving people in at least three ways, each articulating with the concurrent or subsequent objectives and interests of political forestry. First, some residents, notably rural Chinese in Malaya, Sarawak, and Kalimantan, were forced out of the area's jungles. In these places, Chinese were suspected of supporting insurgents and/or of being communist or left-leaning, and thereby of compromising national security. A second strategy involved ordering jungle dwellers to live in consolidated spatial settlements, a strategy known later as "strategic hamlets." Consolidation was thus applied to both Chinese who were forced to live behind barbed wire in Malaya or contained in Sarawak, as well as to upland minorities – tribal peoples – in Thailand and Indonesian Kalimantan, though the strategy's long-term effects on people of these groups differed. A third type of movement was the resettlement of ethnic majorities *into* conflicted jungle zones in large-scale re-settlement or colonization schemes. This was done where the governments believed that colonists would be more dependent on government services, more loyal to the national center, and not supportive of insurgents (Uhlig 1984; Soemadi 1974). Colonists were expected to clear forests for permanent agriculture, changing the environment of the insurgent area as well as the region's ecological makeup.

Regardless of which strategy was adopted – and often all three were deployed – the intention was to divide forests and agriculture into separate territorial-institutional domains of state authority – taking the jungle out of the forest – in order to isolate insurgents from the cover and sustenance provided by the jungle and its inhabitants. Political forests and permanent agriculture were not new technologies of power but served the ruling governments well in these violent border environments.

Racializing insurgent landscapes

Central to counter-insurgency was the minoritization of jungle inhabitants and insurgent populations, and the association of insurgencies with particular racialized populations.[17] Some minority groups were used in counter-insurgency operations, while those who were not were demonized by national governments and militaries. An important but sometimes overlooked point is that so-called tribal peoples (or hill tribes) were remade into national minorities, while lowland "ethnic groups" were constructed as national majorities. Scale mattered, not least because nation-states had just recently come into being as the predominant regional macro-political organization.

In all cases, government authorities and national militaries connected majority and minority racial status with a person or group's alleged political loyalties. Certain ethnic minorities often became suspect national subjects, even if they were full citizens or considered "native." In areas where majority populations lived in both upland and lowland environments, such as in Thailand, uplanders associated with those jungles were more suspect than lowlanders. Association with communist parties and affiliate groups in Malaysia, Indonesia, Thailand or with upland Islamists in Indonesia also rendered individuals living in the vicinities of their strongholds more suspect to national governments. Some ethnic minorities were recruited to aid in the nationalizing projects, or suspected less – the politics of loyalty could shift suddenly, however. Complicating these processes of racializing loyalties and territories was the fact that tribal peoples' purportedly "natural" knowledge of "the jungle" was as important to counter-insurgency operations as it was to resident or mobile insurgents.[18]

Racialization was not a new process: rather, racialized identities were refashioned by military strategists out of pre-existing ideas of subject groups' origins, violent or governable predispositions, territorial histories and presumed associations with insurgent and counter-insurgent forces (see, e.g. Soemadi 1974). Such ideas in some cases had originated with anthropologists, geographers, and customary law specialists (Ellen 1999). What was new was the idea of these groups as *national* minorities, either tribal or formerly alien, and how minority status was seen to relate to national goals.

Racialized ideas were acted upon as if people ascribed ethnic identities possessed certain political characteristics. For example, "tribal" peoples such as Hmong (in Thailand), Karen, Dayaks, and Orang Asli were fashioned as "backward," as well as, in some cases, "innocent," because they lived far from the nation-states' centers and practiced swidden agriculture. But there was also a darker side to what Tsing (1999) called the "green development fantasy" of NGOs in the 1990s who represented forest-based tribal peoples as noble savages with positive ecological sensibilities. That dark side was their representation during and after insurgencies as "savage headhunters" and "wild settlers" who had to be tamed and taught better ways (Peluso 2003a). Paradoxically, many tribal peoples were thus also considered warlike, fierce, and violent with the potential of using their tribal warfare skills to support organized political violence against the urban-based states. Their alleged backwardness in fact fed this representation, following from the idea that they were too primitive to recognize good civilization and reject "bad" alternatives.

The shifting state narratives about forested uplands as spaces inhabited by primitive tribal peoples are thus crucial to unpack for their underlying political content and motives. We are not romantic about the views of those involved in alternative civilizing or state-building projects, who often held similar condescending views of jungle-based people. Leaders, theoreticians, and strategists in these parties were neither "tribal" minorities nor of forest village origins. The treatment of jungle-dwelling people under communism, had it succeeded in Thailand, Indonesia, or Malaysia, may have been as coercive and insistent on removing them from forests as were the national states that succeeded in defeating these insurgencies.[19]

Even more suspect to new national states were rural people of Chinese background self-constructed or ascribed as "Overseas Chinese," "Indonesian Chinese," "Sino-Thai," or locally called by the Chinese dialect or language they spoke most frequently[20] – "Khek" (Hakka), "Foochow," or "Teochieu," among others. In Malaysia and Indonesia, "Chinese" were constructed as "alien" or "migrants." In Malaysia and West Kalimantan, the governments explicitly associated Chinese (as opposed to other ethnic subjects such as Malaysi and Dayaks) with membership in communist parties or groups, although most serious research indicates that some people of nearly all ethnic backgrounds joined or supported communist groups. In the 1950s the Communist Party of Malaya consisted primarily of urban and Chinese intellectuals, and Chinese in rural areas were targeted for resettlement. Many resettled Chinese had moved into rural areas to farm for subsistence during the Japanese Occupation, joining earlier rural and farming Chinese populations (Hack 2001). In rural West Kalimantan, tens of thousands of people were classified as Chinese by the government, and hundreds of thousands had some Chinese ancestry (Heidhues 2003).

A third group of racialized subjects, "national ethnic majorities," presented a more complex picture when comparing across nation-states, especially because of the particularly complex ethnic mixes in Indonesia and Thailand. Javanese constituted a national majority in Indonesia but they were not the only non-tribalized ethnicity – people of various ethnicities associated with Sumatra, Sulawesi, and Bali were considered ethnically different from Javanese but not "tribal."[21] This was in part dependent on whether most ethnic subjects of a particular group professed Islam (or in Bali, Hinduism), as opposed to an animist belief system. Further, because in Indonesia communism was legal and represented by a powerful national party under the first presidential regime (until 1965), it could not be represented as embraced by a single ethnic group; rather, it cut a broad swathe across minority and majority populations. When second president Suharto took over, regional differences shaped the ways adherents to this political ideology were represented. In Java, Bali, and Sumatra, it was clear that people of all sorts of ethnic heritage were communist – struggle was at least ideologically based on class or village lines. In West Kalimantan, however, the military conflated communist sympathies with being both rural and Chinese, even though "Chinese" were as or more likely to be poor as middle class or well off.[22]

Islamist insurgencies in Indonesia, DI/TII, were less racialized, as insurgents and their supporters were mainly part of national or regional ethnic *majorities* in Java, Sumatra, and Sulawesi. In Thailand, the primarily lowland Thais fit a "majority slot," although distinct regional identities in the north, south, and northeast made some people "more" ethnically Thai than others, as described below. In Malaysia, on the other hand, Malays were the dominant national ethnic group, although in East Malaysia (i.e. the Borneo states) this was contentious, as many Malays were migrants from the mainland.

All three nation-states created the conditions for newly racialized landscapes by actively organizing or encouraging the movement of majority or loyal subjects to "remote" jungle areas to cut down forests and convert them to permanent cash cropping. This became a key strategy for pursuing the territorial expansion of national states (De Konick and Dery 1997; Dove 1985). In forest areas where these presumably loyal subjects had been resettled by national policy, and around international border areas where political affiliations had long been mixed and shifting across various political borders, governments constructed majority populations as "needing military protection." This military protection was often pursued by organizing loyal villagers – national minorities or majorities – into self-defense militias such as village scouts or border patrols (e.g. Bowie 1992; Stubbs 1988). Local people in the areas occupied by insurgents were also encouraged to become more tied to central states through incorporation into agricultural development schemes, reforestation of national forests through *taungya*, and other state-sponsored development programs.

Importantly, the landscape effects, property rights, and management goals of counter-insurgency varied. In some cases, counterinsurgency helped to produce forests that were devoid of human settlements – at least from an administrative point of view. This privileged forest resurgence, protection, or extraction. In other cases, counter-insurgency led to the replacement of forests with permanent agriculture, to forestry's detriment and forest decline. The new property rights and state territories – such as forests or industrial agricultural zones – served both accumulation and security purposes. Although this was not the intent, political violence and development represented primitive accumulation at its most basic: states expropriated forest-based (or jungle) subjects' land in the name of national security, reallocating it (often to others) in new forms of property (Glassman 2006) and forced the intensification of market relations through which surplus value could be appropriated. This was later, however.

In the remainder of this section, we present some (necessarily schematic) examples of how the convergences of counter-insurgency and forest-making played out in a few specific sites, and for people of certain ethnic identities. These examples are not meant to be exhaustive, but they will provide examples of different ways that counter-insurgency helped concurrently to produce national states, political forests, and racialized landscapes.

The production of racialized strategic territories in northern borderlands of Malaya

During fieldwork in the northeastern state of Kedah in Malaysia, Vandergeest found forest maps confirmed interview accounts of upper watershed hamlets that were moved to lowland sites during the early 1950s. These hamlets included people classified as Malay (Malay speaking, Islamic), Siamese (Buddhist, Siamese-speaking), and "Sam-Sam" (Siamese-speaking, Islamic). Where the emptied upland areas were not already gazetted as forest reserves, resettlement was accompanied by forest reservation, meant to consolidate the territorial control of the forestry department.

Racial classification did not affect whether people were moved, but it did organize the resettlement process. Ethnic "Siamese" (considered alien populations in Malaya), were contained in fenced camps. "Malays" (glossed as natives and the national ethnic majority) were provided with new villages and land for growing rice and rubber. In a clear example of the making of ethnicity and the racialization of the landscape, Sam-Sam, who were Muslim Siamese speakers, were absorbed into the Malay category when they were moved. They were "made into" Malays through the agricultural practices they were allowed to continue, and through their locations within the new rural landscapes.

After the height of the Emergency had passed, the camps containing Siamese were opened and residents allowed to establish rubber smallholdings on state lands, as had previously been promoted among resettled Malays. Unlike Malays, however, most Siamese never received formal land titles. Their marginal positions were maintained not through violence or coercive movement, but through exclusion from access to the resources of the state, especially the legal recognition of their landholdings through land titles.

The incomplete hegemony of these arrangements has been underlined by older villagers' stories. In villages adjacent to reserve forests in Kedah, interviewees said that displaced villagers continued for decades to travel seasonally to their old village sites to harvest fruit, especially durian from multi-generational trees. These visits ceased only when the fruit trees were submerged by reservoirs from new dams, or claimed by ecotourist resorts located on reservoirs. Today conservationists, government foresters, and their supporters are likely to represent these sites as pristine or conservation forests.

Identifying Chinese as the enemy and enclosing or evicting them from rural "jungles"

During the Malayan and Sarawak "Emergencies" half a million Chinese forest "squatters" were moved into camps called "New Villages" (Stubbs 1988: 286; Sioh 2004). The counter-insurgency link to forestry was different in Malaysia – including both Peninsular Malaya and Sarawak, and Indonesia in Borneo, specifically in West Kalimantan. In Malaysia, the forestry department developed silviculture models specifically for reforesting land where forest cover had been cleared for farming. British forester Wyatt-Smith developed the Malayan Uniform System as a technique for regenerating forests cut by the Japanese occupying army and Malay and Chinese peasants resettled by them during the Occupation (1942–1945). In some areas, the declaration of the Emergency allowed the Forest Department to promote this technique as a scientific rationale for expelling Chinese and other cultivators who occupied forest villages and were suspected of supporting Malayan Communist Party insurgents (Wyatt-Smith 1947, 1949; Ali 1966). These scientific practices were thus both silvicultural management techniques and forest department strategies for reclaiming land as political forests. They were successful because they built on and reinforced counter-insurgency practices.

Similarly, in West Kalimantan, the Indonesian army forcibly engaged "tribes-people," other locals, and non-Chinese migrants to evict long-settled rural Chinese families in the late 1960s, after Suharto's rise to power and the accompanying agrarian and anti-Chinese violence in Java and other parts of Indonesia. All people officially identified as Chinese were glossed as communists or supporters of communist guerillas, many of their village living sites were referred to as "jungles" (*dalam rimba*). They were forced to move to refugee camps and resettlement areas or to find refuge with families in urban areas (Davidson and Kammen 2002; Peluso 2003a, 2003b). Dayaks (various ethno-linguistic groups considered native to Kalimantan), Malays, and other non-Chinese residents were made to prove their loyalties to the Indonesian state by participating in – or not obstructing – these evictions. Some frightened villagers as well as insurgents took refuge or sought new bases in jungle areas closer to the international border with Sarawak. Dayak villagers in particular were forced to support the national military's counter-insurgency during the Indonesian army's subsequent seven years (1967–1974) of jungle operations in the province, mostly by serving as "jungle guides" (Soemadi 1974; Rachman *et al.* 1970). The year the evictions started, 1967, the first *national* Forest Law was established for Indonesia; previously most forested areas had been under the jurisdiction of customary authorities or provincial governors. As national minorities, Dayaks – who had been a provincial majority – now had different political relations with these forests, once they were rendered national. Lines were drawn on maps to create huge national forest territories in West Kalimantan and other provinces, disregarding prior and conflicting customary claims and uses. The

Indonesian army was given a wide swathe of territory (20 kilometers wide) as a security and revenue-producing concession along Kalimantan's 1000-km border with the Malaysian state of Sarawak, expanding the national security role it had been given by first president Sukarno in 1960 (Robison 1986). This extensive border concession under national military control made the violent national state a highly visible material entity in formerly "remote" West Kalimantan.

Planned and spontaneous settlement by "ethnic majorities"

Resettlement and colonialism led to the most diverse outcomes in regard to the choice between turning jungles into political forests, or converting jungles to "permanent" agriculture. West Kalimantan and West Java came to be connected in this way, as many West Javanese (Sundanese) transmigrants were resettled in West Kalimantan. They were (unbeknownst to them) given land in areas that had been forcibly abandoned by Chinese, which were as much inhabited rural areas as jungles (Peluso 2009). The Chinese had left hundreds of thousands of hectares of irrigated and rainfed rice paddies, vegetable garden land and fruit and rubber gardens, more than could be used by locals who remained after the evictions. Transmigration also converted massive amounts of "jungle" to rubber and oil palm production, and transferred property rights from customary to private (Barber and Mathews 2002; Charras 1992; Elmhirst 2004).

In addition, retired or decommissioned soldiers and police were resettled into areas considered dangerous or ongoing security threats *("rawan")* in West Kalimantan. This practice has been used by victorious sides after many wars.[23] These ex-soldiers, as well as the resettlement sites and the new army bases built to accompany new forces stationed there, provided another powerful, everyday symbol of the *national* Indonesian state occupying the former jungles of West Kalimantan, but in a different part of the landscape – heavily populated agroforestry areas (Peluso 2009). The Indonesian military thus gained a symbolic and material West Kalimantan presence at multiple scales: the army fought against insurgency in these jungles (1967–74), the various branches of the military were awarded timber concessions when the forest law was passed (1967–80s), and retired soldiers were given local land to cultivate (1980s–1990s).

In Thailand, agricultural expansion overrode the making of political forests in most areas. Counter-insurgency efforts affected extensive areas that had been or were in the process of being demarcated and gazetted as reserve forests. These were occupied by millions of "spontaneous" migrants, not part of official resettlement programs, but sanctioned by the state (Uhlig 1984; Hirsch 1990; Vandergeest 1996). Although clearing was in violation of forest law, these settlements were condoned and even encouraged by authorities who saw the movement of people into these areas as a way of decreasing forest cover for insurgents, as well as a counter-insurgency strategy aimed at winning over the loyalties of land-poor farmers susceptible to insurgent propaganda. The government also planned and established colonies in forest areas, sponsoring the movement of lowland "Thai" farmers into these areas. Leblond (2009), for example, drawing on his exhaustive research on the question of rural population displacement, describes how the government developed a policy during the 1970s of surrounding insurgent strongholds with deforested land and new villages populated by loyal subjects who received government development programs. The approach was implemented through the Self Defense Border Village project, which established 578 villages between 1978 and 1981 close to borders and communist strongholds around the country as well through smaller royal projects and other programs.

Racialized agricultural conversions in Thailand

Thailand is a useful illustration of the racialization of counter-insurgency operations, as the Thai national state has often been considered less racialized than, for example, Malaysia, where the postcolonial national state has maintained and solidified the racial classifications introduced by the colonial state. The contrasting approaches to counter-insurgency in the North and Northeast of Thailand makes this point. Ethnic "Lao" or "Isan" people, regionally dominant in the Northeast of the country, were considered suspect in terms of *national* loyalty compared to the Central Thai; the region had a history of supporting left-wing politics, and thousands of peasants had joined the jungle-based resistence to the Japanese and Thai alliance during World War II (Somchai 2006:40ff). As Buddhists, however, and wet-rice cultivators who spoke a language close to Thai, they were definitely not politically treated in the same way as upland, tribal peoples of the north. The counter-insurgency/forest management approach was therefore not to resettle them out of forests, but to find better ways of linking them to the urban state center through development projects, and facilitating the expansion of permanent agriculture through roadbuilding and the promotion of upland cash crops linked to international markets. Ways were found to recognize their land rights even in forest reserves (Vandergeest and Peluso 2006b). Not surprisingly then, the most rapid decline of Thailand's forests during this period was in the Northeast. According to the Royal Forestry Department, forest cover in Northeast Thailand declined from about 42 percent in 1961 to just 15 percent in 1985 (Hirsch 1993:55).

In the North of Thailand[24] however, forests were associated with "hill tribes," who were considered much more difficult to enlist into the national civilizing project.[25] Counter-insurgency measures here involved resettling upland ethnic groups into consolidated forest villages, as described above. Hearn (1974:187–188) lists 101 tribal villages that were abandoned or destroyed in northern provinces, and whose occupants were resettled into 13 sites, encompassing around 12,000 people in 1972. Although these resettlement efforts were rapidly abandoned and replaced with the policy described above of surrounding communist strongholds with loyal subjects (LeBlond 2009), the overall effect was in stark contrast to the Northeast: overall there was considerably less settlement of loyal subjects into the forests in the North compared to the Northeast, and the beginning of attempts to limit the access of ethnic minorities (hill tribes) to forests, policies that have since continued and been reinforced as reserve forests were later transformed into national parks and wildlife sanctuaries (Atchara 2009; Sturgeon 2005; Roth 2008). In 1985, a few years after the insurgency ended, forestry department statistics (reproduced in Hirsch 1993:55) showed 50 percent of land area in the North as still under biological forest, down from 69 percent in 1961, compared to the decline from 42 to 15 percent over a similar period in the Northeast. The distinctiveness of approaches to counter-insurgency was not the only reason for these regional differences, as topography (the North being more mountainous) also shaped how people moved. They were, however, major contributing factors to the contrasting outcome for the making of forests in the North and Northeast, and signaled the beginning of the restrictions on "forest farming" (Kunstadter and Chapman 1978) that has drawn the attention of researchers working on the political ecology of forestry in northern Thailand.

We finish this section with a few comments on other ways that forest-based political violence also contributed to the separation of jungle into forest and agriculture. One way was simply that during insurgency and counter-insurgency, forest areas became dangerous places for farmers and forest product (previously called "jungle produce") collectors. For example, on the Indonesian side of Borneo, Dayak villagers said they were afraid to make new swidden fields, fearing Indonesian soldiers from Java and Sumatra would mistake them as rebels. Farmers, especially women, were afraid they might run into combatants (government or oppositional) in the forested areas and stopped grazing cattle and collecting forest products as well. Insurgents, like government soldiers, suspected villagers too. Stories abound in all these study sites about not being able to trust anyone during these times. DI/TII explicitly forbade West Java villagers from burning the forest, afraid that their bases and hiding places would be revealed (Peluso 1992). Moreover, after large tracts of land were allocated to the military in Indonesia, local people were afraid to complain or act if they lost access to customary land, trees, and other forest products.

In sum, through this period of insurgency and counter-insurgency, millions of people moved into and out of jungles. These movements helped set the conditions under which forest departments could subsequently practice forestry and the terms by which they could challenge government claims of exclusive control over political forests. The movements of forest subjects had ecological effects because of changes in everyday and structural forestry practices. This era was also crucial in relation to the refashioning of racialized state subjects, reconstituting their spatial relations to political forests and agricultural areas and their positioning and political relations within the nation-state.

Deploying military resources in jungle emergencies

Our final argument is that conflicts generated by these competing state-building projects drew huge military resources into enhancing surveillance and facilitating state access to and control of forested areas. The effect was to reinforce the coercive power of state forest departments, police, and militaries, and thus their abilities over the long term to enforce the separation of agriculture from forests. We briefly highlight some of the ways that insurgency, militarization, and counter-insurgent programs and practices contributed to transformations in forestry.

A key way that militaries supported forestry was through the intensified surveillance and mapping of forest areas. In Thailand, the Royal Survey Department became in many ways an arm of the US military. The department used aerial photos to produce the well-known series of 1:50,000 maps of all forested areas starting in the 1950s, which were periodically updated based on new aerial photographs. These maps presented topography, vegetation, crops, village locations and so on, and were shared with other government departments, in particular the forestry department, where they became the base maps for forestry work. In particular, they were the base maps used to demarcate reserve forests, with the boundaries of reserve forests often drawn along the contour lines and vegetation zoning (Vandergeest 1996, 2003). By the early 1970s over 40 percent of the terrestrial area of Thailand was demarcated on such maps as reserve forest, with minimal ground checking into local forest use. Similar stories about mapping can be told about Malaysia, Sarawak, and Kalimantan see e.g. (Harper 1997:21; Barr *et al.* 1999).

Rural development in insurgent areas was explicitly a form of counter-insurgency. Field research in the 1990s in Sarawak and West Kalimantan indicated that both the SALCRA scheme for smallholder cash crops in western Sarawak and the distribution of fruit tree seedlings and rubber smallholding projects (PPKR) in West Kalimantan prioritized sites near insurgent bases. In Thailand, the key development program was the Accelerated Rural Development (ARD) scheme, supported by USAID funds (and employing many Peace Corps and CUSO volunteers), in needy provinces – where insurgency was most active. The primary ARD activity was road-building (Muscat 1990).

Roads had multiple purposes, including to provide easier military access for troop and supply movement and surveillance; as integral components of logging operations, forest conversions, and other capitalist projects; and to draw existing populations further into the sphere of central state rule by increasing their access to domestic markets. Roads were often built through reserve forests (Uhlig 1984). Between 1960 and 1980, total road length in Thailand tripled, (Hirsch 1990). It more than tripled in West Kalimantan and Sarawak during the same period. Road-building facilitated "spontaneous" migration, as land-poor farmers flooded to the forests to grow both subsistence and cash crops (maize, cassava, sugarcane) (Uhlig 1984; Hirsch 1990; Cleary and Eaton 1994; Brookfield *et al.* 1995). Road building was critical to transmigration as well as commodity marketing, and of course logging, in West Kalimantan and Sarawak.

Militaries also transferred other technologies and organizational cultures to forestry departments. Helicopters, for example, started out as a technology of war; and became a technology that assisted foresters to monitor forest cover change, rural settlement, the illegal cutting of swidden fields, and generally proved useful for intimidating resident peoples who violated the forest-agriculture boundary (though many continued to do so nonetheless) (Atchara 2009). The organizational structures and institutional patterns of forestry had long imitated the military, as reflected in the territorial structure of forest range management, the rotation of foresters to avoid their becoming too attached to the people in their districts, and in some cases, the arming of forester enforcement units (Kaufman 1960). In some areas, forestry departments and militaries worked together to both control and profit from forest exploitation, as we saw in the example from West Kalimantan, where timber concessions were allocated to PT Yamaker – an army timber concession on the long international border between Indonesian and Malaysian Borneo. Retired military men were (and are) frequently hired by timber companies for security, and both timber companies and other forest product traders generally paid "taxes" to local army bases along their routes from forest to market.

During violence, the deployment of troops – thousands of men – inside and at the edges of the jungles occupied by insurgents kept foresters as well as villagers out – unable to tend their tree or field crops during those times. In Kalimantan, new military bases were built throughout an extensive "border area" that extended south from the international Sarawak border through the city of Singkawang. These symbols of the nation became permanent landscape installments, with territorial jurisdictions over land and forests in the vicinity decades after the end of physical violence.

Even after the insurgents no longer posed a serious challenge to the national states we have discussed here, national security arguments continued to shape the practice of professional forestry in border areas. The fear of further insurgencies helped motivate the reshaping of property rights to land and forest products, the practices around forestry, the location of population settlements and the use of military personnel as private guards for forest enterprises. These practices continued to put foresters, militaries, and big extractive businesses into close connection, and to shape what happened to the forests.

Discussion and conclusion

The overall effect of insurgent political violence and counter-insurgency on forestry practices in Indonesia, Malaysia, and Thailand was to intensify and strengthen the legal and institutional processes used in the making of professional or scientific forestry and forests in Southeast Asia. "Jungles" as theaters of insurgency were tamed through massive rearrangements of property rights, land use zones, vegetative cover, and human settlements. The political violence provided a justification as well as a mechanism – military deployment and tactics – for intensive and extensive national state intervention in landscapes over which it had had only weak hegemonic power. Political violence preceded both forest enclosures and state territorializations.

The period of widespread insurgency we described in southern Southeast Asia was also the period that the FAO was promoting its "forest-for-development" model of professional forestry (Westoby 1987). Forestry for development was generally preceded by enclosure/reservation of forests and the dispossession of rural people from huge tracts of forest lands, except as forest labor. Like forest enclosures and reservation, counter-insurgency operations also aimed to evict people from jungles

– in order to faciliate permanent conversion of the land to industrial agriculture. It is noteworthy that the tactics of counter-insurgency often included government personnel (foresters or military) burning huge tracts of forest to rout out insurgents.

These forms of violence against subjects and forests were followed by apparently more ordered spatial practices – the creation of state territories. As applied in those jungles of Southeast Asia where ruling states did not adopt communist forms, "forestry for development" was not only a strategy for development or forestry, but concurrently for counter-insurgency, nation-state building, and the production of *national* natures. Thus the ideologies and institutional practices associated with the conservation era's romantic notions of preserving "rainforests," "primary forests," and "pristine forests," were both preceded and enabled by this earlier, violent period in which the jungles were made into "primary forests" or divided between political forests and agriculture. These realities of forest history are ignored or forgotten in most contemporary conservation discourse. The peopled jungles of the Cold War era do not fit the notion of pristine environments.

During the Cold War era insurgencies, the jungles of Southeast Asia represented a variety of frontiers: not only those at the edges of "civilization" and national state hegemony, but also the frontiers of brutal extractions of biomass. Contemporary state forestry and the shapes and ecologies of the political forests are as much products of this era as of colonial institutions and discourses. Today, in these countries where centralized, national control was solidified, "jungle" discourses have largely disappeared from references to the managed state forests, nature reserves, and timber concessions that populate the landscape. Use of the term "jungles" continues, however, in reference to those tropical forests where anti-state political violence (insurgency) is occurring in border or other marginal and contested forests. In other words, jungles still exist in certain parts of the Philippines (Mindanao), Burma/Myanmar, and West Papua, to name a few.

Political forests formed of violence are thus like the "imperial debris" recently described by Ann Stoler (2008:193), caught up in "the evasive space of imperial formations past and present as well as the perceptions and practices by which people are forced to reckon with features of those formations in which they remain vividly and imperceptibly bound." The "debris" in this chapter is not of the imperial projects of traditional colonial powers, but re-invention and extension of imperial practices of nation-states working to control national territories by rendering these "natural." We have shown here that jungle counter-insurgency operations were not only concerned with territorial control, but that they also produced racialized subjects connected to the national state and political forests in new ways. Those groups whose loyalty to the central state was most suspect are most likely today to lack formal land rights.

In sum, our major argument has been that it is difficult to understand the shapes and political lives of contemporary forests without understanding their connections to certain kinds of political violence – and that violence itself must be understood in concrete, grounded ways. The "Emergencies" of the Cold War era were qualitatively different than were violent colonial conquests and the structural forestry violence generated in both moments. The outcome is that except in the case of war, insurgency, or adventure tourism, tropical forests are jungles no more

Notes

1 The direct impacts of war on the environment are well-known and have been well-documented. Violence between foresters and forest-based populations has also generated a vast literature.
2 On this, see our earlier work (Peluso and Vandergeest 2001; Vandergeest and Peluso 2006a, 2006b).
3 See also Sioh 1998 and Neumann 2004.
4 The sources of these technologies were largely foreign: the US, Australia, Britain – and in some cases, the Soviet Union and China – setting the stage for new global hegemonies of the post Cold War.
5 On the pitfalls of recognition as national political subjects for indigenous groups, see Povinelli (2002).
6 See, e.g. Pred 2000, Moore 2005, Neumann 2004, Massey 2005.
7 Not all the "geographical scholars" we discuss are in Geography Departments.
8 This is a vast literature. See, e.g. Smith 1991; Neumann 2004; Watts 2003. On forests specifically, see Roth 2007; Potter, 2008; Peluso and Vandergeest 2001.
9 This is a huge literature, for beginners, see, e.g. Hecht and Cockburn 1989; Guha 1990; Peluso 1992; Vandergeest and Peluso 1995a; Bryant 1997; Sivaramakrishnan 1999; Abe *et al.* 2003.
10 For example, see the different versions of the "Small Wars Manual United States Marine Corps," available on the internet (e.g. http://www.au.af.mil/au/awc/awcgate/swm/index.htm); or, on Indonesia, *Fundamentals of Guerrilla Warfare* (1953) by Nasution.
11 Many countries in Central and South America, including Guatamala, Peru, Brazil, and Chile were wracked by similar types of "jungle wars" in the 1950s–1970s.
12 Also during this period of agrarian violence in Indonesia, peasants and landless farmers in Java and Sumatra in communist and socialist groups invaded plantations and private landlords.
13 The latter occurred when second president of Indonesia criminalized communism, and was hunting down both West Kalimantan and Sarawak members of the guerrilla forces trained by Sukarno and his army inside Indonesia (and by the Indonesian army). For a detailed account of these low-impact wars, see e.g. Coppel 1983; Dennis and Grey 1996; Mackie 1974; Davidson and Kammen 2002).
14 These are often the primary subject of political ecologies of forestry (e.g. Guha 1990; Peluso 1992; Bryant 1997; Neumann 1998; Sivaramakrishnan 1999; Roth 2008).
15 Indonesian communism in some ways failed because it organized in rural areas through village patrons, rather than in class-based organizations (Mortimer 1974).
16 Note that not all "jungles" were rendered political forests some were seen as too degraded.
17 In the early twentieth century, colonial administrators and other western observers used the term "race" to refer to different ethnic groups, often classifying them in social evolutionary terms for political purposes. We use the term racialization rather than "ethnicization" because the characteristics ascribed to "races" of people were assumed to be inherent, primordial, genetically transmitted.
18 See e.g. Leary 1995; Endicott 1997; Nasution 1953; Jonsson 2005; Vandergeest 2003; Peluso 2003a).
19 As has been the case in Vietnam and more recently in Laos.
20 Many spoke several dialects in addition to other languages used regionally on a daily basis.
21 Cf Li's (1999) use of "tribal slot" and Michel Trouillot's (Trouillot 1991) now classic "savage slot."

22 See Rachman *et al.*, 1970 on the strategy (Davidson and Kammen 2002).
23 For example, in the US after the revolutionary war, in various Central American countries after the 1980s and 1990s wars; in post-WWII Soviet Union (Brown 1999), ancient and contemporary China (Menzies 1992), and elsewhere.
24 The "north" of Thailand as it is used in Thailand does not include what we have referred to here as "The Northeast."
25 The situation in the north was further complicated by the presence of other military powers interested in opium and the violent politics in Burma, Laos, and China (Sturgeon 2005; McCoy 1972).

Bibliography

Abe, K.I., Wil de Jong, and T.P. Lye. 2003. *The Political Ecology of Tropical Forests in Southeast Asia: Historical Perspectives*. Melbourne, Australia: Trans Pacific Press.

Agrawal, A. and K. Sivaramakrishnan eds. 2001. *Agrarian Environments: Resources, Representations, and Rule in India*. Durham, NC: Duke University Press.

Ali, I. B. H. 1966. "A Critical Review of Malayan Silviculture in the Light of Changing Demand and Form of Timber Utilization." *Malayan Forester* XXIX: 228–238.

Anderson, B. 1991. *Imagined Communities: Reflections on the Origin and Spread of Nationalism*. 2nd ed. London; New York: Verso.

Anderson, D. and Richard H. Grove. 1989. *Conservation in Africa: Peoples, Policies and Practice*. New York, Cambridge: Cambridge University Press.

Atchara, R. 2009. "Constructing the Meanings of Land Resource and a Community in the Context of Globalization." PhD Dissertation, Chiang Mai University, Chiang Mai.

Barber, C. 1989. "State, People and the Environment: The Case of Forests in Java." PhD dissertation, University of California, Berkeley.

Barber, C.V. and E. Mathews. 2002. "State of the Forest, Indonesia." Global Forest Watch Publications. Accessed online June 2009 at http://www.globalforestwatch.org/common/indonesia/sof.indonesia.english.pdf

Barr, C., D. Brown and A. Casson. 1999. Corporate Debt and the Indonesian Forestry Sector. *Unpublished Paper*.

Bowie, K. 1992. "Unraveling the Myth of the Subsistence Economy: Textile Production in Nineteenth-Century Northern Thailand." *The Journal of Asian Studies* 51 (4): 797–823.

Brackman, Arnold C. 1966. *Southeast Asia's Second Front: The Power Struggle in the Malay Archipelago*. New York: Frederick A. Praeger,

Brockington, D. ed. 2002. *Fortress Conservation: The Preservation of the Mkomazi Game Reserve*. Oxford: Oxford University Press.

Brookfield,, Harry, Leslie Potter, and Yvonne Byron. 1995. *In Place of the Forest: Environmental and Socio-economic Transformation in Borneo and the Eastern Malay Peninsula*. Tokyo, New York, Paris: United Nations Press.

Brown, D. 1999. *Addicted to Rent: Corporate and Spatial Distribution of Forest Resources in Indonesia; Implications for Forest Sustainability and Government Policy*. Jakarta: Indonesia-UK Tropical Forestry Management Programme.

Bryant, R. 1997. *The Political Ecology of Forestry in Burma, 1824–1994*. Honolulu: University of Hawaii Press.

—— 2007. "Burma and the Politics of Teak: Dissecting a Resource Curse." In P. Boomgaard and G. Bankoff, eds. *The Wealth of Nature*, eds. London: Palgrave Macmillan.

Buhaug, H. and Paivl Lujala. 2005. "Accounting for Scale: Measuring Geography in Quantitative Studies of Civil War." *Political Geography* 24 (4): 399–418.

Charras, Muriel. 1982. *De la forêt maléfique à l'herbe divine: la transmigration en Indonésie, les Balinais à Sulawesi.* Editions de la Maison des sciences de l'homme. Paris, Ann Arbor, MI: University Microfilms International.

—— 2006. "Westerm Kalimantan in Development: A Regional Disappointment." In G. Smith and H. Bouvier, *Communal Conflicts in Kalimantan: Perspectives from the LIPI-CNRS Conflict Studies Program.* Jakarta: PDII-LIPI, pp. 131–169.

Cleary, Mark and Peter Eaton. 1992. *Borneo: Change and Development.* New York: Oxford University Press.

Collier, P. and Anke Hoeffler, "Greed and Grievance in Civil Wars," *Oxford Economic Papers* 56 (2004), 563–595.

Coppel, C. 1983. *The Indonesian Chinese in Crisis.* Oxford: Oxford University Press.

Davidson, J. and D. Kammen. 2002. "Indonesia's Unknown War and the Lineages of Violence in West Kalimantan." *Indonesia* 73 (April): 1–31.

De Konick, R. and S. Dery. 1997. "Agricultural Expansion as a Tool of Population Redistribution in Southeast Asia." *Journal of Asian Studies* 28.

Dennis, P. and J. Grey. 1996. *Emergency and Confrontation: Australian Military Operations in Malaya* and *Borneo 1950–1966.* St. Leonards, Australia: Allen & Unwin in association with the Australian War Memorial.

Departemen Kehutanan (Forestry Department). 1986. Sejarah Kehutanan Indonesia (Indonesian Forestry History). Jakata: Government of Indonesia.

Dove, M. 1985. "The Agroecological Mythology of the Javanese and the Political Economy of Indonesia." *Indonesia* 39 (April): 1–36.

—— 1993. "Rubber Eating Rice, Rice Eating Rubber." Paper read at Agrarian Studies Seminar, Yale University, New Haven, Connecticut.

Ellen, R. 1999. "Forest Knowledge, Forest Transformation. Political Contingency, Historical Ecology and the Renegotiation of Nature in Seram." In T. Li, ed., *Transforming the Indonesian Uplands,* Amsterdam: Harcourt Press, pp. 131–156.

Elmhirst, R. 2004. "Labour Politics in Migrant Communities: Ethnicity and Women's Activism in Tangerang, Indonesia." In Elmhirst and R. Saptari , eds., *Labour in Southeast Asia. Local Processes in a Globalised World,* London, New York: Routledge Curzon, pp. 387–406.

Endicott, K. 1997. "Review: Violence and the Dream People." *Journal of Asian Studies* 56 (1): 262–263.

Feldman, A. 1991. *Formations of Violence: The Narrative of the Body and Political Terror in Northern Ireland.* Chicago: University of Chicago Press.

Flint, C. R. ed. 2005. *The Geography of War and Peace: From Death Camps to Diplomats.* New York: Oxford University Press.

Forsythe, Timothy and Andrew Walker. 2008. *Forest Guardians, Forest Destroyers: The Politics of Environmental Knowledge in Northern Thailand.* Seattle: University of Washington Press.

Glassman, J. 2006. "Primitive Accumulation: Accumulation by Dispossession, Accumulation by Extra-Economic Means." *Progress in Human Geography* 30 (5): 608–625.

Grove, R. and D. Anderson 1990. *Conservation in Africa: People, Policies and Practices.* Cambridge: Cambridge University Press.

Guha, R. 1990. *The Unquiet Woods: Ecological Change and Peasant Resistance in the Indian Himalaya.* Berkeley: University of California Press.

Hack, Karl. 2001. *Defence and Decolonisation in Southeast Asia: Britain, Malaya and Singapore, 1941–1968.* Richmond, UK: Curzon.

Haraway, D. 1991. "Situated Knowledges." In *Simians, Cyborgs, and Women: The Reinvention of Nature*. London: Routledge.

Harper, T. N. 1997. "The Politics of the Forest in Colonial Malaya." *Modern Asian Studies* 31 (1): 1–29.

Hearn, R. M. 1974. *Thai Government Programs in Refugee Relocation and Resettlement in Northern Thailand*. Auburn, NY: Thailand Books.

Hecht, S. and A. Cockburn. 1989. *The Fate of the Forest: Developers, Destroyers, and Defenders of the Amazon*. London: Verso.

Heidhues, M. S. 2003. *Gold Diggers, Farmers, and Traders in the "Chinese Districts" of West Kalimantan, Indonesia*. Ithaca, NY: Cornell University and Southeast Asian Program Publications.

Hirsch, P. 1990. *Development Dilemmas in Rural Thailand*. Singapore: Oxford University Press.

—— 1993. *Political Economy of Environment in Thailand*. Manila: Journal of Contemporary Asia Publishers.

Jarvie, James, Ramzy Kanaan, Michael Malley, Trifin Roule, and Jamie Thomson. 2003. *Conflict Timber: Dimensions of the Problem in Asia and Africa, Volume II Asian Cases*. Washington: USAID Report prepared by ARD. Accessed online September 2009, at pdf.dec.org/pdf_docs/Pnact463.pdf.

Jonsson, H. 2005. *Mien Relations: Mountain People and State Control in Thailand*. Ithaca, NY: Cornell University Press.

Kaimowitz, D. 2005. *Vital Forest Graphics*. Rome: Food and Agriculture Administration.

Kathirithamby-Wells, K. 2005. *Nature and Nation: Forests and Development in Peninsular Malaysia*. Honolulu, HI: University of Hawaii Press.

Kaufman, Herbert. 1960. *The Forest Ranger: A Study in Administrative Behavior*. Baltimore, MD: Johns Hopkins Press for RFF.

Klare, M. T. 2002. *Resource Wars: The New Landscape of Global Conflict*. New York: Henry Holt Paperbacks.

Kosek, Jake. 2006. *Understories*. Raleigh, NC: Duke University Press.

Kuletz, V. 1998. *The Tainted Desert: Environmental Ruin in the American West*. New York: Routledge.

Kunstadter, P. and E. C. Chapman. 1978. "Problems Of Shifting Cultivation And Economic Development In Northern Thailand." In P. Kunstadter, E. C. Chapman and S. Sabhasri, 3–23. eds. *Farmers in the Forest*. Honolulu, HI: East-West Center and the University of Hawaii Press.

Leary, J. 1995. *Violence and the Dream People: The Orang Asli in the Malayan Emergency, 1948–1960*. Athens, Ohio: Ohio University Center for International Studies, Monographs in International Studies.

Le Billon, P. 2000. "The Political Ecology of Transition in Cambodia 1989–99: War, Peace, and Exploration." *Development and Change* 31 (4): 785–805.

—— 2001."The Political Ecology of War: Natural Resources and Armed Conflicts." *Political Geography* 20 (5): 561–84.

—— 2004. "The Geopolitical Economy of 'Resource Wars.'" *Geopolitics* 9 (1): 1–28.

—— 2007. "Geographies of War: Perspectives on Resource Wars." *Geography Compass* 1/2: 163–182.

—— 2008. "Diamond Wars? Conflict Diamonds and the Geographies of Resource Wars." *Annals of the American Association of Geographers* 98 (2): 345–372.

Leblond, J.P. 2009. "Population Displacement and Forest Management in Thailand." Draft ChATSEA Working Paper, University of Montreal.

Leigh, Michael 1998. "The Political Economy of Logging in Sarawak." In Philip Hirsch and Carol Warren, eds. *The Politics of Environment in Southeast Asia, Resources and Resistance.* London: Routledge.

Li, Tania Murray, ed. 1999. *Transforming the Indonesian Uplands: Marginality, Power, and Production.* Singapore: Harwood Academic Publishers.

Mackie, Jamie A. C. 1974. *Konfrontasi: The Indonesia-Malaysia Dispute 1963–1966.* Kuala Lumpur: Oxford University Press.

Mahmud, Z. B. H. 1979. "The Evolution of Population and Settlement in the State of Kedah." In D. Asmah Haji Omar. ed. *Essays on Linguistic, Cultural and Socio-Economic Aspects of the Malaysian State of Kedah*, Kuala Lumpur: University of Malaya.

Mamdani, Mahmood. 2004. *Good Muslim, Bad Muslim: America, the Cold War, and the Roots of Terror.* New York: Pantheon.

Massey, Doreen. 2005. *For Space.* London: Sage Publications.

McColl, R. 1967."A Political Geography of Revolution: China, Vietnam, and Thailand." *The Journal of Conflict Resolution* 11 (2): 153–167.

McCoy, Alfred W. 1972. *The Politics of Heroin in Southeast Asia: CIA Complicity in the Global Drug Trade.* New York: Harper & Row.

Menzies, Nicholas. 1992. "Strategic space: exclusion and inclusion in wildland policies in Late Imperial China." *Modern Asian Studies* 26 (4): 719–733.

Moore, Donald S. 2005. *Suffering for Territory: Race, Place and Power in Zimbabwe.* Durham, NC: Duke University Press.

Mortimer, R. 1974. *Indonesian Communism Under Sukarno.* Ithaca, NY: Cornell University Press.

Murphy, A. B. 2005. "Territorial Ideology and Interstate Conflict: Comparative Considerations." In C. R. Flint, ed., *The Geography of War and Peace: From Death Camps to Diplomats*, New York: Oxford University Press, pp. 280–296.

Muscat, R. J. 1990. *Thailand and the United States: Development, Security, and Foreign Aid.* New York: Columbia University Press.

Nasution, A. H. 1953. *Fundamentals of Guerilla Warfare.* Jakarta: Indonesian Armed Forces.

Neumann, R. 1998. *Imposing Wilderness: Struggles over Livelihood and Nature Preservation in Africa.* Berkeley, CA: University of California Press.

—— 2001. "The Last Wilderness: Reordering Space for Political and Economic Control in Tanzania." *Journal of the International African Institute* 71:4: 641–665.

—— 2004. "Nation–State–Territory: toward a critical theorization of conservation enclosures." In Richard Peet and Michael Watts, eds. *Liberation Ecologies*, second edition. London: Routledge.

Nevins, J. and N. L. Peluso eds. 2008. *Taking Southeast Asia to Market: Commodities, Nature, and People in the Neoliberal Age.* Ithaca, NY and London: Cornell University Press.

O'Loughlin, J. and Frank Witmer. Forthcoming. "The Localized Geographies of Violence in the North Caucasus of Russia, 1999–2007." *Annals of the Association of American Geographers.*

Osborne, Milton E. 1968. *Strategic Hamlets in South Vietnam: A Survey and Comparison.* Ithaca, NY: Cornell Southeast Asia Program Publications.

Peluso, N. L. 1992. *Rich Forests, Poor People: Resource Control and Resistance in Java.* Berkeley, CA: University of California Press.

—— 1996. "Fruit Trees and Family Trees in an Anthropogenic Forest: Ethics of Access,

Property Zones, and Environmental Change in Indonesia." *Comparative Studies in Society and History: An International Quarterly* 38 (3): 510–548.

—— 2003a. "Territorializing Local Struggles for Resource Control: A Look at Environmental Discourses and Politics in Indonesia." In P. Greenough and A. Tsing, eds. *Environmental Discourses in South and Southeast Asia*, eds. Durham, NC: Duke University Press.

—— 2003b. "Weapons of the Wild: Strategic Uses of Wildness and Violence in West Kalimantan." In C. Slater, ed. *In Search of the Rainforest*. Berkeley: University of California Press, pp. 204–245.

—— 2009. "Rubber Erasures, Rubber Producing Rights. Making Racialized Territories in West Kalimantan, Indonesia." *Development and Change* 40 (1): 47–80.

Peluso, N. L. and E. Harwell. 2001. "Territory, Custom, and the Cultural-Politics of Ethnic War in West Kalimantan Indonesia." In Nancy Peluso and Michael Watts, eds., *Violent Environments*. Ithaca, NY: Cornell University Press, pp. 83–116.

Peluso, N. L. and P. Vandergeest. 2001. "Genealogies of the Political Forest and Customary Rights in Indonesia, Malaysia, and Thailand." *Journal of Asian Studies* 60 (3): 761–812.

Peluso, N. L. and M. Watts, eds. 2001 *Violent Environmennts*. Ithaca, NY: Cornell University Press.

Porritt, V. L. 2004. *The Rise and Fall of Communism in Sarawak, 1940–1990*. Clayton, Victoria: Monash University Press.

Povinelli, Elizabeth. 2002. *The Cunning of Recognition: Indigenous Alterities and the Making of Australian Multiculturalism*. Raleigh, NC: Duke University Press.

Potter, L. 2003. "Forests versus Agriculture: Colonial Forest Services, Environmental Ideas and The Regulation of Land-use Change in Southeast Asia." In Lye Tuck-Po, Wil de Jong, and Abe Ken-ichi, eds. *Political Ecology of Tropical Forests in Southeast Asia*. Kyoto: Kyoto University Press.

Pred, Allen. 2000. *Even in Sweden: Racisms, Racialized Spaces, and the Popular Geographical Imagination*. California Studies in Critical Human Geography, 8. Los Angeles: University of California Press.

Rachman, *et al.* 1970. *Sejarah Singkat Kodam XII Tanjungpura, Kalimantan Barat*. Pontianak.

Reno, W. 1999. *Warlord Politics and African States*. Boulder, CO: Lynne Rienner Publishers.

Robison, Richard. 1986. *Indonesia: The Rise of Capital*. Sydney, London: Allen and Unwin.

Ross, Michael L. 2003. "The Natural Resource Curse: How Wealth Can Make You Poor." In Ian Bannon and Paul Collier, eds., *Natural Resources and Violent Conflict: Options and Actions*. Washington D.C.: World Bank.

Roth, R. 2008. "'Fixing' the Forest: The Spatiality of Conservation Conflict in Thailand. *Annals of the Association of American Geographers*, 98 (2): 373–391.

Rustad, S. C. A., J. K. Rod, W. Larsen, and N. P. Gleditsch. 2007. "Foliage and Fighting: Forest Resources and the Onset, Duration, and Location of Civil War." *Political Geography* 27 (7): 761–782.

Scott, J. 1998. "Freedom and Freehold: Space, People, and State Simplification in Southeast Asia." In K. D. Reid, ed. *Asian Freedoms: The Idea of Freedom in East and Southeast Asia*, Cambridge: Cambridge University Press, pp. 37–64.

Sioh, M. 1998. "Authorizing the Malaysian Rainforest: Configuring Space, Contesting Claims and Conquering Imaginaries." *Ecumene* 5 (2).

—— 2004. "An Ecology of Postcoloniality: Disciplining Nature and Society in Malaya, 1948–57." *Journal of Historical Geography* 30: 729–746.

Sivaramakrishnan, K. 1997. "A Limited Forest Conservancy in Southwest Bengal, 1864–1912." *The Journal of Asian Studies* 56 (1): 75–112.

—— 1999. *Modern Forests: Statemaking and Environmental Change in Colonial Eastern India*. Stanford, CA: Stanford University Press.

Slater, C. 1995. "Amazonia as Edenic Narrative." In William Cronon, ed., *Uncommon Ground: Rethinking the Human Place in Nature*. New York, London: W. W. Norton, pp. 114–131.

Smith, Neil. 1991. *Uneven Development: Nature, Capital and the Production of Space*. Cambridge: Basil Blackwell.

Soemadi. 1974. *Peranan Kalimantan Barat dalam menghadapi subversi Komunis Asia Tenggara: suatu tinjauan internasional terhadap gerakan Komuni dari sudut pertahanan wilayah khususnya Kalimantan Barat*. Pontianak: Yayasan Tanjungpura.

Soepardi, R. 1974. *Hutan dan Kehutanan Dalam Tiga Jaman*. Jakarta: Perum Perhutani.

Somchai, P. 2006. *Civil Society and Democratization: Social Movements in Northeast Thailand*. Leifsgade, Denmark: NIAS Press.

Stoler, A. L. 2008. "Imperial Debris: Reflections on Ruins and Ruination." *Cultural Anthropology* 23: 191–219.

Stubbs, R. 1988. *Hearts and Minds in Guerrilla Warfare: The Malayan Emergency 1948–1960*. Singapore, Oxford: Oxford University Press.

Stump, R. W. 2005. "Religion and the Geographies of War." *The Geography of War and Peace*: 149–173.

Sturgeon, Janet C. 2005. *Border Landscapes: The Politics of Akha Landuse in Thailand and China*. Seattle: University of Washington Press.

Sundberg, Juanita. 2009. "Cat Fights on the Rio and Diabolic Caminos in the Desert: The Nature of Boundary Enforcement in the United States-Mexico Borderlands." Paper presented at Environmental Politics Colloquium, Berkeley, CA, California: University California.

—— 1975. *Whigs and Hunters: The Politics of the Black Act*. London: Allen Lane.

Thongchai, W. 1997. *Siam Mapped: A History of the Geo-Body of a Nation*. Honolulu: University of Hawaii Press.

Trouillot, M. R. 1991. "Anthropology and the Savage Slot: The Poetics and Politics of Otherness." In R. Fox, ed. *Recapturing Anthropology: Working in the Present*. Santa Fe, NM: School of American Research.

Tsing, A. 1999. "Becoming a Tribal Elder and other Green Development Fantasies." In T. Li, ed. *Transforming the Indonesian Uplands*, London: Harwood Academic Publishers.

Uhlig, H. 1984. *Spontaneous and Planned Land Settlement in Southeast Asia*. Hamburg: Institute of Asian Affairs, Giesssener Geographische Schriften

Vandergeest, Peter. 1996. "Territorialization of Forest Rights in Thailand." *Society and Natural Resources* 9: 159–175.

—— 2003. "Racialization and Citizenship in Thai Forest Politics." *Society and Natural Resources* 16 (1): 19–37.

Vandergeest, Peter and Nancy Lee Peluso. 1995a. "Social Aspects of Forestry in Southeast Asia: A Review of Postwar Trends in the Scholarly Literature." *Journal of Southeast Asian Studies* 26: 196–218.

—— 1995b. "Territorialization and State Power in Thailand." *Theory and Society: Renewal and Critique in Social Theory* 24 (3): 385–426.

—— 2006a. "Empires of Forestry: Professional Forestry and State Power in Southeast Asia, Part 1." *Environment and History* 12 (1): 31–64.

—— 2006b. "Empires of Forestry: Professional Forestry and State Power in Southeast Asia, Part 2." *Environment and History* 12: 359–393.

Wainwright, Joel and Morgan Robertson. 2003. "Territorialization, Science, and The Colonial State: The Case of Highway 55 in Minnesota." *Culture Geographies* 10: 196–217.

Watts, Michael. 2001. "Petro-Violence: Community, Extraction, and Political Ecology of a Mythic Commodity." In N.L. Peluso and M. Watts, eds, *Violent Environments*. Ithaca, NY: Cornell University Press.

—— 2003. "Development and Governmentality." *Singapore Journal of Tropical Geography*, 24:1: 6–34.

—— 2004. "Resource Curse? Governmentality, Oil and Power in the Niger Delta, Nigeria." *Geopolitics* 9 (1): 50–80.

Westad, O. A., 2006. *The Global Cold War: Third World Interventions and the Making of Our Times*. Cambridge: Cambridge University Press.

Westoby, J. 1987. *The Purpose of Forests: Follies of Development*. Oxford, New York: Blackwell.

Wyatt-Smith, J. 1947. "Save the Belukar." *Malayan Forester* XI: 24–26.

—— 1949. "Regrowth in Cleared Areas." *Malayan Forester* XIII: 83–86.

13 Mutant ecologies: radioactive life in post-Cold War New Mexico

Joseph Masco

It is a curious phenomenon of nature that only two species practice the art of war – men and ants.

(Norman Cousins, *Modern Man is Obsolete* (1945))

A full-scale nuclear attack on the United States would devastate the natural environment on a scale unknown since early geological times, when, in response to natural catastrophes whose nature has not been determined, sudden mass extinctions of species and whole ecosystems occurred all over the earth. . .It appears that at the outset the United States would be a republic of insects and grass.

(Jonathan Schell, *The Fate of the Earth* (1982))

In the classic Hollywood science fiction film, *Them!* (dir. Gordon Douglas, 1954), Los Alamos weapons science and Cold War logics of "containment" are turned quite sensationally on their heads. Rather than producing international security in the form of a military nuclear deterrent, the American nuclear complex is portrayed as the domestic source of proliferating radiation effects, creating an entirely new ecology of risk in the form of gigantic mutant carnivorous ants. These fantastic creatures are identified, in the film, as the products of the very first atomic explosion in central New Mexico on July 16, 1945. The Trinity Test is portrayed then not as the first triumph of American big science, nor as the technoscientific means of ending World War II, nor as the military foundation of the world's first nuclear superpower. Rather, the first atomic explosion, in this science fiction, is the source of an inverted natural order, in which the smallest of creatures can become a totalizing threat, and where the security state must be deployed to protect citizens from the unintended consequences of nuclear science. *Them!* engages a new kind of nuclear fear in 1954, one based not on the apocalypse of nuclear war but on the everyday transformation of self and nature through an irradiated landscape. Remembered today mostly for its McCarthy-era theatrics in which the giant ants play a thinly veiled allegory for the communist "menace," the film more subtly presents a devastating critique of U.S. nuclear policy at the very height of the Cold War: it argues that on July 16th 1945 Americans entered a post-nuclear environment of their own invention. From this perspective, the nuclear apocalypse is not in the future – a thing to be endlessly deterred through nuclear weapons and international relations – it is already here, being played out in the unpredictable movement of radioactive materials moving through bodies and biosphere.

Them! arrived on American movie screens in June of 1954, just three months after Los Alamos scientists conducted the largest thermonuclear explosion of the Cold War at Bikini Island in the Marshall Islands. Detonating with 2.5 times its expected force, the "Bravo" event produced a 15-megaton yield and vast atmospheric fallout. The Bravo test ultimately contaminated 50,000 square miles of the Pacific with "serious to lethal levels of radioactivity" (Weisgall 1994: 305). Among the exposed were 223 indigenous residents of Rongerik, Rongelap, Ailinginae and Utirik atolls as well as the 23-member crew of the Japanese fishing boat, *The Lucky Dragon*, which was over 80 miles from ground zero at the time of detonation. Thus, as American theatergoers flocked to *Them!* in the summer of 1954, making it one of the most successful films of the year, news reports were simultaneously following the progression of radiation sickness among the Marshall Islanders and *Lucky Dragon* crew – educating many Americans, for the first time, to the biological effects of radioactive fallout. This graphic documentation of the ecological costs of nuclear testing, as brute reality as well as cinematic fantasy, worked to transform America's nuclear program for many individuals from an exclusively military project to a global environmental threat. After Bravo, the bomb was increasingly recognized to be both an explosive and, in the form of fallout, a biological weapon

Figure 13.1 Bravo test

in the U.S., challenging the nature of the "experiment" being conducted by Los Alamos weapon scientists. For if each U.S. nuclear detonation advanced the potential of the bomb as a military machine, each test also added to the global burden of radioactive elements in the biosphere, overturning the "national security" logics of the U.S. nuclear arsenal by introducing the possibility of cellular mutations in plants, animals, and people on a global scale.

As the first U.S. popular culture text to engage the bomb not as a military weapon but as an ecological threat, *Them!* is worth revisiting in Post-Cold War America. The film is important not only because it reveals a moment when the U.S. nuclear arsenal was not yet a normalized (all but invisible) aspect of everyday life, but also because it is the *ur*-text for an ongoing fascination with mutation in American popular culture, an important cultural legacy of the Manhattan Project. Although it appears today as a form of atomic kitsch, the film (an academy award winner for its one special effect: the giant ants) is played straight, and remains a compelling textual effort to assess the "newness" of the atomic age. *Them!* begins, quite hauntingly, as a crime story: The police encounter a young girl, wandering the New Mexican desert alone in her bathrobe, too traumatized to speak (the image of a post-nuclear survivor). Discovering a series of bizarre and violent murders in the area, including the girl's parents, the police struggle to make sense of crime-scene evidence (buildings destroyed from the inside out, recurring traces of sugar and formic acid, and an apparent lack of motive). The police soon call in the FBI, but these domestic agents of the security state are equally limited in their ability to assess the "crime," also unable to make the imaginative leap required to see mutant nature as the cause. Frustrated by a lack of fingerprints, for example, the police and FBI agents stare without recognition at a strange impression found in the earth. The plaster cast of the impression reveals the footprint of a giant ant, constituting a criminal signature literally too large for the police to comprehend. Nuclear nature simply baffles; as one policeman puts it "lots of evidence, loaded with clues, but nothing adds up." The problem here is that the crimes are "unnatural" by the standards of pre-nuclear America, making the first problem of the nuclear age one of linking perception and imagination in a world operating by new horizons of possibility. A team of entomologists from the Department of Agriculture eventually identify the footprints, and lead the police back through the "crime scenes" looking for evidence of a natural order transformed. In the fallout zone from the 1945 Trinity test, an area untouched by people in nine years, they discover strange mound formations that signal the arrival of a new species and trigger the first confrontation with the giant ants. The film thus suggests that the nuclear transformation of everyday life can occur at any time, and anywhere, even in the silence of an eerie, and seemingly empty, desert.

Them! both deploys and ridicules the military logics of containment by asking: If the giant ants are a crime, then who is responsible? The film enacts a split vision, both demonizing the ants as an external other, while recognizing that they are a creation of the U.S. security state. Thus, the terrible joke embedded in the title of the film – "Them!" – which suggests that the agents of destruction are foreign born rather than domestic, "theirs" rather than "ours." The film ultimately argues that the dispersal of nuclear materials in the environment (recognized as a global phenomenon by 1954) is the source of a new kind of nature, mutant, wild, and *un*containable by the state. As imagined here, the U.S. nuclear complex is responsible not only for new technologies of mass death but also for producing new kinds of mutant life, as species are reinvented at the genetic level. As nuclear allegory and ecological critique, *Them!* also implicitly argues that human beings are not only responsible for creating a mutant ecology via the bomb, they are also part of this ecology, producing a future that is as unpredictable at the level of genetic stability as it is at the level of international relations.

Them! is the cinematic instantiation of a larger cultural discourse in the U.S. about the bomb, in which nuclear critics have deployed insects as a means of engaging the philosophical status of the nuclear age. From Norman Cousins 1945 essay "Modern Man is Obsolete" written days after the atomic bombings of Hiroshima and Nagasaki to Jonathan Schell's 1982 portrayal of a post-nuclear American republic of "insects and grass," insects have been used to articulate a "species" knowledge in relation to the nuclear. The "nature" of nature is interrogated in these discourses, as the power of atomic energy, the "purity" of ecosystems, and the adaptability of certain organisms to a radioactive environment, are positioned against human "nature." Cousins' argument that only "men and ants" make war asks if it is a biological imperative to organize conflict in both species (as highly organized but ultimately mindless beings), while Schell argues that humans are too fragile a species to survive a nuclear war, and that the only victors would be the insects that could withstand and adapt to a radioactive environment. Both authors argue that the destructive power of the bomb demands social evolution, and deploy insects as a mirror to humanity. Edmund Russell has tracked this historical impulse to link people and insects in a remarkable analysis of chemical weapons and pest control in twentieth-century America. Tracking the technoscientific, organizational, and ideological tools used in military and public health campaigns against human and insect "enemies," he documents a structural interaction between species under the concept of "extermination," concluding that "war and control of nature co-evolved: the control of nature expanded the scale of war, and war expanded the scale on which people controlled nature" (2001:2). Technoscience militarizes nature in these discourses, enabling a dual deployment of social evolution and biological extinction, the focal points of a new kind of modernity. In other words, the atomic bomb produces not only new understandings of self, nature, and society but also (as *Them!* argues) initiates a profound mutation in each of these terms.

Consider, for example, the concept of "background radiation," which references the baseline level of radiation considered to be inherent in the environment by federal authorities. The background radiation figure is the amount of radiation the average American receives in a given year from all sources; it is also the standard with which U.S. industrial radiation exposure rates are measured. The current background radiation rate for U.S. citizens is 360 millirems per year. Of this, 300 millirems come from "naturally" occurring sources, such as cosmic rays, radon, radiation from the surface of the earth, and from potassium-40 in our bodies. The remaining 60 millirems come from the cumulative atmospheric effects of industry

– including nuclear medicine, nuclear power, and nuclear weapons testing (Wolfson 1991:60–63). What now constitutes the "background" field for all studies of radiation effects is a mix of naturally occurring and industrial effects. More specifically, the trace elements of Los Alamos weapons science now saturate the biosphere creating an atomic signature found in people, plants, animals, soils, and waterways. The Manhattan Project not only unlocked the power of the atom, creating new industries and military machines, but it also inaugurated a subtle but total transformation of the biosphere. But if nature entered a new kind of nuclear regime in 1945, then how should we now assess that transformation? After all, the very idea of a background radiation standard is to establish a norm, a new definition of the "natural" in which the past effects of the nuclear complex are embedded as a fundamental aspect of the ecosystem. To appreciate the full scope of this nuclear revolution, we need to examine the effects of the bomb not only at the level of the nation-state but also at the level of the local ecosystem, the organism, and ultimately, the cell.

The background radiation rate constitutes an average and thus does not apply to any specific individual. The true evaluation of nuclear risk is tied to specific exposures rather than the background radiation count (which, although measurable, constitutes a negligible health risk). Makhijani and Schwartz, for example, identify seven classes of people negotiating health risks from U.S. nuclear production (1998: 396):

(1) Workers in uranium mines and mills and in nuclear weapons design, production and testing facilities;
(2) armed-forces personnel who participated in atmospheric weapons testing;
(3) people living near nuclear weapons sites;
(4) human experiment subjects;
(5) armed forces personnel and other workers who were exposed during the deployment, transportation and other handling and maintenance of weapons within the Department of Defense;
(6) residents of Hiroshima and Nagasaki in August 1945; and
(7) the world's inhabitants for centuries to come.

The world's inhabitants for centuries to come. The enormous difference in the types and degrees of exposures among these populations demonstrates both the generality and specificity of the nuclear age: exposures are simultaneously collective (involving everyone on the planet) and highly individualized (involving specific classes of people – soldiers, miners, nuclear workers). While we all have trace elements from the Cold War nuclear project in our bodies, no two exposure rates are identical, as geographical location, occupation, and nuclear events (whether from nuclear industry, atmospheric nuclear tests, or accidents such as Chernobyl) combine with individual physiology and specific ecosystems to define actual rates and degrees of risk. Nevertheless, if we were able to track back in time and space, following the trajectory of the various chemicals and nuclear materials now in each of our bodies, one subset of these industrial signatures would lead back to Los Alamos and the Cold War national security project, offering a different vantage

290   *Joseph Masco*

point from which to assess the nuclear age. From this perspective, America's nuclear project has witnessed the transformation of human "nature" at the level of both biology and culture, leading to the formation of new kinds of risk societies, unified not by national affiliation, but by exposure levels, health effects, and nuclear fear.

These ever-present signatures of the nuclear security state constitute, for the vast majority of people, a theoretical rather than a known health risk. However, while studies of the survivors of Hiroshima, Nagasaki, the Marshall Islands, and Chernobyl, as well as of nuclear workers, have produced a detailed scientific understanding of the effects of high levels of radiation exposure, the effects of low-level radiation remain a subject of intense scientific debate. It exists, as Adriana Petryna (2002) has put it, at the level of "partial knowledge," making the challenge of the nuclear age as much the regulation of uncertainty as the documentation of biological effects. This uncertainty is intensified by the specific attributes of radiation-induced illness, which includes a displacement in time (sometimes occurring decades after exposure) and a potential to be genetically transferred across generations. Recognizing the subtle but totalizing scope of the nuclear transformation of nature – the dispersion of plutonium, strontium, cesium, and other elements into the biosphere – challenges the traditional concept of a "nuclear test," which in Los Alamos has referred most prominently to the detonation of a nuclear device. For how does one define or limit the scope of the

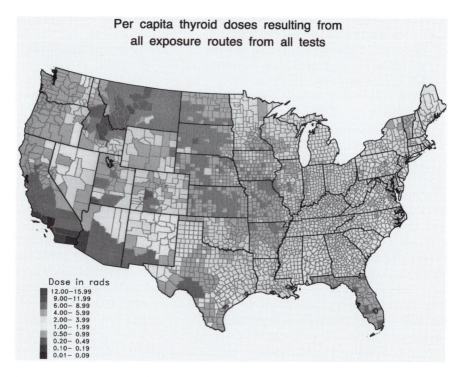

Figure 13.2 Long-term stewardship chart

nuclear laboratory when its trace elements can be found literally everywhere on the planet? Thus, while Los Alamos scientists worked through the Cold War to perfect nuclear weapons through the Cold War as the core technology in a "closed world" system of military command, control, and surveillance (see Edwards 1996), their testing regime also transformed the biosphere itself, turning the earth into a vast laboratory of nuclear effects that maintain an unpredictable claim on a deep future. The world produced by the bomb is structured by its totalizing scale (the entire planet) and by more localized, multigenerational effects that are highly changeable, rooted in any given moment as much in ambiguity or latency as in material fact. The 24,000-year half-life of plutonium, for example, presents a multi-millennial colonization of the future, requiring a different temporal analytic for investigating radioactive ecologies.

To this end, I propose extending our theorization of the complexity of nature-culture forms via the concept of "mutation." A mutation occurs when the ionization of an atom changes the genetic coding of a cell, producing a new reproductive outcome. As cells replicate over time, mutagenic effects can have three possible outcomes: (1) evolution, or an enhancing of the organism through a new adaptation to the environment; (2) injury, such as cancer or deformity; or (3) genetic noise, that is, changes that neither improve nor injure the organism but can still affect future generations. A concept of mutation implies, then, a complex coding of time (both past and future); it assumes change, but it does not from the outset judge either the temporal scale or the type of change that will take place. It also marks a transformation that is reproduced generationally, making the mutation a specific kind of break with the past that reinvents the future. Engaging the U.S. nuclear project through the lens of mutation privileges not only the institutional and technoscientific networks needed to construct the bomb but also the long-term social and environmental effects of the production complex itself. The ecological effects of atmospheric nuclear testing, for example, may not be fully realized for decades, and an understanding of their cultural effects requires an investigation into the different conceptions of nature that inform local forms of knowledge.

Thus, while the Cold-War American nuclear project has not yet produced any giant ants, it has distributed new material and ideological elements into the biological bodies of citizens, and the social body of the nation, that continue to proliferate, promising unpredictable outcomes. As such, the Manhattan Project remains an unending experiment: Nuclear war is still possible today, just as the biosphere and specific social orders continue to be transformed by the accumulating effects of (Post)-Cold War military nuclear science. While each U.S. citizen negotiates the traces of nuclear weapons science in their bodies and biosphere – making each of us real or potential mutants – the nuclear future remains highly mobile. Consequently, the remainder of this chapter investigates debates and practices involving new "species" logics in the nuclear age, examining how the pursuit of security through military technoscience has raised questions about the structural integrity of plants, animals, and people. As we shall see, the nuclear saturates both environments and social imaginations in New Mexico, revealing mutant ecologies subject to new possibilities.

Radioactive natures: life in the wildlife/sacrifice zone

At the end of the Cold War, the U.S. nuclear complex formally occupied a total continental landmass of over 3,300 square miles, involving 13 major institutions and dozens of smaller production facilities and laboratories (O'Neill 1998:35). These production sites were predominantly located in isolated, rural areas as a complex form of domestic development. Huge new industrial economies were created in Oak Ridge (Tennessee), Hanford (Washington), and Los Alamos (New Mexico) in 1943 and later in Aiken (South Carolina), Amarillo (Texas), Idaho Falls (Idaho), Rocky Flats (Colorado), and at what became the Nevada Test Site (see Hales 1997; O'Neill 1998). It was in these mostly rural, nonindustrial locations that nuclear materials were mass-produced, nuclear weapons were built and tested, and nuclear waste was stored, fusing local ecologies and local communities with the American nuclear project. The internal logics of nuclear development required deliberate acts of territorial devastation, producing an archipelago of contaminated sites stretching across the continental United States from South Carolina to Nevada, from Kentucky to Washington, and from Alaska to the Marshall Islands. This "geography of sacrifice," as Valerie Kuletz (1998) has called it, is currently estimated to entail a $216–400-billion environmental restoration project for those sites that can, in fact, be "remediated," and it is likely to cost more than the Cold War nuclear arsenal itself (see Schwartz 1998; U.S. DOE 1995a, 1995b). Nuclear security has required complex new forms of internal cannibalism, as both the biology of citizens and the territories of the state encounter an array of new nuclear signatures after 1943.

In the post-Cold War period, the U.S. nuclear complex has implicitly recognized these transformations through a new type of territorial reinscription. On October 30, 1999, for example, Secretary of Energy Bill Richardson announced the formation of a 1000-acre wildlife preserve within a 43-square mile territory of Los Alamos National Laboratory (LANL). The new White Rock Canyon Preserve was singled out by the DOE as a "unique ecosystem" one that is "home to bald eagles, peregrine falcons, southwestern flycatchers, 300 other species of mammals, birds, reptiles, and amphibians, as well as 900 species of plants" (U.S. DOE, LAAO 1999:1). As Secretary Richardson explained:

> How fitting that we are here today at Los Alamos, the place that witnessed the dawn of the atomic age. . .In places of rare environmental resources, we have a special responsibility to the states and communities that have supported and hosted America's long effort to win the Cold War – and we owe it to future generations to protect these precious places so that they can enjoy nature's plenty just as we do. Los Alamos's White Rock Canyon is such a place, an able bearer of New Mexico's legacy of enchantment. After today, it will be more so as we celebrate the reunification of land and community.

We celebrate the reunification of land and community. The "wildlife preserve" as a concept forwards a claim on purity, marking specific ecologies worth preserving as precious resources in a "state of nature." What can such a claim mean, however, in the context of a U.S. nuclear site? Richardson's appeal to a "legacy of enchantment" as well as to the reunification of land and community in New Mexico comes after a decade of intense environmental politics concerning the Cold War legacies of nuclear weapons work at Los Alamos. The post-Cold War period began in New Mexico with the near simultaneous announcements of a moratorium on nuclear weapons tests and the designation of 2,200 contaminated sites within Los Alamos National Laboratory, requiring an estimated cleanup of over $3.3 billion (U.S. DOE 1995a:xiv). While many New Mexicans discovered the scale of Cold War nuclear research at Los Alamos through its environmental costs, community groups throughout northern New Mexico began mobilizing for health studies as well increased surveillance of water, soil, and air quality. The reunification of land and people proposed by the "wildlife preserve" recognizes the unique cultural investments of Pueblo and Nuevomexicano communities in the area now occupied by Los Alamos. However, the discourse of "preservation" enabling such recognition can only do so by ignoring the long-standing practices of environmental ruin, informing past and present research at the laboratory.

This ideological project to link the "national security" offered by the atomic bomb during the Cold War to sustaining the biodiversity of U.S. territories, however, forwards a deep structural contradiction. The effects of nuclear production have transformed the global environment, making the biosphere itself a postnuclear formation. Since the trace elements of atmospheric fallout are now ubiquitous in soils and waterways, flora and fauna, the "nature" of wildlife as a concept has changed in the nuclear age. If exposure is now a general condition – a question of degree rather than kind – then what does it mean to promote such images of survival in the midst of contamination?

This recuperation of "nature" within post-Cold War debates about the environmental and health dangers of nuclear production articulates a new form of state territoriality. In the continental United States alone, the DOE has recently transformed over 175,800 acres of land by legislative fiat from industrial nuclear sites to wildlife preserves. Carved out of the vast security buffer zones established around nuclear sites, most of these areas were fenced off in the middle of the twentieth century and isolated from human contact during the Cold War. Consequently, these sites were among the most heavily fortified wilderness areas in the world. By presenting these sites as untouched in over 50 years, the DOE seeks to redefine the value and object of that military fortification, replacing nuclear weapons systems with biodiversity as the security object of the nuclear state. This suturing together of wildlife preserve and national sacrifice zone has become an expansive post-Cold War project.

At the Savannah River Site, which produced plutonium and tritium for the U.S. nuclear arsenal, 10,000 acres (of the 200,000-acre nuclear facility) became the Crackerneck Wildlife Management Area and Ecological Reserve in 1999 (U.S. DOE, SROO 1999). Celebrating some 650 species of aquatic life found on the site, the DOE presented a remarkable image of biodiversity to the public. DOE representatives failed to mention, however, that the unusually healthy alligators and rather large bass fish found at the Savannah River Site are also unusually radioactive (Associated Press 1999). Their bodies contain Cesium 137, a byproduct of nuclear material production on the site, which is home to five nuclear reactors. The Savannah River Site now presents a uniquely modern contradiction. The site maintains a massive environmental problem in the form of 34 million gallons of high-level radioactive waste, a multi-millennial challenge to the future, but it has been rescripted by the nuclear state as an ecological reserve preserved, as the DOE notes, for "future generations."

At the Idaho National Engineering and Environmental Laboratory (INEEL), 74,000 acres are now included in the Sagebrush Steppe Ecosystem Reserve. The DOE has devoted this preserve to the protection of some 4,000 species of plants and 270 species of animals – including the ferruginous hawk, the pygmy rabbit, and Townsend's big-eared bat (U.S. DOE, INEEL 1999). Inaugurating the reserve, Secretary Richardson remarked (ibid.):

> The Department of Interior estimates that 98 percent of intact sagebrush steppe ecosystems have been destroyed or significantly altered since European settlement of this country. Because the INEEL has been a largely protected and secure facility for 50 years, it is still home to a large section of unimpacted sagebrush habitat. Our action today will help preserve for future generations one of the last vestiges of this important ecosystem.

INEEL – a largely protected and secure facility. With 52 nuclear reactors, and 11 gigantic tanks filled with 580,000 gallons of high-level nuclear waste, INEEL is redefining the definition of "protected" and "secure" – as well as "impact" and "risk" – for distant future generations. Townsend's big-eared bat and the pygmy rabbit may have gained new state recognition via the reserve, but their new status is primarily a bureaucratic one and does not address the mobility of animals, ecosystems, and radionuclides between territories identified as wildlife reserves and nuclear production sites.

The hard insight informing these new wildlife preserves is that isolation from human traffic provides an enormous ecological benefit: human contact is more immediately toxic for many ecosystems than are radioactive materials. The DOE wildlife reserve/sacrifice zone dual structures seems to argue, however, that nuclear materials can be kept in place and that the border between preserve and wasteland can be effectively patrolled over millennia. This logic is trumped most convincingly at the Hanford Reservation in Washington State, which produced plutonium for the U.S. arsenal from 1945–1992 and is now recognized as the most seriously polluted site in the United States. The DOE has recently devoted 89,000 acres of Hanford's 540 square miles to preserving the long-billed curlew, Hoover's desert parsley, and Columbia yellow cress (U.S. DOE, PNNL 1999). However, mulberry trees on the Hanford Reservation have been showing increasing amounts of strontium-90 over the last decade (Lavelle 2000); and the Russian thistle plant has recently created a new kind of environmental hazard: the radioactive tumbleweed (Associated Press 2001). The Russian thistle shoots its roots down 20 feet into the earth, sucking strontium-90 and cesium into its system from contaminated areas.

The head of the plant eventually breaks off to become a windblown radiation source. Hanford now spends millions of dollars each year managing this form of contamination and has crews armed with pitchforks patrolling the reservation in trucks to wrangle the radioactive weeds. This inability to enforce the distinction between wilderness and wasteland was further dramatized at Hanford in 1998, when fruit flies landed in liquid radioactive material and carried contamination far and wide over the next weeks, requiring nothing less than a $2.5 million dollar DOE cleanup operation (Stang 1998).

Radioactive tumbleweeds, contaminated fruit flies, and toxic alligators – these are all survivals of the Cold War nuclear project, as well as new forms of nuclear nature. Adjacent to each of the DOE wildlife preserves, however, are sites that are not just minimally radioactive according to federal standards, but rather present such profound environmental hazards that they will need to be fenced off and monitored for, in some cases, literally tens of thousands of years. These sites represent Cold War survivals of another kind. Despite the rhetorical and institutional effort to find areas of "purity" within the ecology of the nuclear complex, the broader context involves a massive state-sponsored territorial sacrifice during the Cold War that has been wildly productive in specific areas. The U.S. nuclear complex could not have produced 70,000 nuclear weapons from 1943–1992 without favoring industrial production over environmental concerns. Just as the current background radiation rate normalizes the atmospheric effects of above-ground nuclear testing as an aspect of nature, the new wildlife zones offer an image of nature created through nuclear politics and radioactive practices. The wildlife preserve is thus an exception that proves the rule within the U.S. nuclear complex. Despite the new bureaucratic recognition of the ferruginous hawk, the pygmy rabbit, and the larkspur, the division between normal, abnormal, and pathological is being redefined in these nuclear sites, as contaminated nature is recognized to be not only valuable and robust, but to greater or lesser degrees, ever-present. In other words, the experimental projects that produced and now maintain the bomb have collectively turned the entire biosphere into an experimental zone – one in which we all live – producing new mutations, as we shall now see, in both natural and social orders.

Environmental sentinels, or the militarization of the honey bee

The radioactive future of the Cold War nuclear complex is already mutating in the post-Cold War period, producing a complex mobilization of future generations, technoscience, and state institutions. The DOE has not only offered up zones of conservation to future generations but also acknowledged that as many as 109 sites within the nuclear complex are too contaminated to remediate. The challenge of what to do with these radioactive sites over decades, centuries, and in some cases, millennia, is now articulated through a new discourse of environmental surveillance and control known as "long-term stewardship." The DOE (2001) defines the project as:

> The Long-Term Stewardship Program will maintain and continuously improve protection of public health, safety, and the environment at a site or portion of

a site assigned to DOE for such purposes. This mission includes providing sustained human and environmental well-being through the mitigation of residual risks and the conservation of the site's natural, ecological, and cultural resources. Mission activities will include vigilantly maintaining "post-cleanup" controls on residual hazards; sustaining and maintaining engineered controls, infrastructure, and institutional controls; seeking to avoid or minimize the creation of additional "post-cleanup" long-term stewardship liabilities during current and future site operations; enabling the best land use and resource conservation within the constraints of current and future contamination; and periodic re-evaluation of priorities and strategies in response to changes in knowledge, science, technology, site conditions, or regional setting. The Long-Term Stewardship Program will coordinate activities to identify and promote additional research and development efforts needed to ensure this protection and to incorporate new science and technology developments that result in increased protection of human health and the environment and lower costs.

Sustained human and environmental well-being through the mitigation of risk. The Long-Term Stewardship Program approaches the radioactive and chemical legacies of Cold War nuclear production as a bureaucratic, as well as technoscientific, problem. Promising an increasingly intimate interaction with contaminated sites, the Long-Term Stewardship Program hopes to minimize future environmental effects by systematically deploying as yet undeveloped technologies. This is a utopian program that imagines perfect management of Cold War nuclear waste and

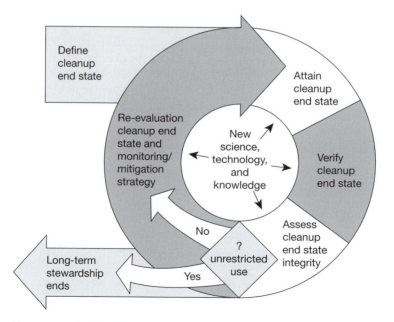

Figure 13.3 NCI fallout map

contaminated sites for millennia – despite the prior 50 years of environmental neglect.

Creating "sustained human and environmental well-being" in a postnuclear environment, however, requires a complex new form of governmentality. For Foucault (1991), governmentality is the focus of the state on policing its population to improve the health and well-being of its citizens. In long-term stewardship, the logic of national security is inverted; the threat of foreign arsenals and armies are replaced by an internal discourse of contamination and territorial colonization. In this context, governance means protecting citizens from the industrial effects of the nuclear security state, thus redrawing the lines between policing and welfare. However, it is not clear how "environmental well-being" can or will be defined. The Department of Energy cannot return ecosystems to a preindustrial, prenuclear state. Rather, "clean-up" here means meeting U.S. regulatory standards, which are dependent on expected land use. The hope of the Long-Term Stewardship Program is that, through surveillance and applying cutting-edge science to the environmental legacy of the Cold War, a kind of ecological stasis can be achieved in the near term, as science improves over time to solve the problems posed by radioactive contamination and waste. However, in recognizing that some sites are too damaged to treat effectively, the program also reveals that the Cold War maintains a powerful claim on a deep future. With budget projections currently made out only to the year 2070, the DOE estimates that the program will require $100 million *per year* simply to maintain the 109 long-term stewardship sites for an indefinite future (U.S. DOE 2001:108).

If the wildlife zone is one new form of nuclear nature, the long-term stewardship site is another, with an equally deep claim on future generations. Indeed, in orienting scientists, technologies, and communities around long-term stewardship sites, the DOE is also creating long-term stewardship communities, producing entirely new ecosocial orders. To make this point, we do not have to look thousands or even hundreds of years into the future. One long-term stewardship site in Los Alamos is known as Area G, which has been the laboratory's primary nuclear waste site since 1957. Area G is a 100-acre facility located on one of the finger-like mesas that make up the Pajarito Plateau. Low-level radioactive waste (consisting mostly of objects contaminated during laboratory operations), as well as significant quantities of plutonium 239 and uranium 238 from nuclear weapons research, is stored in 500-foot long pits and in deep shafts. While inventories have been carefully documented since 1988, few records were kept for the period 1957–1971, and poor records for the period 1971–1988. The incomplete knowledge of what is in Area G is important because just to the east of the site is the town of White Rock (population 6,800), while immediately north is San Ildefonso Pueblo territory. Pueblo members collect plants and hunt game in the shadow of Area G, as well as maintain shrines and sacred sites in the area. A recent laboratory "performance assessment" concludes that Area G will be completely full by 2044, initiating a new kind of territorial project (Hollis 1997:10):

Active institutional control will continue for a period of 100 years (between 2047 and 2146). During institutional control, site access will be controlled, environmental monitoring will be performed, and closure cap integrity will be maintained. After the institutional-control period, it is assumed the site will be maintained by the DOE or its equivalent for as-yet undefined industrial uses. This industrial-use period is assumed to prevail for the 900 years remaining in the compliance period (between 2147 and 3046).

The 900 years remaining in the compliance period (between 2147 and 3046). Evaluating the exposure risks to future populations along a variety of intrusion scenarios, the report confirms that the Manhattan Project inaugurated a new ecological regime on the Pajarito Plateau, one which is now intimately involved with negotiating the 24,000-year half-life of plutonium and other nuclear materials (see Rothman 1992 and Graf 1994). Currently evaluating risk only on a 1,000-year time frame, Area G is nonetheless one instantiation of a larger Cold War nuclear legacy that the discourse of long-term stewardship rhetorically seeks to contain using rational technoscientific measures.

The Area G Performance Assessment concludes, "The ability to contain radioactivity locally depends largely on nature, while the ability to prevent intrusion depends solely on man." It therefore assumes from the start that "current natural conditions will prevail" and "a government entity will maintain the site and control access to it" for the next 1,000 years (Hollis 1997:16). Both nature and the state are, for the sake of the study, assumed to be stable entities across the next millennium, even as the evidence of the last 50 years shows a dramatic change in both. Indeed, more subtle changes are already shaping the nuclear future of the Pajarito Plateau, offering a new state of nature, more mutant than stable. Plumes of tritium contamination as well as chemical residues from high explosives are already leaking from Area G, demonstrating that the geology of the Pajarito Plateau is more permeable than previously assumed. Thus, even as the performance assessment assumes a forever-vigilant state agency to watch over a stable ecosystem at Area G, environmental surveillance is revealing a more mobile ecological formation. Indeed, surveillance itself has become the basis for new kinds of nature.

Consider the role now played by the Italian honeybee (*Apis mellifera*) at Area G. As a creature that flies over a wide area foraging for pollen and nectar in flowers and then returns to a fixed location (the hive) to produce honey, the honeybee is a natural environmental surveyor. Los Alamos scientists have demonstrated that the honeybee is particularly sensitive to tritium, a radioactive substance used in nuclear weapons to enhance the size of the explosion and that is notoriously difficult to contain. Deploying the honeybee as an environmental tool since the late 1970s, scientists have documented increasing tritium contamination rates at Area G through the 1990s (Fresquez *et al.* 1997). This instrumentalization of the honeybee takes more than one form at Los Alamos, but in the context of Area G, it reveals a profound transformation in ecological regimes. Neighboring Pueblo communities identify mesa tops as areas of particular cultural importance, containing shrines and sacred sites that participate in a different conception of nature. Pueblo cosmology has traditionally worked, not to deploy nature as a technoscientific object, but to integrate Pueblo members into the local ecology (see Ortiz 1969).

Within Eastern Pueblo cosmologies, the bee plays a crucial role in pollinating plants and is both a symbol and an agent for life itself; consequently, pollen figures prominently in ceremonies of purification and seasonal renewal. The Manhattan Project colonized this ecological regime with one that focuses on the techno-scientific deployment of nature. The value of the bee, in this new context, is no longer as a life-giving entity but as a toxic being, marking the transformation of the plateau from a wild space of nature to a new kind of mutant ecology.

While specific animal forms are being deployed – and reinvented – to shape environmental politics in post-Cold War Los Alamos, a more subtle aspect of the Manhattan Project has been to transform regional human populations into radiation monitors. Activist groups spent much of the 1990s pushing for environmental impact studies and increased regulation of the laboratory, helping to produce a cross-cultural regional dialog about the environmental consequences of nuclear weapons research at Los Alamos. Concurrently, LANL scientists, Pueblo representatives, as well as officials from the Bureau of Indian Affairs, each began conducting independent tests of air, water, soil, plants, and animals in the region, not only to define the level of risk to Pueblo citizens living adjacent to the laboratory but also to confirm the accuracy of LANL science. The Pueblos of Jemez, Cochiti, Santa Clara, and San Ildefonso have begun training new generations of youth as environmental scientists to prepare them to take over responsibility for monitoring the environmental effects of the laboratory. Thus, communities throughout the region – LANL scientists, Los Alamos community members, Native Americans, Nuevomexicanos, and antinuclear activists – all now claim the title of "environmentalist," maintaining deeply felt, if asymmetrical, investments in the Pajarito Plateau. However, while each of these populations is committed to preserving the regional ecology, their cultural understandings of that ecology are construed on radically different terms.

As New Mexicans took an increasingly public interest in LANL's environmental standing after the Cold War, many also played the unwitting role of environmental test subjects. New Mexicans did so at two levels: first, as workers at the laboratory who were monitored for radiation exposures on the job, and second, as regional populations who (often unwittingly) participated in the Los Alamos Tissue Analysis Program, an effort started in the 1950s to track radiation exposures via tissue sampling. In the late 1990s, relatives of 407 individuals who had tissue samples taken during autopsies in Los Alamos and regional hospitals brought a class-action lawsuit against the laboratory. The multimillion-dollar settlement acknowledged that informed consent was not received from family members during these autopsies. Workers in the laboratory as well as residents of Northern New Mexico have thus been part of a larger environmental monitoring project for decades – similar to the bees – but, in this case, their own bodies have been placed in the role of "environmental sentinel." In this sense, tracking radionuclides through the biosphere and specific bodies in Northern New Mexico has become an expanding project for all concerned. The medical knowledge produced by these efforts, however, remains partial and controversial. The four-fold elevated presence of thyroid cancer in Los Alamos discovered in the 1990s might simply be an effect, for example, of the intensity of the screening regime in Los Alamos hospitals (Athas 1996). Nevertheless, while the long-term health effects of nuclear production at Los Alamos remain controversial at the level of technoscience, there is no doubt of the effect they have had on social imaginations in northern New Mexico. Illnesses throughout the region are attributed to the laboratory, revealing another aspect of the nuclear reinvention of nature.

The social logics of mutation

While interviewing Los Alamos employees who believed their health had been damaged on the job, I was told repeatedly about a videotape reported to document hazardous work conditions at Area G. For these workers, the tape held the promise of standing as evidence in future legal proceedings, a means of making visible to the outside world the everyday practices that were usually shielded by gates, security, and the power of the nation-state. A former Area G worker, who was concerned about his health and did not believe in the veracity of the cumulative radiation badge measurements recorded in his Los Alamos medical file, invited me to view the videotape in his home. As I watched, I was confronted with a complex textual record of mutation. The tape was originally made by Los Alamos personnel to document efforts to consolidate space at Area G for the accruing nuclear waste from laboratory operations. The banality of worker job descriptions is soon ruptured, however, when a tractor accidentally punctures a partially buried barrel of nuclear waste. The narrative then shifts from recording the formal statements of workers during the handling of the ruptured barrel to informal moments with the work crew playing to the camera. Eventually, the multiracial workforce splits along racial lines, as the white program managers don anti-contamination gear to test the drum for radionuclides, while the Nuevomexicano and Pueblo workers remain in normal work clothes. The manual labor of digging up and moving barrels of radioactive waste takes place underneath the deep blue New Mexican sky with a ferocious wind that completely covers workers in dust from the site. My host claimed that the dust from the waste site might well have contaminated workers, and then explained to me how easily the radiation monitors could be turned off at Area G to allow such exposures to go unrecorded.

The videotape reveals the difficult work conditions and physical labor needed to move drums of nuclear waste, but the novel presence of the camera also becomes central to the recording: the workers not only practice describing their jobs prior to formal taping and then deal with the accident, while being taped, but they also mug for the camera. Midway through the video, my host interrupts to tell me that he knows what happened to Karen Silkwood, the Kerr-McGee whistleblower who died mysteriously in a car crash in 1974. Her organs were sent to Los Alamos for analysis as part of the tissue registry program but were then mysteriously lost. He tells me that her organs were placed in a laboratory refrigerator, which subsequently failed, and was then dumped at Area G, packed full of the damaged organs of U.S. nuclear workers. Area G becomes, in his presentation, not merely an ongoing health threat to current workers but also literally a grave, a site where the human evidence of radiation exposures is buried as industrial waste. He hopes that the videotape can help reveal this fact, documenting for an outside world the ongoing biological sacrifice of nuclear workers. Twenty minutes into the videotape, the scene shifts to the office spaces at Area G, where the camera operator discovers and then plays with the mirror function on the video camera to produce a series of special effects. For the next 20 minutes of tape, he entertains his fellow workers – by giving them a third eye, or merging their foreheads into giant mutant forms, or giving them tails, while laughing hysterically at the visual results. The videotape that begins with the serious work of nuclear waste disposal, in other words, shifts to a literal discourse of mutation, one that visually transforms each Area G worker into a monstrous being. The Area G workers I spoke with focused more on the official acts documented in the first half of the videotape, than on the cultural logics and fears revealed in the second half. But the videotape records not only the everyday practices at Area G, the brute work of moving nuclear waste around and the precariousness of containment, but also a surreal form of nuclear play that displays workers not as potential mutants but as present ones – linked by tails, misshaped heads, and multiple eyes.

The Area G videotape ends on an equally jarring note, as it cuts from the play of mutation at the nuclear waste site to a garage somewhere in the northern Rio Grande Valley, where a Nuevomexicano relative of the camera operator (who has taken the camera home) stands stiffly and without emotion in the center of the screen, playing ranchero music on an accordion. This eruption of the nonnuclear everyday into the narrative of Area G is a reminder of the multiple cultural worlds informing life in northern New Mexico that are linked both formally and informally to the nuclear project at Los Alamos. The Area G videotape reveals the radical transformation of the region into a nuclear economy: It documents the burying of nuclear waste on the plateau, permanently transforming the ecology of that space. It also documents the mobilization of whole communities that are now devoted simply to monitoring and working with the nuclear waste produced by America's national security regime, and ultimately, it demonstrates the fears of mutation that permeate workers' psyches, underscoring the psychosocial effects of living within a nuclear ecology. These forces are not static, but rather highly mobile, making it impossible to discuss the regional effects of the Manhattan Project without taking into account how material realities fuse with sociocultural logics and nuclear fear. A political ecology of the bomb that investigates the interaction between regimes of nature reveals the American nuclear project to have been ecologically trans-formative and multigenerationally productive: it has reinvented the biosphere as a nuclear space, transformed entire populations of plants, animals, insects, and people into "environmental sentinels," and embedded the logics of mutation within both ecologies and cosmologies. The giant cinematic ants of 1954 have, in other words, been replaced now by far more subtle and serious forms of life defined by the ambiguities and dangers of inhabiting specific radioactive spaces, mutant ecologies which now present an ever evolving biosocial, political, and ethnographic terrain.

References

Associated Press 1999 "Biological Bounty at Former Nuclear Bomb Factory" *Associate Press*, June 24.

—— 2001 "Getting Rid of Radioactive Weeds" *Associate Press*, May 4.

Athas, William F. 1996 *Investigation of Excess Thyroid Cancer Incidence in Los Alamos County*. Santa Fe, NM Department of Heath.

Cousins, Norman 1945 *Modern Man is Obsolete*. New York: Viking Press.

Edwards, Paul N. 1996 *The Closed World: Computers and the Politics of Discourse in Cold War America*. Cambridge, MA: MIT Press.

Foucault, Michel 1991 "Governmentality" in Graham Burchell, Colin Gordon and Peter Miller (eds) *The Foucault Effect: Studies in Governmentality*. Chicago: University of Chicago Press.

Fresquez, P.R., D.R. Armstrong, and L.H. Pratt 1997 *Tritium Concentrations in Bees and Honey at Los Alamos National Laboratory: 1979–1996*. Los Alamos, NM: Los Alamos National Laboratory.

Graf, William L. 1994 *Plutonium and the Rio Grande: Environmental Change and Contamination in the Nuclear Age*. New York: Oxford University Press.

Hales, Peter Bacon 1997 *Atomic Spaces: Living on the Manhattan Project*. Urbana, IL: University of Illinois Press.

Hollis, Diana 1997 *Performance Assessment and Composite Analysis of Los Alamos National Laboratory Material Disposal Area G*. Los Alamos, NM: Los Alamos National Laboratory.

Kuletz, Valerie 1998 *The Tainted Desert: Environmental and Social Ruin in the American West*. New York: Routledge.

Lavelle, Pat 2000 *Facing Reality at Hanford*. Seattle, WA: Government Accountability Project.

Makhijani, Arjun and Stephen I. Schwartz 1998 "Victims of the Bomb" in Stephen I. Schwartz (ed.) *Atomic Audit: The Costs and Consequences of U.S. Nuclear Weapons Since 1940*. Washington, DC: Brookings Institution Press.

O'Neill, Kevin 1998 "Building the Bomb" in Stephen I. Schwartz (ed.) *Atomic Audit: The Costs and Consequences of U.S. Nuclear Weapons Since 1940*. Washington: Brookings Institution Press.

Ortiz, Alfonso 1969 *The Tewa World: Space, Time, Being, and Becoming in a Pueblo Society*. Chicago: University of Chicago Press.

Petryna, Adriana 2002 *Life Exposed: Biological Citizens after Chernobyl*. Princeton, NJ: Princeton University Press.

Rothman, Hal 1992 *On Rims and Ridges: The Los Alamos Area Since 1880*. Lincoln: University of Nebraska Press.

Russell, Edmund 2001 *War and Nature: Fighting Humans and Insects with Chemicals From World War I to Silent Spring*. Cambridge: Cambridge University Press.

Schell, Jonathan 1982 *The Fate of the Earth*. New York: Knopf.

Schwartz, Stephen I. (ed) 1998 *Atomic Audit: The Costs and Consequences of U.S. Nuclear Weapons Since 1940*. Washington, DC: Brookings Institution Press.

Stang, John 1998 "Tainted Tumbleweeds Concern Hanford" *Tri-city Herald*. December 27.

U.S. Department of Energy (DOE) 1995a *Estimating the Cold War Mortgage: The 1995 Baseline Environmental Management Report*. Washington, DC: U.S. Government Printing Office

——1995b *Closing the Circle on the Splitting of the Atom*. Washington: U.S. Government Printing Office.

—— 2001 *Long-term Stewardship Study. Washington: U.S. Government Printing Office.* Available online at: http://lts.apps.em.doe.gov/mission.asp (accessed 15 October 2003).

U.S. Department of Energy, Idaho National Engineering and Environmental Laboratory (DOE, INEEL) 1999 "Energy Department, Bureau of Land Management Create Sagebrush Steppe Reserve." News Release, July 19: 1–2.

U.S. Department of Energy, Los Alamos Area Office (DOE, LAAO) 1999 "Richardson Announces 1,000 Acres at Los Alamos National Laboratory to Protect Wildlife." News Release, October 30:1–2.

U.S. Department of Energy, Pacific Northwest National Laboratory (DOE, PNNL) 1999 "Ecology Reserve: A Haven for Plants and Animals." Backgrounders, August: 1–2.

U.S. Department of Energy, Savannah River Operation Office (DOE, SROO) 1999 "Department of Energy Teams With State on Dedication and Management of 10,000-Acre Crackerneck Wildlife Management Area and Ecological Reserve." News Release, June 24: 1.

Walker, J. Samuel 1999 *Permissible Dose: A History of Radiation Protection in the Twentieth Century*. Berkeley: University of California Press.

Weisgall, Jonathan M. 1994 *Operation Crossroads: The Atomic Tests at Bikini Atoll.* Annapolis, MI: Naval Institute Press.

Wolfson, Richard 1991 *Nuclear Choices: A Citizen's Guide to Nuclear Technology.* Cambridge, MA: MIT Press.

Part V

Fuelling capitalism: energy scarcity and abundance

14 Past peak oil: political economy of energy crises

Gavin Bridge

Introduction

Energy is one of modernity's fundamental mediums and metrics. Defined as the capacity to do work, energy is the productive force at the heart of many economic, social and environmental changes associated with modernist transformation. Social mastery of the earth's energies is also one of the most potent of modernity's ideological tropes. Measures of national development centre on the growing availability, accessibility, and efficiency of energy over time: bodily calorie intake, installed electricity generating capacity, kilometers of paved roads – these and others are used to distinguish whether, when, and where modernity's fuse has been lit. Governments seeking short cuts to modernity plough resources into nuclear power, large-scale hydro, and rural electrification as a means of spreading the light of development across a territory, drawing citizens out of the metaphorical darkness of tradition, and forging a national imaginary (Coronil 1997, Nye 1998).

Energy, then, is one of the principal components of modernity in both a statistical and an ideological sense. Figure 14.1 evidences the strong, positive correlation between energy consumption and economic output for around sixty countries. A similar association is found at the world scale, where economic output and total commercial primary energy supply have both risen about 16-fold in the last 100 years (Smil 2005: 65). Beyond graphs like Figure 14.1, ethnographies of daily life reveal how access to energy transforms experience and expectations at a personal level: comparisons of geographical mobility, or the effort expended on household chores show how increased energy availability can underwrite far-reaching domestic, urban and regional transformations. And for each person who has experienced such changes first hand, many others aspire to an energy-intensive modernity in which technologies such as the car, the fridge and the light bulb diminish distance, forestall the seasons, and render irrelevant the earth's rotation.

Given the centrality of energy to modernity, what is the meaning of energy *crisis?* Popular use of this powerful couplet denotes more than the technical failure of a provisioning system. "Energy crisis" strikes at the heart of modernist transformation: it suggests the moment when the grand arc of human progress stalls, and the lights of prosperity flicker and dim. Crisis signals a moment of collective existential doubt, a historic and geographic conjuncture when the ability of energy

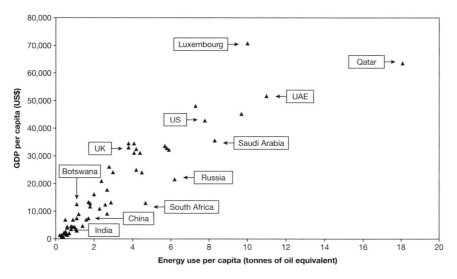

Figure 14.1 The relationship between GDP and energy consumption (data for 2005)

Assembled by the author from data sourced from World Resources Institute (2010).

consumers to continue to summon energy in familiar ways is suddenly – and perhaps permanently – called into question. "Energy crisis," then, is the existential terror of an empty tank and flaccid fuel lines: modernity, fused (Calvino 1974).

From energy crises to peak oil: anatomy of a chapter

This chapter moves from the general to the specific, engaging energy crisis as a general concept before turning to its specific materialization in "peak oil." Having highlighted the tight linkages between energy and modernity, the next section discusses the origins of "energy crisis" as a particular way of thinking about the availability of energy in a modern, high-energy society. It is followed by a brief examination of the specificity of fossil fuels and, in particular, the historical distinctiveness of liquid hydrocarbons. In the remaining sections, the phenomenon of "peak oil" is critically examined as a contemporary manifestation of "energy crisis." With a twist on its main title, the chapter seeks to move "past peak oil" in the sense that it advances a critique of peak oil's core claim that geological limits are now the primary driver of oil's availability, and that an imminent peak in global oil production constitutes a crisis. It takes issue both with the popular fixation that peak oil is the Achilles heel of a modern high-energy society, and the critical left's recent flirtation with peak oil as a crisis of capitalism.

The critical review of peak oil proceeds via four arguments: first, that peak oil's claim about an imminent provisioning crisis is misplaced, curiously irrelevant and ultimately conservative, because it identifies crisis not in the social relations of the energy system but in physical depletion; second, that critical political economy has been too ready to embrace peak oil as a symptom of a "fossil fuel mode of production" in crisis; third, that the inequities and developmental and environmental distortions of high-energy, fossil-fuel-dependent societies constitute an on-going "crisis" even in the absence of supply interruptions, so that "crisis" is a normal state of affairs; and fourth that, through the politicization of some of these inequities and distortions, the production and consumption of conventional oil will indeed peak (and is already doing so in some countries). Importantly, however, the peak will be a "demand-side" phenomenon, as societies seek to re-allocate and prevent some of the social and environmental costs associated with conventional oil. In sum, the chapter displaces "energy crisis" from a conventional bourgeois concern with supply disruption to show how crises arise from the material forms produced by fully functioning, fossil energy provisioning systems.

Naming scarcity: energy crisis as oil shortage

"Energy" and "crisis" first became closely associated in the popular imagination following rapid rises in the price of oil in 1973 and 1979, associated with restrictions by key oil exporting states on the rate at which oil flowed to major markets. The conflation of energy with oil was no accident, given the dominant role of oil in the transportation, power and heating sectors of most industrial economies and, in particular, the United States. But the energy crisis was about more than a belated recognition by U.S. policy-makers of their dependence on imported oil. Gasoline lines and rationing cut Americans to the quick: these were material realities that confronted a national ideology of perpetual progress built on energy abundance. By shaking a cherished "faith that the days of our children would be better than our own," the energy crisis precipitated what President Carter would come to call a national "crisis of confidence" (Carter 1979).

Since the 1970s, energy crisis has been widely applied to describe a variety of short-term "supply squeezes" that threaten the reproduction of socio-economic relations. Classic examples in industrial, high-energy economies include strikes by coal miners and railway workers cutting the flow of coal to power stations, and international disputes over the transit of natural gas that interrupt a significant proportion of national supply. Occasionally scarcity reflects absolute shortage, and "fuel famine" conditions can prevail: coal famines were a feature of urban life well into the twentieth century in Europe and the United States, particularly in winter when ice blocked the ports and snow impeded rail movements of coal from the mines to cities. More often, however, scarcity is a market condition: shortage is experienced as a rise in the price of energy inputs that impinges on conventional rates of use and excludes a greater number of the poor from access to energy supplies. In either case, however, interruption of supply signals a deeper socio-political crisis: a breakdown in the "normal" processes and conventions through which techno-social networks are reproduced and prevailing socio-economic relations sustained.

By temporarily dispossessing those accustomed to energy abundance, sharp rises in the price of fuel can expose the more permanent, structural social inequalities that govern access to energy. In doing so, they can reveal geographies of deprivation that are obscured from those able to command energy during the "normal" operation of energy provisioning systems (Buzar 2007). In this way, classic supply-squeezes can reveal a very different notion of crisis: the chronic exclusion of many people from participation in the high-energy society. Crisis here is not an interruption to the normal state of affairs, but the conditions of inequality that enable some to command abundance while others go without. This alternative definition manifests itself at different spatial scales. Within industrial economies, for example, the phenomenon of "fuel poverty" highlights how access to energy in nominally "high-energy societies" is differentiated by income and quality of housing stock. At the global scale, an estimated 1.6 billion people have no access to electricity and around 2.4 billion – over a third of the world's population – rely on biomass as a primary source of energy, such as crops residues, wood and charcoal (Cleveland 2007). In many places in sub-Saharan Africa, for example, biomass accounts for over 75 percent of the energy consumed in the course of daily life (Mwampamba 2007). Addressing "energy crisis" in these contexts means significantly expanding access to energy and, in particular, improving access to the high-grade energy sources of electricity and liquid fuels. In the popular pairing of "energy" and "crisis," then, one finds an exquisite combination of abundance and scarcity – of enfranchising optimism and existential fear – in which can be glimpsed one of the driving dialectics of modernity.

The language of "energy crisis" has made a comeback in recent years. Rising energy prices, awareness of climate change and recognition of the vulnerabilities of the contemporary energy system have conspired to make redefining the social relationship to energy one of the "grand challenges" of the twenty-first century. In national and international policy arenas, the search for a "new energy paradigm" (Helm 2007) is conceived explicitly as a problem of social reproduction: that is, as a question of how to change the material basis through which a society reproduces itself while preserving social structure and function. The ambitious goal is to effect a "rapid transformation to a low-carbon, efficient and environmentally benign system of energy supply" (IEA 2008). At the centre of this problematic is management of an extended, complex and unruly *hydrocarbon commodity chain* that stretches from the upstream processes of fossil fuel extraction and processing to the downstream processes of distribution, consumption, and the sequestration of carbon (Bridge 2008). Like other modern energy crises, it is the rate of flow through this hydrocarbon chain that sets the bounds of possibility, and around which various utopian and dystopian visions of energy futures currently circulate.

At the *downstream* end of the hydrocarbon commodity chain – where hydro-carbons are combusted and carbon-dioxide released – the gap between rates of carbon emission and carbon sequestration is problematized through the science and discourse of greenhouse gas accumulation and climate change. At the *upstream* end of the chain, there is an insurgent skepticism about the ability to expand or even sustain the current rate at which conventional oil is extracted worldwide (approxi-mately 85 million barrels per day). This skepticism is expressed in the model and narrative of "peak oil," which posits that the rate at which petroleum liquids can be pumped from the earth is nearing its global maximum. Peak oil is a classic specter of scarcity, a "hysterical concern with machine fodder" for an industrial system dependent on high energy and material throughputs (Illich 1973). In this sense, peak oil is like other incarnations of eco-scarcity that have haunted the development of the modern material economy, such as national panics over timber famines in the US and Europe at the beginning of the twentieth century, shortages of critical industrial minerals in the post-war period, and neo-Malthusian pronouncements regarding global food supply and population. Unlike the energy crisis of the 1970s, however, peak oil roots the origins of scarcity firmly below ground, rather than in the above-ground, social realm. As we shall see shortly, peak oil proclaims a particular form of "energy crisis" that is global in scope, geological in origin, and which takes the form of a permanent reduction in the rate at which conventional oil can be extracted.

Oil: the precarious incumbent

Six generations ago, oil was a bit player in an emerging lubricant market where it competed with the rendered bodies of whales, lizards, and fish. Oil now pulses through the lives of most people on earth. Profoundly distinctive, modern industrial economies are high-energy societies that have shattered historic "solar-income budget constraints" (Daly 1974, cited in Clark and York 2005: 406; McNeill 2000, Smil 2005). They bear little resemblance to the low-energy "biological old regimes" that characterize the bulk of human history, in which all energy was "derived directly from the sun: human and animal muscle power, wood, flowing water, and wind" (Marks 2002, cited in Huber 2009a; Hall *et al.* 2003: 318). However, like those previous regimes, energy's role in contemporary social metabolism is just as central, where social metabolism refers to the flows of matter and energy that enable social reproduction (Foster 1999). It is only by tapping and converting a variety of extrasomatic sources of energy over time that societies have modified environments, expanded resource bases, and generated the energy and economic surpluses that enable advanced social divisions of labor (Martinez-Alier 1987, Price 1995, Crosby 2006).

Over the last 200 years the shape and form of the material economy has been tightly linked to the discovery and development of very dense reservoirs of energy in the form of fossil fuels. The significance of these fuels is their capacity to generate tremendous energy "surpluses" (the net return on energy invested in their extraction) compared to renewable energy sources (Cleveland 2007). The raiding of fossil energy stocks in this period has enabled those who command such energies to have a scale and intensity of influence on biogeochemical flows that, in many cases, now exceeds that of natural systems (Crutzen 2002, Dalby 2007). The effective subsidies that fossil fuels provide have also underpinned huge gains in labor productivity (as fuel-based machines substitute for human labor), an unprecedented concentration of productive powers that enable massive economies of scale, and a deepening of the exploitation of nature. As such, the application of ever-greater fossil fuel inputs has been a primary means by which prodigious increases in the production of renewable and non-renewable resources have been maintained over the last 200 years, in the face of declining quality of raw materials and the exhaustion of localized stocks. Ecological economists point out how the much-vaunted productivity of agricultural and industrial systems in value terms is inversely proportional to their productivity in energy terms: rising agricultural yields have been secured only through substantial fossil fuel inputs, for example (Martinez-Alier 1987). In this way "energy crisis" – in the sense of an impending failure to provision industrial societies with the abundance of energy to which its machines and infrastructures have become accustomed – is the mother of all scarcities, a sputtering of the elixir of perpetual expansion by which economies have procured more from less, expanding resource output even while drawing on poorer and poorer quality materials (Bridge 2009).

The step change in available energy enabled by the development of fossil fuels is described by the sociological concept of "metabolic rift," a rupture in the historical patterns of material exchange through which societies reproduce themselves (Foster 1999, Clark and York 2005). Rift here refers both to changes in the scale and geography of physical flows – a massive net transfer of carbon from lithospheric stocks to the atmosphere through the combustion of coal, oil and gas, for example – and to substantial shifts in the geographical organization of economic activity and social life. Unlike wind, water or other solar-derived sources of energy, fossil fuel energies can be expanded and made to flow at will, enabling the realization of economies of scale. They are mobile in a way that wind and water power are not, and allow an unprecedented geographical concentration of production (Huber 2009a). The reliance on machines and large-scale infrastructure for the release of fossil energies also creates conditions for the concentration of capital. Capitalism pre-dates the widespread use of coal in industrial production, but the productive possibilities of fossil energies have given industrial capitalism many of its distinctive social and geographical forms so that it makes sense to talk of a "fossil fuel mode of production." Huber (2009a), for example, argues that by decisively shifting productive forces from human labor to machines, fossil fuels generalized the conditions for a class monopoly over the means of production. And in the sphere of circulation, fossil fuels overcame "the biological constraints of transporting goods" and became a primary means for expanding markets and reducing the costs of circulation.

Although coal remains a cornerstone of the world's primary energy supply, it is oil that has underpinned the modernization of agriculture, driven an unprecedented expansion in personal mobility, and underwritten global economic integration via international trade over the past 60 years (nearly two-thirds of the world's refinery output is used in transportation (Smil 2008: 15)). Energy historians describe a coal-to-oil transition in the twentieth century, in which coal's share of the world's primary energy supply fell from 75 percent in 1900 to less than 23 percent in 2000, although absolute output of coal rose nearly 6-fold in this time (Smil 2005: 13; Podobnik 2005). The significance of this transition in energetic terms is that it moved the material base of the economy up the so-called "resource pyramid," from a lower-grade source of energy (coal) to a higher-grade (liquid fuels, derived from crude oil). With the highest energy content per unit volume, liquid fuels occupy the pinnacle of the fossil-fuel resource pyramid (McCabe 1998).

Oil's high energy density and liquid properties provide "peculiarly prolific capacities for enhancing productivity" and have made it the pre-eminent *social* hydrocarbon of the twentieth century (Huber 2009a: 106). While the mix of energy sources varies widely at a regional or national scale – and one can find starkly different energy systems co-existing in places with marked social inequalities – oil has become embedded in social institutions and material infrastructure to a peculiar degree. The sheer ubiquity of oil means that it is a commodity like no other: oil's applications are more numerous, more intimate and more transformative of social life than those associated with coal, steel or uranium. Through the extensive socio-technical networks associated with mobility and mass consumption, oil has permeated throughout social and cultural life during the twentieth century. In North America, and to a lesser extent in Europe, oil was integral to the Fordist/Keynesian solutions to a crisis of under-consumption that manifested itself in the Great Depression. Through expanding automobility, suburbanization, and development of a mass market for consumer durables, demand for oil was effectively built – via physical engineering and socio-cultural convention – into the social and geo-graphical structure of urban, national and international economies (Huber 2009b, Mitchell 2009a).

The high-energy, oil-fuelled economy, then, is a heavily incumbent energy system with a degree of political, economic and social embeddedness that makes it difficult to dislodge (Kern and Smith 2008). Yet this sprawling, self-reproducing system perches on the lofty heights of the resource pyramid, its material magnificence dependent on the continuing support of a narrow range of top-quality hydrocarbons. It is this ungainly, potentially precarious state of affairs that provides the context for concerns about peak oil.

Peak oil: a crude crisis

Peak oil is a proposition about the relationship between the rate at which conventional crude oil is currently taken from the ground and the rate at which it can be extracted in the future. Although peak oil has developed a wide and popular following since the late 1990s – and to some observers has become a "catastrophist cult" (Smil 2006a) – its narratives and models are strikingly coherent having been developed through the activities of the Association for the Study of Peak Oil (ASPO) and its network of national organizations (Bridge and Wood 2010). Peak oil consists of three distinct but sequential claims: (1) that the extraction rate for petroleum liquids (i.e. for conventional oil) is nearing its global maximum; (2) that the imminent peaking of global production marks the onset of a permanent shortage relative to demand; and (3) that permanent shortage conditions change the structural characteristics of the international oil market in ways that will be felt as an "oil shock" – a hike in price that marks "the end of cheap oil" (Campbell and Laherrère 1998). Thus one of the early proponents of peaking concludes that the next "oil shock" will be "paralyzing and permanent" and "will not be solved by any redistribution patterns or by economic cleverness, because it will be a consequence of pending and inexorable depletion of the world's inexpensive conventional crude oil supply" (Ivanhoe 1995: 88).

Underpinning these claims is an application to the global scale of the "Hubbert model" – a statistically-based model of oil depletion developed for U.S. oil production by a Shell geologist, Marion King Hubbert, in the 1950s (Deffeyes 2001). The model indicates that, for any large region, unrestrained oil production will peak when about half the reserve is gone. By comparing reliable figures on oil extraction worldwide over the last 150 years with the best assessments of remaining reserves, global applications of Hubbert's model indicate a maximum rate of extraction sometime between 2010 and 2025, depending on how conservative an estimate of reserves one adopts. Supporting evidence comes from the reduced rate at which large fields are being discovered (despite significant improvements in exploration technology), from decline curves/depletion rates at existing large fields, and from the recent history of major oil provinces (like the North Sea) which reproduce the peaking that Hubbert correctly predicted would occur in the United States in the 1970s.

The eponymous "peak" – the inflection point of the global oil supply curve – is significant because it marks the point beyond which aggregate extraction cannot be increased. Peak oil is not, then, the same as an "oil crunch," a *reversible* supply squeeze arising from a combination of historic investment rates in the oil sector and unexpectedly rapid increases in demand. The importance of the peak, in other words, is that it marks a permanent departure from the long experience of rapidly expanding oil production and consumption in the twentieth century. Aleklett (2006: 10) neatly captures both the scale and sense of historic inversion associated with the onset of a peak in production: whereas 50 years ago global consumption stood at 4 billion barrels per year, and new exploration yielded 30 billion barrels per year, now consumption worldwide is 30 billion barrels per year, and only 4 billion barrels per year are added via exploration. Post-peak is a supply-constrained world in which oil is no longer a standard commodity: rather than being determined by the marginal costs of supply (give or take some fluctuation), oil prices are a function of the economic and strategic value of oil to consumers.

A uncomfortable embrace: misplacing crisis in depletion

The relationship of resource and environmental geography to statements of physical resource scarcity and "limits talk" is complex, and reveals a continuing struggle between two different epistemological traditions that make up contemporary "resource geography." On the one hand, resource geographers and some political ecologists have found gainful employment inventorying and cataloguing natural resources and their rates of consumption at local, state, federal, and international scales in order furnish some kind of answer to contextual questions about environmental and resource limits. On the other hand, political ecology and resource geography also have a robust tradition of critical inquiry which examines how ideas about nature – and nature's limits – are integral to the production of social differentiation and domination (Harvey 1974, Peluso and Watts 2001). For this critical tradition, "natural resources" and "resource constraints" are problematic claims that express positional – as opposed to universal – appraisals of the meaning, significance and value of the non-human world. The notion of "natural limits," for example, has been used to legitimate regressive social policies that deny rights and freedoms to less powerful groups, that curb redistributive ambitions, and which regulate social behavior in the name of saving the earth. This critical epistemological tradition has broadly taken an anti-naturalist position when it comes to resource scarcity: it argues that environmental conditions are under-determined by nature, and that the criteria for deciding among different environmental futures come from within society rather than being imposed by natural limits.

It surprising, then, to find that peak oil's claims are accepted by many writers and thinkers in the critical left-tradition. One finds in their engagement with peak oil a disturbing flirtation with the "future-as-disaster" that is a consequence of having abandoned a well-honed critical position on the production of scarcity. This capitulation to "limits talk" crops up in a number of places, including, somewhat surprisingly, in the work of ecological Marxists. Altvater (2007: 46, 55), for example, sees peak oil as a sort of judgment day for "fossil capitalism": "the peak," he argues "and thus the limits, of oil production have a major effect on the capitalist accumulation process" and "the fuel driving capitalist dynamics is running out." In an extensive and largely affirming review of the peaking literature, Foster (2008: 7) posits that peak oil may represent a "global turning point" because of the way "an imminent peak in conventional oil. . .strikes at the lifeblood of the existing capitalist economy." Although he explicitly guards against a charge of environmental determinism, the peak oil condition is presented throughout as an external, geologically-based constraint, a primary influence on the rate at which oil can be made to flow from the ground, and the stage for a new round of geopolitical struggle. In other analyses – including Harvey's (2003: 23) *New Imperialism* – an increasing scarcity of oil becomes the explanatory context for oil imperialism: a generalized scarcity argument that propels a "bidding war" or "resource war" where notions of natural scarcity and national security interweave (Caffentzis 1992). Other authors pick up peak oil's arguments for different reasons: in recent work on the politics of calculation around oil and climate, for example, Mitchell (2009b) rests his careful analysis of the production and proliferation of uncertainty on peak oil's premise that the moment of maximum extraction is imminent. One finds, then, a range of writers and thinkers who are usually suspicious of natural limits echoing peak oil's core claims that (1) oil is increasingly scarce; (2) the constraints on its availability are now primarily geological; and (3) emergence of these constraints drives social and political action.

Perhaps part of what is going on here is a strategic determinism about peak oil: that is, a calculated recognition that arguments about imminent scarcity can ignite a deeply-rooted fear of running out, and provide a politically effective lever with which to drive economies away from oil. As important, however, to this consideration of oil's physicality is an effort to admit some of the materialities of oil into analysis – to engage "the properties of oil itself" (Mitchell 2009a: 401) –

and make a space in accounts of oil's political economy for the non-human. Such a motive and ambition are welcome, of course, as they are essential to developing a more thoroughly materialist, political-ecological account of the worlds produced in and through oil. Peak oil's strong sense of fixed geological constraints governing oil's future availability, however, is a crude means of introducing materiality. It snaps analysis back into a naturalist position that forecloses arguments about the social organization of oil production, the limited time-horizon of a market-based energy policy, or the ways in which the social metabolism of oil via the hydrocarbon commodity chain has effectively excluded many of oil's social and environmental costs from the calculus of market price.

The rush to embrace peak oil as a convenient example of capitalism's ecological contradictions is as surprising as it is politically problematic. In reaching too quickly to claim peak oil for its own, recent work on the critical left lets slip a rich body of empirically-grounded arguments about the fundamentally social origins of oil's availability, including historical accounts of the strategies used to organize scarcity in oil markets in the face of prodigious abundance. Peak oil hands the prevention of abundance/organization of scarcity over to geology, effectively depoliticizing the investment strategies of oil firms and the actions of resource-holding states. Empirical observation and carefully-crafted claims about the influence of "above ground" conditions on oil's availability – cartelization, the investment strategies of international oil companies, and resource nationalism – are cast aside as being of secondary importance compared to the new strictures of "below-ground" limits. The problem here is that for critical political economy the creation of scarcity is the heart of capitalist society, whereas for peak oil it is a natural condition.

The materialities of oil

Peak oil subscribes to a simple view of materiality as geological fixity: the rigid confines of the empty barrel. The aim of this penultimate section is to take forward the notion that "the properties of oil itself" matter by showing how the diverse materialities of the hydrocarbon commodity chain are much more expansive than those imagined by peak oil. Materiality here provides a way to capture and express the physicality of oil as a complex and combustible hydrocarbon but without attributing intrinsic qualities and causality to a substance whose forms, functions and geographies are defined socio-culturally, at least in part (Bakker and Bridge 2006). The materialities of oil, then, refer to the biophysical characteristics and material forms of oil as it flows in and through society and the way these are productive of particular forms of social relations.

From social metabolism to materialities

This section highlights five of the more significant of oil's diverse materialities. Each illustrates how some of the most troublesome social relations associated with the hydrocarbon commodity chain are not independent of oil's diverse materialities but bear their imprint (cf. Prudham 2005). Through these selected cases, the primary "energy crises" of the age of oil are shown to be related not to physical depletion – *pace* peak oil – but to struggles over the distribution of economic value and socio-ecological costs of oil production and consumption. *First,* consider oil's properties as a liquid fuel and its superior capacities for work and transformation. In this usage materiality is very similar to the Marxian concept of "use values" where it refers to the qualities of materials and their usefulness to society. The point, here, however, is that these use values are neither uniform nor universal but vary geographically as part of a landscape of uneven development (Bakker 2004). The utility of gasoline, for example, is geographically and historically specific, so that the rate of gasoline consumption is not a fixed law based on its energy-to-weight ratio, but is related to patterns of use (Benton 1989). The "scarcity" of oil, then, is not a generalized physical shortage but a shortage relative to modes of living: as with other resources, oil's scarcity "presupposes certain social ends" (Harvey 1974, Bakker 2004).

Venerable and significant as this observation may be, it nonetheless tends to land with dull philosophical thud. Its implication, however, is that geographical variation in the utility of oil underpins oil trade, giving rise to an uneven landscape of oil creditor states (net exporters) and oil debtor states (net importers). Fewer than 30 countries are net exporters, and only a handful hold the majority of known conventional reserves. Accordingly, the hydrocarbon commodity chain is structured by a central tension between the interests of exporting and importing states, a tension that takes shape in the reciprocal flows of oil and finance between them. The vulnerabilities and strategic opportunities created by these flows of oil and money are at the core of international geopolitics, and structure the domestic politics of large exporting and importing states alike. Oil prices – a conventional measure of scarcity – can be read as an indication of the balance of power between oil exporting countries (symbolized institutionally by OPEC) and oil importers (institutionalized in the form of the International Energy Agency), with periods of relatively high prices corresponding to moments when exporters have the upper hand (Mommer 2001). It is here that peak oil's myopic fixation on "below-ground" constraints becomes clear. Its focus on rising oil prices and the deteriorating reserve/production ratios of many international oil companies ignores how the last decade has seen a significant swing back towards a "proprietorial" model of resource governance. This is manifested in, for example, the renegotiation of access agreements in Venezuela, Russia and Bolivia and a form of resource nationalism that sees decisions about the rate of production increasingly determined by the interests of resource-holding states.

Second, oil may have useful properties but, as a hidden, flammable and explosive liquid, it is difficult to appropriate and commodify. Popular imagery aside, crude oil does not occur as large, neatly-defined underground lakes, but as dispersed concentrations of variable chemical and physical quality trapped among ancient sediments. Some of the reservoirs targeted by international oil companies are far from markets and deeply buried – significant recent discoveries in the deepwater Gulf of Mexico are over 35,000 feet (approximately 9km) below the surface. Conventional oil's liquid character assists its transportation on the surface (it may be pumped in pipelines, for example), but underground its flow character makes it a fugitive resource. The migratory character of oil complicates the process of capture (particularly in large oil fields where there are multiple owners) because the actions of neighboring property owners are not independent. This has led historically to the development of distinctive rules of property in an effort to prevent ruinous competition (Libecap and Wiggins 1984). Commercial realization of oil's value, then, requires capturing a fugitive fluid of generally awkward physical and chemical properties, corralling it to the exclusion of others, and standardizing variations in chemical composition and flow rate so that it may be channeled in an orderly stream over long distances. Oil is not readily a "co-operative" commodity (Bakker 2004) and its proliferation as such is a remarkable feat of science, capital and law (as accounts like Yergin's (1992) *The Prize* and Chernow's *Titan* (1998) illustrate).

One indication of crude's recalcitrance when it comes to the commodity form is that the barriers to entry in the upstream oil sector are high: capital and technological requirements are forbidding, and oil production is dominated by a limited number of large firms. In turn, this organizational structure exerts an influence on the sort of oil reservoirs that are targeted: larger firms need to locate larger bodies of oil in order to replace reserves depleted by production, and large fields are far less numerous than smaller ones. This pattern has intensified following a wave of corporate mergers and acquisitions during the relatively low-price, lean years of the 1990s to produce an upstream industry structure dominated by a few companies that need to locate very large bodies of oil. Any shortage of drilling targets is, then, in part a function of the organizational structure of the industry. That apparently "geological" criteria like the type and size of reservoir are not independent of the way the industry is organized has at least two implications: the rate at which oil arrives at the surface is, to a significant degree, a function of the exploration and development strategies adopted by a few large firms; and competition to secure the requisite "giant" fields propels firms into frontier-type environments around the world, creating conflicts over access to resources and socio-ecological impacts of extraction that suggest strongly – and to no-one more so than those inhabiting the supply zones of the global economy – that maintaining a "normal flow regime" produces states of crisis.

Third, the materialities of oil also extend to the "geography of holes" – that is, to the specific material form that oil extraction takes in the field and, in particular, to the contrast between geographies of territory and extraction. An oil well represents a discrete, molecular point of access rather than a contiguous territorial claim. Compared to the expansive spaces of forestry or agriculture, for example – where production and the generation of value is diffused across a broad surface – the extraction of oil occupies a point in space rather than a laminar, extensive presence. In ecological economics, this highly concentrated spatial form is regarded as significant because oil fields have some of the highest power densities (measured in Watts per square meter) of any energy natural or anthropogenic feature on earth: compared to biomass or other renewables, an oil well produces an enormous flow of energy from a very small area (Cleveland 2007). Oil extraction, therefore, has a punctuated and discontinuous geographical expression that does not coincide with notions of national territory or development. The discontinuous geographies of extraction – their "molecular" rather than "territorial" logic – means that a principal axis of competition is the struggle to secure exclusive control over specific pieces of ground, rather than expanding a territorial domain. The extractive landscape is one of discrete spatial monopolies – patchworks of oil concessions that codify a logic of holding ground and "securing the hole," in which power comes not from the administration of territory but from the ability to control specific patches of ground. Not only does this produce in oil a particular logic of violence and possession that confounds notions of the modern state, justice and democracy (Watts 2004). It also contributes to the conspicuous failure of oil to drive nationally-based development in many parts of the world, and to the crisis of the "resource curse" – the persistence of poverty and deprivation amid enormous natural resources.

Fourth, the geographies of resource access described above mean that oil emerges into the social realm in highly concentrated, tightly-regulated streams. Oil wells are, in this regard, vertical analogues of the chokepoints and bottlenecks conventionally associated with oil shipment, concentrations of flow that generate significant opportunities for control. This vertical and horizontal chokepoint geography creates a high potential for price instability because of the relative ease – imagined or real – of shutting off significant sources of supply (via, for example, interruptions to production in the Niger Delta, hurricanes in the Gulf of Mexico, or periodic shifts in the level of geopolitical tension in the Persian Gulf). Analysts of short-term oil market shifts frequently allude to the bottleneck geographies of oil supply and, in doing so, indicate the tremendous opportunities such imaginary geographies of vulnerability create for capturing value through trading future claims to oil (i.e. oil futures) and via financially hedging against shifts in the price of oil. Since the creation of oil futures markets in the 1980s, the financialization of oil has played an increasingly important role in the production and trade of physical oil, as well as within international political economy more generally (Labban, forthcoming). Trade in oil futures changes the presumed causal relationship between physical oil supply and the market price, and seriously challenges peak oil's narrative that high oil prices and price volatility signal the inflection point of the production curve (Labban, forthcoming). By focusing on the material vulnerabilities of oil supply – and the way perceptions of supply vulnerability and risk create the conditions for the accumulation of fictitious capital via futures contracts – it is possible to see how expectations of crisis are integral to the financialization of oil. Rather than high oil prices being evidence for a geologically-determined crisis of supply, peak oil provides a crisis narrative about the fundamentals of oil supply and demand that fuels the financialization of oil.

Fifth, the materialities of oil extend to the systematic "leakage" of carbon at various points along the hydrocarbon commodity chain, via which fossil stocks of carbon are transferred from the lithosphere to atmosphere. This is not limited to cases of technical rupture – such as burst pipelines or tanker spills – but extends to the institutions of the high-energy society that make it possible for the full social effects of oil's use not to register in the price paid by consumers. The social and geographical displacement of pollution associated with oil is well-catalogued in studies of urban air quality and refinery emissions, and in accounts of environmental destruction during the extractive phase. These indicate how trade in oil is a process of ecologically unequal exchange, in which highly-ordered, high-value energy sources accrue to those with wealth and power while the socio-ecological costs of tapping, refining and using highly concentrated energy sources are displaced onto others (Martinez-Alier 2002). Since the early 1990s, the primary target of scientific and popular concern is the emission to the atmosphere of carbon dioxide from fossil fuel combustion. By taking advantage of the dispersal and mixing capacities of the atmosphere, the negative social and environmental effects of oil use can be made to fall outside the hydrocarbon commodity. Because it enables market prices for oil to be some way below oil's full social cost, the availability of the atmosphere as an unpriced external "vent" for the metabolized waste products of oil ensures oil is artificially "cheap," exerting a powerful influence on the rate at which oil flows through economies. From this perspective, peak oil's concern with an impending slow-down in the rate of oil extraction is curiously irrelevant: the primary challenge for the twenty-first century when it comes to oil (and other fossil fuels) is to slow the rate at which fossil carbon is mobilized and released to the atmosphere. *Contra* the claims of peak oil, the problem is not one of trying to get more oil (or coal or gas) out of the ground, but of finding ways to keep it shut in.

Conclusion

A conventional understanding of "energy crisis" centers on the prospect of sudden and prolonged shortage due to a failure of institutional structures that, thus far, have proven capable of collecting, converting, and distributing energy in ways that reproduce prevailing socio-economic and environmental forms. The narrative of "peak oil" re-articulates this long-standing fear, by pointing to the moment when the "immutable physics" of oil reservoirs will kick in to constrain the ambitions of a modern, high-energy society based upon expanding the throughput of conventional oil (Campbell 1998). I have argued in this chapter that the fossil-based model of social metabolism is indeed in crisis, but not because of a geologically-driven supply-side failure as imagined by the proponents of peak oil. I have suggested that the "crises" of the fossil-fuel mode of production lie not in any post-peak apocalypse, but in the everyday "normal" operation of the contemporary oil economy. The degree to which oil has become embedded in economic and social life means these crises take on a variety of forms, and I have illustrated just a handful of the different ways in which the hydrocarbon commodity chain may be considered to be "in crisis" even in the absence of significant disruptions to supply. From this perspective, crisis is the political expression of contradictions that are inherent to the social metabolism of energy and matter.

Critical geography possesses a set of well-honed tools for analyzing scarcity and, in that sense, the apparent embrace of peak oil by some commentators on the critical left is surprising. In an effort to redress the crude naturalism of peak oil, this chapter has adopted a "new materialist" understanding of scarcity (Castree 2009). It has sought to specify the materialities of oil and carbon, and the way in which these material forms give a structure to the social relations that surround the extraction, distribution and consumption of oil, in order to understand why and how oil is made simultaneously abundant and scarce over time.

In contrast to peak oil's proposition about natural limits, energy futures "past peak oil" will be socially rather than geologically determined. Oil may be incumbent, and the challenges of moving "beyond oil" should be not under-estimated. But there are encouraging signs that some of oil's "normal" contradictions are becoming increasingly politicized – around climate change, the development effects of oil extraction, and "energy security" for example – so that the political economy of energy is being reshaped now more than at any time since the early 1970s. This struggle to define the contours of a new energy regime is expressed in contemporary policy discourses about energy transition and low-carbon pathways, as well as in alternative formulations like "décroissance" and "energy descent," among others (Latouche 2009, Odum and Odum 2001). What is clear, however, is that one cannot pin hopes on peak oil's recessionary limb as a means of addressing the contradictions of oil: fossil fuels – and conventional oil – are likely to remain dominant sources of energy for a long time yet, because there is "no readily available non-fossil energy source that is large enough to be exploited on the requisite scale" (Smil 2006b). Significant increases in the use of coal, gas and oil are likely, particularly in developing economies. In fact, in economic terms there is "no urgency for an accelerated shift to a non-fossil world: fossil fuel supplies are adequate for generations to come; new energies are not qualitatively superior, and their production will not be substantially cheaper" (Smil 2006b: 23). Any transition away from fossil fuels, therefore, will be historically unique in that transition will be towards lower quality, more costly resources and largely as a political response to recognition of the incumbent energy system's social and environmental costs (Smil 2006b, Cleveland 2007).

Articulation of the contradictions inherent to the production and use of fossil fuels is currently most advanced around climate change and, in particular, in calls by "low-energy" countries in the global South for climate justice. The failure to secure binding agreements on carbon emissions at the Copenhagen Climate Change Conference in 2009, however, revealed something of the geographical political economy of fossil energy use. Copenhagen demonstrated that climate change is part of a broader crisis of social metabolism that revolves around geographically uneven rates of hydrocarbon throughput. The strong linkage between energy availability and human development (Figure 14.1) means that *expanding* the provision of high quality energy sources – such as liquid fuels and electricity – remains a priority for large numbers of people in the global South (Cleveland 2007). Technological innovation and investment in new, low-carbon energy infrastructures will certainly form a substantial part of any new energy paradigm. The analysis of "energy crises" offered here, however, suggests that because a progressive energy future will necessarily include a substantial role for fossil fuels, an important goal should be to reform the institutions of the hydrocarbon commodity chain as part of constructing a more sustainable and just energy future "past the peak."

References

Aleklett, K. 2006. "Oil: A Bumpy Road Ahead" *World Watch Magazine* 19(1): 10–12.

Altvater, E. 2007. The Social and Natural Environment of Fossil Capitalism. In L. Panitch and C. Leys (eds), *Coming To Terms with Nature: Socialist Register*. New York: Monthly Review Press.

Bakker, K. 2004. *An Uncooperative Commodity: Privatizing Water in England and Wales.* Oxford: Oxford University Press.

Bakker, K. and G. Bridge. 2006. "Material Worlds? Resource Geographies and The 'Matter of Nature'" *Progress in Human Geography* 30(1): 1–23.

Benton, T. 1989. "Marxism and Natural Limits: An Ecological Critique and Reconstruction" *New Left Review* 178: 51–86.

Bridge, G. 2008. "Global Production Networks and the Extractive Sector: Governing Resource-based Development" *Journal of Economic Geography* Vol. 8 (3): 389–419

—— 2009. "Material Worlds: Geographies of Natural Resources" *Geography Compass* 3(3): 1217–1244.

Bridge, G., and A. Wood. 2010. "Less is More: Spectres of Scarcity and the Politics of Resource Access in the Upstream Oil Sector" *Geoforum* 41(4): 565–576.

Buzar, S. 2007. *Energy Poverty in Eastern Europe: Hidden Geographies of Deprivation.* Aldershot: Ashgate.

Caffentzis, G. 1992. The Work/Energy Crisis and the Apocalypse. In Midnight Notes Collective (ed.), *Midnight Oil: Work, Energy, War 1973–1992.* New York: Autonomedia.

Calvino, I. 1974/1996. The Petrol Pump. In *Numbers in the Dark and Other Stories.* New York: Vintage Books.

Campbell, C. J. 1998. "Running Out of Gas: This Time the Wolf is Coming" *The National Interest* 51: 47–55.

Campbell C. J. and J. H. Laherrère 1998. "The End of Cheap Oil" *Scientific American* 278, 80–85.

Carter, J. 1979. Presidential Speech to the Nation, Washington D.C. July 15th 1979. Available online http://www.pbs.org/wgbh/amex/carter/filmmore/ps_crisis.html

Castree, N. 2009. "Who's Afraid of Charles Darwin?" *Geoforum* 40(6): 941–944.

Chernow, R. 1998. *Titan: The life of John D. Rockefeller, Sr.* New York: Vintage.

Clark, B. and R. York 2005. "Carbon Metabolism: Global Capitalism, Climate Change and the Biospheric Rift" *Theory and Society* 34(4): 391–428.

Cleveland, C. 2007. Energy Transitions Past and Future. In Cutler J. Cleveland (ed.) *Encyclopedia of Earth.* Washington, D.C.: Environmental Information Coalition, National Council for Science and the Environment. Available online at http://www.eoearth.org/article/Energy_transitions_past_and_future

Coronil, F.1997. *The Magical State: Nature, Money and Modernity in Venezuela.* Chicago: University of Chicago Press.

Crosby, A. 2006. *Children of the Sun: A History of Humanity's Unappeasable Appetite for Energy.* New York: W.W. Norton.

Crutzen, P. 2002. "Geology of Mankind" *Nature* 415, p. 23.

Dalby, S. 2007. "Anthropocene Geopolitics: Globalisation, Empire, Environment and Critique" *Geography Compass* 1(1): 103–118.

Daly, H. 1974. "The Economics of the Steady State" *American Economic Review* 64 (2): 15–21.

Deffeyes, K. 2001. *Hubbert's Peak: The Impending World Oil Shortage.* Princeton and Oxford: Princeton University Press.

Foster, J. B. 1999. "Marx's Theory of Metabolic Rift: Classical Foundations for Environmental Sociology" *The American Journal of Sociology* 105(2): 366–405.

Foster, J. B. 2008. Peak Oil and Energy Imperialism. Monthly Review 60(3). Available online at http://www.monthlyreview.org/080707foster.php

Hall, C. A. S., Pradeep Tharakan, John Hallock, Cutler Cleveland and Michael Jefferson. 2003. "Hydrocarbons and the Evolution of Human Culture" *Nature* 426 (6964): 318–322.

Harvey, D. 1974. "Population, Resources and the Ideology of Science" *Economic Geography*, 50, 256–377.

—— 2003. *The New Imperialism*. Oxford: Oxford University Press.

Helm, D. (ed). 2007. *The New Energy Paradigm*. Oxford: Oxford University Press.

Huber, M. 2009a. "Energizing Historical Materialism: Fossil Fuels, Space and The Capitalist Mode of Production" *Geoforum* 40(1): 105–115.

—— 2009b. "The Use of Gasoline: Value, Oil, and the 'American way of life'" *Antipode* 41(3): 465–486.

Illich, I. 1973. *Energy and Equity*. Paris: Le Monde.

International Energy Agency. 2008. *World Energy Outlook*. Paris: International Energy Agency.

Ivanhoe L. F. 1995. "Future World Oil Supplies: There is a Finite Limit" *World Oil* October: 77–88.

Kern, F., and A. Smith. 2008. "Restructuring Energy Systems for Sustainability? Energy Transition Policy in the Netherlands" *Energy Policy* 36(11): 4093–4103.

Labban, M. 2010. "Oil in Parallax: Scarcity, Markets and the Financialization of Accumulation" *Geoforum* 41(4): 541–552.

Latouche, S. 2009. *Farewell to Growth*. Cambridge: Polity Press.

Libecap, G., and S. Wiggins. 1984. "Contractual Responses to the Common Pool: Prorationing of Crude Oil" *American Economic Review* 74: 87–98.

McCabe, P. 1998. "Energy Resources – Cornucopia or Empty Barrel?" *Association of American Petroleum Geologists Bulletin* 82(11): 2110–2134.

McNeill, J. R. 2000. *Something New Under the Sun: An Environmental History of the Twentieth Century*. New York and London: Norton.

Marks, R. 2002. *The Origins of the Modern World: A Global and Ecological Narrative*. Lanham, MD: Rowman and Littlefield.

Martinez-Alier, J. 1987. *Ecological Economics: Energy, Environment and Society*. Oxford: Blackwell.

—— 2002. *The Environmentalism of the Poor: A Study of Ecological Conflicts and Valuation*. Cheltenham: Edward Elgar.

Mitchell, T. 2009a. "Carbon Democracy" *Economy and Society* 38(3): 399–432.

—— 2009b. "Future Measures: How To Calculate – For a World Without Oil" Annual Public Lecture, ESRC Centre for Research on Socio-Cultural Change, University of Manchester. May 14 2009.

Mommer, B. 2001. *Global Oil and the Nation State*. Oxford: Oxford University Press.

Mwampamba, T. 2007. "Has the Woodfuel Crisis Returned? Urban Charcoal Consumption in Tanzania and its Implications to Present and Future Forest Availability" *Energy Policy* 35(8): 4221–4234

Nye, D. 1998. *Consuming Power: Social History of American Energies*. Cambridge: MIT Press.

Odum, T., and E. Odum. 2001. *A Prosperous Way Down*. Boulder, CO: University Press of Colorado.

Peluso, N. and M. Watts (eds) 2001. *Violent Environments*. Ithaca, NY: Cornell University Press.

Podobnik, B. 2005. *Global Energy Shifts: Fostering Sustainability in a Turbulent Age*. Philadelphia: Temple University Press.

Price, D. 1995. "Energy and Human Evolution" *Population and Environment* 16(4): 301–320.

Prudham, S. 2005. *Knock on Wood: Nature as Commodity in Douglas Fir Country*. New York: Routledge.

Smil, V. 2005. *Energy at the Crossroads: Global Perspectives and Uncertainties*. Cambridge, MA: MIT Press.

—— 2006a. "Peak Oil: A Catastrophist Cult and Complex Realities" *World Watch Magazine* 19(1): 22–24.

—— 2006b "21st Century Energy: Some Sobering Thoughts" *OECD Observer* 258/9. December 2006.

—— 2008. *Oil: A Beginners Guide*. Oxford: One World.

Tabb, W. 2006. Resource Wars. *Monthly Review* 58 (8): 32–42.

Watts, M. 2004. "Antinomies of Community: Some Thoughts on Geography, Resources and Empire" *Transactions of the Institute of British Geographers* 29(2): 195–216.

World Resources Institute 2010. Climate Analysis Indicators Tool (CAIT). On-line database available at http://cait.wri.org/. World Resources Institute, Washington D.C.

Yergin, D. 1992. *The Prize: The Epic Quest for Oil, Money and Power*. New York: Free Press.

15 The geopolitics of energy security and the war on terror: the case for market expansion and the militarization of global space

Mazen Labban

In 2006, the prestigious journal *Foreign Policy* and the Center for American Progress surveyed "America's top foreign policy experts" to find out how the United States was faring in the war on terror (*Foreign Policy* 2006). The survey was motivated by an apparent contradiction between the Bush administration's declaration that the "war on terror is being won" and its warning that "another attack is inevitable." What is remarkable about the survey is that 82 percent of the experts interviewed by *Foreign Policy* agreed that "Becoming less dependent on foreign sources of energy will strengthen national security." What is even more remarkable is that 90 percent of the public agreed with the experts on the threat to national security from dependence on foreign sources of energy – the only matter of national security on which both the general public and the experts agree. Those experts saw that the "single most pressing priority in winning the war on terror" resided in ending the dependence of the US on foreign oil, more pressing than "killing terrorist leaders," "promoting democracy in the Muslim world" and "stopping the proliferation of nuclear weapons." Sixty-four percent of those experts believed (in 2006) that US energy policies have actually made things worse as US oil imports appeared to fund terrorism.

US dependence on foreign oil appears to undermine national security at a more immediate, material level. Two-thirds of the "highest echelons of America's foreign policy establishment" believe that direct attacks on energy infrastructure will be the most likely "method" employed by terrorists to attack the US. Yet, because the actual incidence of terrorist attacks has declined since 2001, presumably because of the "success" of the war on terror, energy and security experts have resorted to the *threat* of terrorism to justify the military protection of the global oil supply network. The mere threat of a terrorist attack, the argument goes, is enough to spike oil prices because the decline in the expansion of oil production capacity behind rising global demand has created a very tight market. This has largely resulted from the investment strategies of state-owned companies that continue to block the flow of capital into expanding reserves and production. This structural condition has magnified the effect of any potential disruption, or threat of disruption, to the supply of oil – effects that would not be confined to the US, but would have global repercussions. The threat of terrorism thus requires the *securitization* of global space in its entirety as a preventive measure against the risk of terrorist attacks:

uncertainties and contingencies of the future are lived as permanent, pervasive risk in the present (see Martin 2007; Žižek 2005). This threat of "subjective" violence, the particular and visible terrorist attack, justifies the global expansion and extension of the less visible objective and "systemic" violence of the "smooth functioning" of militarized global capitalism (see Žižek 2008).

The current discourse of energy security goes beyond the question of US dependence on foreign oil and the financing of terrorism. It embodies a larger project of a structural transformation of the world energy market and the militarization of global space, justified by imminent threats to *global* energy security. The energy security discourse, emanating largely from the US, identifies three trends from which stem threats to global energy security: resource nationalism in oil producing countries and its effects on the decline in global oil reserves and production capacity; growing demand for imported oil in developing countries and the global expansion of their state-owned companies into the international energy market; and the threat to the disruption of oil supply and sudden hikes in oil price from attacks on the energy infrastructure by terrorist organizations or "hostile nations." Taken together, those threats undermine energy security understood in its traditional sense as reliable oil supplies at reasonable and stable prices, which could lead to global economic recession and international instability. They could be contained, or averted, according to the expert discourse by the expansion of open markets to allow the free flow of commodities and investment capital; the erection of multilateral cooperative agreements to coordinate the energy policies of major oil consumers; and the military protection of critical infrastructure and maritime trade routes through US-led collective security agreements.

Those elements of energy security are deeply entwined and they cohere around tensions between questions of US "homeland security" and the security of the global energy market. I examine those tensions as they manifest themselves in the hegemonic discourse on energy security in the US. I focus on the US because this is largely where concerns about energy security in relation to terrorism and resource nationalism dominate the public discourse. By hegemonic discourse I mean the discourse emanating from civilian and military departments of government; influential research and policy centers, such as the Council on Foreign Relations and the Baker Institute, and advisory councils such as the National Petroleum Council and the Energy Security Leadership Council; periodicals and other publications associated with those institutions; and individual "experts" who have rotated among, or held simultaneous positions in government, think tanks, private consulting firms and oil companies. First, I briefly examine energy security in relation to changes in the world oil market since the crises of the 1970s; then I examine the tension between (US) energy independence and (global) energy security. The first two sections form the background to the rest of the chapter, which discusses, in some detail, the twin dangers of resource nationalism and physical attacks on the energy infrastructure, and their implications on the world market and collective security, respectively.

Energy security and the liberalization of the oil market

As evident in its historical trajectory, the term "energy security" can be endowed with any meaning, depending on the political expediency of the moment. The concept, as security of energy supplies from politically or geologically induced disruptions, developed in the nineteenth century with the development of mechanical warfare (see Farrell *et al.* 2004). Energy security became synonymous with diversification of energy sources following the historic decision of the British navy to switch from coal to petroleum to fuel its ships in the rivalry with Germany on the high seas. In the US, energy security retained its military significance through the 1950s. It referred to the protection of domestic oil production as a matter of military preparedness, i.e. to ensure the availability of adequate supplies in the case of war (Jaffe and Soligo 2008). It was not until the Libyan oil shock of 1971, and especially the Arab oil embargo of 1973, that energy security became synonymous with independence from "foreign" oil imports, i.e. from outside the "western hemisphere," to cushion against shocks from sudden disruptions and sudden increase in price. Those threats prompted the US to engage its Cold War allies in a security framework that ensured "policy coordination" and collaboration among the industrialized countries and prevented a scramble for supplies in the event of disruptions. Its key components were the International Energy Agency (IEA) and the Strategic Petroleum Reserve (SPR) in the US and similar stockpiles in Europe and Japan.

The liberalization of the oil market in the 1980s, however, and the expansion of trade in crude oil on the spot market, outside long-term contracts with fixed prices, provided an additional measure of supply security. Oil could now be obtained on the open market according to prices determined by "market fundamentals" and relatively immune to the monopoly power of OPEC, whose members in effect had started trading their oil on the spot market to benefit from higher oil prices. The introduction of oil futures and other derivatives on the New York Mercantile Exchange in 1983, and on the International Petroleum Exchange (London) in 1988, and the boost in energy derivates trading in commodity markets throughout the 1990s, expanded the oil market further and provided hedging opportunities against price risk. Financial markets effectively separated the movement of oil prices from the exchange of physical deliveries in physical markets, at the same time that they made the oil price more susceptible to "non-market" events that could potentially threaten physical supplies (see Labban 2010). Paradoxically, the security of oil supplies grew with the volatility of the oil price.

Oil, in fact, remained abundant throughout the following two decades, because of hoarding by oil companies to hedge against, or speculate on, price increases in the future, and because of the slowdown in global demand as a result of the slowdown in economic growth in the industrialized countries. Moreover, the establishment of the IEA and SPR made the economies of the OECD more "resilient" to disruptions of energy supplies (Yergin 1991). The development of further mechanisms increased the "security margin" of the industrialized countries. Those included the expansion and development of domestic production;

diversification of supply sources (geographically) and types of fuel (increased use of nuclear energy, liquefied natural gas, clean coal, renewables, etc.); promotion of conservation and energy efficiency (see Yergin 1988; 2006; 2008; Morse 1991; Bahgat 2003; Yetiv 2004). Concerns about the security of supplies subsided further as high oil prices in the 1970s made production in "marginal" regions profitable; oil production expanded outside OPEC, in the North Sea and the Gulf of Mexico particularly. OPEC appeared to have lost its role as "key player in the political economy of oil" and to have plunged into a process of "institutional decay" – horizontal ties among OPEC members appeared to give way to "new alignments among oil producers and consumers" (Morse 1991; see also Morse 1986). Moreover, the expansion of some OPEC members into the downstream sector deepened their interest in steady and stable supply to consumer markets and reduced their "incentives for politically engendered price increases" (Yergin 1988). With the further liberalization of the global economy in the 1990s OPEC appeared practically dead (see Morse 1999).

The collapse of the Soviet Union in 1991 brought with it promises of new sources of energy that could further weaken OPEC's hold on the market. Much was made of the hydrocarbon potential of the Caspian Sea Basin and the prospect of more reliable energy supplies and, more importantly, direct investments in upstream production in the former Soviet Union, including Russia (see Morse and Richard 2002). The apparent alliance between the US and Russia in the war on terror following September 2001 improved the prospect of Russia displacing OPEC as a reliable energy supplier and potential swing producer. Former Soviet republics, especially those eager to become US clients, were also eager to open their territories to western oil companies. More realistic and accurate assessments of Caspian hydrocarbon reserves, however, burst the inflated Caspian bubble (see Labban 2009). The expansion of state control over the Russian oil industry, on the other hand, and the souring of relations between Russia and the US following the invasion of Iraq in 2003 abolished the prospect of opening the Russian upstream sector to western oil companies. Even worse, Russia appeared to gravitate closer to OPEC both in its energy policies an in the current attempt to form a natural gas cartel similar to OPEC.

Resurgence of concern with energy security and dependence on foreign oil after the relative calm of the 1990s came on the heels of several events in oil producing countries almost coinciding with each other: strikes by oil workers in Venezuela; attacks on oil infrastructure in Nigeria; the invasion of Iraq in 2003, which on one hand removed more oil from the market but on the other hand threatened to spread political instability to other oil producing countries in the Middle East; mounting threats of military strikes against Iran following the latter's resumption of its nuclear energy program; the resurgence of resource nationalism and the expansion of state control over the oil industry in Russia, Venezuela, and elsewhere; disruptions from natural disasters; several (failed) terrorist attacks on oil production facilities and export infrastructure.[1] The security of global oil supply seemed threatened by potential disruptions from any one of this variegated set of events, which when taken together magnified the threat even more, especially in the context of a tight market. Low investment in expanding production capacity throughout much of the 1990s left the market with little spare capacity to meet rising demand following global economic recovery and to prevent sudden increases in the price of oil. Tight markets with little spare capacity imply more vulnerability to disruptions, accidental or intended, and it is in this context that the contemporary notion of energy security, and threats to energy security, must be understood.

Managing market security, managing global interdependence

No one in the community of experts on energy security and policy would seriously consider US independence from foreign oil. Those who contemplate US self-sufficiency in energy acknowledge that this is not achievable in the foreseeable future, i.e. the next two decades (cf. Deutch and Schlesinger 2006). Nonetheless, for most analysts dependence, as long as it is not concentrated on one geographical source and type of energy, does not necessarily constitute a threat to energy security. Conversely, energy independence does not necessarily guarantee energy security: energy supplies may remain vulnerable to all sorts of disruptions, despite self-sufficiency. In a fundamental way, however, the question of territorial self-sufficiency has become irrelevant since it does not account for the degree to which the territorial economy of the US has become interdependent with the global economy and the global energy market. The effect of energy disruptions on the US economy is not directly determined by the access of the US to energy resources or self-sufficiency in energy supplies, but is mediated by the effect of the global energy market on global economic growth.

The underlying problem is growing *global* dependence on imported oil. There is one energy market, and disruptions in one place would reverberate throughout the whole market. For the same reasons, however, development of resources in any particular place would have beneficial effects on the whole market. Accordingly, energy independence is not only unrealistic but also dangerous – it is the obverse of resource nationalism and would hinder investment in the development of global energy supplies. Indeed, strengthening international trade and investment in energy would enhance the security of the US: "There can be no U.S. energy security without global energy security" (NPC 2007). Thus, the question is not how to end US dependence on imported oil – an unreasonable objective – but how to *manage* its consequences (Bahgat 2003; Yergin 2005; Deutch and Schlesinger 2006; West 2008). The task is to derive energy security from managing interdependence in the world energy market and the global system of international relations as a whole. As Yergin (2007) put it in his testimony to Congress, "Secession is not an option":

Energy security inevitably exists in a larger context of overall security and international relationships. In a world of increasing interdependence, energy security will depend much on how countries manage their relations with one another, whether bilaterally or within multilateral frameworks.

Management of growing interdependence is necessitated by two major sources of tension within the global energy system: a) growing resource nationalism and the closure of the upstream sector in oil producing countries to western major oil companies; b) growing competition among consumers, especially the developing countries of the periphery whose state-owned companies do not simply compete for oil resources across the globe but represent a potential threat to the very structure of the energy market. The global energy market system would be aggravated further in the event that state-owned companies from producing and consuming countries join their "state-orchestrated strategies" into an "axis of oil" that challenges the "rule-based international order" for energy trade and investment (Leverett and Noël 2006). The security of the energy market could be ensured through the management of two sets of relations to prevent such an alliance against market forces: a) relations between consumers and producers, mainly between OECD and OPEC and other producing countries with state-owned companies (Russia); b) relations among consumers, within the OECD but especially between the OECD and the developing economies of the periphery with state-owned oil companies (China and India).

Resurgence of resource nationalism and the return of OPEC

The significance of contemporary resource nationalism can only be understood against the ideology of the free market that consolidated throughout the 1990s. Economic liberalization appears to have extended the principles of free trade and economic integration into energy markets, providing "ready access to resources and the efficient application of investment capital, technology and management by an internationally competitive petroleum industry" (NPC 2007: 216; see also Morse 1999). Oil consuming countries benefited from a larger number of market participants, market determined prices and free access to oil supplies, while the governments of oil producing countries also benefited from access to foreign capital (Baker Institute 2008). Energy and finance circulated freely in unfettered markets until those came under attack by resurgence of OPEC at the end at the 1990s and the expansion of state-owned companies riding the wave of resources nationalism with the economic recovery and high oil prices of the early 2000s.

Until the price collapse of 1998, OPEC was traditionally divided between "doves" and "hawks," those who advocated lower prices and higher export volumes (Saudi Arabia and Kuwait) and others, with smaller reserves and lower production capacity, who pursued higher revenues from higher prices (Iran and Venezuela) (Yetiv 2004). By the end of the decade, however, OPEC as a whole came to favor higher prices, adopting a "more radical, confrontational developing world approach that favors revenues over other issues" (Morse and Jaffe 2005). Saudi Arabia, until then the swing producer that secured oil supplies in case of price rise or interruptions of supplies, now supported higher prices over volume. Years of market gluts and struggles for market share within OPEC and between OPEC and non-OPEC producers had inhibited investment in exploration and production. Despite recovery of world oil demand and prices in 2001, however, OPEC kept constraints on the expansion of production capacity and investment in additional supplies, presumably to maintain high oil prices. As a result, OPEC spare production capacity declined from 5 million barrels per day in 1998 to 0.7 million bbl/d in 2005. This left OPEC with little ability to meet rising demand from economic recovery and the industrial growth in the periphery. Imminent oil scarcities and constraints on oil production in the near future would derive from a combination of "above ground" factors, namely OPECs "developing world approach" and the inefficiency of its members' state-owned companies, not from geological limits.

The barrier to the free movement of financial and technical capital is not specific to OPEC, however, but is a problem posed by the nature of state-owned companies. As the wave of resource nationalism spreads across most oil and gas exporting countries outside the OECD, the barriers to the movement of capital proliferate. What makes matters worse is the expansion of state-owned companies, from producer and consumer countries alike, into *international* energy markets, exporting their investment strategies abroad and placing further constraints on the expansion of market practices. The resurgence of OPEC and the global expansion of state-owned companies are seen to pose threats to the structure of the global energy market, and only as such to the energy security of particular consumers. This threat can be averted by expanding and strengthening the market itself. As Yergin (2007; see also Yergin 2008) declared on many occasions, "markets *themselves* need to be recognized as a source of security." Markets are believed to provide security for two reasons: a) At the level of physical supplies, "large, flexible, well functioning" markets react to the movement of price and can absorb shocks and respond to disruptions "more quickly," "more efficiently and effectively" than controlled and regulated markets. The fewer the barriers and regulations that constrain the free movement of (physical and financial) resources, the more easily supplies and prices can adjust to market forces outside the control of any one group of producers – as long as trade is safe and secure. (Yergin 2006; 2007; Yetiv 2004; see also Morse 1999). On the other side, in "flexible, well-functioning markets" consumers would adjust demand and consumption according to the movement of price, thus encouraging "economy and innovation in the supply and demand for energy" (Jaffe and Soligo 2008: 54). b) From a financial perspective, open markets allow the free flow of investment capital into the upstream sector, which is essential for growth in production to meet growing demand when necessary. As global markets become more dependent on production from complex and more remote places, development of new resources, expansion of production capacity and the infrastructure necessary to bring resources to market becomes more costly and dependent on a "stable and attractive investment climate" (NPC 2007; see also Yergin 2007).

Despite exaltations of the virtues of the market, there is near consensus in the expert community that the market alone cannot "automatically deliver the best outcome," not because of a flaw in the free market itself, but because its "proper functioning" is distorted by OPEC and because state-owned oil companies presumably operate according to non-market rules.[2] As the two sections that follow demonstrate, the proposed solutions to avert these threats center on the simultaneous liberalization of the energy market and the integration of oil consumers, and potentially oil producers, into liberal market mechanisms through multilateral agreements and international institutions. The primary purpose of those mechanisms is to open the upstream sector of oil producing countries to foreign investment and to prevent state oil companies of consuming countries from pursuing bilateral, state-to-state deals outside the market. In what follows, we first analyze the argument against state-owned companies and then look at the problem that their entry into the market poses.

State-owned companies: the limit to market expansion

The threat that state-owned companies pose to global energy security derives from the constraints they place on the flow of financial capital and high technologies into the upstream sector. The limits on investment in oil producing countries derive from two factors. First, stringent investment rules repel the foreign capital required to expand production capacity, which leaves oil producers with scant resources and backward technologies at a time when oil production has become more complex and costly. Infrastructure and production capacity in productive regions deteriorate while the capital and technology of transnational oil majors is forced to migrate to less productive and more costly regions, imposing therefore additional costs on the market. Second, as generators of the larger part of the national income, state-owned oil companies' earnings must contribute to the national budget, thus removing finances generated by oil revenues from the cycle of reinvestment in oil production:

> To the extent that NOCs [national oil companies] must meet national socio-economic obligations such as income redistribution, over-employment, fuel price subsidization, and industrial development, NOCs have fewer incentives or resources for reinvestment, reserve replacement, and sustained exploration and production activity.. . .The tendency of NOCs to focus on socio-economic activities other than oil field maintenance and expansion is partly responsible for the slow pace of resource development relative to the rapid rise in global demand and could mean that new production will not materialize to meet rising oil requirements in the future, leaving major oil consuming nations with a scarcity of fuel.
>
> (Jaffe and Soligo 2008: 12)

State-owned companies, moreover, are presumably "less responsive to market price signals," thus they deprive the market of potential increase in oil supply even when oil prices increase. Not only state-owned companies, but also consumers in some oil producing (and oil consuming) countries are "less responsive to market price signals" because of subsidized fuel consumption. Subsidies, it turns out, have several detrimental effects on the security of the energy markets. In the first place, fuel subsidies maintain high demand even when prices in the international market are high, because they shield consumers from the "true market costs of fuel consumption" and make domestic consumption less responsive to the movement of price. Because consumers in countries with no subsidies continue to pay "artificially" inflated higher prices, they effectively finance end-user consumption through government subsidies. When these subsidies are in oil-exporting countries, they make less oil available for export, thus depriving the international oil market of oil supplies and potentially driving international prices even higher, at the same time that they channel more funds away from reinvestment in expanding production (ESLC 2008: 106; see also Baker Institute 2008: 16; Deutch and Schlesinger 2006: 8). In summary, spending oil revenues on social welfare, income redistribution, and industrial development in oil exporting countries leaves the governments of oil producing countries with little investment capital, which increases the cost of energy while putting the security of the market at risk. The risk to long-term energy supplies is magnified by the closure of oil producing countries to foreign capital.

The global expansion of state-owned companies into global energy markets

If state-owned companies in producer countries pose a problem to the market by placing barriers to its expansion, state-owned companies from consumer countries threaten the market by their very expansion into it. The rising demand for oil in China and India has led to the "going out" of their state companies to "lock up" resources where major oil companies cannot go. Not everyone perceives this "going out strategy" as a stress on resource availability. On the contrary, this expansion may prove beneficial to world energy markets as investments by state-owned companies from developing countries expand global oil and gas reserves. Moreover, if those companies invest in places where transnational companies cannot invest (because of sanctions) or would not invest (because of high costs), then such investments would end up subsidizing consumption in the US and elsewhere. According to current estimates, China for example imports about 10 percent of the oil produced by Chinese oil companies abroad; the rest is sold on the world market (Victor 2007; see also Yergin 2008). Accordingly, the OECD and the IEA should encourage foreign investments by Chinese and Indian companies rather than treat them as a threat to energy security.

The real threat to energy security of investments from developing countries derives from the form of investment, or the "mercantilist approach" to securing foreign investments in energy production. On one level, this approach places transnational oil majors at a competitive disadvantage. China and other developing countries have sought bilateral agreements that attach extra-commercial provisions to investments in energy production. Those include aid, soft loans, investment in non-oil transportation and telecommunication infrastructure, and other commitments that transnational oil companies would not be able to offer to the governments of oil producing countries, whose expectations may change accordingly. State-owned companies, moreover, have direct access to cheaper capital from their governments, making their costs of investment more competitive compared with private oil companies that have to borrow capital from capital markets (see Leverett and Noël 2006; Deutch and Schlesinger 2006;

NPC 2007; Yergin 2007). On a broader level, the "Chinese approach" would be detrimental to the world market if it were to become the model for other consumers to emulate. More important than the hostility between producers and consumers, or among consumers competing for access to resources, is the potential "hostility to globalization" that could "fracture the global trading system" into myriad bilateral and preferential agreements that would undermine the "proper functioning" of the world energy market (Deutch and Schlesinger 2006; NPC 2007). Indeed, the risks of the "mercantilist" approach transcend its effect on the energy market: state-owned companies, of oil producing and consuming countries, could become an alternative source of international funding and investment capital that could upset the geopolitical-financial order centered on the IMF and other institutions dominated by the North Atlantic. On this view, Venezuela's financing Argentina's exit from IMF indebtedness is seen as a case of using oil revenues for "political pur-poses that harm U.S. interests," while Chinese investments in African countries is believed to weaken the leverage of international financial institutions and western governments in improving the prospects of "good governance" and "sound fiscal policies" in developing countries.

Managing energy security: open markets and multilateral cooperative agreements

The goal of the campaign against resource nationalism is not to abolish state-owned companies, but to establish mechanisms that would integrate them into the world economy according to market rules. The primary task, presumably, is to improve the performance of state-owned companies and make it conducive to the expansion of production by exposing them to domestic competition. Governments of oil producing countries should be encouraged to promote "private sector investment, joint ventures, and technology transfer" (Deutch and Schlesinger 2006). Energy should become central to the existing "international economic architecture." Multilateral agreements could curb the spread of resource nationalism by bringing "the rules of global oil trade and investment in harmony with rules governing trade in manufacturing and services" (Baker Institute 2008: 17; see also Morse and Jaffe 2005: 91; Luft 2008):

> Access to consuming country markets and preferential trade status should be linked in some measure to oil-producing states' energy sectors delivering more liberalized policies toward investment in their oil resources. This would mean . . . discriminating more actively against those countries that do not permit foreign investment in their energy resources and that limit their exports to manipulate prices.

In other words, consumer countries could use the WTO, IMF and other institutions to prise oil-producing countries open to foreign oil companies. This is not an easy task, as the promotion of free and open markets abroad comes against persisting protectionism against foreign investment in the US and the EU (see Labban 2008:

53, 81–82). Similarly, the campaign against cartels comes against the campaign for "consumer cooperation." For all its emphasis on fair competition, the hegemonic policy discourse is averse to the growing competition for oil and gas resources among consumers, which drives prices higher and allows oil producers to manipulate the competition to their interest. Cartel behavior on the part of the producers, however, justifies a consumers' cartel; but whereas the former appears to undermine the security of the energy market, the latter appears to guarantee it, as it paradoxically promotes market principles: "The United States and other large consuming countries, if banded together, can do a great deal more to enhance the institutional mechanisms that favor markets over political intervention by producers" (Baker Institute 2008: 17). Here also energy could become part of the existing "international economic architecture." New competitors from developing economies could be engaged in the "global network of trade and investment" to transform potential geopolitical rivalry not only into benign economic competition but also into "more durable cooperation" that also ensures the newcomers own energy security (Yergin 2007; 2008; Leverett and Noël 2006).

The engagement of developing major consumers, and potentially energy suppliers such as Russia and Brazil, entails one of three forms of integration: either an extension of the IEA, or the creation of a "cooperative arrangement within the IEA," or the creation of parallel cooperative arrangements comprising developing countries to collaborate with the IEA. Such cooperative agreements serve as a "deterrent" to the monopoly power of OPEC by ensuring a "harmonized energy policy" among oil consumers and by expanding the "IEA-coordinated stockpiling system" to China and India and other major consumers to absorb shocks from sudden disruptions (Baker Institute 2008; Deutch and Schlesinger 2006; see also Lugar 2006). The OECD established the IEA in 1974 ostensibly for this purpose. But there is another aspect of the IEA that lends more significance to the prospect of extending it to the developing economies of the periphery. The IEA, the brainchild of Henry Kissinger and in circulation within the US administration since 1969, was intended as a mechanism that would "coordinate" the energy policies of its members, i.e. to prevent bilateral relations between members of the OECD and oil producing countries independent of the US.[3] Since the 1980s, the IEA became increasingly an instrument for expanding free markets: opening the Japanese market for oil products; removing subsidies over European coal; and so on. Thus, extending the IEA or similar mechanism to newly industrializing countries would achieve two simultaneous goals. It would promote "market-based approaches to energy" and encourage new members "to view efficient markets as the best way to obtain resources" (NPC 2007; Victor 2007). Equally important, engaging developing countries into multilateral energy frameworks would align their energy policies with those of the US and "enable closer monitoring of their compliance with international agreements" (NPC 2007). Membership in the IEA or the "development of new regional energy security arrangements" would encourage energy producers and consumers such as China and Russia to "define their goals in manners compatible with US objectives" (Morse and Jaffe 2005).

The war on terror and the militarization of global space

The military hegemony of the US played a crucial role in the development of liberal capitalism after the Second World War. Military alliances, such as NATO, upheld US hegemony over its Cold War allies by integrating them within asymmetrical structures that set limits to their political autonomy while expanding the reach of the US military on behalf of global capitalism as well as a matter of national security. This simultaneity continues to shape contemporary US military strategy, at least at the level of self-representation. Just as the security of the international system has depended on US military power, the security of the US, according to the US Defense Department's *National Defense Strategy* of 2008, is "tightly bound up with the security of the broader international system." Providing "enduring security for the American people" entails the promotion of "collective democracies" and fostering "a world of well governed states that can meet the needs of their citizens and conduct themselves responsibly in the international system" (DoD 2008: 1; see also ESLC 2008: 14). The ultimate task of the US military forces is not only to defend the national territory but also to improve "the capacity of the international system itself to withstand the challenge posed by rogue states and would-be hegemons" (DoD 2008: 6). In matters relating to energy security, the deployment of US military power serves to secure access to the "global commons" and the "flow of energy resources vital to the world economy" (US DoD 2008: 16).

The militarization of global space is an essential aspect of a properly functioning world market, as long as the US "maintains overwhelming superiority in military and, in particular, naval power" (Jaffe and Soligo 2008: 2, 54). As the world (energy) market expands, threats to its smooth functioning increase, their sources and origins becoming more diverse. Global economic interdependence has created "a web of interrelated vulnerabilities" and has spread risk across global space, increasing the uncertainty about the extent of crises and their speed in spreading across space. This risk is compounded by the development of non-conventional warfare – chemical, biological, nuclear weapons – with potential "catastrophic capabilities." In this new strategic environment, as with any cautious trader on Wall Street, military planners must "develop the military capability and capacity to hedge against uncertainty" (US DoD 2008). Hedging against geographical uncertainty, however, is as crucial as hedging against future uncertainty: the "spreading web of globalization" makes it imperative "to defend the homeland by identifying and neutralizing threats as far from our shores as possible" (United States Navy 2007). This defense of "the homeland in depth" involves on one hand the protection of the global energy infrastructure and trade networks against terrorist (and pirate) attacks and, on the other, the engagement of rising powers in military-to-military cooperative agreements, especially in the naval sphere.[4]

Terrorism and the protection of energy infrastructure

Energy security intersects with (the discourse on) terrorism on two levels: finance and the physical infrastructure. The question of financial flows to terrorist organizations is particularly difficult, since one of the main strategies in the global war on terrorism is to cut the material and financial support of terrorist groups and the institutions that support and propagate the extremist ideology that underpins them. Global dependence on imported oil, however, has kept oil prices high, and the higher the oil price the higher the revenues of oil exporting states, including those that support terrorism, *no matter where oil is purchased*. What complicates matters, from a geopolitical perspective, is that one of the culprit states, Saudi Arabia, is one of the staunchest US allies from which the US imports about 1.5 million barrels per day. Although it is practically impossible to determine the extent to which oil revenues finance terrorist organizations, it has become an article of faith in the policy circles of the US that part of the petrodollars flowing to Saudi Arabia (and other Gulf states) finds its way through various channels to the "jihadist movement": Saudi Arabian charities, governmental and nongovernmental institutions, and private individuals, have allegedly financed terrorist organizations (directly or through financing Islamic fundamentalism) and have fueled "conflicts from the Balkans to Pakistan" (Luft 2008; see also Woolsey 2002; Wirth *et al.* 2003; Bahgat 2004; Friedman 2006; Deutch and Schlesinger 2006; Byman 2008a; 2008b; Baker Institute 2008; Jaffe and Soligo 2008; Luft 2008).

The relation between oil revenues and the incidence or threat of terrorism, however, is neither direct nor necessarily linear. Terrorist activities are "relatively low-cost endeavor[s]" and terrorist networks do not require large capital outlays, as with standing armies, but are organized in a manner "to make the most of scant funds." (Islamic) fundamentalism, or "ideological extremism," moreover, is not dependent on the price of oil and is not likely to move with it. Al-Qaeda flourished when oil prices were at their historical lowest in the 1990s, relying on sponsors such as Sudan and Afghanistan, poor countries by any standard. A decline in oil prices and state revenues might play into the hands of terrorist networks by undermining the stability of the authoritarian governments that depend on high oil revenues for their stability (see Victor 2007; Taylor and Van Doren 2008). Indeed, the objective of attacks on the energy infrastructure promoted in speeches by al-Qaeda leaders is both to undermine the economic basis of the regimes of oil producing countries in the Middle East while also raising the price of oil for western consumers.

At this second level, at the level of threat to the physical infrastructure, energy security encounters the problem of terrorism still with physical and financial dimensions. Quite paradoxically, although the energy market has become more stable and resilient than it was thirty years ago, it has also become more physically vulnerable and financially volatile. As the international energy market expands and production becomes more dependent on more remote and difficult to mine resources, the transportation and distribution infrastructure expands and its vulnerabilities multiply. The vulnerability of the network derives not only from its vastness, however, or the (physical) concentration of the infrastructure, but also from its connectivity: disruptions of supply in one place might create shocks at the regional, or even global scale. Successful attacks on the infrastructure can disrupt the physical flow of resources and create (short-term) shortages. But a local attack can also have "wide-ranging consequences" by creating financial volatility in the oil market, precisely because of the over-sensitivity of financial markets to perceived repercussions of non-market events. Financial markets are as vulnerable to actual attacks or threats of attack as to failed attacks and perceived or imagined threats of attack; indeed, the ramifications of a perceived event can exceed the actual significance of the event itself (see Cook 2008; see also Labban 2010). A hike in price therefore could result from a successful or failed attack, or even the mere threat of an attack on oil infrastructure. On this logic, there is no such thing as a failed terrorist attack if the goal is to raise oil prices or cause panic in oil markets. Even if a terrorist attack fails to halt the delivery of oil to western markets, an increase in oil price would still accelerate the "transfer of wealth" from oil importing countries to oil exporting countries – allegedly a "central part of [Al-Qaeda's] plan" (Luft 2008). If the problem of financing terrorism with oil revenues is taken into account (above), a vicious cycle of high oil prices and disruptions to the supply of oil appears to loom in the near future: high oil prices increase the revenues of states financing terrorists whose attacks on the infrastructure would raise prices even higher, leading to higher revenues and more funding for terrorist organizations, and so on.

The flexibility and unpredictability of terrorist organizations exacerbate the threat to the network further: protection of one part of the network shifts their intentions elsewhere. The implication is that the only way to secure the (global) energy infrastructure is to protect it in *its entirety*: "to prevent or respond to physical threats or attacks on the *entire supply chain*" (Yergin 2005: 57). Protection of the (domestic) energy infrastructure gained increasing significance with the rise of "catastrophic terrorism" in the 1990s, culminating in the attacks on New York in 2001. Although the power outage in New York in 2003 and Hurricane Katrina in 2005 were not the works of Al-Qaeda or any other terrorist group, they underscored further the vulnerability of the domestic energy grid and supply network. The protection of critical infrastructure, or CIP, soon became a central element in ensuring homeland security. The infrastructure could become the object of terrorist attacks, or terrorists could use the infrastructure as a "weapon of mass destruction," by releasing the energy stored in it to inflict damage on life and property with very little input. Attacks could be carried *on* the infrastructure or *through* the infrastructure (Bajpai and Gupta 2007; see also Farrell *et al.* 2004; Yergin 2005).

The attention given by Al-Qaeda to oil prices and oil production and transportation facilities after 2004 globalized the problem of protecting the energy infrastructure. It was no longer enough to protect the concentrated infrastructure in the US, points of production, refining, and so on, but the whole network at the global scale. This implies the protection of millions of oil wells and offshore extraction platforms, vast pipeline networks, more than four thousand tankers, "chokepoints" and maritime routes, pumping stations, gathering and distribution plants, terminals and storage facilities, refineries, petrochemical plants – in addition to the electronic and computerized control systems, and networks covering space and the ocean floor, that manage the operations of this vast material network and that are also vulnerable to cyberattacks. The materiality of the homeland thus extended through the physical and cyber web of the global infrastructure that connects energy markets, consumers, and producers. Protection of the homeland required protecting the vast infrastructure though which vast amounts of key resources – material, cyber and financial – must remain in constant flow, which amounts to the militarization of global space in its entirety and through all its layers.

Several CIP measures were put in place by the Clinton administration in the late 1990s (Farrell *et al.* 2004: 440–441). The most significant bureaucratic development, however, was the creation of the Department of Homeland Security in 2002. Homeland Security Presidential Directive 7 (HSPD-7) established the "policy for enhancing 'protection of the Nation's CI/KR [Critical Infrastructure/Key Resources]' and mandate[d] a national plan to actuate that policy"; the Homeland Security Department's *National Infrastructure Protection Plan* (NIPP) meets the mandate set forth in (HSPD-7).[5] The NIPP recognizes from the beginning that measures taken to protect "the Nation's CI/KR" do not end "at a national border" and must therefore require international coordination and cooperation, among governments and private interests (US DHS 2006: 51–52).[6] The Department of Homeland Security, according to NIPP, collaborates with the Departments of Justice, Commerce, Defense, Treasury, and especially the Department of State, "to reach out to foreign countries and international organizations to strengthen the protection of U.S. CI/KR," by integrating CIP programs into the international agreements that govern international markets (US DHS 2006: 5, 18, 19, 22; see also ESLC 2008: 31, 108–109).[7]

The rationale for fusing homeland security with international coordination derives from the fact that the domestic energy infrastructure is interconnected to varying degrees with global energy, transportation and other infrastructures – "systems of systems" on which depend the "Nation's safety, security, prosperity, and way of life."[8] These systems "must be protected both at home and abroad"; risk must be managed "as far as possible outside the physical borders of the United States" (US DHS 2006: 55, 124; see also United States Navy 2007; US DoD 2008). The international vulnerabilities of the Nation's infrastructure, "vulnerabilities based on threats that originate outside the country," derive from three characteristics of this infrastructural interconnection. First, at the most immediate and material level, the domestic energy infrastructure is vulnerable to cross-border interdependencies, beginning with the extension of the infrastructure into Canada and Mexico. At a more general level, it owes this vulnerability to the cross-sector and cross-border integration of the energy infrastructure into international and global markets – thus, financial services, transportation, and telecommunications all become components of the energy infrastructure. Second, also at a material but less immediate level, energy sectors in the US may depend on "inputs that are not within the control of U.S. entities." Government functions and facilities may be "directly affected by foreign-owned and -operated commercial facilities." The functioning and operation of the domestic energy infrastructure, therefore, depends to a certain extent on decisions made outside the US, by foreign entities. Finally, the third characteristic that lends the Nation's infrastructure international vulnerability is the ownership by private corporations from the US and the federal government of "a significant number of facilities located outside the United States that may be considered CI/KR." The NIPP does not illustrate, but we can speculate that very few overseas assets owned by the US government or private corporations would not be considered "critical" or "key" to the security of "the Nation."

Conclusion: energy security as preemptive aggression

Energy security is a global question. We looked at it from the standpoint of the United States because the discourse of energy security, since its inception in the 1970s, made energy security synonymous with US national security, and made the energy security of the US synonymous with the security of the global market. US energy security is construed as necessary for global economic stability as much as global energy security is essential for the US economy and national security. Any interruption in the flow of oil to the US – any threat to US national interests – will bring about global economic disorder. This is the double blackmail of the globalism embodied in the energy security discourse emanating from the US: everybody gains from increased energy security at the global scale; everybody suffers from threats to US energy security and the vulnerability of the homeland to disruptions in energy supply. An expert expressed this danger on the eve of the Arab oil embargo (1973) as the "the dangers to world harmony and peace – that lie in a situation of growing American dependence on external energy sources, especially Middle East oil" (Wilson 1973). More recently, a former commander in chief of the US Pacific Command assured readers of the preeminent policy journal *Foreign Affairs* that there is "far smaller" risk to the interruption of the international maritime flow of oil than generally assumed because of the preponderance of US naval power, adding however that only the US has sufficient naval power to "seriously disrupt oil shipments": "Today, the U.S. Navy has no rivals in its capacity to impose and sustain . . . blockades" over "a large area of water over a long period of time," and the US will employ naval blockades "when necessary." The world benefits from the naval hegemony of the US as long as other countries "do not endanger Washington's vital interests" (Blair and Lieberthal 2007).

According to the hegemonic discourse, military and economic threats to global energy security could be hedged by the simultaneous expansion of an open market, managed by multilateral agreements that guarantee the uninterrupted circulation of material and financial resources, and military forces through existing formal alliances such as NATO and informal arrangements such as the Global Maritime Partnership initiative, in a manner that coincides with the national interest of the US. Terrorism is one threat that could be eliminated in the (very) long run by the expansion of free market democracy and in the short run by military expansion and the arming and training of local police and military forces in vulnerable regions. The threat of intensified competition for hydrocarbon resources, especially from developing countries bent on modernizing their militaries and developing blue water naval forces, and that could result potentially in armed conflict, could be contained by integrating those countries in collective energy and collective security arrangements that align their energy policies with those of the US; their state-owned companies would eventually behave according to market-based rules and participate in the international market to the benefit of the market as a whole. Ultimately, the threat of resource nationalism, the source of current and potential oil scarcities, can be resolved by promoting "good governance" and creating "hospitable investment climates" in oil producing countries, to channel national revenues and open the space to foreign capital for investment in the expansion of reserves and production. It is this last threat, the threat from resource nationalism, which appears to magnify the effects of the other two by creating resource scarcities and maintaining a tight energy market. Forecasts about imminent scarcities and potential attacks on the network or resource wars bring imagined crises from the future into the present and project them across global space in its entirety, making the expansion of a militarized global market necessary and urgent. In the name of global energy security, military and market aggression becomes an act of defense, a hedging strategy against threats in space and time, preempting presumed future threats before they become present, taking the fight to the enemies "around the world so we do not have to face them here at home." Acts against aggressive expansion in the name of security – from terrorism to resource nationalism – become the justification and legitimation of the necessity of this act of preemptive aggression.

Notes

1   The most prominent were the attacks on the French tanker *Limburg* near Yemen (6 October 2002); Yanbu' and al-Khobar, in Saudi Arabia (1 and 29 May 2004); the Abqaiq refinery in Saudi Arabia (24 February 2006); attacks on oil refineries and storage facilities in Yemen (15 September 2006; 8 November 2007).

2   Few analysts, however, are keen on adhering to radical free market principles and have argued against any form of market intervention or management, much less building up military presence in oil producing regions. Such measures would undermine energy security, because they would prevent the market from functioning according to its own principles. This is largely the position of experts associated with the libertarian CATO institute. See Gholz and Press 2007; Gordon 2008; Taylor and Van Doren 2008.

3   The European Economic Community and the Arab League established the Euro-Arab consultative council in July 1974 and (with Japan) sought bilateral agreements with OPEC, offering also manufacturing exports, economic and technical assistance, and recycling of petrodollars. The IEA guaranteed against an independent European energy policy, at the same time that it preserved the special relation between the US and Saudi Arabia. When Japan sought independent oil trade with Kuwait, it was threatened with embargo against Japanese manufactures in Europe; European companies came under similar threats by the Reagan administration when Western European countries pursued gas contracts with the Soviet Union in 1985.

4   Naval security has gained increasing significance with the attack on the French tanker *Limburg* offshore from Yemen in 2002 and the increase in piracy in the Red Sea in recent years. The apparent modernization of the Chinese navy and potentially the Indian and Russian navies has pressed the case for naval cooperation, with Iran some times construed as the potential threat to maritime energy security other than nonstate groups. On the threat of Iran to maritime energy security, see Brumberg *et al.* 2008. On the threat of piracy to maritime trade, see Luft and Korin 2004. On naval cooperation on matters related to energy, see Blair and Lieberthal 2007; see also Deutch and Schlesinger 2006; ESLC 2008.

5   US DHS 2006. This plan is to be updated every three years, and it was updated in 2009. DHS 2009 varies very little from the plan of 2006, but interestingly places more emphasis on "partnering" with the private sector and adds the aim to be "cost-effective and to minimize the burden on CI/KR owners and operators." The plan also requires sector specific agencies to produce sector specific plans (SSP), such as the Department of Energy's plan, produced in May 2007 and will be reissued in 2010 (see US DoE 2007).

6   Note that the "homeland" already implies a space extending beyond the territorial US. According to the Homeland Security Act (2002), the homeland includes, besides the fifty states and all tribal territories, American Samoa, Guam, Northern Mariana Islands, Puerto Rico, the Virgin Islands, and "any possession of the United States."

7   Quite significantly, among the "international outreach programs," including strengthening existing bilateral and multilateral partnerships and the development of new international institutions to protect the global energy infrastructure, the NIPP includes the promotion of a "global *culture* of physical and cyber security" and the adoption of activities "designed to improve the protection of U.S. CI/KR overseas, as well as the reliability of international CI/KR on which this country depends" (US DHS 2006: 56, 128, 130; emphasis added). In material terms, this implies that the Departments of State and Justice devise programs "to train national police and security forces to defend and secure energy infrastructure in key countries." Even more: "The State Department should improve its capacity to intervene diplomatically" where threats to the energy security arise from local conflicts, and to collect intelligence regarding potential conflicts that could affect US energy security – including "intelligence on national oil companies and their reserves in order to allow policymakers to make better decisions about future alliances and the nation's strategic posture on energy suppliers" and to prevent "dangerous geopolitical tensions" that may arise from uncertainty about reserves. See also ESLC 2008: 11, 112–113.

8   From a security perspective, systems are more critical to the energy sector than individual assets, because of the interdependence of energy systems in the energy sector and between the energy sector and other sectors. See US DoE 2007: 23.

References

Bahgat, G. 2003. *American Oil Diplomacy*. Gainesville, FL: University Press of Florida.
—— 2004. "Terrorism and Energy: Potential for a Strategic Realignment" *World Affairs* 167 (2) pp. 51–58.
Bajpai, S. and J.P. Gupta. 2007. "Securing Oil and Gas Infrastructure." *Journal of Petroleum Science & Engineering* 55 (1/2) pp. 174–186.
Baker Institute. 2008. *The Global Energy Market: Comprehensive Strategies to Meet Geopolitical and Financial Risks – the G8, Energy Security, and Global Climate Issues*. Baker Institute Policy Report No. 37. Houston, TX: The James A. Baker III Institute for Public Policy, Rice University.
Blair, D. and K. Lieberthal. 2007. "Smooth Sailing: The World's Shipping Lanes are Safe." *Foreign Affairs* 86 (3) pp. 7–13.
Brumberg, D., J. Elass, A. Myers Jaffe, and K. B. Medlock III. 2008. *Iran, Energy and Geopolitics*. Working Paper Series: The Global Energy Market: Comprehensive Strategies to Meet Geopolitical and Financial Risks. Houston, TX: The James A. Baker III Institute for Public Policy, Rice University.
Byman, D. L. 2008a. *The Changing Nature of State Sponsorship of Terrorism*. Washington, D.C.: The Saban Center for Middle East Policy at the Brookings Institution. Analysis Paper No. 16, May 2008.
—— 2008b. "Rogue Operators." *The National Interest* 96, pp. 52–59.

Cook, D. 2008. *Oil and Terrorism*. Working Paper Series: The Global Energy Market: Comprehensive Strategies to Meet Geopolitical and Financial Risks. Houston, TX: The James A. Baker III Institute for Public Policy, Rice University.

Deutch, J. and J. R. Schlesinger. 2006. *National Security Consequences of U.S. Oil Dependency*. Independent Task Force Report No. 58. Washington, D.C.: Council on Foreign Relations.

Energy Security Leadership Council (ESLC). 2008. *A National Strategy for Energy Security: Recommendations to the Nation on Reducing U.S. Oil Dependence*. Washington, D.C.: Energy Security Leadership Council.

Farrell, A. E., H. Zerriffi, and H. Dowlatabadi. 2004. "Energy infrastructure and security." *Annual Review of Environment & Resources* 29 (1) pp. 421–469.

Foreign Policy 2006. "The Terrorism Index," July/August, pp. 48–55.

Friedman, T. L. 2006. "The First Law of Petropolitics" *Foreign Policy* 154, pp. 28–36.

Gholz, E. and D. G. Press. 2007. *Energy Alarmism: The Myths that Make Americans Worry about Oil*. Policy Analysis No. 589. Washington, D.C.: CATO Institute.

Gordon, R. C. 2008. *The Case Against Government Intervention in Energy Markets Revisited Once Again*. Policy Analysis No. 628. Washington, D.C.: CATO Institute.

Jaffe, A. M. and R. Soligo. 2008. *Militarization of Energy – Geopolitical Threats to the Global Energy System*. Working Paper Series: The Global Energy Market: Comprehensive Strategies to Meet Geopolitical and Financial Risks. Houston, TX: The James A. Baker III Institute for Public Policy, Rice University.

Labban, M. 2008. *Space, Oil and Capital*. London and New York: Routledge.

—— 2009. "The struggle for the Heartland: Hybrid Geopolitics in the Transcaspian." *Geopolitics* 14 (1) pp. 1–25.

—— 2010. "Oil in Parallax: Scarcity, Markets and the Financialization of Accumulation." *Geoforum* 41 (4) pp. 541–552.

Leverett, F. and P. Noël. 2006. "The New Axis of Oil." *The National Interest* 84, pp. 62–70.

Luft, G. 2008. *Oil and the New Economic Order*. Washington, D.C.: Institute for the Analysis of Global Security.

Luft, G. and A. Korin. 2004. "Terrorism Goes to Sea." *Foreign Affairs* 83 (6) pp. 61–71.

Lugar, R. G. 2006. "The New Energy Realists." *The National Interest* 84, pp. 30–33.

Martin, R. 2007. *An Empire of Indifference: American War and the Financial Logic of Risk Management*. Durham and London: Duke University Press.

Morse, E. L. 1986. "After the Fall." *Foreign Affairs* 64 (4) pp. 792–811

—— 1991. "The Coming Oil Revolution." *Foreign Affairs* 69 (5) pp. 36–56.

—— 1999. "A New Political Economy of Oil?" *Journal of International Affairs* 53 (1), pp 1–29.

Morse, E. L. and A. M. Jaffe. 2005. "Opec in Confrontation with Globalization," in J. H. Kalicki and D. L. Goldwyn (eds.) *Energy & Security*. Washington, D.C. Woodrow Wilson Center Press, pp. 65–95.

Morse, E. L. and J. Richard. 2002. "The Battle for Energy Dominance." *Foreign Affairs* 81 (2) pp. 16–31.

National Petroleum Council (NPC). 2007. *Hard Truths: Facing the Hard Truths about Energy*. Washington, D.C.: National Petroleum Council.

Taylor, J. and P. Van Doren. 2008. "The Energy Security Obsession." *The Georgetown Journal of Law & Public Policy* 6 (2) pp. 475–485.

United States Department of Defense (DoD). 2008. *National Defense Strategy*. Washington, D.C.: Department of Defense.

United States Department of Energy (DoE). 2007. *Energy: Critical Infrastructure and Key Resources Sector-Specific Plan as Input to the National Infrastructure Protection Plan.* Washington, D.C.: Department of Energy.

United States Department of Homeland Security (DHS). 2006. *National Infrastructure Protection Plan.* Washington, D.C.: Department of Homeland Security.

—— 2009. *National Infrastructure Protection Plan: Partnering to Enhance Protection and Resiliency.* Washington, D.C.: Department of Homeland Security.

United States Navy. 2007. *A Cooperative Strategy for 21st Century Seapower.* Washington, D.C. Department of the Navy.

Victor, D. G. 2007. "What Resource Wars." *The National Interest* 92, pp. 48–55.

West, J. R. 2008. "Running on Empty." *The National Interest* 95, pp. 4–6.

Wilson, C. L. 1973. "A Plan for Energy Independence." *Foreign Affairs* 51 (4) pp. 657–675.

Wirth, T. E., C. B. Gray, and J. D. Podesta. 2003. "The Future of Energy Policy." *Foreign Affairs* 82 (4) pp. 132–155.

Woolsey, R. J. 2002. "Defeating the Oil Weapon." *Commentary* 114 (2) pp. 29–33.

Yergin, D. 1988. "Energy Security in the 1990s." *Foreign Affairs* 67 (1) pp. 110–132.

—— 1991. *The Prize.* New York: Simon and Schuster.

—— 2005. "Energy Security and Markets" in J.H. Kalicki and D.L. Goldwyn (eds.) *Energy & Security.* Washington, D.C. Woodrow Wilson Center Press, pp. 51–64.

—— 2006. "Ensuring Energy Security." *Foreign Affairs* 85 (2) pp. 69–82.

—— 2007. "The Fundamentals of Energy Security." Testimony to the Committee on Foreign Affairs, US House of Representatives. *Hearing on Foreign Policy and National Security Implications of Oil Dependence*, March 22.

—— 2008. "Energy Under Stress." in K. M. Campbell and J. Price (eds.) *The Global Politics of Energy.* Washington DC: Aspen Institute.

Yetiv, S. A. 2004. *Crude Awakenings.* Ithaca, NY: Cornell University Press.

Žižek, S. 2005. *Iraq: The Borrowed Kettle.* London: Verso.

—— 2008. *Violence: Six Sideways Reflections.* New York: Picador.

Part VI

Blue ecology: the political ecology of water

16 Commons versus commodities: political ecologies of water privatization

Karen Bakker

The World Water Forum – a global gathering held every three years to debate the world's most urgent water issues – is intended to be a solemn affair. But protests invariably disrupt the proceedings. The meeting in the Dutch city of The Hague in March 2000 was typical: as Egypt's Minister of Public Works and Water Resources began to give his inaugural speech, two audience members – one male, one female – suddenly appeared on stage. In full view of the audience of dignitaries and government ministers, the protesters approached the presidential table, removed their clothes and handcuffed themselves together. Strategically scrawled on their bodies were the words "No to Water Privatization" and "Yes to Water as a Human Right."

Meanwhile, protesters in the audience (discreetly chained to their seats) shouted slogans accusing governments of colluding with private water companies to profit from the world's water resources. Some of their concerns related to the support given by conference organizers to private water companies, and to their links with development organizations in favor of water privatization, including the World Bank. But protestors' slogans also targeted governments responsible for environmentally destructive and socially inequitable water management.

The security guards were quickly overwhelmed, and the meeting ground to a halt. The protesters' message, captured by bemused journalists, was clear: water privatization had to be stopped, and government management of the world's water had to be dramatically reformed. But the Ministerial Declaration issued a few days later took no heed of these demands: the world's governments voiced support for private water management, but made no mention of the human right to water, nor of the protesters' demands for environmental and social justice.

The events in The Hague are but one example of the fierce protests against water privatization which have taken place around the world over the past decade.[1] Opponents have deployed court cases, referendums, street protests, and media-savvy campaigns in their struggle to safeguard "public" water, targeting private water companies and the aid organizations and governments that support them. These protests have arisen in response to the recent, dramatic increase in private sector activity in the control and management of water supply systems. During the 1990s, some of the world's largest multinationals (Bechtel, Enron, Vivendi) began expanding their operation and ownership of water supply systems on a global scale:

the largest (Veolia, a Vivendia subsidiary) now has nearly 140 million water and sanitation customers worldwide. At the same time, many governments tried to treat water resources as commodities: pricing water as an economic good, and creating private water rights and water markets.

This has generated fierce controversy. Proponents of privatization assert that private companies will perform better: they will be more efficient, provide more finance, and mobilize higher-quality expertise than their government counterparts. Supporters also often argue that private involvement will facilitate broader reforms – such as the treatment of water as an economic good – that are required in order to ensure environmentally friendly outcomes such as water conservation and the reduction of pollution. These arguments rest on the claim that government management of urban water supply is beset by several inter-related problems: low coverage rates, low rates of cost-recovery, low tariffs, under-investment, deteriorating infrastructure, over-staffing, inefficient management, and unresponsiveness to the needs of the poor. This hotly disputed litany of supposed government woes has dominated the pro-water supply privatization development discourse over the past decade, and is often summed up by the label "government failure" (see, for example, World Bank 1993). From this perspective, it is unethical not to involve private companies if they can perform better than governments at providing water, particularly to the poor.

In contrast, opponents of privatization argue that government-run water supply systems, when properly supported and resourced, are more effective, equitable, and responsive, have access to cheaper forms of finance (and thus lower water tariffs), and can perform just as well as their private sector counterparts (Lobina and Hall 2008). Opponents warn of the negative effects – social and environmental – of private ownership and management of water resources and water supply systems. Those who reject privatization also argue that it is unethical to profit from water, a substance essential for life and human dignity: Harvey, for example, characterizes privatization of water supply as one example of "accumulation by dispossession" – the enclosure of public assets by private interests for profit, resulting in greater social inequity (Harvey 2003). Some go further, and argue that environmental protection and water conservation should be fostered through an ethic of water use, whether based on a spirit of solidarity, environmental consciousness, eco-spirituality, or traditional water use practices (Petrella 2001; Shiva 2002).

This has obvious parallels with debates in many other sectors (from health care to housing). But water privatization inspired particularly fierce protest and had become, by the end of the 1990s, one of the most controversial issues debated in international development and environmental management circles (see Birdsall and Nellis 2003; Castro 2008; Davis 2005; Hall *et al.* 2005; Prasad 2006). Why would this be the case? One reason is that water fulfills multiple functions and is imbued with many meanings. Water is simultaneously an economic input, an aesthetic reference, a religious symbol, a public service, a private good, a cornerstone of public health, and a biophysical necessity for humans and ecosystems alike. It should thus come as no surprise that protests against water privatization have united a strikingly diverse range of movements: unions, environmentalists, women's groups, fair trade networks, alternative technology advocates, religious organizations, indigenous communities, human rights organizations, anti-poverty and anti-globalization activists, and development watchdogs. United in politically suggestive coalitions, these groups protest both privatization and the broader water governance reforms with which it is associated.

Another reason for the fierceness of protests against water privatization is the fact that fresh water is a "final frontier" for capitalism. Essential for life and (at least in the case of drinking water) non-substitutable, water throws up challenging barriers – technical, ethical, and political – to private ownership and management. The water privatization debate is thus a microcosm of contemporary debates over the roles of states and markets, and over the acceptability and efficacy of markets and private ownership as solutions to the world's putative environmental crises.

Debating privatization

Debates over privatization conventionally pit partisans of classic forms of government intervention against "neoliberals," whose reformulation of the role of the government emphasizes the need for selective regulation by the state, rather than direct state provision of public services. Much of the debate between opponents and proponents of water supply privatization (and "free market environmentalism" more generally) hinges on the role and extent of states versus markets (or the "public" versus "private" sphere). A range of political economic arguments thus typically dominates water privatization debates, as with debates over privatization more generally.

Debates over water privatization also have an environmental dimension. Indeed, the arguments of water privatization proponents are perhaps best captured by the term "free market environmentalism": a mode of resource regulation that offers hope of a virtuous fusion of economic growth, efficiency, and environmental conservation (Anderson and Leal 2001). This has much in common with debates over neoliberalism – a set of ideas that have dominated public policy debates since the 1970s, at the core of which is the doctrine that market exchange is an ethic in itself, capable of acting as a guide for all human action (Harvey 2005). But "free market environmentalism" also captures the set of concerns over local and global environmental crises which rose to the fore in the 1970s. The 2008 edition of the Worldwatch Institute's influential *State of the World* report (subtitled *Innovations for a Sustainable Economy*) is a classic example of the resulting convergence of environmental and economic concerns: advocating the use of markets for a wide range of environmental purposes, the report argues that markets will play a critical role in environmental conservation, while environmental issues simultaneously "rewrite the rules" for capitalism (WRI 2008). Supporters of "free market environmentalism" (or "green neoliberalism," as its opponents term it) argue that, through establishing private property rights, employing markets as allocation mechanisms, and incorporating environmental externalities through pricing, environmental goods will be more efficiently allocated, thereby simultaneously addressing concerns over environmental degradation and inefficient use of resources. In short, markets will be deployed as the solution to, rather than the cause of, environmental problems.

In response, opponents of water supply privatization often frame free market environmentalism as a form of "green imperialism" or "green neoliberalism." They point to studies that have demonstrated the limits, unexpected consequences and impacts of neoliberalizing nature in a broad range of historical and geographical contexts (Heynen and Robbins 2005; Himley 2008; Mansfield 2008). They argue that environmental degradation (an inevitable by-product of capitalism) may be mobilized as opportunities for continued profit (O'Connor 1996), but that the involvement of private companies will not necessarily ensure an overall improvement in environmental quality; on the contrary, companies are likely to engage in cost-cutting measures detrimental to the environment.

The view from communities and non-governmental organizations in developing countries, often rooted in indigenous water use practices, is often overlooked, but offers another important perspective. The community perspective offers a cultural as well as political economic critique of private and government provision of water. For many urban residents, community (rather than government of private providers) plays an important role in enabling or constraining access to water supply on a daily basis. Community action is of great importance, as environmental concerns are central to the livelihoods of the urban poor. Poor environmental quality is costly, in both health and economic terms. This is as true for lack of access to safe drinking water as it is for many other issues. And water carries an additional set of threats for the urban poor, who live within the interstices of the city (in floodplains and along riverbanks, on steeply eroded slopes and marshy land). In the zones where the poor settle, water is often a threat to physical safety, both in terms of flooding and poor water quality. Environment-related water concerns are thus an imperative, and not a luxury, for the urban poor, and privatization is merely one of the many water-related issues with which they may grapple on a daily basis.

Privatization: the solution to the global water crisis?

This latter point raises one of the key dimensions of the water privatization debate: proponents often justify privatization on the grounds that private companies can solve the world's urban water supply "crisis." The main features of this crisis are well known: the most recent estimates suggest that over 1 billion people are without access to "basic" water supply, pegged at 20 liters per person per day (although the World Health Organization recommends 100 liters per person per day – and North Americans use 3 to 4 times as much) (WHO 2003). This situation has persisted despite the fact that supplying water has been high on the agenda of the international community since the United Nations Water and Sanitation Decade (1981–90), during which bilateral aid and multilateral finance were directed toward water supply projects in unprecedented amounts. At the end of the decade, more people (in absolute terms) were being supplied with "improved water supplies" than ever before, yet in many countries this failed to keep pace with population growth (WHO 1992; and J. Bartram, WHO, personal communication). Moreover, these efforts inadequately coped with the acceleration of rates of urbanization over the latter half of the twentieth century. The scale of the problem is seemingly beyond even our best efforts: even the ambitious Millennium Development Goals only call for a reduction by 50 percent of those without sustainable access to safe drinking water.

The world's water crisis is thus very real. But we should nevertheless be skeptical about the rhetorical uses to which the specter of "crisis" can be put (particularly when it serves as a justification for privatization). It should not prevent us from questioning the ways in which human action contributes to, or even creates lack of access to water. For example, we might explore how urban water management often amounts to the "production of thirst," in which cultural norms, political commitments, and the seemingly mundane practices of water managers combine to exclude the poor from accessing water (Kooy and Bakker 2008a, 2008b).

This claim of generalized water scarcity deserves close scrutiny. Water is a resource mobilized by humans on a massive scale. Humans withdraw 5,200 cubic kilometers – or 5.2 trillion metric tons – of water annually (10 percent of total surface runoff) (Gleick 1993). These global figures mask, of course, the high degree of spatial and temporal availability of water; withdrawals are greater than 50 percent of runoff in some regions (North Africa, Central Asia, South-Western US, South-Eastern England). In these regions, water quantity, and in more humid regions, declining availability due to declining water quality, are the causes of the scarcity experienced by humans. Scarcity, in other words, is socially produced (sometimes termed "second-order" scarcity). That our awareness of scarcity is growing is a signal not of *absolute* scarcity, but of *relative* scarcity, due to factors such as increasing pollution, population density, and water use per capita.

The increasing awareness of scarcity is in part attributable to the growth in second-order (i.e. human-produced) scarcity, and also to the appeal of the often implicit Malthusian-style assumptions about the "limits to growth." This is particularly but by no means exclusively applicable to some environmental groups, for whom an ethic of care has been gradually displaced by a discourse of sustainability and associated logic of compensation in their convergence with economists in advocating an ethic of efficiency in resource management. The shift from state to market, and from a focus on water supply to water conservation, are thus often intertwined in current water policy debates. Conservation should here be read in both a political economic and ecological sense: as the preservation of capitalism as a socio-economic system; and as the prioritization of environmental conservation, as both market opportunity and strategic necessity, enacted by both private companies and the state. From this perspective, water marketization is part of a more generalized transition to a new mode of resource regulation (Table 16.1).

The implementation of the state hydraulic paradigm was concretized (at least in most OECD countries) in the post-war period during which the state undertook to provide those services necessary to capital accumulation that were assumed to be unfeasible for the private sector. In many European countries, and in contrast to the nineteenth century, the state entered into the business of water supply,

Table 16.1 State hydraulic versus market environmentalist modes of water regulation

	State hydraulic	*Market environmentalist*
Economic regulation	Command-and-control	Market-based instruments
Resource management	Growth-oriented, supply-driven	Scarcity-responsive, demand-led
Network manager	State	Market
Primary goals	Universal provision; quantity	Efficiency; quality
Provision ethos	Service	Business
Consumer identity	User	Customer
Method of charging	Unmetered	Metered
Raw water	Resource – subsidized or free	Environment – abstraction priced
Water supply pricing	Social equity (ability to pay)	Economic equity (benefit principle)

Source: Adapted from Bakker 2004

approaching water supply provision as a welfare service, and developing water resources as necessary and strategic factors of production. In most industrialized countries, and in many urban zones of developing countries, public ownership of utilities and development of resources was underpinned by a model of social welfare in which state provision of an expanded sphere of "public goods" was thought to be in the general economic and social interest (Graham and Marvin 2001). This mode of regulation (variously termed Keynesian, Fordist, or social welfarist) was expressed in different ways in different countries. In the United Kingdom, the entire water use and wastewater disposal cycle was brought under centralized public ownership. In France, private companies continued to operate as service providers to infrastructure-owning municipalities. In Spain, investment in water resources was crucial to the Franco's project of agricultural modernization, and the state assumed complete control over surface water resources across all sectors (Bakker 2002).

In those countries where a high degree of control was assumed by the state, continued public provision of this resource was, by the end of the twentieth century, being undermined by the contradiction which beset public goods provision more generally: the continued legitimacy of the state was dependent upon the satisfaction of expectations that it had itself sanctioned, but which threatened to undermine either environmental sustainability (both in terms of degradation of quality and in terms of second-order (human-created) scarcity) and/or economic competitiveness (Bakker 2004). The global crisis of the "administered" mode of regulation beginning in the final quarter of the century both contributed to, and was exacerbated by the breakdown of the state hydraulic paradigm. In the case of the water sector, the macro-economic crisis of the state justified under-investment in infrastructure and services; the lack of public finance and the resulting decline in service provision standards (declining quality or quantity, rising prices) undermined the legitimacy of the state as service provider, in turn providing justification for marketization. By the end of the century, governments (often spurred by multilateral financial agencies and bilateral aid organizations) were actively encouraging private companies to take over the management of water supply systems. Rural areas and outlying (or "peri-urban") settlements attracted little interest from private companies, as their small scale and low densities reduced profitability potential. Large urban centers were the focus of attention. By the late 1990s, many capital cities of developing countries had committed to "private sector participation" contracts, from Buenos Aires to Jakarta, Manila to Casablanca. It is largely in urban areas, in other words, that private companies have expanded their operations over the past two decades. Of the world's total population, estimates suggest that only 3 percent are supplied via private operators, although this figure is much higher in some countries (Winpenny 2003). But when we look at cities, and particularly large cities, the picture changes: perhaps 20 percent of the world's urban population are supplied by the private sector, amounting to hundreds of millions of customers, most of whom became clients of private companies in the past two decades.[2] Given that the prospect of profitability generally increases with the size of the urban area (because of important economies of scale), the urban bias of private sector participation is unsurprising. And, as industry rhetoric frequently claims, the global water crisis is a global opportunity for private capital. Urban water supply is thus the primary battleground over which water supply privatization is fought.

The rise and (partial) retreat of the private sector in water supply

The range of Private Sector Participation options advocated during the 1990s arose as a response to widely recognized challenges facing the water sector. Many of these challenges were summarized in numerous reports, such as the World Bank's 1992 Buky Report and the United Nations World Water Assessment Programme Reports (World Bank 1992a, 1992b; UNWWAP 2003, 2006). The now-familiar litany of water sector challenges (contested by some proponents of public water supply, which argue that many water utilities are well-performing) includes low coverage rates, indiscriminately low tariffs (which encourage wasteful water use, and/or render difficult the recovery of costs from tariff revenues), under-investment, deteriorating infrastructure, over-staffing, inefficient management, and unresponsiveness to the needs of the poor. Views on the causes of these failures differed: privatization proponents argued that they were due to "state failure," whereas opponents of privatization argued that these failures could have been prevented by better public management and stronger development lending practices.

This debate over the causes of "failure" in the water sector continues. But in the 1990s, the so-called Washington Consensus and new commitments to private sector development on the part of a range of multilateral financial institutions and bilateral donors drove market environmentalist reforms in water supply (and to a lesser extent irrigation), in line with other sectors (Table 16.2) (Bangura and Larbi 2006; Miller-Adams 1999; OECD 1995; Williamson 2000).These reforms typically adopt a dual strategy of (i) commercialization of water management (through the introduction of commercial principles, such as recovery in water management); and/or (ii) introduction of private actors, either via private sector management of water supply networks (through a range of privatization contracts), or, much more rarely, water resources rights and markets (regarding the range of private sector participation contracts, see World Bank 1997. See also Cullet 2009; Dinar 2000). These reforms occurred in many countries, and in some cases were supported by donors, or were imposed through loan conditionalities, generating controversy (Conca 2006; Grusky 2001; Hall and de la Motte 2004). Additional, sector-specific factors included the drive of some private water firms to internationalize and the "demonstration effect" of private sector participation in water supply in some countries, such as Chile, France, and the United Kingdom and France (Bakker 2004).

Table 16.2 Resource management reforms: examples from the water sector

Category	Target of reform	Type of reform	Example drawn from the water sector
Resource management institutions	Property rights	Privatization (enclosure of the commons or asset sale)	Introduction of riparian rights (England (Hassan 1998); or sale of water supply infrastructure to private sector (England and Wales (Bakker 2004))
	Regulatory frameworks	De-regulation	Cessation of direct state oversight of water quality mechanisms (Ontario, Canada (Prudham 2004))
Resource management organizations	Asset management	Private sector 'partnerships' (outsourcing contracts)	French municipal outsourcing of water supply system management to private companies (Lorrain 1997)
	Organizational structure	Corporatization	Conversion of business model for municipal water supply: from local government department to a publicly owned corporation (Amsterdam, The Netherlands (Blokland *et al.* 2001))
Resource governance	Resource allocation	Marketization	Introduction of a water market (Chile (Bauer 1998))
	Performance incentives/ sanctions	Commercial- ization	Introduction of commercial principles (e.g. full cost recovery) in water management (South Africa (McDonald and Ruiters 2005))
	User participation	Devolution or Decentralization	Devolving water quality monitoring to lower orders of government or individual water users (Babon River, Indonesia (Susilowati and Budiati (2003))

Source: Adapted from Bakker 2007.

But hopes that private sector participation, particularly via the concession model, could mobilize significant financing for the water sector have largely not materialized. In the early to mid-1990s, private sector investment in the water sector increased rapidly. However, total private sector investment in the water supply and sanitation sector peaked in 1997, and has declined since then (Marin and Izaguirre 2006; PriceWaterhouseCoopers 2003). This was in part structural, as the Asian, Russian, and Argentine financial crises, as well as the bursting of the hi-tech bubble, reduced the overall availability of private sector finance. Cancellations across all utility sectors increased after 2000, including water (which had a higher cancellation rate, in terms of total investment flows, than other utility sectors, due to the cancellations of several large contracts) (Annez 2006; Izaguirre 2005; Marin 2009; see also OECD 2000). As the World Bank acknowledged in its response to the Camdessus/World Panel on Financing Water Infrastructure Panel in 2003, private sector finance had provided a relatively small proportion of total capital needs in the sector, and the declining trend in private sector investment was likely to continue (World Bank/IMF 2003). By 2004, the authors of a World Bank report concluded that: "expectations of private sector participation in the financing of infrastructure needs were overoptimistic" (Briceño-Garmendia, *et al.* 2004).

This suggests that the conventional concession model adopted for private sector provision has failed. Some problems which affected concessions – such as strategic under-bidding, fraught contract renegotiations, poor information on asset conditions, and a lack of effective regulation – were not necessarily inherent to the concession model, and might have been addressed through improving regulatory models and contract negotiation processes. But the risks stemming from the substantial investment obligations associated with concessions were significant, and in some cases insurmountable. Exchange rate (currency) risk was significant in cases of substantial hard (foreign) currency investments. "Revenue risk" was also significant: this risk of revenue shortfalls often stemmed from low collection rates; but at times was also related to political influence on tariff-setting (and a consequent failure for tariffs to rise to levels necessary to ensure revenues sought by the private sector). A key conundrum for private operators was that network extensions were difficult to finance given high capital costs (including profit margins), on the one hand, and the high proportions of poor customers with low "ability-to-pay," on the other.[3] Proponents of privatization point to cases such as Casablanca, in which private involvement resulted in significant increases in coverage, but general trends suggest that the hope that private sector concession contracts could make significant headway in providing water to the unserved ("addressing the backlog," in development parlance) was largely not met.

In short, the concession model of water privatization has not met the expectations of proponents in terms of development objectives, such as mobilizing financing or solving the world's water supply crisis through supplying the urban poor (Castro 2008; Davis 2005; Prasad 2006). But it would be a mistake to assume that the private sector had retreated; rather, companies have reconfigured their strategies, focusing on management contracts (which carry lower risk) and on higher-income cities and countries (Marin 2009). Public statements by senior executives of water supply services firms reveal a retreat from earlier commitments to pursuing private sector opportunities in water globally, with senior figures publicly acknowledging high risks and low profitability in supplying the poor (Robbins 2003). Some international financial institutions have begun officially acknowledging the limitations of the private sector (UNDP 2006; ADB 2003a; IMF 2004). High-profile cancellations of water supply concession contracts – including Atlanta, Buenos Aires, Jakarta, La Paz, and Manila – seem to bear out the hypothesis that water presents difficult and perhaps intractable problems for private sector management. The multinational private sector has retreated from supplying water to poor communities in lower-income countries, but continues to pursue aggressive growth strategies elsewhere.

Indeed, the total population served by private water operators has continued to increase – although the annual rate of new contract awards (particularly large concession contracts) has reduced (Marin 2009). These observations are not universally true, as the level of privatization activity has been spatially variable; not all countries have experienced a decline. Concession contracts were most rare in regions where overall investment was lowest (sub-Saharan Africa), reflecting an investment bias towards East Asia (particularly China) and Latin America (WDM 2005; Saghir 2006). Within countries, companies tend to concentrate on wealthier areas and neighbourhoods, particularly where governments do not take an active stance in promoting supply for the under-privileged. In short, private water firms have tended to concentrate on the higher end of emerging markets, in which they prefer lower-risk contracts. This spatial bias, it should be noted, is equally true of donor funding for water and sanitation, as lower-income countries tend to be under-funded (WHO 2008). Some argue that these trends imply a worrisome (but expected) strategy of "cherry-picking" on the part of private companies, which opponents of privatization believe to be objectionable given a perception that international donors unreasonably favor private operators (for example, in accessing finance). Others argue that these trends represent the end of oligopoly and, as such, a welcome maturation of the sector (WDM 2005; Marin 2009).

As the anecdote at the outset of this chapter suggests, private companies also strategically retreated from water supply in response to sustained political protest (Hall *et al.* 2005). Protests took a variety of mechanisms: court cases (which occurred in several of The Water Dialogues-participating countries); campaigns to pressure governments into canceling contracts; and street protests (in rare cases leading to contract cancelation or expropriation). These protests brought together a broad range of groups, representing a range of interests, including organized labor, consumers, environmentalists, women's groups, and religious organizations (Bennett *et al.* 2005; Morgan 2005; Olivera and Lewis 2004).

In some instances, protests were framed in terms of an outright rejection of private sector participation; in other instances, specific issues were disputed, such as tariff increases. Although the range of grievances varies, concerns over the impacts of privatization on equity (fairness in pricing and access to services), employment, and working conditions predominate. Environmental concerns and nationalist rejection of "foreign" ownership and management of water supply utilities have been less frequently represented than the concerns of consumers and unions, although in many instances a nascent "red–green" alliance has cohered around the anti-privatization agenda (as in Cochabamba, Toronto, Jakarta, and Vancouver). The views of protesters are, of course, disputed by privatization proponents, but by 2005, even the World Bank acknowledged the fact that "the frequent (not inevitable) result [of water privatization] was popular protests, dissatisfied governments, and unhappy investors" (World Bank 2005). This had the effect of raising the costs of private sector participation, and, according to some observers contributed to reduced private sector interest in long-term concessions (given the greater exposure of this type of contract to political risks).

Protesting privatization: the commons and the "human right to water"

For many anti-privatization activists, water privatization is an act of what David Harvey terms "accumulation by dispossession": the appropriation, for profit, by the private sector of both the natural environmental "commons" and of public goods created (and/or subsidized) by the state (Glassman 2006; Harvey 2003). This act of dispossession is emblematic of "globalization from above," with the negative consequences that this process is assumed to entail (Barlow and Clarke 2003; Bond and Dugard 2008; McDonald and Ruiters 2005; Petrella 2001; Shiva 2002). According to its opponents, the involvement of private companies invariably introduces a pernicious logic of the market into water management, which is incompatible with guaranteeing citizens' basic right to water. Accordingly, most anti-privatization activists reject the conceptual bases upon which proponents of privatization base their claims. First, they do not agree that water should be priced at its full cost; nor do they agree that this will lead to more equitable outcomes. They reject the claim that private management, on a for-profit basis, will improve performance; and they equally reject that claim that accountability to customers and shareholders is more direct and effective than the attenuated political accountability exercised by citizens via political representatives (Johnstone and Wood 2001; Rogers and Hall 2003; Shirley 2002; Winpenny 1994).

Underlying this stance is the assumption that public goods are inviolably collective, and that water, as the biophysical and spiritual basis of life, is the public good *par excellence*. Defending the sphere of public goods – a reserve of natural and social capital supportive of communities but under threat as a potential source of private profit – is the broader task that anti-privatization activists see themselves as carrying out. The notion that water has an economic *value* is admitted by some (although not all); but the notion that water is therefore an economic *good* is rejected by most of the anti-privatization advocates. They argue, instead, that water is a "commons" (and, according to some, a "public trust" – a concept not entirely compatible with the commons). This ambivalence is not exclusive to activists. For example, the language of the EU Water Framework Directive, arguably the most sophisticated piece of water legislation in the past decade, captures this ambivalence neatly. The directive's first article (subjected to protracted wrangling as it was being drafted) reads: "Water is not a commercial product *like any other* but, rather, a *heritage* which must be *protected, defended, and treated as such*" (European Union 2000; emphasis added).

Should we accept that water is a commodity? A mainstream reaction would be to argue that, while water has economic value, it cannot simply be defined as a commodity, for both moral and technical reasons. A more radical critique offered by anti-privatization scholars and activists frames privatization as part of a broad-based intensification of the "accumulation" of nature under capitalism, in which non-monetary values and community management are displaced by economic values and private-sector-controlled (or market-mimicking) ownership, management, and allocation. Market-mimicking techniques, from this perspective, are not the benign, neutral management techniques presented in most of the contemporary water management literature. Resistance to privatization is thus, in many instances, also linked to resistance to markets (and capitalism) in a broader sense. As such, anti-privatization campaigns and protests are emblematic of Polanyi's "double movement," in which the attempt to dis-embed the economy from society (and from social constraints on markets) provokes a corresponding attempt to re-embed private sector activity within society, through re-imposing social constraints on markets and private firms. As a result, a broad range of "alternatives" to privatization has appeared within recent years (Table 16.3).

Table 16.3 is intended to be useful in structuring our analyses of activism and advocacy. For example, activists often describe water as a "commodity" in contrast to water as a "human right." But Table 16.3 suggests that this is misleading, insofar as the term "commodity" refers to a property rights regime applicable to resources, and human rights to a legal category applicable to individuals. The more appropriate, but less widely used, antonym of water as a commodity would more properly be a water "commons."

Similarly, the typology has had significant implications for anti-privatization campaigns around the world. Many of these campaigns have focused on a human right to water as a means of "defending" water against privatization. But this characterization has had ambiguous results, as we shall see, in part because the notion of water as a human right is not necessarily incompatible with water privatization.

Campaigning for the human right to water

The international campaign for a human right to water has grown enormously over the past decade (Gleick 1998. Morgan 2004; WHO 2008). Beginning with a set of declarations by activists, a broad-based campaign has emerged bringing together development-focused aid watchdogs (such as the UK's World Development Movement), mainstream international organizations (such as the World Health Organization), human rights organizations (notably Amnesty International), environmental groups (the Sierra Club, one of the largest U.S. activists), consumer groups (such as Ralph Nader's Public Citizen), and more radical anti-privatization and alter-globalization groups (such as South Africa's Anti-Privatization Forum and the Council of Canadians), as well as networks of influential international actors (such as Mikhail Gorbachev's Green Cross and Ricardo Petrella's Group of Lisbon).[4]

Table 16.3 Neoliberal reforms and alter-globalization alternatives

Category	Target of reform	Type of reform	Alter-globalization alternative
Resource management institutions	Property rights	Privatization	• Mutualization (re-collectivization) of asset ownership (Wales (Bakker 2004)) • Communal water rights in village 'commons' in India (Narain 2006)
	Regulatory frameworks	De-regulation	• Re-regulation by consumer-controlled NGOs such as 'Customer Councils' in England (Franceys 2006; Page and Bakker 2005)
Resource management organizations	Asset management	Private sector 'partnerships'	• Public-public partnerships (e.g. between Stockholm's water company (Stockholm Vatten) and water utilities in Latvia and Lithuania (PSIRU 2006) • Water cooperatives in Finland (Katko 2000))
	Corporatization	Organizational structure	• Low-cost, community owned infrastructure (e.g. Orangi Pilot Project, Pakistan (Zaidi 2001))
Resource governance	Resource allocation	Marketization	• Sharing of irrigation water based on customary law ('usos y costumbres') in Bolivia (Trawick 2003)
	Performance incentives/ sanctions	Commercial-ization	• Customer Corporation (with incentives structured towards maximization of customer satisfaction rather than profit or share price maximization (Kay 1996))
	User participation	Devolution or Decentralization	• Community watershed boards (Canada) (Alberta Environment 2003) • Participatory budgeting (Porto Alegre, Brazil) (TNI 2005)

Source: Adapted from Bakker 2007.

Activists have also focused on country-specific campaigns for constitutional and legal amendments, notably in Uruguay, where a national campaign resulted in 2004 in a national referendum on the human right to water, and a resulting constitutional amendment declaring water access a human right. Similar efforts on the right to water subsequently emerged elsewhere in Latin America (particularly Bolivia, Colombia, Ecuador, El Salvador, and Mexico).

As the campaign for the human right to water has gathered momentum, activists have gained support from mainstream international development agencies including the World Health Organization, the United Nations Development Program, and the United Nations General Assembly, whose president appointed a high-profile anti-privatization activist (Canadian Maude Barlow), as Senior Advisor on Water in late 2008. This follows the Comment issued by the UN Committee on Economic, Social and Cultural Rights in 2002, asserting that every person had a right to "sufficient, safe, acceptable, physically accessible, and affordable water" (UN ECOSOC 2002; Hammer 2004).[5] The Committee declared that the right to water was implied in the International Covenant on Economic, Social and Cultural Rights, and that states party to the Covenant (which entered into force in 1976) have the duty to realize, without discrimination, the right to water. Importantly, the Committee argued that realization of the right to water is linked to the realization of many other rights, including the rights to food, health, adequate housing, the right to gain a living by work and the right to take part in cultural life. And in 2008, the UN Human Rights Council adopted a resolution reconfirming the obligations of governments to ensure access to safe drinking water and sanitation, and establishing an Independent Expert on related human rights obligations. This means that the UN human rights system now has a separate mechanism exclusively dedicated to issues related to the right to water and sanitation.

These developments are celebrated by advocates, who argue that the human right to water is necessary for several reasons, most notably the non-substitutability of drinking water, rendering this right "essential for life." Some would contend that the right to water already implicitly exists; for example, it is recognized through legal precedents when courts support the right of non-payment for water services on grounds of lack of affordability (UNWWAP 2006). In addition, other human rights that are explicitly recognized in the UN Conventions are predicated upon an assumed availability of water (e.g. the rights to food, life, and health).

Explicit recognition of the right to water would strengthen its basis in international law, supporters argue, with several advantages. First, new legal avenues would be created enabling citizens to compel states to supply basic water needs. Second, the human right to water offers stronger protection than the current essential services designation, at least when it comes to state provision (Vandenhole and Wielders 2008). Third, the human right to water would enable the fulfillment of other related rights. Finally, the human right to water fulfills basic principles of human rights law: "equal citizenship," a "social minimum," "equality of opportunity," and "fair distribution" (UNDP 2006).

But opponents have pointed out the potential difficulties in implementing a right to water. These concerns echo general criticisms of the rights-based approach to development: advocating a human right to water is unhelpful, because it "belabors the obvious, and ignores what is difficult" (Brooks 2008). Critics point to a range of difficulties: the lack of precision of the definition of the human right to water (how much water, in practical terms?); accountability; pricing (should water be free?); environmental issues (could governments use the human right to water to justify environmentally damaging water projects?); and potential conflict with existing systems of water rights (particularly "informal" water rights, which are often territorially based, unrecognized in formal law, and crucial importance to indigenous peoples) (see, for example, Boelens 2008, 2009). Indeed, some indigenous rights activists fear that a human right to water will provide additional leverage for states intent on wresting control of water resources from local communities. This is echoed by the arguments of scholars who critique the Western bias (or Eurocentrism) of human rights, which (from their perspective) stems from an individualistic, libertarian philosophy that is culturally conditioned, rather than universal in nature (see, for example, Boelens and de Vos 2006). As such, they say, human rights are of limited applicability in non-Western societies. In the case of water, for example, complex cultural relationships with hydrological landscapes imply highly differentiated (and usually collective) sets of water rights only imperfectly captured, and indeed sometimes threatened by the notion of water as a human right (Boelens 2009). Others respond that the human right to water, like other rights, is an important mechanism for individuals to make against states in all societies, and a critical tool for the fight for social justice, regardless of cultural context.

The foregoing criticisms carry significant weight. But from the perspective of anti-privatization campaigners, the most devastating criticism of the human right to water is the possibility that it will not foreclose private sector management of water supply systems. Current interpretations suggest that private sector provision is compatible with the human right to water (Langford 2005). The most important piece of international jurisprudence to date (the 2002 Comment by the Committee on Economic, Social and Cultural Rights discussed earlier) takes an agnostic approach on questions of private sector participation, acknowledging that states have a responsibility to ensure the human right to water, but refraining from commenting on the legitimacy of different political economic models of provision. The Comment also avoids imposing direction obligations on multinational water corporations (see Langford 2006). Indeed, the shortcomings of "human right to water" anti-privatization campaigns became apparent following the Kyoto World Water Forum in 2003, as proponents of private sector water supply management began speaking out in favor of water as a human right. Senior water industry representatives identified water as a human right on company websites, in the media, and at high profile events such as the Davos World Economic Forum.[6] Right-wing think tanks such as the Cato Institute backed up these statements with reports arguing that "water socialism" had failed the poor, and that market forces, properly regulated, were the best means of fulfilling the human right to water (see, for example, Segerfeldt 2005). Non-governmental organizations closely allied with private companies, such as the World Water Council, also developed arguments in favor of water as a human right (Dubreuil 2006a, 2006b). Shortly after the Kyoto meeting, the World Bank released a publication acknowledging the human right to water (Salman and McInerney-Lankford 2004). And less than a year after the Uruguayan Parliament approved the Constitutional amendment to a human right to water, the government passed an executive resolution allowing the private companies that had signed concession contracts prior to the referendum to continue.

Private companies agree that water is a human right, which they are both competent and willing to supply – but only if risk-return ratios are acceptable, a condition which cannot be met by most communities.

This outcome is not unexpected by critics of human rights doctrines operating from what we might term a radical left tradition, which posits that human rights are compatible with capitalist political economic systems. They would point out that many citizens of capitalist democracies accept that commodities are not inconsistent with human rights (such as food and shelter), but that some sort of public, collective safety net must exist if these rights are to be met for *all* citizens. This requirement is true for housing and food (as inadequate as these measures may be in practice). Full privatization may be inconsistent with a human right to water unless it is coupled (as it is in England) with a universality requirement (laws prohibiting disconnections of residential consumers), and with strong regulations for price controls and quality standards.[7] But private sector *participation* in water supply certainly fits within these constraints. In other words, rooted in a liberal tradition that prioritizes private ownership and individual rights, the current international human rights regime is flexible enough to be fully compatible both with private property rights and with the private provision and operation of infrastructure, whether for water or other basic needs. This "tyranny of rights," as Kneen terms it, has important limitations in the defense of the public domain (whether conceived of as a "commons," "public goods," or the classic state provision of services (Kneen 2009)). In short, human rights are individualistic, anthropocentric, state-centric, and compatible with private sector provision of water supply; and as such, a limited strategy for those seeking to refute water privatization or the application of market principles to water management. Pursuing a human right to water as an anti-privatization campaign thus makes three strategic errors: conflating human rights and property rights; failing to concretely connect human rights with different service delivery models; and thereby failing to foreclose the possibility of increasing private sector involvement in water supply.

Conclusions: what might a political ecology approach contribute to the debate?

The water privatization debate speaks to broader debates over the respective roles of states, markets, and communities in economic life, our collective response to environmental crises, and the role of civil society (or the "public sphere") in adjudicating questions of social and ecological justice. As suggested in this chapter, the rise and (partial) retreat of water privatization over the past two decades is illustrative of the tendency of the "accumulation of nature" under capitalism to marginalize disempowered communities (the "unserved poor"). This suggests that the debate over the world's water crisis, as with environmental problems more generally, is a redistributive struggle with issues of economic and political power at its core. From this perspective, the debate over the human right to water is highly salient, as it is illustrative of some of the pitfalls and issues that have arisen in ongoing debates over the distributive implications of free market environmentalism.

What might a political ecological analysis contribute to this debate? First, my analysis emphasized the environmental dimension of water supply privatization – an issue often glossed over in contemporary debates. A political ecological analysis widens the lens to issues of *reproduction* (rather than issues of production, from a traditional political economic perspective). In doing so, political ecological analyses tend to focus on the materiality (or "biophysicality") of specific resources in all of their particularities and complexities (rather than focusing on some idealized, general "nature"). This, in turn, helps us better identify and explain the various types and degrees of barriers to commodifying and capturing differential rents from different resources. In particular, a political ecological perspective can help us understand why water privatization is re-emerging at the turn of the twenty-first century despite the fact that water remains a liminal resource for capitalism. This is particularly the case when political ecological work acknowledges the co-production of socio-economic and environmental change thereby, confronting the issue of agency – of both humans and non-humans – in a way that much political economic research does not. In short, a political ecological analysis helps us to reframe the question of privatization in two ways: to examine privatization as an ecological as well as socio-economic phenomenon; and to integrate an analysis of privatization with an understanding of the simultaneous and often overlapping roles played by government, private, community, *and* non-human actors.

This leads to my second point: that a political ecological approach can contribute to conceptual analyses of, and activism on, alternatives to privatization. Political ecology wrestles simultaneously with questions of social justice and environmental justice, and thus approaches the impacts of water privatization and commer-cialization rather differently than a strictly political economic perspective. This is because political ecology not only begins from the assumption that socio-economic and environmental change are co-produced, but also broadens the set of actors – non-humans, as well as humans – who are considered both as objects of study, and also as holders of legitimate claims to equitable treatment. Privatization and commercialization of water often occur together with a simultaneous commodi-fication and (re)valorization of the environment – prioritizing environmental protection over consumer's ability to pay, or industrial demands for water. The "free market environmentalist" paradigm, when applied to water management, produces clear gains for the environment in some cases; hence the frequent dis-agreements between environmental groups and consumer groups in contemporary debates over water privatization. The human right to water, for example, has been the subject of wary disinterest (and in some cases resistance) on the part of environmental groups.

Third, and closely linked to the previous point, political ecology provides an alternative vantage point (although not one which political ecologists always employ) from which to evaluate the role of the state, particularly in redistribution. My claim here is not that the "interrogation of the state" is a move unique to political ecology. Rather, political ecologists approach the state in a manner somewhat distinct from political economists. More precisely: the "retreat of the state" is a very ambivalent process when its environmental impacts (rather than the redistribution of the social surplus to humans) are considered. The state has in some cases rationally administered massive environmental degradation and systematic under-provision of environmental goods. Some of the great gains in human welfare during the twentieth century associated with the "state hydraulic paradigm" were made at the expense of the environment – with the state temporarily devolving costs onto the environment in what might be termed an "ecological fix" (Bakker 2004, 2009). Attitudes toward the state become more ambivalent (and the conflation of "state" with "public" interest more obviously erroneous) when one factors the environment into the redistributive equation. This is particularly relevant to "developing" countries, where community-led resource management remains widespread and in many cases a more viable option to state-led development models – more accurately described, in many cases, as the territorialization of state power through an imposition of control over local resources.

More generally, acknowledging the critical role of the state in resource allocation allows us to transcend the public/private binary often invoked in debates over resource privatization, and to appreciate the active, strategic role of the state in market environmentalism: not a "retreat" but a repositioning of the state – as an active agent in the transition from a "state hydraulic" to "market conservation" mode of water supply regulation. This interpretation of market environmentalism as a process actively led by the state raises an important question: *why* would the state seek to cede water management functions to the private sector? This question can only be answered in specific contexts. A political ecological framework sets the stage for a constructive response, reminding us that water privatization reconfigures the relationships between the state, the market, our water environments, and one another.

Notes

1   The definition of privatization is disputed. Some favour a broad definition, which includes a broad range of processes ranging from commercialization, corporatization, liberalization, private sector participation, and asset sale. Others use a precise definition, reserving the term "privatization" for the sale of assets to the private sector. In this chapter, the term privatization is used in the former sense, to refer to the participation of private businesses in the full range of water services and sanitation activities, including full privatization, divestiture, concessions, lease/affermage, management and service contracts, consulting services, and public-private partnerships with NGOs. This is technically inaccurate, but a useful shorthand.
2   As of January 2006, 11 percent of the world's large cities (population greater than 1 million) had private sector participation contracts in place (Lobina and Hall 2008). As with many other aspects of the debate over privatization, disagreement frequently arises about how to measure the precise number of customers of private sector companies. Perhaps 10 percent of cities larger than 300,000 in lower- and middle-income countries have private sector participation (2006 data). These tend to be larger cities, representing approximately 20 percent of the world's urban population. The figure of 3 percent is cited in Winpenny 2003.
3   The term ability-to-pay is the subject of dispute. The fact that households without network connections pay more per unit volume of water than households with network connections is widely recognized, as is the benefit of extending water networks to those

households. The dispute arises when the amounts currently paid by unconnected households are used as a proxy of "ability-to-pay." Critics argue that this assumption is false, as consumers in this situation are "price-takers" rather than "price-makers." Those who take this latter view favour the use of income-related thresholds (e.g. 5 percent of income) for water services.

4   These declarations include the Cochabamba Declaration, the Group of Lisbon's Water Manifesto (Petrella 2001), and the Declaration of the P8 (the world's poorest 8 countries, organized as a counterpart to the G8) at their 4th Summit in 2000. Campaigns include the UK-based "Right to Water" (www.righttowater.org.uk), the "Octubre Azul/Blue October," the Canada-based "Friends of the Right to Water," the U.S.-based "Water for All," and the "Green Cross" campaign for an international convention on the right to water (http://www.watertreaty.org).

5   ECOSOC 2002. The Committee on Economic, Social and Cultural Rights (CESCR) is the body of independent experts that monitors implementation of the International Covenant on Economic, Social and Cultural Rights by its "States parties."

6   Veolia's French-language website states, for example: "L'eau est considérée à la fois comme un bien économique, social, écologique et comme un droit humain." [Water is considered an economic, social, and ecological good as well as a human right]. Accessed at: http://www.veoliaeau.com/gestion-durable/gestion-durable/eau-pour-tous/bien-commun. See also Antoine Frérot's comments (at the time, the Director General of Veolia, one of the largest private water companies in the world) during the Open Forum on "Water: Property or Human Right?" at the 2004 Davos Forum. Accessed at: http://gaia.unit.net/wef/worldeconomicforum_annualmeeting2006/default.aspx?sn = 15810

7   As recognized by the UN Committee in its General Comment on the human right to water, which observes that, in permitting third parties (such as the private sector) in addition to state actors to supply water, an additional burden is placed upon regulatory frameworks, including "independent monitoring, genuine public participation, and imposition of penalties for non-compliance" (ECOSOC 2002, article 24).

References

ADB, 2003. "Beyond Boundaries: Extending Services to the Urban Poor." Manila: Asian Development Bank, 2003.

Alberta Environment. 2003. *Water for Life: Alberta's Strategy for Sustainability* Edmonton: Alberta Environment.

Anderson, T. and D Leal. 2001. *Free Market Environmentalism.* New York: Palgrave Macmillan.

Annez, P. 2006. "Urban Infrastructure Finance from Private Operators: What Have We Learned from Recent Experience?" World Bank Policy Research Working Paper 4045. Washington: World Bank.

Bakker, K. 2002. "From State to Market: Water *Mercantilización* in Spain. *Environment and Planning A* 34: 767–790.

—— 2004. *An Uncooperative Commodity: Privatizing Water in England and Wales.* Oxford: Oxford University Press.

—— 2005. Neoliberalizing Nature? Market Environmentalism in Water Supply in England and Wales. *Annals of the Association of American Geographers* 95(3): 542–565.

—— 2007. "The "Commons" versus the "'Commodity': 'Alter'-Globalization, Anti-Privatization and the Human Right to Water in the Global South." *Antipode* 39(3): 430–455.

—— 2009. Commentary: Neoliberal Nature, Ecological Fixes, and the Pitfalls of Comparative Research. *Environment and Planning A* 41, 1781–1787.

Bangura, Y. and G. Larbi. 2006. *Public Sector Reform in Developing Countries*. London: Palgrave Macmillan.

Barlow, M. and Clarke, T. 2003. *Blue Gold: The Fight to Stop the Corporate Theft of the World's Water*. New York: Stoddart.

Bauer, C. 1998. "Slippery Property Rights: Multiple Water Uses and The Neoliberal Model in Chile, 198–1995." *Natural Resources Journal* 38, 109–154.

Bennett, V., S. Dávila-Poblete and M. Nieves Rico (eds.) 2005. *Opposing Currents: The Politics of Water and Gender in Latin America*. Pittsburgh: University of Pittsburgh Press.

Birdsall, N. and J. Nellis. 2003. "Winners and Losers: Assessing the Distributional Impacts of Privatization." *World Development* 31, no. 10: 1617–1633.

Blokland, M., O. Braadbaart, and K. Schwartz (eds). 2001. *Private Business, Public Owners*. The Hague: Ministry of Housing, Spatial Planning, and the Environment.

Boelens, R. 2008. *The Rules of the Game and the Game of the Rules. Normalization and Resistance in Andean Water Control*. Dissertation. Wageningen University, Wageningen, Netherlands.

—— 2009. "The Politics of Disciplining Water Rights." *Development and Change* 40, no. 2: 307–331.

Boelens, R. and de Vos, H. 2006. "Water Law and Indigenous Rights in the Andes" in *Cultural Survival Quarterly* 29, no. 4: 18–21.

Bond, P. and Dugard, J. 2008. "Water, Human Rights and Social Conflict: South African Experiences." *Law, Social Justice & Global Development.* 1, 1–21.

Briceño-Garmendia, C., A. Estache and N. Shafik 2004. "Infrastructure Services in Developing Countries: Access, Quality, Costs, and Policy Reform." World Bank Policy Research Paper, WPS 3468, 2004.

Brooks, D. B. 2008. "Human Rights to Water in North Africa and the Middle East: What is New and What is Not; What is Important and What is Not." In *Water as a Human Right for the Middle East and North Africa*, edited by A.K. Biswas, E. Rached, and C. Tortajada. New York, NY: Routledge, 19–34.

Castro, J. E. 2008 "Neoliberal Water and Sanitation Policies as a Failed Development Strategy: Lessons from Developing Countries" *Progress in Development Studies* 8, no.1: 63–83.

Conca, K. 2006. *Governing Water: Contentious Transnational Politics and Global Institution Building.* Boston: MIT Press, 2006.

Cullet, P. (ed.) 2009. *Water Law, Poverty, and Development: Water Sector Reforms in India*. Delhi: Oxford University Press.

Davis, J. 2005. "Private-Sector Participation in The Water and Sanitation Sector." *Annual Review of Environment and Resources* Vol. 30, pp.145–183.

Dinar, A. 2000. "The Political Economy of Water Pricing Reforms." Washington: World Bank, and Oxford: Oxford University Press, 2000.

Dubreuil, C. 2006a. *The Right to Water: From Concept to Implementation*. Marseilles: World Water Council.

—— 2006b. *Synthesis on the Right to Water at the Fourth World Water Forum, Mexico.* Marseilles: World Water Council.

ECOSOC 2002. *General Comment 15*. Geneva: United Nations Committee on Economic, Social and Cultural Rights.

European Union. 2000. Water Framework Directive 2000/60/EC, Official Journal L327s.

Franceys, R. 2006. "Customer Committees, Economic Regulation and the Water Framework Directive." *Journal of Water Supply: Research and Technology – Aqua* 6, 9–15.

Glassman, J. 2006. "Primitive Accumulation, Accumulation by Dispossession, Accumulation by 'Extra-Economic' Means." *Progress in Human Geography* 30, no. 5: 608–625.

Gleick, P. 1993."Water and Conflict: Fresh Water Resources and International Security." *International Security* 18, no. 1: 79–112.

—— 1998. "The Human Right to Water." *Water Policy* 1: 487–503.

Graham, S. and S. Marvin. 2001. *Splintering Urbanism: Networked Infrastructures, Technological Mobilities and the Urban Condition.* London: Routledge.

Grusky, S. 2001. "Privatization Tidal Wave: IMF/Bank Water Policies and the Price Paid by the Poor." *Multinational Monitor* 22, no. 9.

Hall, D. and R. de la Motte. 2004. *Dogmatic Development: Privatisation and Conditionalities in Six Countries*, PSIRU report for War on Want, London: Public Services International Research Unit.

Hall, D., E. Lobina and R. de la Motte. 2005. "Public Resistance to Privatisation in Water and Energy." *Development in Practice* 15, no. 3–4: 286–301.

Hassan, J. 1998. *A History of Water in Modern England and Wales*. Manchester: Manchester University Press.

Hammer, L. 2004. "Indigenous Peoples as a Catalyst for Applying the Human Right to Water." *International Journal on Minority and Group Rights* 10: 131–161.

Harvey, D. 2003. *The New Imperialism*. Oxford: Oxford University Press.

—— 2005. *A Short History of Neoliberalism*. Oxford: Oxford University Press.

Heynen, N. and P. Robbins. 2005. "The Neoliberalization of Nature: Governance, Privatization, Enclosure and Valuation." *Capitalism, Nature, Socialism* 16(1): 5–8.

Himley, M. 2008. "Geographies of Environmental Governance: The Nexus of Nature and Neoliberalism. *Geography Compass* 2(2): 433–451

IMF. 2004. "Public-Private Partnerships." Washington: International Monetary Fund (Fiscal Affairs Department), 2004.

Izaguirre, K. 2005. "Private Infrastructure: Emerging Market Sponsors Dominate Private Flows. Public Policy for the Private Sector. Note No. 299." Washington: World Bank, 2005.

Johnstone, N. and L. Wood. (eds.) 2001. *Private Firms and Public Water. Realising Social and Environmental Objectives in Developing Countries*. London: International Institute for Environment and Development.

Katko, T. 2000. *Water! Evolution of Water Supply and Sanitation in Finland from the mid-1800s to 2000*. Tampere: Finnish Water and Waste Water Works Association.

Kay, J. 1996. "Regulating Private Utilities: The Customer Corporation." *Journal of Co-operative Studies* 29(2), 28–46.

Kneen, B. 2009. *The Tyranny of Rights*. London: Earthscan.

Kooy, M. and K. Bakker. 2008a. "Splintered Networks? Urban Water Governance in Jakarta" *Geoforum* 39(6), 1843–1858.

—— 2008b. Technologies of Government: Constituting Subjectivities, Spaces, and Infrastructures in Colonial and Contemporary Jakarta *International Journal of Urban and Regional Research*, 32(2), 375–391.

Langford, M. 2005. "The United Nations Concept of Water as a Human Right: A New Paradigm for Old Problems?" *International Journal of Water Resources Development*. 21, no.2: 273–282.

—— 2006. "Ambition that Overleaps Itself? A Response to Stephen Tully's Critique of the General Comment on the Right to Water." *Netherlands Quarterly of Human Rights*. 24(3), 433–459.

Lobina, E. and D. Hall. 2008. The Comparative Advantage of the Public Sector in the Development of Urban Water Supply *Progress in Development Studies* 8(1) 85–101.

Lorrain, D. 1997. Introduction – The socio-economics of water services: the invisible factors. In *Urban Water Management – French Experience Around the World* edited by D. Lorrain. Paris: Hydrocom, pp. 1–30.

McDonald, D. and G. Ruiters. 2005. *The Age of Commodity: Water Privatization in Southern Africa*. London: Earthscan.

Mansfield, B. (ed.) 2008. *Privatization: Property and the Remaking of Nature–Society Relations*. Oxford: Blackwell.

Marin, P. 2009 *Public-private Partnerships For Urban Water Utilities: A Review of Experiences in Developing Countries*. Washington, DC: World Bank.

Marin, P. and A. Izaguirre. 2006. "Private Participation in Water: Toward a New Generation of Projects?" *Gridlines* 14. Washington: World Bank, Public Private Infrastructure Advisory Facility.

Miller-Adams, M. 1999. *The World Bank: New Agendas in a Changing World*. New York: Routledge.

Morgan, B. 2004. "The Regulatory Face of the Human Right to Water." *Journal of Water Law* 15, no. 5: 179–186.

—— 2005. "Social Protest against Privatization of Water: Forging Cosmopolitan Citizenship?" In *Sustainable Justice: Reconciling International Economic, Environmental and Social Law*, edited by M-C. Cordonier Seggier and C.G. Weeramantry, Chapter 21. Boston: Martinus Nijhoff Publishers.

Narain, S. 2006. *Community-led Alternatives to Water Management: India Case Study*. Background paper: Human Development Report 2006. New York: United Nations Development Programme.

O'Connor, J. 1996. "The Second Contradiction of Capitalism." In *The Greening of Marxism*, edited by T. Benton. London: Guilford, 197–221.

OECD. 1995. "DAC Orientations for Development Cooperation in Support of Private Sector Development." Paris: Organization for Economic Cooperation and Development,.

—— 2000. *Global Trends in Urban Water Supply and Waste Water Financing and Management: Changing Roles for the Public and Private Sectors*. Paris: Organisation for Economic Cooperation and Development,.

Olivera, O. and T. Lewis. 2004. *Cochabamba! Water War in Bolivia*. Boston: South End Press.

Page, B. and K. Bakker. 2005. "Water Governance and Water Users in a Privatized Water Industry: Participation in Policy-Making and in Water Services Provision—A Case Study of England and Wales. *International Journal of Water* 3(1): 38–60.

Petrella, R. 2001. *The Water Manifesto: Arguments for a World Water Contract*. London: Zed Books.

Prasad, N. 2006. "Privatisation Results: Private Sector Participation in Water Services after 15 Years." *Development Policy Review*, 24, no. 6: 669–692.

PriceWaterhouseCoopers. 2003. "Comparative Review of IFI Risk Mitigation Instruments and Direct Sub-Sovereign Lending." London: PricewaterhouseCoopers Securities LLC, World Bank, and Bank-Netherlands Water Partnership.

Prudham, S. 2004. "Poisoning the Well: Neoliberalism and the Contamination of Municipal Water in Walkerton, Ontario." *Geoforum* 35: 343–359.

PSIRU 2006. *Public–Public Partnerships as a Catalyst for Capacity Building and Institutional Development: Lessons from Stockholm Vatten's Experience in the Baltic Region*. University of Greenwich: Public Services International Research Unit.

Robbins, P. 2003. "Transnational Corporations and the Discourse of Water Privatization." *Journal of International Development* 15(8): p. 1073–1082.

Rogers, P. and A. Hall. 2003. "Effective Water Governance." TEC Background Papers, no. 7. Sweden: Global Water Partnership Technical Committee.

Saghir J. 2006. *Public-Private Partnerships in Water Supply and Sanitation: Recent Trends and New Opportunities.* Paris: OECD Global Forum on Sustainable Development, November 2006.

Salman, S. and S. McInerney-Lankford. 2004. *The Human Right to Water: Legal and Policy Dimensions.* Washington, D.C.: World Bank.

Segerfeldt, F. 2005. *Water for Sale: How Business and the Market Can Resolve the World's Water Crisis.* Washington, D.C.: Cato Institute.

Shirley, M. 2002. *Thirsting for Efficiency: The Economics and Politics of Urban Water System Reform.* Washington, D.C., World Bank.

Shiva, V. 2002. *Water Wars: Privatization, Pollution and Profit.* London: Pluto Press.

—— 2003. *Human Development Report 2003: Millennium Development Goals: A Compact for Nations to End Human Poverty.* New York: United Nations Development Program, Human Development Report Office.

Susilowati, I. and L. Budiati. 2003. "An Introduction of Co-Management Approach into Babon River Management in Semarang, Central Java, Indonesia." *Water Science & Technology* 48(7), 173–180.

TNI, 2005. *Reclaiming Public Water: Achievements, Struggles and Visions from Around the World.* Amsterdam: Transnational Institute.

Trawick, P. 2003. "Against the Privatization of Water: An Indigenous Model for Improving Existing Laws and Successfully Governing the Commons." *World Development* 31(6), 977–996.

UNDP. 2005. *Human Development Report 2005: International Cooperation at a Crossroads: Aid, Trade and Security in an Unequal World.* New York: United Nations Development Programme, Human Development Report Office,.

—— 2006. *UN Human Development Report 2006: Beyond Scarcity: Power, Poverty, and the Global Water Crisis.* New York: United Nations Development Programme, Human Development Report Office.

UN ECOSOC. "Economic, Social and Cultural Rights: The Right to Food." Report Submitted to the 59th Session of the Commission on Human Rights, by the Special Rapporteur on the Right to Food, Jean Ziegler, in accordance with Commission on Human Rights resolution 2002/25." E/CN.4/2003/54: UN Economic and Social Council, 2003.

UNWWAP. 2003. "Water for People, Water for Life." Rome, FAO.

—— 2006. "Water: A Shared Responsibility." Rome, FAO.

Vandenhole, W. and T. Wielders. 2008. "Water as a Human Right – Water as an Essential Service: Does it Matter?" *Netherlands Quarterly of Human Rights.* 26(3), 391–424.

WDM. 2005. "Dirty Aid, Dirty Water: The UK Government's Push to Privatise Water and Sanitation in Poor Countries." London: World Development Movement.

—— 2006. "Out of Time: The Case for Replacing the World Bank and IMF." London: World Development Movement.

—— 2007. "Going Public: Southern Solutions to the Global Water Crisis." London: World Development Movement.

Williamson, J. 2000. "What Should the World Bank Think About the Washington Consensus?" *World Bank Research Observer* 15, no. 2: 251–264.

Winpenny, J. 1994. *Managing Water as an Economic Resource*. London: Routledge.

—— 2003. "Financing Water for All: Report of the World Panel on Financing Water Infrastructure." Geneva: World Water Council/Global Water Partnership/Third World Water Forum.

World Bank. 1992a. "Water Supply and Sanitation Projects: The Bank's Experience: 1967–1989." Washington, D.C.: World Bank Operations and Evaluation Department.

—— 1992b. "Effective Implementation: Key to Development Impact. Report of the Portfolio Management Task Force." Washington, D.C.: World Bank.

—— 1993. *Water Resources Management: Policy Paper*. Washington, D.C.: World Bank.

—— 1997. *Toolkits for Private Participation in Water and Sanitation*. Washington, D.C.: World Bank .

—— 2005. "Infrastructure Development: The Roles of the Public and Private Sectors: World Bank Group's Approach to Supporting Investments in Infrastructure. World Bank Guidance Note." Washington, D.C.: World Bank.

World Bank/IMF 2003. *Implementing the Word Bank Group Infrastructure Action Plan (With Special Emphasis on Follow-up on the Recommendations of the World Panel on Financing Water Infrastructure*. Washington, DC: World Bank and International Monetary Fund Development Committee, 2003–0015.

WHO. 1992. *The International Drinking Water Supply and Sanitation Decade: End of Decade Review*. Geneva: World Health Organization.

—— 2003. *The Right to Water*. Geneva: World Health Organization.

—— 2008. *UN-Water Global Annual Assessment of Sanitation and Drinking Water. Pilot Report: Testing a New Reporting Approach*. Geneva: World Health Organization.

WRI. 2008. *Seeding the Sustainable Economy: State of the World Report 2008*. Washington: World Resources Institute.

Zaidi, A. 2001. *From Lane to City: The Impact of the Orangi Pilot Project's Low Cost Sanitation Model*. London: WaterAid.

17 The social construction of scarcity: the case of water in western India[1]

Lyla Mehta

The global politics of water scarcity

Water scarcity has emerged as one of the most pressing problems in the twenty-first century. It is estimated that 2.7 billion people will face water scarcity by 2025 (UN 2003). Against a growing alarmism of "water wars," several global agencies, national governments and NGOs have been concerned with emerging water "crises" and potential water conflicts (e.g. UN 2003; FAO 2003). For example, projections of water supply and population growth rates are predicting a dark scenario of the future: while the average per capita supply of water will decrease by one-third by 2025, water use will increase by about 50 percent during the same period (Vision 21 2000).

There are several polarized views regarding how the water needs for present and future generations of a country and region can be met (see Mehta 2005). Until a few decades ago, the large dam[2] was universally considered to be the panacea for water scarcity. The proponents of large dams tend to downplay the social and environmental costs of large dams against the benefits of hydropower and irrigation (British Dam Society 1999). These views have been contested by a world-wide constituency comprising academics, scientists and members of voluntary agencies who have highlighted the problems of involuntary resettlement and environmental damage due to large dams (e.g. Goldsmith and Hildyard 1992; McCully 1996). The World Commission on Dams (WCD) in 2000 concluded that while dams have made a considerable contribution to human development, in too many cases unacceptable costs have been borne in social and environmental terms. The Commission also argues that often water and energy needs can be met through alternative solutions that would fare better than dams on equity grounds (WCD 2000).

There are still very polarized views on large dams and the role that they play in mitigating water scarcity. To some extent, the controversies around large dams represent the dilemmas around the "very meaning, purpose, and pathways for achieving development" (Alhassan 2009: 151, and Roy 1999.). Ten years on from the WCD process, there is still no consensus regarding the place of large dams in development and water resource management, and there are still many reservations to the conclusions of the WCD, especially in dam-building nations such as India, China and Turkey. Is this because large dams are urgently required to solve the problems in water-needy areas, or is it because of questions concerning a wider political economy? Is there a need to investigate the relationship between discourses of water scarcity and the vested interests in large-scale development projects? Are there different and more locally appropriate ways to view and address water scarcity? This chapter addresses these complexities surrounding water scarcity by taking the case of Kutch, a semi-arid to arid region in western India and its relationship with the controversial Sardar Sarovar (Narmada) Project (SSP), a controversial dam under construction in western India. By drawing on water scarcity in Kutch, India, the chapter argues that scarcity is not a natural condition. Instead, it is usually socially mediated and the result of socio-political processes. Often water scarcity tends to be naturalized and its anthropogenic dimensions are whitewashed. The paper argues that it is wrong to conceive of water scarcity only in absolute or volumetric terms. Instead there is an urgent need to link water scarcity with wider socio-political and institutional processes. It is also important to distinguish between the biophysical aspects of scarcity that are lived and experienced differently by different people and its "constructed" aspects.

Conceptual points of departure

The scarcity of water and drought are complex phenomena that can be analyzed differently from social, political, meteorological, hydrological and agricultural perspectives. However technical and popular understandings of water scarcity have tended to be simplistic (Falkenmark and Chapman 1989). There has been the tendency to direct attention to the lack of supply of water due to natural forces rather than look at human-induced land and water use practices and at socio-political considerations. Real causes of scarcity can be obscured leading to inappropriate solutions. This is because most academic and policy portrayals of water scarcity focus on the finite nature of water supplies (e.g. Shiklomanov 1998). Countries are also classified according to a "water stress index" on the basis of their annual water resources and population (see Falkenmark and Widstrand 1992) and water scarcity scenarios are created for groupings of countries or regions based on projections of future water demands and needs (e.g. Rosengrant *et al*. 2002). While there is some acknowledgement of the differences between water shortages – which refer to physical amounts – and water scarcity – (which could be a social construct or the result of affluence, lifestyle choices and expectations – (see for example Winpenny in FAO, n.d.), largely most of the literature focuses on volumetric and physical measures, especially with respect to both a growing population and competing demands for water.

Why does it matter? Four important implications arise through these conventional and sometimes problematic framings of scarcity. One, the scarcity of essential goods is often used to argue for the need for markets and institutions that are needed to mediate the transactions of scarce or "economic goods" (such as water and land) which are made the objects of property (Xenos 1987). In the water sector, this line of thinking was endorsed at the Dublin conference in 1992, where water was recognized as having an economic value in its competing uses. Through the 1990s, thus, water reform processes have instituted controversial pricing and cost recovery mechanisms as well as the institutionalization of formal tradable rights to water, in order to facilitate the emergence of water markets. These debates however highlight the economic aspects of water, rather than focus on symbolic and cultural aspects. However, the declaration of water as an economic good still remains highly controversial in the water domain because it blanks out the multi-faceted nature of water which my study demonstrates (see also Mehta 2004).

Two, scarcity underpins much thinking around conflicts arising through competing claims around scarce resources. In the water domain, the popular assertion of global water wars is well known. More generally, in the 1990s there was a surge of literature positing links between natural resource scarcity and violent conflict. For example, work by Thomas Homer-Dixon has argued, for example, that declining environmental resources such as clean water accompanied by large-scale movements of people and the resultant economic deprivation would all lead to conflicts and resource wars (1994). By drawing on a series of case studies from around the world, this work demonstrates that environmental scarcity plays an independent role in causing conflict. Even though other factors – such as ideology, power relations, unequal property rights – matter too, they are subordinate to environmental scarcity, which may be the causal factor. Examples where, despite scarcity, cooperation rather than conflict occurred are largely ignored. The focus on environmental scarcity as a causal variable tends to ignore other explanatory variables (see Peluso and Watts 2001). Scarcity is not seen as the result of powerful actors getting away with resource appropriation and thus enhancing degradation. Moreover, "the politics of distribution disappear into the environmental scarcity concept" (Hauge and Ellingsen 1998). Finally, ways in which scarcity can encourage co-operation are neglected, as has been ably demonstrated in the large body of approaches grounded in Common Property Resources (CPR) Theory (e.g. Ostrom 1990; Bromley and Cernea 1989; Wade 1988 and Berkes 1989). Finally, scarcity is a concept that can provide meta-level explanations for a wide range of phenomena over which humans ostensibly have no control, and science and technology are evoked as the panaceas. For example, most policy interventions in developing countries still focus on supply solutions for dealing with increased water demand. These include large dams, the extra-basin transfer of water along with small-scale solutions such as rainwater harvesting.

This paper takes the view that it is important to link scarcity debates with socio-political perspectives that engage with discourses and contestations around scarcity. Socio-political perspectives of scarcity draw on a variety of disciplinary approaches including political ecology (Blaikie and Brookfield 1987; Bryant 1992; Peet and Watts 1996; Forsyth 2003; Robbins 2004) and Foucauldian discourse analysis. Here it is important to examine how scarcity is perceived by different actors and the extent to which the definition is context-bound. It is also necessary to look at contexts of scarcity in a wide range of areas: the household, community, state and world. Thus, I am concerned with both examining contested meanings of scarcity as well as how scarcity is created through the politics of allocation and through competing claims and conflicts over water resources. The former emphasizes the need to understand how scarcity is constructed discursively; the latter is concerned with the intricate web of power and social relations governing access to and control over water at the macro and micro levels.

The case of Kutch

The crescent-shaped peninsula of Kutch is the largest district in Gujarat and has an area of 45,612 sq. km constituting 23 per cent of the state. Kutch is like an island as it is bound by the sea in the South and West and by the Ranns (salt marshlands) in the East and North. Apart from its very heterogeneous social and ethnic composition, the region has nine ecological zones (Gujarat Ecology Commission 1994).

Kutch has an arid to semi-arid type of climate. Temperature ranges from 45 degrees centigrade in the summer to two degrees in winter. Humidity and evapotranspiration are high throughout the year. In some areas, groundwater supplies are abundant, but increasingly the levels are dropping. Overexploitation of the aquifer combined with sea water ingression has led to salinity in the water and soils and a sinking water table. The groundwater table sinks at a rate of a meter a year and in two *talukas* in the district fall under the over-exploitation category (Gujarat Ecology Commission 1994: 14).

Rainfall is erratic and variable and averages about 350 to 370mm. There is high regional variation, ranging from 440mm in southern Kutch to 338mm in western Kutch (Raju 1995: 10). It only rains a few days a year (15 on an average), with significant intra-district variations. In official discourse, Kutch is considered drought-prone, with droughts taking place every 2–3 years. Scarcity conditions in Kutch are often attributed to dwindling rainfall (Mehta 2005). However, this is a myth, both in Kutch as well as in other parts of the world (cf. Falkenmark *et al.* 1990). Rainfall data of the past 60 years prior to 1997 indicates that, while there have been erratic variations in the quantity of rainfall, there is no evidence to suggest that precipitation rates have changed. A t-test, comparing the rainfall in Kutch over 30 years (1968–1997) with the previous 30 year period (1938–1967), revealed no significant difference ($t_{obt.} = -.28$, $p > 0.05_{2\text{-tail}}$ see Sinclair 1998). Inference tests using rainfall data for the *talukas* of Abrasa, Bhuj and Rapar over a longer period (120 years) were conducted to compare rainfall differences between four 30-year periods (1878–1907, 1908–1937, 1938–1967 and 1968–1997). A repeated measures analysis of variance revealed no significant differences over these periods (Sinclair 1998).

Kutchi identity is moulded around water, or the lack of it. Villagers across the length and breadth of the district say that the lack of water is the cause of their misery, the depopulated villages and mass migration out of Kutch. Water scarcity is attributed to low rainfall, ever-decreasing rainfall and perennial droughts. There is a wide-spread belief in Kutch that, due to the harsh climate, erratic water supply, declining groundwater sources and frequent droughts, the only solution is to get water from the rivers of Gujarat (Kutch Development Forum 1993). That is why all hopes are often pinned on the controversial Sardar Sarovar Project (SSP) under construction on the River Narmada in Gujarat. The planned 163-meter dam is part of the ambitious Narmada Project, which comprises two mega and several large dams. The SSP is also made out to be Gujarat's lifeline (see Raj 1991) and is also made out by many to be *the* only hope for Kutch.[3] At the time of writing, the dam's height is at about 110 meters. Its construction has been very slow owing to cost and time overruns and due to a highly dynamic protest movement that has been highlighting, amongst other things, the severe problems concerning resettling and rehabilitating the communities affected by the dam.

The politics of scarcity: Kutch and the SSP

Plans to provide water for Kutch from the river Narmada have a long history and are no less complicated than the history of the Sardar Sarovar dam itself. Though the project was conceived almost a century ago, actual work has been stalled due to inter-state conflicts such as the height of the dam, the extent of submergence and the sharing of benefits. Different committees were set up to resolve all these inter-state conflicts such as the Narmada Water Disputes Tribunal of 1979. Kutchis maintain that the state of Gujarat did not represent their interests adequately and were biased in favor of obtaining benefits for Central Gujarat, which has far better water endowments. As a result, during the course of several rounds of negotiations, Kutch successively lost out. From an original plan of three canals, the Narmada Water Disputes Tribunal sanctioned only the canal along the coast in 1979. Instead of allowing for the irrigation of 9.45 lakh acres of land in Kutch, only 95,000 acres of land were to get irrigation (Kutch Development Forum 1993). In this way less than two per cent of Kutch's area stands to benefit from the Kutch Branch Canal. Largely, all over Kutch and other parts of Gujarat it is large farmers and the agro-industrial lobbies that stand to gain the most, especially in Central Gujarat (see Mehta 2005). By contrast, as I discuss soon, the interests of poorer groups, such as dry land cultivators and pastoralists, are neglected.

The present plan envisages a canal of 200 kilometers in Kutch passing through a tiny coastal strip in Eastern and Southern Kutch. Not all of this area is considered to be drought-prone. Much of this strip is rich in groundwater endowments and is part of the belt that has experienced the green revolution in Kutch. The industrial belt of Kutch situated in the Kandla-Gandhidham area is also located in the command area. Thus, the needs of industrial residents and rich farmers may be met more than those of needy farmers in other drought-prone areas. It is ironic that water-hungry Kutch should be used by the dam proponents to justify the project, especially as it does not appear as though Kutch is likely to benefit significantly from the project, at least in terms of irrigation, the stated aim of the project. In 2004 a drinking water pipeline brought Narmada water to Samakhiyari village in Kutch, some 600 km northwest of the Sardar Sarovar dam in Gujarat state. This is the Narmada Pipeline Project designated to deliver safe drinking water to roughly 20 million people in more than 8,000 villages in Gujarat. In 2010 when I returned to Kutch, I found that piped water connections to many households had reduced women's daily trudge for water, even though water regularity continued to be a problem (see also Talati *et al.* 2004). They are, however, still waiting and hoping for the promise of irrigation water.

Largely, the propaganda machinery used by the state as well as decades of political promises have succeeded in "manufacturing" perceptions or myths that reinforce the bounty that is supposed to be the SSP. Here I borrow Herman and Chomsky's concept of "Manufacturing Consent" (1994). In a book with this title they describe the role of the media in "manufacturing" consent and describe how support is mobilized for special interests that dominate the state activity. In Gujarat, the state has *"manufactured"* one dominant perception of water, namely, the Narmada project as the single solution. In doing so, political and business interests all over the state are being served.[4] The project is also legitimized in the name of the water-hungry in drought-prone Gujarat and Kutch. Additionally, the discourse on water resources management is hegemonized by this one project. The focus on externally supplied water has prevented water-harvesting schemes from gaining widespread acceptance in Gujarat. Officials of the Gujarat State Land Development Corporation (GSLDC) feel that their work is marginalized in water resources departments in Kutch and in Gandhinagar. Their efforts are stymied due to the state-wide obsession with the Narmada project and they feel that their work is not taken seriously (Mehta 2005). Villagers in the research village also echo these sentiments. Every year they watch helplessly as water flows unchecked into the Rann due to Kutch's topography. Thus it is necessary that water is sufficiently tapped through rainwater harvesting and catchment area treatment instead of all attention being focused on the large project.

In the late 1990s, Kutchis had largely bought into the grand narrative of this "water wonder" of the SSP. They felt that it would solve all their problems and make up for the injustices of climate and history, since the people felt betrayed that they have constantly had to live with a series of broken promises. The widespread "manufactured" nature of debates around the SSP also helped obscure and whitewash the anthropogenic nature of scarcity to which I now turn.

The anthropogenic dimensions of scarcity

While the actual volume of water bestowed by the Rain God might not have changed, the severity of drought or scarcity is felt more acutely today than in the past. This manifests itself in concrete and biophysical dimensions. Scientists and local people maintain that the intensity of drought has increased (cf. Murishwar and Fernandes 1988). There are several factors at play. The first factor is increasing *devegetation*, which has certainly taken place due to an increase in commercial logging activities in the last five decades. Prior to 1948, areas known as *Rakhals* were set aside, where tree cutting and grazing were prohibited. Despite their elitist nature, the *Rakhals* were successful in experiments concerning the types of trees suitable for Kutch's unique requirements, and considerable forest cover was created (Rushbrook Williams 1958: 29). After 1948, these institutional restrictions ceased to exist, and there was a boom in unchecked logging. This has had serious repercussions on the vegetational cover of Kutch. The wild growth of *Prosopis Juliflora* has also led to loss of grass cover and the undermining of indigenous tree species. Moreover, it is believed that *Prosopis Juliflora* neither attracts rain nor gives moisture to the soil even though it might conserve water within its own system. Bad water management practices have also played a role in vegetational reduction. The world famous grasslands in northern Kutch, for example, have suffered considerably due to the damming of Kutch's northern rivers. The damming stopped the annual inundation and natural fertilization by the silt traditionally brought by the rivers. The grasslands are now dependent only on rainfall for their rejuvenation.

Another dimension to anthropogenic scarcity is the *overexploitation of ground-water aquifers*. Access to and control over groundwater in Kutch is marked by tremendous inequality. In my research village, "higher castes" such as the Rajputs and Jadejas comprise less than 30 per cent of the population but they control about 65 per cent of the land. They also own most of the wells in the village. Well ownership goes hand in hand with land ownership. Those who have access to land, control the water below them. The rich irrigators in rural areas (popularly known as "water lords") are often responsible for depleting vast amounts of groundwater resources. These water lords overcome groundwater constraints by their willing-ness and financial ability to invest in yearly or even monthly well-digging, -broadening and -deepening operations. They are also successful at circumventing legislature and making the best of institutional loopholes. The groundwater crisis, hence, is not just one of dwindling water levels, but instead a crisis of access and control over scarce resources.

However, in popular discourses promoted in the media and by politicians, the anthropogenic dimension of water scarcity is obscured (Mehta 2005). The culpability of large farmers, bad water management practices and state policies is denied. The story of "dwindling rainfall" obscures the fact that water has been misused and legislation is constantly circumvented. The power of the water lords remains unquestioned and their greed is exonerated. The water problem is seen as "natural," something beyond human agency, even though rainfall and drought patterns are characterized by high uncertainty and variability. Projects such as the SSP are evoked as the only solution to set right what nature has ostensibly disturbed.

Local experiences of scarcity

I now turn to village level experiences of water scarcity by drawing on findings from a village which I call Merka in eastern Kutch. The village is situated in the potential command area of the SSP. It has been declared a "no source" village by the state, which means that existing water supplies in the village are not sufficient to provide water to its population. Water is, thus, supplied by the Gujarat Water Supply and Sewage Board, either by tanker or by pipeline.

Merka is a multi-caste village. Caste is the basis for most social interactions and also plays a crucial role in local water resources management practices. Merka's castes range from the erstwhile feudal lords (*Jadejas*) to *Rjputs* (warrior castes), pastoralists (*Rabaris, Bharvads*) and the *Dalits* (formerly known as Harijans or "untouchables"). Sources of water comprise tanks around the village where rainwater is collected, wells with groundwater and *virdas*, holes in the river bed.

Institutional arrangements around water

In Hindu and village cosmology, water is considered pure and holy. It is considered to have a cleansing and purifying effect and is revered by all. Religious and caste-based institutions provide rules of purity and pollution dictating whose water can be drunk, whose should be avoided and who should fetch the water. Water is used as a metaphor to accentuate differences and social distance between the groups in the village. Declarations of difference between communities are based on whether *the other's* water can be drunk or not. Even though state-based institutions prohibit water-based discrimination, the "higher" castes still insist on discriminatory practices. However, these rules and restrictions are often bent or even totally dropped under certain circumstances. For example, during drought periods "higher" castes do not hesitate to drink water from *Dalit* wells. Thus even caste-based institutions display a certain degree of flexibility during times of drought. High caste villagers explain this in the following way: sub-terrain water is the same everywhere; it becomes differentiated only when it acquires the attributes of the user. Thus, according to village logic water in a well used by *Dalits* is not impure, but the water in a Dalit's house is. This perception allows for flexibility in the otherwise strict caste-based water institutions.

This discussion reveals that water as a natural resources has symbolic, cultural and spiritual dimensions and highly differentiated in its use in local contexts. Even though water is used as a metaphor to express difference, water-related rules and practices are sometimes bent and dropped. Official water resources management discourses (such as those endorsed in the 1992 Dublin principles) tend to focus on the material values of water. But merely viewing water through an economic lens can undermine its embeddedness in the everyday symbolic, cultural and social contexts within which people live their lives. In doing so, water is robbed of its multifaceted meanings.

Merka's social fabric is very heterogeneous and differentiated. Dominant castes still enjoy most control over the village's natural resources. Most of the land is under the control of the *Jadejas* and the *Rjputs*. One *Rjput* clan owns over half the irrigated land. Even though their former glory may have declined, the erstwhile feudal chiefs, the *Jadejas*, exercise de facto control over the village commons, even though these lands officially come under the jurisdiction of the state.

Formal institutional arrangements which are created by the state or by extension workers tend to neglect the differentiated nature of community. Water-directed interventions in Merka are usually directed towards and brokered by a few dominant elites, usually male leaders from the high castes. They are the ones who benefit from irrigation schemes, drought-relief programmes and other state-directed interventions. It is assumed that these leaders will speak with one voice for the whole village and that they are interested in collective benefits for all.

Traditional power structures sometimes override the more recent state-driven institutions which aim to create an equitable use of land and water resources. For example, it is the uncodified customary arrangements that tend to prevail over state tenure arrangements in land arrangements. High caste families still continue to own hundreds of acres of land in the village despite state-introduced ceiling acts.

Often formal institutional arrangements tend to reinforce the position of the traditional elites. For example, drought relief schemes encourage the rent-seeking activities of the elites. They ensure that the power status quo remains unchallenged. Planners often assume a homogeneous village, forgetting the different goals and priorities of the different village members.

In Merka, institutions governing water use are highly differentiated and often serve to reinforce dominant power and social relations. In some parts of the village, tanks are often the only water sources and are central to the lives of the people. They are used for bathing, drinking, watering livestock and, in some cases, irrigation. Until recently, tank management was the responsibility of the rich and powerful, who would pay for their upkeep. Tank management went hand in hand with the notions of blessing and benediction. Hence, tank cleaning and management activities are considered to generate, an important form of symbolic capital (cf. Bourdieu 1977) in the community. The gains arising out of tank management are therefore not just material but also symbolic, such as reward in the after-life and prosperity for one's descendants. By enhancing the power and status of tank benefactors, indigenous institutions thus reinforce the power and prestige of the rich and powerful in the community. In the past few decades, state-sponsored drought relief programmes have increasingly assumed responsibility for tank maintenance with the aim of drought proofing the area and eliminating water scarcity. Contrary to the popular view that these have displaced local initiatives, informal arrangements to manage tanks still exist. As and when the need arises local collections are initiated and tanks are de-silted. These activities do not proceed according to fixed rules, but instead have an *ad hoc* character and are rooted in religious practices and beliefs, instead of merely in natural resource management practices.

Living with scarcity

In rural Kutch, the outcome of every year is uncertain. Periods of abundance are interspersed with periods of dearth and impoverishment. Rainfall is largely characterized by uncertainty and can be seen to be *"regularly irregular."* What are the institutional arrangements that deal with this uncertainty and scarcity? Livelihood strategies display a high degree of flexibility. Let me begin with dry land agriculture and pastoralism and the links between the two.

Dry land agriculture employs a wide range of risk minimization strategies, such as the spreading of land assets over different land parcels distributed over a variety of soil types. Decision-making regarding field preparation is often an innovative response to an ever-changing environment. For example, if villagers sense a lean year, they are likely to plant drought-hardy crops. If the year appears promising they invest in cotton. Crop-related decisions are not just dependent on exogenous factors such as the rainfall. Personal need, practicalities and collegiality towards field neighbors are also important factors. Thus, agricultural practices are flexible responses to situations at a given time and given place. They are adaptations to the year, particular soil conditions and to highly specific contingencies arising within the social world. For example, it is usual to confer with field neighbors and collectively negotiate on crops to be grown in a particular vicinity. To borrow Paul Richards's useful analogy, all these factors make agriculture in Kutch an ongoing performance which is a "sequential adjustment to unpredictable conditions" (Richards 1989: 41). Clearly of course, not all cultivators have uniform strategies. Large landowners with irrigation facilities enjoy the maximum buffer against uncertainty. By contrast, dry land cultivators and marginal farmers face the knocks of scarcity more.

The same resource base is also used by herders, given that the livestock-based economy has always been one of the most important sources of livelihood in Kutch. Kutch's semi-arid to arid type of climate encourages a vegetation of short annual grasses ideal for livestock rearing. The pastoralists are usually sedentary but during lean years migration is a necessity, given the uncertainty of rainfall and forage availability in the village environs. Those with large herds can afford to migrate for about 400 kilometers. Migration thus allows pastoralists with large herds to adapt to a variable and heterogeneous environment. Due to this mobility they can exploit and access different social and ecological patches across the range. One always hopes, quite literally, that the grass is greener on the other side. The institutional arrangements need to be highly flexible and adaptable and entail constant decisions and responses to "here and now" contingencies. Each site has its own set of forage opportunities and restrictions. The water situation is always different, as is the reception from the host community. Survival is only possible due to constant adaptation and *ad hoc* arrangements. Those with fewer animals (under 100) cannot afford to migrate and have to make do with locally available grasses.[5]

Migratory pastoralism is possible only due to the wide support and social networks spread out over a wide area, indicating the embeddedness of institutions in wider social structures. These social networks include kinship ties amongst other pastoralists but also reciprocal relationships with farmers that have been built over several generations. The relationship between cultivators and pastoralists, who use the same resource base, has largely been synergistic. Landowners appreciated the manure provided by the pastoralists, and they were allowed to pitch camp on fallow or harvested fields during their migratory routes. Recently, however, changes in agricultural patterns have made the relationships less symbiotic, with pastoralists losing out. State policies and interventions have tended to offer agricultural subventions to cultivators and have led to the introduction of double and triple cropping. The migration of pastoralists is actively discouraged, with pastoralists being fined or areas being sealed off. There are no state policies in Kutch directed towards pastoralists or for the protection of CPRs. This has led to a general lack of appreciation of the diverse ways in which different resource users use the same land and CPR resources. It has also led to a general undermining of the institutional flexibility displayed by cultivators and pastoralists as they adapt their livelihoods to deal with uncertainty and led to a general worsening of ties between the two groups.

Of course, the livelihood strategies in drought-prone Merka are not only very diverse. They also depend on people's occupational status and wealth assets. Rich irrigator families have the financial clout to dig wells and grow fodder crops that ensure fresh feed for their cattle, sometimes in the most extreme drought conditions. They do not suffer tremendously due to the hardships of drought. There is no change in their diet, and milk continues to be drunk by all members, including women and girls. Drought for this family means fewer yields and fewer profits, which mean not having flowing cash which they would use to build a house or celebrate a wedding. In no way does drought entail misery or loss.

By contrast, drought means debt, hardship and a somewhat reduced intake of milk and milk products for poorer households who earn most of their money through seasonal labor and share-cropping. A poor *Dalit* family in 2000–2001 reported how the failure of rains meant falling into the pernicious trap of being indebted to the money-lender. The intake of milk produce is drastically decreased and the dependence on casual labor and state-sponsored relief measures is strong. They also could not irrigate one of their fields near the dam because the rich irrigators had used up all the water in the tank. Since their relationship with them was one of one of patronage and dependence, they could not be overtly critical of them. Thus, scarcity and drought mean different things to different resource users, and their experiences and perceptions are largely linked with people's wealth, assets and social positioning.

I have experienced drought in Merka several times. While local people are often weighed down by fodder scarcity, low agricultural yields, debts and *complain* about the lack of flowing cash money and few or no off-farm employment facilities, there is also the relative normalcy of drought, no matter how difficult and hard:

> We are used to drought. Two years are bad and one year is good. This is our life. When it's bad we disappear away from the village. When the rains come, we race back. This is our home and we are happy here.

But this acceptance of the cyclical nature of drought and scarcity may not always persist. Even the highly adapted, flexible and diverse livelihood strategies of both cultivators and pastoralists will not always be able to withstand the problems of dwindling groundwater aquifers, devegetation, soil degradation and the lack of grass cover. There are limits to local resilience. I do not want to overly glorify "adapting to and living with scarcity and uncertainty." However understanding their dynamics will help planners and policy-makers overcome their "dry land blindness" and promote interventions that contribute to mitigating scarcity, instead of naturalizing it.

Largely, planners have not built on local people's coping strategies *vis-à-vis* scarcity. Instead of promoting dry land agriculture or agro-pastoralist occupations they have neglected them. They do not view scarcity as a temporally bound phenomenon. Instead, Kutch is made out to be permanently drought-prone and cursed by scarcity. State-sponsored water interventions have not succeeded in mitigating scarcity. In fact, some of them have exacerbated the water problems in certain areas, making scarcity indeed ever present and all pervasive. Planners also have idealized views of local communities and local institutions. These flawed interventions arise because of the prevailing world-views and experiences of policy-makers and their dry land blindness, and because of institutional weaknesses in water management programmes. As long as this situation persists, scarcity and its accompanying "scarcity industry" will remain an all-pervasive feature of life in Kutch.

Discussion

This paper has argued that scarcity is not necessarily "natural." Instead, it refers to a concrete period of dearth either of water, milk or fodder, which is felt acutely by the human and livestock population in rural areas. Several strategies, rooted in local knowledge systems and practices, exist to cope with seasonality and uncertainty, and rural livelihoods have adapted to the variable and uncertain nature of Kutch's rainfall. The coping strategies against scarcity are highly differentiated. The wealthy of the village tend to have the most options and can resort to a wider range of coping strategies than the poor. To a certain extent, social forms of differentiation such as caste, historical legacies and gender legitimize the unequal access to and control over scarce resources. These are the "lived and experienced" aspects of scarcity.

Powerful discourses of scarcity have largely served the interests of powerful people (e.g. politicians, business constituencies and irrigators). They have obscured the fact that there is highly unequal access to and control over land and water resources in Kutch. They also succeed in essentializing scarcity in Kutch and making it seem as "natural," thus ignoring its anthropogenic nature. Scarcity is also used to legitimize the controversial SSP by evoking notions of its bounty and potential contribution to Gujarat's development. But this consensus has largely been "manufactured" due to the socio-political processes discussed. These are the "constructed" or "manufactured" aspects of scarcity. Thus, there emerges the need to analyze water scarcity at two levels: One, at the discursive level where scarcity is "constructed" and two, at the material level as a biophysical problem where it is lived and experienced differently by different people.

The case study highlights several wider lessons for debates concerning so-called environmental crises. One, there are problems in merely focusing on the use or material values of resources or property. As demonstrated, water in Kutch has symbolic and cultural meanings that are not captured in policy debates. Two, technological "solutions" to scarcity such as large dams are not neutral. Instead, they are contested, as the SSP case demonstrates. Three, conflicts around resource use may not be merely due to "scarcity." Instead, conflicts emerge due to unequal access to and control over resources. This is because local "users" have diverse and sometimes conflicting interests over property and resources (cf. Li 1996; Mosse 1997), as was demonstrated the Kutch case. Four, as discussed in the introduction, the case study highlights how scarcity can be examined through socio-political perspectives that engage with discourses and contestations around scarcity. Thus, socio-political perspectives of scarcity focus on an analysis that is both discursive *and* materialist (cf. Escobar 1996 and Yapa 1995), where the nexus of power, ideas and social relations is the centerpiece of enquiry. Such an analysis tries to marry an ecological phenomenon (i.e. a shortage of food/water, etc.) with political economy. For example, Yapa talks of "discursive materialism" (Yapa 1995; Yapa 1996) where the focus is not just on the social or material or discursive, but on all three. The historian Ross (Ross 1996) distinguishes between socially-generated scarcity (insufficient necessities for some people and not others) versus absolute scarcity (insufficient resources, no matter if equitably distributed). Similarly, my work has distinguished between "lived/experienced" scarcity (something that local people experience cyclically, due the biophysical shortage of food, water, fodder, etc.) and "constructed" scarcity (something that is manufactured through socio-political processes to suit the interests of powerful players – in this case the dam-building lobby and the rich irrigators and agro-industrialists). I also demonstrated how the discursive nature of manufactured scarcity often exacerbates biophysical scarcity. Clearly there is the constructivist dilemma. To cast everything as "socially and politically constructed" could in some ways deny the existence of a "real" ecological crisis around water, food, land, and so on. Constructivists could be accused of fiddling while Rome burns (Ross 1996). But we still need to be aware of the dangerous implications of scarcity politics and the ways in which scarcity is deployed to colonize the future in different ways (see Mehta 2010). It is thus important to maintain a focus on materialistic aspects and on how resource shortages and ecological degradation are often a result of the uneven social measures that manufacture scarcity all over the world for the economic and political gain of powerful interests.

Conclusion

I have used the case of Kutch to highlight the multifaceted nature of scarcity and how it is socially and politically constructed to meet certain ends. By taking the case of water, the chapter questioned conventional understandings of water scarcity. It argued that water scarcity is not necessarily a given, but instead has both "lived/experienced" as well as "constructed" elements. Institutional perspectives have played in a key role in moving away from alarmist portrayals of scarcity and property rights by demonstrating how local people can manage and live with scarcity. Still, to be true to women's and men's everyday realities, they need to be complemented by analyses that locate property rights within wider historical, cultural and socio-political processes that marry both discursive and materialist analyses.

Notes

1 The empirical material presented in this paper draws on Mehta 2005 and 2007 and on research conducted in Kutch which I have been visiting since 1995. I am as always grateful to the people of Merka and Kutch for their friendship, warmth and hospitality. The usual disclaimers apply.

2 According to the World Commission on Dams, there are currently over 800,000 dams in the world, of which 45,000 are large. A large dam has a wall height of more than 15 meters.

3   For reasons of space, it is impossible to provide all the details on the project (see Mehta 2005 for more details on Kutch and the SSP and Morse and Berger (1992) and Fischer (1995) for details on the Narmada project and its controversies).

4   For example, the Gujarat government has been promoting industries coming up along the "Golden Corridor," largely situated in the SSP's command in Central Gujarat. It has attracted investments worth Rs 75,000 crores for this purpose (for further details see Mehta 2005).

5   Merka's pastoralists belong to the *Rabari* and *Bharvad* communities and comprise about 17% of the village population. About 70 percent of them claim to be landless, making livestock their chief form of wealth and property. Together they own only about 7 percent of Merka's land (see Mehta 2005 for more details).

References

Alhassan, H. S. (2009). "Viewpoint – Butterflies vs. Hydropower: Reflections on Large Dams in Contemporary Africa," *Water Alternatives* 2(1): 148–160.

Berkes, F. (1989). *Common Property Resource: Ecology and Community-Based Sustainable Development*. London, Belhaven Press.

Berry, S. (1989). "Social Institutions and Access to Resources." *Africa* 59(1): 41–55.

—— (1987). *Property Rights and Rural Resource Management: The Case of Tree Crops in West Africa*. Boston, MA, Boston University, African Studies Centre.

Blaikie, P. and H. Brookfield (1987). *Land Degradation and Society*. London and New York, Routledge.

Bourdieu, P. (1977). *Outline of Theory of Practice* (translated from the French by Richard Nice). Cambridge, Cambridge University Press.

British Dam Society (1999). Press statement at the Reuters/IUCN press conference on the World Commision of Dams, London, July.

Bromley, D. and M. Cernea (1989). *The Management of Common Property Natural Resources: Some Conceptual and Operational Fallacies*. Washington, The World Bank.

Bryant, R. (1992). "Political Ecology. An Emerging Research Agenda in Third-World Studies." *Political Geography* 11(1): 12–36.

Cleaver, F. (1998). "Moral Ecological Rationality, Institutions and the Management of Common Property Resources." *Development and Change,* 31 (2): 361–383.

Escobar, A. (1996). "Constructing Nature: Elements for a Poststructural Political Ecology." In *Liberation Ecologies: Environment, Development, Social Movements*. R. Peet and M. Watts (eds). London and New York, Routledge: 46–68.

FAO (Food and Agriculture Organisation) (2003). "No Global Water Crisis – But Many Developing Countries will Face Water Scarcity," *FAO News Stories*, 12 Mar, Rome, available at www.fao.org/english/newsroom/news/2003/15254-en.html (accessed March 21, 2003).

Falkenmark, M. and T. Chapman (eds). (1989). *Comparative Hydrology. An Ecological Approach to Land and Water Resources*. Paris: UNESCO.

Falkenmark, M., and C. Widstrand (1992). "Population and Water Resources: A delicate Balance," Population Bulletin, Population Reference Bureau.

Fallkenmark, M., J. Lundqvist, and C. Widstrand. (1990). "Coping with Water Scarcity. Implications of Biomass Strategy for Communities and Policies." *Water Resources Development* 6(1): 29–43.

Fischer, W. (ed.). (1995). *Towards Sustainable Development: Struggling over India's Narmada River*. Armonk, M.E. Sharpe,

Forsyth, T. (2003). *Critical Political Ecology*. London: Routledge.

Goldsmith, E. and N. Hildyard (eds.). (1992). *The Social and Environmental Effects of Large Dams. Volume III. A Review of the Literature*, Wadebridge, Cornwall, UK, Wadebridge Ecological Centre.

Gujarat Ecology Commission, (1994). *Current Ecological Status of Kachchh*. Vadodara, GEC.

Hauge, W. and T. Ellingsen (1998). "Beyond Environmental Scarcity: Causal Pathways to Conflict." *Journal of Peace Research* (Special Issue on Environmental Conflict) 35(3): 299–317.

Herman, Edward and Noam Chomsky. (1994). *Manufacturing Consent: The Political Economy of the Mass Media*. London, Vintage.

Homer-Dixon, T. F. (1994). "Environmental Scarcities and Violent Conflict: Evidence from Cases." *International Security* 19(1): 5–40.

Kutch Development Forum (1993). *A Demand for Review of Water Allocation to Kutch. Recommendations, Observation, Extracts. From Report of Review Group on Narmada.* Bombay, KDF.

Li, T. M. (1996). "Images of Community: Discourse and Strategy in Property Relations." *Development and Change* 27: 501–527.

McCully, P. (1996). *Silenced Rivers: The Ecology and Politics of Large Dams*, London: Zed.

Mehta, L. (2004). *Financing Water for All: Behind the Border Convergence in Water Management*. IDS working paper 233. Brighton, Institute of Development Studies

—— (2005). *The Politics and Poetics of Water: The Naturalisation of Scarcity in Western India*. New Delhi, Orient Longman.

—— (2007). "Whose Scarcity? Whose Property? The Case of Water in Western India." *Land Use Policy* 24: 654–663.

Mehta, L. (ed.) (2010). *The Limits to Scarcity: Contesting the Politics of Allocation*. London: Earthscan.

Morse B. and T. Berger (1992). *Sardar Sarovar: Report of the Independent Team*. Ottawa, Futures International Inc.

Mosse, D. (1997). "The Symbolic Making of a Common Property Resource: History, Ecology and Locality in a Tank-Irrigated Landscape in South India." *Development and Change* 28: 467–504.

Murishwar, J. and W. Fernandes (1988). "Marginalisation, Coping Mechanisms and Long-term Solutions to Drought." *Social Action*, 38: 162–178.

Ohlsson, L. and A. Turton (2000). *The Turning of a Screw: Social Resource Scarcity as a Bottle-neck Adaptation to Water Scarcity*. Stockholm Water Front – Forum for Global Water Issues, No. 1. Stockholm: Stockholm International Water Institute (SIWI)

Ostrom, E. (1990). *Governing the Commons: The Evolution of Institutions for Collective Action*. New York, Cambridge University Press.

Peluso, N. and M. Watts (eds) (2001). *Violent Environments*. Ithaca, NY, Cornell University Press.

Peet, R. and M. Watts (eds.) (1996). *Liberation Ecologies: Environment, Development, Social Movements*. London and New York, Routledge.

Potanski, R. and W. M. Adams (1998). "Water Scarcity Property Regimes and Irrigation Management in Sonjo, Tanzania." *Journal of Development Studies* 34(4): 86–116.

Raj, P. A. (1991). *Facts: Sardar Sarovar Projects*. Gandhinagar, Narmada Nigam Limited.

Raju, K. C. B. (1995). "Strategies to Combat Drought in Kutch." Workshop on Strengthening of Community Participation in Disaster Reduction and Role of NGOs. 13–15 January 1995, New Delhi. Mandvi: Vivekanand Research and Training Institute.

Richards, P. (1989), "Agriculture as a Performance." In R. Chambers, A. Pacey and L.A. Thrupp (eds.). *Farmer First. Farmer Innovation and Agricultural Research.* London: Intermediate Technology, 39–42.

Robbins, P. (2004). *Political Ecology: A Critical Introduction.* Oxford: Blackwell Publishing.

Rosegrant, M. W., X. Cai, and S. A. Cline (2002). *World Water and Food to 2025: Dealing with Scarcity*, Washington: International Food Policy Research Institute (IFPRI).

Ross, A. (1996). "The Lonely Hour of Scarcity." *Capitalism, Nature, Socialism* 7(3): 3–26.

Roy, Arundhati (1999). *The Cost of Living.* New York: Modern Library.

Rushbrook Williams, L. F. (1958). *The Black Hills. Kutch in History and Legend: A Study in Indian Local Loyalties.* London, Weidenfeld and Nicolson.

Shiklomanov, I. A. (1998). *World Water Resources. A New Appraisal and Assessment for the 21st Century.* Paris, UNESCO. Available at http://unesdoc.unesco.org/images/0011/001126/112671eo.pdf

Shiva, V. (2002). *Water Wars: Privatization, Pollution and Profit.* Cambridge, MA, South End Press.

Sinclair, A. (1998). "A Statistical Analysis of Rainfall Data in Kutch." Mimeo. Brighton, Institute of Development Studies, University of Sussex.

Talati, J., M. Dinesh Kumar, and D. Shah. (2004) *Quenching the Thirst of Saurashtra and Kachchh Regions through Sardar Sarovar Project.* IWMI-Tata Water Policy Program, Annual Partners' Meet. Gujarat: International Water Management Institute (IWMI).

UN. 2003. *Water for People, Water for Life: UN World Water Development Report* (WWDR). UNESCO and Berghahn Books.

Vision 21, (2000). *Vision 21: A Shared Vision for Hygiene, Sanitation and Water Supply and A Framework for Action. Also Forming the Water for People Component of the World Water Vision.* Geneva, Water Supply and Sanitation Collaborative Council. Available online at http://www.wsscc.org/ (accessed March 2009).

Wade, R. (1988). *Village Republics: Economic Conditions for Collective Action in South India.* Cambridge, Cambridge University Press.

Winpenny, J. T. (n.d.). "Managing Water Scarcity for Water Security," prepared for FAO (Food and Agriculture Organisation) as part of an e-mail conference, available from: http://www.fao.org/ag/agl/aglw/webpub/scarcity.htm

Wolfe, S. and D. B. Brooks (2003). "Water Scarcity: An Alternative View and its Implications." *Natural Resources Forum*, 27(2): 99–107.

World Commission on Dams (2000). *Dams and Development. A New Framework for Decision Making. The Report of the World Commission on Dams.* London: Earthscan.

Xenos, N. (1987). "Liberalism and the Postulate of Scarcity." *Political Theory* 15(2): 225–243.

Yapa, L. (1995). "Building a Case against Economic Development." *Geo-journal* 35(2): 105–118.

—— (1996). "What Causes Poverty? A Postmodern View." *Annals of the Association of American Geographers* 86(4): 707–728.

Part VII

Biopolitics and political ecology: genes, transgenes, and genomics

18 Governing disorder: biopolitics and the molecularization of life

Bruce Braun

> When Spinoza says we do not know what a body can do, this is practically a war cry.
>
> (Deleuze 1990: 255)

Introduction

In what ways can it be said of the molecularization of life that it has made our biological existence a political concern in new ways? Nikolas Rose (2001) gives us one answer.[1] With advances in molecular biology, genetics and biochemistry, he argues, we have come to understand the body in terms of its genetic inheritance, with important implications for how we are governed, and the ways in which we govern ourselves (see also Novas and Rose 2000). Like his colleague Sarah Franklin, and the anthropologist Paul Rabinow (who uses slightly different terminology – biosociality – to understand similar practices of bodily self-regulation and management), Rose understands this to entail a shift within the biopolitical regimes of modernity, from political rationalities directed toward the management of risk at the level of populations, to the individual management of the genetic risks peculiar to one's own body, or what he calls "ethopolitics."[2]

Rose's answer has considerable merit. Unlike conservative reactions to technological change that are based on notions of an essential human nature, or that rely on a sharp distinction between the biological and the technological body, Rose understands bodies as composite entities, at once biological, technological and political. Hence, recent advances in genetics and biotechnology do not register a decisive break with a prior human essence; they partially constitute what it means to live our humanity today. To use familiar terms, they are part of the ongoing history that is our "species being," and open possibilities for new "forms of life" in the future (see also Thacker 2005). Like others working in this tradition, Rose also recognizes that technological advances in biomedicine have transformed social identities, given rise to new forms of political association, and opened new circuits of capital. Indeed, as numerous commentators have noted, in the wake of new reproductive technologies, stem cell research, and other biotechnological advances, there have emerged countless new forms of "genetic citizenship," by which individuals and groups have made their biological existence a matter of ethical concern and a basis for political action.[3]

In important respects analyses like those by Rose, Franklin and Rabinow have become the dominant story about how life has been brought into law and politics in the molecular age. But do such accounts fully exhaust how the relation between our biological existence and our political existence is now lived? While it is certainly true that molecular biology, genetics and biochemistry have spatialized our bodies in novel ways, and, in conjunction with shifts in governance, have transformed our ethical and political relations to them, is ethopolitics the only game in town? Perhaps better, for *whom* is the molecular age an *ethopolitical* age that is defined, and experienced, primarily as a matter of choice and the individual management of risk? Despite the appeal of Rose's powerful account of ethopolitics, there may be reasons to read it with caution, from his somewhat limited account of how the body has been "molecularized" and the political rationalities arrayed around it, to his complete erasure of sovereign power in favor of what he sees as pervasive forms of pastoral power.

In this chapter I call both into question. On the one hand, I will argue that Rose relies on a singular, and somewhat simplistic, account of what has transpired with the rise of molecular biology and genetics; namely, that the body has come to be figured in terms of a genetic code that belongs to the individual alone, at once its own property, and that which forms the basis for its "life." For Rose, the individual self and the genetic body coincide; the body is conceived as a bounded entity whose molecular existence is *internal* to it (albeit open to technological modification), and ethics and politics comes to focus on the relation between the somatic individual and its possible biotechnological futures. Yet this is hardly the only way that molecular biology knows bodies. Alongside the genomic body, and at times overlapping with it, can be found another, post-genomic body which is *also* understood at the molecular scale, but considered instead in terms of its *displacement* within wider molecular fields. From this perspective, bodies are understood less in terms of their intrinsic genetic essence – the fantasy of one's genetic code carried on a flash drive – and more in terms of global economy of exchange and circulation, where the body is "thrown" into a chaotic and unpredictable molecular world filled with emergent yet unspecifiable risks. Far from a stable molecular life internal to the bounded body, to be managed and potentially improved, this account gives us a *precarious* body immersed in what Barnard Vallat, Director General of the World Organization for Animal Health (OiE) has called the "great biological cauldron" of the twenty-first century, where biology is significant for its potential to vary, and where the future is less about "care of the self" than it is about imminent catastrophe.[4]

This molecularized body, I will argue, has become the site of very different political rationalities, gathered around the concept of "security," which find no place in Rose's ethopolitical account of the molecular age. Rather than accept a simple and singular account of the molecular age in which biopolitics morphs into ethopolitics, I will ask whether it may be necessary to trace the ways that biopolitics has merged with geopolitics, and the ways that the government of "life" has revealed itself to be intimately related to the exercise and extension of sovereign power. We are perhaps left not with opposed understandings of how our biological existence is related to our political existence – one that draws on notions of governmentality, and another that focuses on sovereignty – but with the task of understanding how the two are related.

From biopolitics to ethopolitics: intrinsic molecularization and the individual management of risk

> The question of what a body can become is an open question
> (Ansell-Pearson 1999: 13)

Let me turn first to Rose's account of ethopolitics. A key point of reference for Rose is the work of Michel Foucault (1977, 1979, 1991). As is well known, Foucault argued that in the seventeenth and eighteenth centuries political power came to interest itself less in decisions over life and death, or control over territory – the traditional concerns of the sovereign prince – and more with the management of "life," that is, with ordering and enhancing the vital or productive processes of human existence. While Foucault used the general term "biopower" to designate those new forms of power that took the capacities of bodies and conduct of individuals as their concern, he at times distinguished between two more specific forms. He used the term "anatomo-politics," for instance, for those disciplinary techniques that sought to maximize the body's forces and integrate them into efficient systems, such as through proper training, or through rationally organizing workplaces, armies and domestic economies. The term "biopolitics," on the other hand, he used to designate those political technologies that took the biological existence of the *nation* as their object, understood as a "population" imbued with mechanisms of life – birth, morbidity, mortality, longevity, vitality – and knowable in terms of statistical norms. Examples of the latter included public health, town planning, and hygienics, each of which conjoined state science (demography, vital statistics, administration) with forms of self-regulation to bring about the normalization of life processes.[5]

Whether or not this resulted in the replacement of sovereign power (power to *take* life or *let* live) by more dispersed forms of disciplinary or pastoral power (power to *make* live or *let* die), has been widely debated, but these debates seem to concern Rose less than the possibility that Foucault's account of biopolitics has become horribly outdated.[6] As Rose (2001: 1, 13) puts it, "the truth regimes of the life sciences have mutated" and with these changes "biopolitics has merged with ethopolitics." Rose's description of these mutations is worth quoting at length, since it forms the basis for all that follows:

> The body that twentieth-century medicine inherited from the nineteenth cen-
> tury was visualized via a clinical gaze, as it appeared in the hospital, on the
> dissection table and inscribed in the anatomical atlas. The body was a vital
> living system, or a system of systems. The skin enclosed a "natural" volume
> of functionally interconnected organs, tissues, functions, controls, feed-
> backs, reflexes, rhythms, circulations and so forth. This unified clinical body
> was located within a social body made up of extra-corporeal systems – of

environment, of culture – also conceptualized in terms of large scale-flows – of air, water, sewage, germs, contagion, familial influences, moral climates and the like. Eugenic strategies took their character from this way of linking the individual and the social body. The genetic body differs on all counts from this eugenic body. Most notably, it is conceived on a different scale. In the 1930s, biology came to visualize life phenomena at the submicroscopic region – between 10^{-6} and 10^{-7} cm. Life, that is to say, was molecularized. This molecularization was not merely a matter of the framing of explanations at the molecular level. Nor was it simply a matter of the use of artifacts fabricated at the molecular level. It was a reorganization of the gaze of the life sciences, their institutions, procedures, instruments, spaces of operation and forms of capitalization. . .life was imagined as sub-cellular processes and events, controlled by a genome which is neither diagram nor blueprint but a digital code written on the molecular structure of the chromosome. This is "the language of life" that contains "the digital instructions" that make us what we are.

(Rose 2001, 13–14)

In other words, when it comes to how we understand biological life today, molecular biology and genetics have replaced physics and chemistry, and with this shift, we have witnessed new ways of conceiving and acting upon bodies.

Later we may have reason to question whether such a sharp epochal shift has indeed occurred, and whether the body is "molecularized" today in such a singular fashion and with such singular effects. Here we need only note that Rose's position has become commonplace in the social sciences and humanities today. As the story is usually told, with such things as DNA diagnostic tools, automated gene-sequencing computers, and data-mining and gene-discovery software, we can at once encode, recode and decode biological materials, translating "wet" DNA (physical samples), for instance, into "dry" DNA (information), which can subsequently be manipulated and ultimately reassembled in new form (see Thacker 2005). For Thacker (2005) and Waldby (2000) this encoding, recoding and decoding has allowed for ever tighter links between biology and capital, as biological matter is translated into mobile and fluid networks of information that can be owned, bought and sold as intellectual property. For Rose, on the other hand, the most significant effects of this shift appear to be found in how we understand and govern bodies and their possibilities. Gene therapy treatments – the novel idea that we can actively transform the genetic material in a living being – and nanotechnologies – the construction of organic and non-organic objects, molecule by molecule – are seen to give us a biological body which is understood less in terms of fate and more in terms of the management and pre-emption of risks, even as something that can potentially be improved. In Rose's (2001: 16, 17) words:

Life now appears to be open to shaping and reshaping at the molecular level: by precisely calculated interventions that prevent something happening, alter the way something happens, [or] make something new happen in the cellular processes themselves.

The result is that our *ethical* relation to our bodies has changed, for dilemmas about what we are, what we are capable of, what we may hope for, now have a molecular form. Translated into the language of biopolitics, Rose argues that it is increasingly our cellular existence – "life itself" – and not just our conduct, which has become subject to what Foucault called "technologies of self."

Ethopolitics, then, is the name Rose gives to this new ethical-political relation to our bodies, which are now defined in terms of open-ended futures. But there is more to Rose's account than merely a shift in the target of political rationalities from the behavior of bodies to their actual make-up; for Rose ethopolitics also relates to crucial changes in the relation between the individual and the state. Rose develops this point in response to critics of biotechnology, for whom the molecularization of life is inescapably haunted by eugenics. With our new-found capacity to diagnose genetic conditions in embryos, for instance, we can now make choices about whether to continue a pregnancy, or to accept an embryo for implantation in IVF therapies, based upon the knowledge of future risks. For a number of critics this has raised the unsettling possibility of political rationalities directed toward eliminating "taints or weaknesses" in populations, based on some bodies being calculated to have less biological worth than others. This discomfort should come as little surprise; as we are all too aware from events in the twentieth century, biopolitical projects can just as readily invest in the life of the collective body through purging "defective" bodies as they can through improving, training, or selecting, "healthy" ones.[7]

It is partly in response to these anxieties that Rose spells out his account of a historical shift from a biopolitics of populations to an ethopolitics characterized by the individual management of the "somatic" self. While he readily agrees that political rationalities are still organized around "risks" to health, he claims that the nature of these political rationalities has changed in such a way that eugenics is no longer the threat it once was. Biopolitical practices in the past, he argues, were directed toward improving the *national* stock, and took two forms which contained the potential for eugenics: hygienics, which was concerned with maximizing the health and productive powers of the national body in the present; and the regulation of reproduction, which was concerned with improving the national stock by eliminating risks to its well-being in the future. These were matters of concern for state policy, as well as for individuals who understood their biological lives (and the lives of their children) in terms of an ethical responsibility to the national body, thus blurring the boundary between coercive and voluntary eugenics.

The present age, Rose argues, is markedly different. To begin, it is not at all apparent that we are still in an age where the state seeks to take charge of "the lives of each in the name of the destiny of all" (Rose 2001: 5). In other words, for Rose the idea that the state should coordinate and manage the affairs of all sectors of society – that it should attach importance to the "fitness" of the national body *en masse* – has fallen into disrepute, since the question of "fitness" is no longer framed in terms of a struggle between national populations, but instead posed in economic terms, such as the cost of days off from work that are caused by ill health. Hence, when it comes to national health, the state seeks to "enable" or "facilitate" the health of individuals, rather than govern bodies in any direct way. The difference between "old" eugenics and what some have today labeled "liberal" eugenics, then, can be seen as the difference between state-led programs that in the past sought to produce a particular population with particular traits and capabilities, and the ethical decisions of individuals in the present, who are exercising "choice" in reproductive matters. Although forms of pastoral power clearly shape these reproductive choices, the state remains neutral. For Rose, this is a crucial difference, and symptomatic of a larger shift whereby health is increasingly a matter of individual, rather than state, responsibility, and citizens are asked to take responsibility for securing their own wellbeing, through such things as purchasing private health insurance, being informed citizens, actively investigating health conditions, joining with others in support groups, contributing to lobby groups and seeking genetic counseling.

It is here, *at the intersection of the molecularization of life with the individualization of risk*, that Rose locates ethopolitics as the dominant biopolitical regime of the present. Within such a biopolitical order, he argues, individuals are presented with new ways of rendering their bodies to themselves in thought and language, making judgments about them, and ultimately acting upon them, whether these decisions are based on DNA samples from amniotic fluid, in the case of reproductive health, or susceptibility to Alzheimer's, due to the presence or absence of particular genes. Thus, the individual who "takes responsibility for her health" is at the same time the individual who thinks her body through its "genetic inheritance," an inheritance to be managed wisely or potentially improved. For Rose, this government of the genetic self is thus decidedly *not* about following general programs but about understanding and making wise choices about the risks that are peculiar to one's self. Risk becomes "individualized"; the individual becomes "intrinsically somatic"; and ethical practices "increasingly take the body as a key site for work on the self" (Rose 2001: 17).

Within the social sciences and humanities this formulation of the biopolitical present predominates, evident in a great deal of work on the social and cultural aspects of biomedicine and biotechnology. From anthropologists and sociologists, for example, we learn that the molecularization of life and the individualization of risk have given rise to new forms of identity and sociality around disease and risk.[8] Individuals are said to increasingly recognize the "self" as the bearer of this or that genetic risk, around which daily routines and future plans must be prudently organized. Likewise, researchers have begun to attend to the myriad of ways that our genetic lives are lived, and ethical decisions about "life itself" are made, within complex networks of activists, scientists, doctors, politicians and corporate interests that are clustered around particular "risks."[9] In many of these accounts the internet looms large, providing novel possibilities for the sharing of biomedical knowledge and life experience among lay advocates, scientists and clinicians, and for forging translocal communities around particular genetic identities.[10] These new deterritorialized "body-geographies" can be seen to challenge local cultures of health and local etiologies of disease, while also providing space for the proliferation of alternative body-knowledges, or for the emergence and organization of new demands on state and capital by individuals and collectives. For Rose and Novas (2004), such practices provide further evidence of the "making up" of the biological citizen from below, rather than the shaping of citizens by the disciplinary power of the state.

We should not take this to mean that power-relations are absent from ethopolitics. Indeed, one of the crucial questions to emerge from Rose's account is precisely what it means to "exercise choice" in the self-management of the body. What defines "choice" for the ethopolitical subject? And who *is* this subject who understands their body in such terms? Drawing upon Dean's discussion of the formation of neo-liberal subjects, we might begin with an initial observation that with the shift to private health insurance and away from the providential state we are in a sense compelled to be subjects who "make choices" about our health.[11] As Deborah Heath, Rayna Rapp and Karen-Sue Taussig (2004) put it: we are asked to be good genetic citizens, which is to say that we are *obliged* to wisely manage our lives through exercising choice. In the absence of other options for securing health, such as those provided by a providential state, *we must make our biological life our life's work.* But this presents us with a further range of problems. On the one hand we are faced with growing populations – undocumented workers, the working poor – who are excluded from this ethopolitical order; that is, those who are denied the political right to health, or who lack the resources that might enable them to "choose." For these subjects the biological self is a *precarious* entity – bare life, exposed to death – rather than an object for personal reconstruction. On the other hand we find that as soon as we look carefully at the social and medical field on which the "somatic" self exercises choice, we find it delimited by numerous parameters: not only traversed by countless forms of pastoral power – all those "professionals of vitality," counselors, therapists and ethicists, not to mention geneticists and physicians, who are there to guide our decisions – but shaped by what Catherine Waldby (2000) has called the production of "biovalue."

Indeed, if molecular biology and genetics have reconfigured the body in terms of information, and if the ethical care of the self occurs *within* this field of informatics, then the question arises of what *sort* of bioinformation is being produced, to what end, and for whom? It is no secret that the driving force behind bioinformatics today is finance capital, such that the future of any given field of research, the sequencing of this or that genome, or the data-mining of this or that genomic database, more often than not flourishes or perishes depending on stock values, and those stock values, in turn, are tied to the actual – or proclaimed – successes or failures of research. Moreover, research is most likely to occur if results can be transformed into products (genetic-based drugs or therapies, for instance), or if it can be mobilized as part of some product development pipeline. Not just anyone can participate in building this informational field, despite the organizing of advocacy groups and online medical communities. As Eugene Thacker (2005) notes, life at the molecular level is only knowable through complex and expensive apparatuses – electron microscopes, ultracentrifuges and x-ray diffraction – and through the expensive, computer-driven analysis of gene-banks. The development of cures and preventive practices is exorbitantly expensive and inaccessible to non-specialists. And the patenting of bio-information means that the right to use such information is constrained by property and law. Not only does this modern day form of enclosure mean that the field of "choice" is circumscribed, but bio-technology's high-capitalization and specialization means that immense challenges stand in the way of any sort of informed critique and public debate.

As Rose (2001: 20) notes, Foucault famously argued that "medical thought [has been] fully engaged in the philosophical status of man." If so, the philosophical status of the human is shaped today as much by the calculations of entrepreneurs as by the decisions of researchers, doctors or patients. Not only does this point out the bankruptcy of much of what travels as "bioethics" – a professional field which always seems to arrive too late, *after* biomedicine, biotechnology and finance capital have ushered in the future, and thus can act only to incorporate new biotechnological realities within law – it also suggests limits for the sort of generalized govern-mentality or self-management that Rose assumes defines our biological existence today, for it becomes impossible to reduce the biopolitical field to the actions of citizens alternately empowered or ensnared in webs of pastoral power.

This does not mean that Rose's account is without merit, or that it fails to cast light on meaningful questions about how individuals negotiate their biological lives in the molecular age. To the extent that the biological lives of affluent members of Western societies have come to be understood in terms of ethopolitics, where the molecularization of life coincides with the individualization of risk, the politics of "life itself" has increasingly come to turn on a set of ethical questions about what a body can do, and a set of political questions surrounding how the body's capacities can be increased, such as through the *recognition* of certain genetic conditions or the *establishment* of institutional forms and legal frameworks that might enable individuals to maximize their genetic potential. Indeed, although he doesn't put it in these terms, Rose's account brings us face to face with the question of democracy; for if "being" has neither fixed form nor determined end, and if what it means to live our humanity is the outcome of politics rather than something given in advance, then it would seem that our most pressing need is for a political order that *corresponds* to this new corporeal order, where the flourishing of life (and the technoscientific practices that facilitate it) is not determined by rates of return on investments, or constrained by law, property, and nation, but is open to an ontological and ethical play beyond the current integration of life and law.[12]

Biology, virtuality, security: extrinsic molecularization and the geopolitics of "life itself"

> Bodies-in-formation betray a virtual potential towards becoming dangerous and so our politics of security are progressively becoming a virtual security politics.
>
> (Dillon 2003: 531)

The concept of ethopolitics gives us a powerful way to think about ethical-political projects of self-formation in the molecular age. But is Rose's genomic body the *only* form that the molecularized body takes today? And is ethopolitics the only game in town? Or, stated differently, is political power in the molecular age primarily pastoral in nature?

Two aspects of Rose's account of mutations in the truth regimes of the life sciences bear further scrutiny. The first is his faith in epistemic shifts. For Rose, the genetic body of the twenty-first century "differs on all counts" from the clinical or eugenic body of the nineteenth century. But does it? Can such a bold proclamation of a new epoch be sustained? The second has to do with *how* Rose imagines that the body has been "molecularized." While it may be true that we now visualize life phenomena at the submicroscopic region, is the biological life of the body conceptualized only in terms of one's "genetic inheritance" and its technological improvement? Is the body really the bounded and autonomous entity that Rose makes it out to be, constituted only in terms of an *internal* genetic essence that contains its future within it?[13]

While it may be true that in industrialized liberal democracies this model of the "somatic" self holds sway, there is another dimension to the molecularization of life that has received far less attention. This has to do with the conceptualization of the body in terms of its *displacement* within wider molecular fields. That is, at the same time that molecular biology and genetics have given us a body known at the molecular scale, and thus made the physical mechanisms of "life" available to political and economic calculation in new ways, they have also, in conjunction with the sciences of immunology and virology, given us another way to conceive of our biological existence, no longer in terms of a self-contained body whose genetic inheritance is to be managed and improved, but in terms of a body embedded in a chaotic and unpredictable molecular world, a body understood in terms of a general economy of exchange and circulation, haunted by the specter of newly emerging or still unspecifiable risks.[14] For every story in the U.S. media that speaks breathlessly of advances in stem cell research and gene therapy, or that worry over the "post-human" futures these might usher into being, we find two or three other stories that speak ominously of migrating birds and backyard chickens, and that mix together Vietnamese peasants, influenza viruses and homeland security. This conjunction of biopolitics and geopolitics, of the molecularized body and the question of biosecurity, finds no place in Rose's ethopolitics, but merits equally close attention. By tracing its tangled threads, we may find ourselves faced with very different political rationalities, no longer framed in terms of the governmentalization of "life itself," but in terms of the extension of forms of sovereign power by which life is ever more tightly integrated with law.

Before turning to these, we should note that there is a long history of thinking the body in terms of exchange and circulation, well documented by historians and epidemiologists. During the plague epidemics of the fourteenth century, for instance, "public health" involved the quarantining of people and goods suspected of harboring infectious diseases. Urban renewal campaigns of the early twentieth century understood disease to spread through food, water and waste, and today the ongoing AIDS pandemic has brought renewed attention to the porosity of human bodies. There is an equally long history of public health linked explicitly to security. As Nicholas King (2003) notes, the Center for Disease Control (CDC) in the United

States began during the Second World War as an effort to investigate and control infections among soldiers, and to keep malaria from spreading to the armed forces from its "reservoir" in the civilian population of occupied countries.

From its inception, then, public health has taken the body to be a *geopolitical* body. Yet in important respects the present moment is more than bare repetition. Two developments hold particular significance. On the one hand we are witness to a set of geographical and historical transformations that have come to be discussed under the rubric of "globalization." These are commonly taken to include the liberalization of markets, the unprecedented mobility of capital and goods, the extension of global supply chains, the transformation of ecosystems and the growth of international travel and tourism. Within these accounts, the present moment is understood in terms of the collapse of space and time (or the "folding" of topological space-time). On the other hand, molecular biology, in conjunction with virology and immunology, has given us ways of conceptualizing bodies in terms of their *molecular geographies* – in terms of networks and pathways, movements and exchanges – with the sort of detail and complexity unimaginable in fourteenth-century Venice, or at the time of Typhoid Mary, or even during the influenza pandemic of 1918. In what follows I propose that one way we can understand the recent emergence of "biosecurity" as a political concern is to acknowledge that it corresponds to a particular way in which the molecularized body has been apprehended within globalization.[15] Stated in slightly different terms, we might propose that, in contrast to "ethopolitics" which names a form of neoliberal governmentality that comes into being at the intersection of the molecularization of life and the individualization of risk, "biosecurity" today names a set of political responses within globalization that take the *unpredictability* of molecular life – its "virtual potential" or "waywardness" – as their own justification and in such a way that "security" appears the only viable political response.

We can explore this proposition further by reference to the recent avian influenza scare. Highly pathogenic avian influenza (HPAI) is the name given by virologists to a number of influenza A viruses that cause high mortality and morbidity, both in animal populations (especially domestic poultry), and among infected humans. Of most concern in recent years has been a particular type of HPAI virus known as H5N1, one of a number of "emerging infectious diseases" that have received growing attention since the mid-1990s.

It is the virtual potential of viral life that has received the greatest attention. Today, molecular virologists tell us that viruses consist of genetic material encased in surface proteins that stick out from the viral envelope (see Figure 18.1). One type of proteins – Hemagglutinin (HA) proteins – determine how, and whether, a virus can penetrate human cells. Once in a cell the virus can replicate. Neuraminidase (NA) proteins, on the other hand, determine the exit strategy, i.e. whether the replicated viral matter can escape the cell to infect other cells. Viruses are named on the basis of these proteins. The designation H5N1, then, corresponds to the type of HA and NA proteins found in the virus. At the time of the avian flu scare, it was believed that among influenza A viruses there existed 16 HA

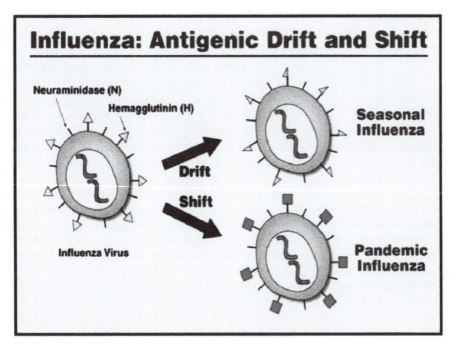

Figure 18.1 Influenza

subtypes and 9 NA subtypes. Of the 16 HA sub-types, all were thought to infect birds while only some were known to infect humans.

Except in rare cases, H5 viruses have not infected humans. It is even more rare for an H5 virus to be transmitted from one human to another. As we heard repeatedly, however, the problem with influenza viruses was that they mutate at an alarming rate, caused by changes to the viral genome, which is segmented into eight separate RNA molecules. Mutations were thought to occur in two ways, genetic drift and genetic shift, where the former referred to a slow process of mutation that occurs as "errors" are introduced into replicated genetic sequences, and the latter referred to the much more rapid reassortment of genes that is thought to occur when different viruses come into contact with each other, such as when an H5N1 virus and a H3N2 virus are found in the same host. Changes to the genome are of great significance, virologists believe, because they can result in changes to the HA and NA proteins.[16]

The mutation of HPAI viruses are said to have two potential outcomes of great significance for humans. Genetic reassortment, and accompanying changes to HA proteins, can potentially change the *transmissibility* of a virus, giving it the capacity to infect new hosts. Mutation can also change the *virulence* of the virus, and thus affect morbidity and mortality rates in the infected population. The widely reported worry was that with the right kind of mutation, H5N1 could mutate into a virulent form that was transmissible between humans, plunging the world into a catastrophic global pandemic.

For our purposes, what is of interest in the story of H5N1 is less the accuracy of biomolecular immunology's account of viral mutation, or the probability of whether a pandemic will occur (for reasons that will soon become clear, the latter is a matter of speculation rather than prediction) than *how* this understanding of molecular life, and the discourse of "emerging infectious diseases" that emerged with it, has transformed our understanding of our own biological existence and given rise to new forms of political rationality. We can begin with the simple observation that with increased focus on such things as avian flu, the ebola virus, mad cow disease, and other zoonotic diseases, molecular life has been recoded as inherently unpredictable, as always carrying within it the potential to vary. Human life, in turn, is understood to be thrown into, or exposed to, this molecular world of chaotic change. Far from a self-contained body with a clear genetic code – the fantasy of the essential "self" stored as information on a flash drive – what we find in the medical and political discourse of "emerging infectious diseases" is a body that is radically open to the world, thrown into the flux of an inherently mutable molecular life where reassortment is not what we control, but what we fear.

This post-genomic world is not understood in terms of one's genetic inheritance – nor is it primarily about "care of the self" or "genetic citizenship" – it is instead understood in terms of a global economy of circulation and exchange that at once precedes and transcends the individual body. By this account biomolecular life is not governed by fixed taxonomies or known in terms of genetic essences; it is instead a dynamic world characterized by novel combinations, where entities jump between bodies and cross between species, and where "life itself" continuously confronts us with the new and the unknown. The philosopher Brian Massumi (1993: 11) has succinctly captured the temporal and affective dimensions of this epistemic shift:

> Viral or environmental. . .these faceless, unseen and unseeable enemies operate on an inhuman scale. The enemy is not simply indefinite (masked or at a hidden location). In the infinity of its here-and-to-come, it is elsewhere, by nature. It is humanly ungraspable. It exists in a different dimension of space from the human, and in a different dimension of time;. . .The pertinent enemy question is not who, where, when, or even what. The enemy is a what not; an unspecifiable may-come-to-pass, in another dimension. In a word, the enemy is virtual.

For Massumi the "virtual" has a precise meaning, taken from Henri Bergson and Gilles Deleuze. It refers not to a non-existent, or immaterial entity, as in popular usage, but to a potentiality that is present in every situation. Unlike the "possible," which is opposed to the real, the virtual *is* real, which is to say that it exists as a non-actualized potential in the present. It is at once immaterial yet real, abstract yet concrete, containing possible futures that are already with us, but which remain ungraspable.

To relate this to H5N1, we might say that the virtual has to do with all the potential mutations that could occur, given what the virus presently is, and the heterogeneous associations into which it may enter. This molecular future is immanent in the present although it cannot be known in advance.[17]

What is the significance of recoding molecular life in terms of its indeterminacy? Most immediately, it transforms our relation to the future, which is, in a sense, already with us. This biological future is radically open, of the same nature as a throw of the dice, full of surprises and unexpected forms (Davies 2007). Yet, as evident in the discourse of "emerging infectious diseases," this future can also be defined in terms of the imminence of a generalized, yet nondescript catastrophe. As bird flu "expert" Michael Osterholm is fond of putting it in relation to the next pandemic: "it's not if, but when."[18] The present is populated by unknown and unknowable risks; we don't know what comes next, but it could be bad.

It is not difficult to see how the indeterminacy of molecular life can be articulated with fear and dread. Nor does it take much imagination to see how an understanding of globalization, which frames the present in terms of the collapse of time and space, might further augment this sense of biological terror. As food networks become increasingly complex and global, this story goes, the molecular geographies that constitute our biological existence are changing in both speed and scale. Air travel is likewise said to give "biological emergence" new urgency, as planes cross between continents far quicker than the incubation period for many pathogens. To borrow language from the medical anthropologist Nicholas King (2002b), the lesson that we were all asked to learn from the SARS crisis of 2003 was that networks were dangerous: through a specific configuration of live animal markets, migration and air travel, the biological existence of people in Singapore and Toronto had become intimately connected in "real time" to the lives of wild bats and civet cats in China.[19] The articulation of biology-as-virtuality with globalization, then, is said to raise the level of urgency and uncertainty; in this fearsome new world of global networks dangerous fragments circulate and recombine in novel ways, threatening our bodies and identities.

The rhizomatic movements of animals only added to the problem. Not only did molecular life "shift" and "drift" and transportation technologies and global trade collapse space and time; birds flew (see Figure 18.2). As was widely reported in western media, it was widely suspected that at least some wild birds had the capacity to carry HPAI without showing symptoms.[20] To use the familiar epidemiological metaphor, wild birds composed a "silent reservoir" of viruses – a faceless, unseen and unseeable enemy – where the distinction between friend and enemy was rendered indistinct. Indeed, the metaphor of "reservoir" was too static for this world of emergence; because birds migrated, they formed unpredictable reservoirs that dispersed and moved about: now in China, then in Turkey, now mixing with these flocks, now with those, forming amorphous transnational networks that respected no borders, and that were visible only in their effects.

It is not difficult to see how this view of molecular life could be taken up in a political register. Although he did not have birds in mind, Donald Rumsfeld in 2003 captured brilliantly what was at stake in a world of virtual risks. As he put it in the context of the "war on terror":

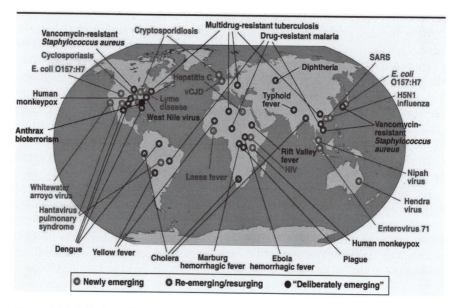

Figure 18.2 Infectious diseases

> There are known knowns. These are things we know that we know. There are known unknowns. That is to say, there are things that we know we don't know. But there are also unknown unknowns. These are things we don't know we don't know.[21]

In a similar way biology-as-virtuality comes to matter – politically and economically – in terms of emergence and its unpredictable spatio-temporalities; that is, in terms of those biological "unknown unknowns" that in an age of globalization could appear from anywhere, and threaten to bring about catastrophic effects, a point driven home to Congress in 2004 by Anthony Fauci, director of NIAID, who presented the world in terms of a complex cartography of "emergence" (see Figure 18.3).[22]

In important respects "biosecurity" today names an answer to the problem of the mutability and unpredictability of biological life within a political-economic order that is premised upon global economic integration. But biosecurity in practice may be more difficult than it may appear, for how does one bring the "unspecifiable" future-to-come within the realm of economic and political calculation? By definition the virtual is incalculable. No algorithm can exhaust its possibilities. Nor can it be incorporated into the probabilistic calculations of insurance. As Melinda Cooper (2006) notes, when biology comes to be known in terms of "emergence" the future can only be "speculative" and political calculation must become "future-invocative," actively intervening *within* the disorder of biological life in order to produce a desired end.

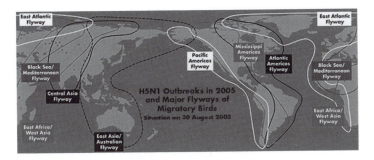

Figure 18.3 Cartography of emergence

Today, the answer to the problem of "unknown unknowns" takes two distinct forms: the speculative act of *pre-emption*, which takes as its target potential rather than actual risks, and the political technology of *preparedness* (or "vital systems security"), which seeks to order life in such a way as to make it resilient in the face of unexpected perturbations (see Lakoff 2007). In terms of the former, Cooper (2006) argues that these "speculative biological futures" have become an important basis for the integration of biomedicine and defense in the United States, including the re-emergence of biological weapons research.[23] What interests me here, however, are the ways in which such future-invocative acts of biological pre-emption are played out globally in a geopolitical register, in ways that extend forms of sovereign power. We can trace this across a variety of practices, beginning with the multiplication and expansion of surveillance networks, which in recent years have increased in number, scale, and complexity. These include "early warning" systems whose objective is to identify and contain "outbreaks" before they threaten the security of either the state or commerce. Many of these are data-mining operations. The Global Public Health Intelligence Network (GPHIN), for instance, is an Internet-based multilingual early-warning tool that continuously scans global media sources – news wires, web sites, blogs – for information about disease outbreaks. ProMED is similar, but more extensive, adding reports of diseases among farm animals and plants to GPHIN's focus on humans. Reports are often contributed by ProMED-mail subscribers, who act as informal "eyes and ears" for the network. The goal of these surveillance networks is nothing less han an unlimited, unending examination of global populations, tied, in turn, to rapid response teams administered by the WHO, which arrive at "hot spots" within 24 hours to do on-the-spot investigations, confirm diagnoses, help with patient management and ultimately contain any outbreak (see also King 2002a). These teams carry with them sophisticated field offices with "robust" information technology and communications set-ups (satellite telephones, radio communications, and field video-conferencing capacity); and, much like the U.S.'s global network of military bases, have pre-positioned specialized protective equipment and medical supplies located at strategic sites around the world.

All of the above are tied to emerging laboratory networks that are linked in real time. These laboratory networks – like the OiE's OFFLU and the WHO's Global Influenza Program – are the diagnostic and research arms of this integrated system, with collaborating centers that circle the globe. At least in principle, these collaborating centers agree to share knowledge, disseminate strains, facilitate the movement of reagents needed for laboratory research, develop databases of virus genomes through gene sequencing and so on.[24]

Such networks are not necessarily bad in and of themselves; they must be evaluated in terms of their effects. The first thing to note is that these networks are increasingly global in reach. As Nicholas King (2002a) comments, we are long past the days of nationally bounded surveillance systems whose goal was to monitor and protect the population of any particular state. What we are witnessing instead are familiar techniques of medical surveillance multiplied globally, where the monitoring of individual bodies in specific places is augmented by the surveillance of the global population in the de-territorialized space of informatics, databases and the Internet. Today these surveillance networks are being extended to animal populations, including wildlife, as animals are reclassified as "biohazards," both to each other, and to humans. The US Department of the Interior, for instance, began sampling migrating birds in the Pacific Flyway, adding to the growing influenza gene bank maintained by the NIH, and with the US Departments of Agriculture, and Health and Human Services, began an interagency strategic plan for the early detection of HPAI. The European Union did the same, while the UN's FAO, in conjunction with the USGS, unveiled an ambitious program that would fit wild birds with tiny backpacks carrying communication technologies linked to a system of radio beacons and satellites, in order to collect and disseminate real time migration data to ecologists, virologists and epidemiologists around the world.[25] While the image of birds carrying backpacks is somewhat comical, it represents both the capacity and the desire to extend the unending examination of global populations across the animal kingdom in order to govern the globe as a single, integrated biological system from which risks might emerge.

Ultimately, these surveillance systems seek to manage a set of spatio-temporal problems. On the one hand, they are about early detection and rapid response, continuously striving to reduce the time between detection, diagnosis and action in order to contain outbreaks, and to accelerate the production of antivirals and vaccines needed to protect more distant populations. Analogies to fighting forest fires abound. At another level they are about anticipating the future through the development of immense gene banks of influenza viruses – more than 900 at last count – which can be quickly mined for relevant data by corporations, research labs and state agencies, in the race to discover and patent pharmaceutical solutions. In important respects, then, these networks presume that the answer to "biology-as-virtuality" is technology – better surveillance, better laboratories, better vaccines – and their advocates frame "emergence" as a logistical problem that demands a technological answer, rather than an existential problem that requires a philosophical response, or a social or an economic problem that demands a political solution.

Perhaps most important for my purposes, these networks involve efforts by states to act extraterritorially. This *geopolitical* dimension was made explicit in the August 2003 statement of Tommy Thompson (2003), the former director of Heath and Human Services (HHS) in the United States. Although it begins by framing health security as a "humanitarian" concern – about "America's mission of compassion abroad" – it became clear that "health security" was to be about something quite different:

> As secretary of Health and Human Services it is my privilege to run a department that performs a critical role in America's mission of compassion abroad. *Public health knows no borders and no politics.* In recent memory alone, we have seen AIDS leap from Africa into our own cities; we have seen severe acute respiratory syndrome (SARS) spread with shocking rapidity from southern China to North America; we have seen the West Nile virus somehow cross the Atlantic and begin a slow spread across our continent; and we have seen that a key to controlling tuberculosis in the United States is controlling it in potential visitors to and from abroad.

It should come as no great surprise that even in its post-Westphalian manifestation, public health remains a geopolitical exercise concerned with the sanctity of borders, dangerous migrations and foreign risks. What *has* changed under the regime of "biosecurity" is the *geography* of health security, for in an age of globalization, it is not enough to protect borders, the fight must be taken "over there," before it "reaches here." Like the war on terror, amorphous viral networks required a global strategy of pre-emption. And for such a strategy America needed allies: other countries, and international organizations such as the WHO, FAO and OiE.

> Indispensable to our public health efforts, then, is the cooperation, leadership, and engagement of our partner nations. The United States can lead and contribute to the cause of global health, but cannot accomplish its mission alone. A prime example of our cooperation with fellow nations was seen in our response to the SARS epidemic. To fight this disease, U.S. health officials cooperated with and worked in places like China, Singapore, Thailand, Taiwan, and Vietnam. *We swiftly undertook several measures designed to turn the tide and defeat the epidemic before it became a serious threat on U.S. soil.*
>   (Thompson 2003, emphasis added)

As the CDC put it in 2000, in an age of global networks, "it was far more effective to help other countries control or prevent dangerous diseases at their source than try to prevent their importation" (quoted in King 2002a).

Thompson's comments remind us that "security" – even in a biological or medical register – is a geopolitical discourse that simultaneously names the enemy and that which is to be protected from it. With this in mind we may wish to modify our definition of biosecurity further, as the term given to a set of political technologies that seek to *govern biological disorder* in the name of a *particular community*, through acts that are *extraterritorial*. Or to say this differently, biosecurity under the auspices of the CDC and HHS retains the ideal of territoriality while simultaneously seizing on de-territorialization as the solution (King 2002a). Indeed, US public health policy, much like its foreign policy, has abandoned isolation in favor of (forced) integration, premised on the idea that the world can be split into a "functioning core" of liberal peace, and a "non-integrated gap" within which emerging threats must be suppressed.[26] Biosecurity is often framed in much the same way, as about *penetrating into* this gap, and *re-organizing it internally* so as to minimize risk.[27] Indeed, one of most celebrated aspects of surveillance networks such as GPHIN is precisely that it can reach into the sovereign space of other states, gathering biological and health information from a variety of sources, often informal or unofficial, and thereby bypass "uncooperative" states that might otherwise not wish to share such information.[28]

While these interventions are frequently posed in terms of "enlightened self interest" – the idea that by acting extraterritorially to achieve security at home, we benefit others too – it is crucial to attend carefully to what these "actions abroad" entail. Most visible in the news media has been the "culling" of bird populations, often resisted by local communities who have much to lose, but encouraged through promises of compensation. Less visible has been the collection of viral genomes from countries like Vietnam and Indonesia by institutions like the CDC, resulting in fierce struggles over property rights. Even less visible have been attempts to internally reorganize social, cultural and biological practices, what the development industry likes to call "capacity building." USAID, for instance, has begun to put money into improving states' diagnostic capacities and the integration of local laboratories into global networks.[29] Money has also flooded into training professional or paraprofessional workers charged with "modernizing" agricultural practices by reordering village spaces and introducing "biosecurity" practices into village life – essentially "investing in life through and through" at the village scale. The WHO, for instance, printed "biosecurity" guides for health workers and development NGOs that outlined specific spatial practices targeting so-called "sector 3" and "sector 4" livestock, and which arranged different forms of human-animal association in terms of lowest to highest biosecurity. These efforts to shrink the "non-integrated gap" through "harmonizing" practices raises countless questions, from the further integration of law with life, to the displacement of local epistemologies and local etiologies of disease, to the social, cultural and economic displacements that accompany new husbandry practices.

Not surprisingly, such attempts to enclose the biological life of villages turn on distinctions between the normal and the pathological, troped as a distinction between the "modern" and the "primitive." Indeed, in recent years nothing has signified "backward" more than the image of peasant children playing with domestic birds, whether alive or dead, the surest sign of the "pre-modernity" and "naivety" of the Third-World villager, whose practices are ultimately a threat to us all. Indeed, newspapers in the United States have been full of stories about the owners of fighting cocks in Indonesia sucking blood from wounded birds and the Vietnamese delicacy of blood pudding, while the Turkish press has focused on the husbandry of Kurdish peasants, who yet again threatened the nation, this time through their improper relations with animals.

Conclusion

If "security" is a political discourse that justifies new forms of sovereign power by placing the actions of the state "outside" politics, then biosecurity risks doing much the same, justifying a continuous *state of emergency* at the level of political life by reference to a continuous *state of emergence* at the level of molecular life. We might conclude, then, that biosecurity names much more than a set of political technologies whose purpose is to govern the disorder of biological life; it increasingly names a global project that seeks to achieve *certain* biomolecular futures by pre-empting others, and does so in part by reconfiguring in other places relations between people, and between people and their animals. Biosecurity weds biopolitics with geopolitics.

We are perhaps now in a position to bring the two halves of this chapter together. In what ways can it be said of the molecularization of life that it has made our biological existence a political concern in new ways? For Nikolas Rose, the molecularization of life has brought us to a new moment in the history of biopolitics, one in which bodies are understood in terms of their "genetic inheritance," the management of risk is individualized, and the make-up of our bodies, and not just their conduct, has become the subject of technologies of self. In this ethopolitical regime biopolitics is understood in terms of governmentality, and politics takes as its concern the recognition of genetic conditions and the mobilizing of resources in their name.

But this is not the only way in which the molecularization of life has been apprehended. If we attend to the global biopolitics of biosecurity we find a quite different relation between the biological and the political. On the one hand, the "genetic inheritance" of the "somatic" self comes to be replaced by "precarious" bodies inhabiting "virtual" biologies. On the other hand, forms of pastoral power recede while new forms of sovereign power appear. But how are we to understand the relation between the two? At the very least, we must see Rose's ethopolitics as something more particular and less universal, as perhaps a form of biopolitics within globalization that is specific to the zone of "liberal peace" in the affluent spaces of the West. But more important, we must ask whether the conditions of possibility for ethopolitics – for secure bodies that are open to "improvement" – include the extension of sovereign power elsewhere in the name of biological security. It is not just that the global South is taken to lie outside the technoscientific and cultural networks that compose the ethopolitical in Rose's account, but that arguably biological existence *there* is increasingly subject to projects that seek to pre-empt risk through new forms of sovereign power. We are faced with the troubling thought that in the molecular age, what appears to us in terms of an ethics of "care of self" and as a pressing problem of democracy, may appear to others as yet another expression of empire.

Acknowledgments

This essay is a revised version of an article that first appeared in *Cultural Geographies*. It has benefited from the helpful comments of audience members at the annual Cultural Geographies lecture, held at the Annual Meetings of the Association of American Geographers in Chicago, as well as comments received at the School of Anthropology, Geography and Environmental Studies at the University of Melbourne, the School of Philosophy at the University of New South Wales, and the PhD Program in Earth and Environmental Sciences at the Graduate Center, City University of New York.

Notes

1   These arguments are developed further in Rose (2007).
2   See Franklin (2006), Franklin *et al.* (2000) and Rabinow (1997).
3   On genetic citizenship, see Heath *et al.* (2004), and Rose and Novas (2004).
4   Vallet made his comments at the International Symposium on Emerging Zoonoses, Atlanta, 22–24 March, 2006.
5   This making a political object of "life" also provided means of resistance to techniques of power, often through the language of "rights." The "right" to a healthy body, for instance, emerged *as a right* in conjunction with a biopolitical regime that took the health of populations as their concern. That the language of rights is *internal* to biopolitics, rather than modifying or contesting it from the outside, is certainly one of the most far-reaching conclusions that emanates from Giorgio Agamben's understanding of sovereign power. See Agamben (1998).
6   Agamben (1998) has famously argued that the production of the biopolitical body was the "original activity of sovereign power," and locates this politicization of biological life as far back as Aristotle's distinction between *zoe* (biological life) and *bios* (political life, or a qualified way of life). In Foucault's work the relation between biopower and sovereign power was never clearly articulated, existing as a sort of "vanishing point" in his writing. Catherine Mills (2007) notes that in *History of Sexuality*, Foucault suggests that sovereignty is replaced by biopower, while in his essay "Governmentality," he suggests that discipline, government and sovereignty coincide historically.
7   Habermas (2003) sets out the problem in somewhat different terms, as contradicting the Enlightenment ideal of autonomy, and thus the ethical self-understanding of the species. With pre-implantation genetic screening, the child *has determined for it* its biological future, with the risk that "we may no longer see ourselves as ethically free and morally equal beings guided by norms and reason" (p. 41).
8   For examples, see Heath *et al.* (2004: 152–156).
9   See in particular Rose and Novas (2004).
10   See Parr (2002), Heath *et al.* (2004). Indeed, so important has the Internet become for the politics of health that *how* these virtual spaces are constructed and negotiated by different individuals and social groups, from accreditation organizations like URAC to individuals and social groups with differing ability to negotiate what is often a bewildering array of information, is of growing concern.
11   For Dean (1999), neoliberal governmentality rests on the production of a particular kind of subject who understands themselves in terms of making (wise) choices, and thus in terms of individual responsibility.
12   For more on this last point, see Mills (2007).
13   Within the terms of Rose's ontology, "subjects" are relational, but the "genetic" body *becomes so* only in contact with biotechnology.

14 Post-genomic microbiology not only suggests that there are much longer "sequential chains" involved in genetic events, but, as Melinda Cooper (2006: 117) notes, it has increasingly posited a "co-evolution" of humans and microbes, such that "We are literally born of ancient alliances between bacteria and our own cells; microbes are inside us, in our history, but are also implicated in the continuing evolution of all forms of life on earth."

15 Bingham and Hinchliffe (2008) identify three different ways that "biosecurity" has entered the political agenda: in terms of invasive species, foodborne illnesses and infectious diseases. Each of these names a different set of material conditions, and ethical and political practices. My focus on infectious diseases – and the political rationalities organized around it – is meant to call attention to emergent forms of sovereign power.

16 Indeed, officials from the WHO believe that there are multiple "strains" of H5N1 currently in circulation, with each strain containing HA and NA proteins with slightly different amino-acid sequences.

17 For more on the virtual, see Massumi (2002).

18 Quoted in "US bird flu scenario eyed," *Associated Press* (21 Sept 2005). Accessed at http://www.cbsnews.com/stories/2005/09/21/health/main870945.shtml, February 2006.

19 See also the collection of essays in Ali and Keil (2008).

20 This was a highly contested claim, criticized by some for deflecting attention away from industrial agriculture.

21 Rumsfeld first made these comments at a Defense Department briefing on Feb 12, 2002.

22 NIAID is the acronym for the National Institute of Allergies and Infectious Diseases, a branch of the NIH.

23 See also Zeese (2006).

24 These networks have been limited by conflicts over intellectual property, with the USA, and the CDC in particular, receiving considerable blame for failing to fully share information. Other countries – Indonesia in particular – have worried about neocolonial forms of bioprospecting and genetic enclosure around infectious disease research.

25 The FAO-USGS program began during summer 2006 with wild swans moving between Mongolia and Eurasia. See "Satellites help scientists track migratory birds" , FAO Newsroom. http://www.fao.org/newsroom/en/news/2006/1000388/index.html, accessed 10 September 2006.

26 See Barnett (2003). For a critical response see Roberts *et al.* (2003).

27 As explained by the OiE's Bernard Vallat: "One country with weaknesses in veterinary services [is] a threat to all," while capacity building is an "international public good." Comments presented at the 2006 International Symposium on Emerging Zoonoses, Atlanta, 22–24 March 2006.

28 The goal of "reaching into" the sovereign space of the nation-state, or "improving" their reporting to global institutions is the explicit aim of the OiE's WAHIS Web application. As Bernard Vallat (2005) explains,

> The active search and verification procedure for unofficial information from various sources that was introduced in 2002 has become more and more effective each year. Its results have improved the exhaustiveness of the OIE's information in general, and the credibility of official information from certain Member Countries in particular.

29 Interview with Scott Dowell (CDC), 2006 International Symposium on Emerging Zoonoses, Atlanta, 24 March 2006.

References

Agamben, G. 1998: *Homo Sacer: Sovereign Power and Bare Life*, translated by D. Heller-Roazen (Stanford, CA: Stanford University Press).

Ali, H. and Keil, R., eds., 2008: *Networked Disease* (Oxford: Blackwell).

Ansell-Pearson, K., 1999: *Germinal Life: The Difference and Repetition of Deleuze* (London: Routledge).

Barnett, T., 2003: "The 'Core' and the 'Gap'," *Providence Journal-Bulletin*, 7 November.

Bingham, N. and Hinchliffe, S., 2008: "Mapping the Multiplicities of Biosecurity," in A. Lakoff, and S.Collier, eds., *Biosecurity Interventions* (New York: Columbia University Press).

Cooper, M., 2006: "Pre-empting Emergence: The Biological Turn in the War on Terror," *Theory, Culture & Society* 23, 115–137.

Davies, G., 2007: "The Funny Business of Biotechnology: Better Living Through Comedy," *Geoforum*, 38, 221–223.

Dean, M. 1999: *Governmentality: Power and Rule in Modern Society* (Thousand Oaks, CA: Sage).

Deleuze, G., 1990: *Expressionism in Philosophy: Spinoza* (New York: Zone Books).

Dillon, M., 2003: "Virtual Security: A Life Science of (Dis)order," *Millennium: Journal of International Studies* 32, 531–558.

Foucault, M. 1977: *Discipline and Punish: The Birth of the Prison* (New York: Vintage Books).

Foucault, M. 1979: *The History of Sexuality*, Volume 1 (New York: Pantheon Books)

Foucault, M. 1991: "Governmentality," in G. Burchell, C. Gordon and P. Miller, eds., *The Foucault Effect: Studies in Governmentality* (Chicago: University of Chicago Press), 87–104.

Franklin, S., 2006: *Born and Made: An Ethnography of Pre-Implantation Genetic Diagnosis* (Princeton: Princeton University Press).

Franklin, S. Lury, C. and Stacey, J., 2000: *Global Nature, Global Culture* (London: Sage Publications 2000).

Habermas, J., 2003: *The Future of Human Nature* (Cambridge, Polity Press).

Heath, D., Rapp, R. and Taussig, K., 2004: "Genetic Citizenship" in D. Nugent and J. Vincent, eds, *A Companion to the Anthropology of Politics* (Malden, MA: Blackwell), 152–167.

King, N., 2002a: "Security, Disease, Commerce: Ideologies of Postcolonial Global Health" *Social Studies of Science* 32, 763–789.

King, N., 2002b: "Dangerous Fragments" *Grey Room* 7, 72–81. See also the collection of essays in A. Harris and R. Keil, eds. *Networked Disease* (Oxford, Blackwell, 2008).

King, N. 2003: "The Influence of Anxiety: September 11, Bioterrorism, and American Public Health," *Journal of the History of Medicine*, 58, 433–441.

Lakoff, A., 2007: "Preparing for the Next Emergency," *Public Culture,* 19, 247–771.

Massumi, B., 1993: *The Politics of Everyday Fear* (Minneapolis, MN: University of Minnesota Press)

Massumi, B., 2002: *Parables for the Virtual: Movement, Affect, Sensation* (Durham NC: Duke University Press).

Mills, C. 2007: "Biopolitics, Liberal Eugenics and Nihilism," in S. DeCaroli and M. Calarco, eds., *Sovereignty and Life: Essays on the Work of Giorgio Agamben* (Stanford, CA: Stanford University Press, 2007), 180–202.

Novas, C. and Rose, N., 2000: "Genetic Risk and the Birth of the Somatic Individual," *Economy and Society* 29, 485–513.

Parr, H. 2002: "New Body-Geographies: The Embodied Spaces of Health and Medical Information on the Internet" *Environment and Planning D: Society and Space* 20, 73–95.

Rabinow, P., 1997: *Essays in the Anthropology of Reason* (Princeton, NJ: Princeton University Press).

Roberts, S., Secor, A. and Sparke, M., 2003: "Neoliberal Geopolitics" *Antipode* 35, 886–897.

Rose, N., 2001: "The Politics of Life Itself," *Theory, Culture and Society* 18, 1–30.

Rose, N., 2007: *The Politics of Life Itself: Biomedicine, Power, and Subjectivity in the Twenty-First Century* (Princeton, NJ: Princeton University Press).

Rose, N. and Novas, C., 2004: "Biological Citizenship," in A. Ong and S. Collier, eds., *Global Assemblages: Technology, Politics, and Ethics as Anthropological Problems* (Malden, MA: Blackwell), 439–463.

Thacker, E. 2001: "Diversity.com/Population.gov." Walker Art Center Web Gallery, http://www.walkerart.org/archive/5/AF73755568388DC9616B. Accessed 3/2/2006.

Thacker, E., 2005: *The Global Genome: Biotechnology, Politics and Culture* (Cambridge, MA: MIT Press).

Thompson, T., 2003: "Public Health Knows no Borders," *U.S. Foreign Policy Agenda*, 8. http://usinfo.state.gov/journals/itps/0803/ijpe/pj81thompson.htm. Accessed 19 Jan. 2006.

Vallat, B., 2005: "Entering a New Era: The Birth of the WAHIS Web Application," OiE, December 2005 (http://www.oie.int/eng/Edito/en_edito_mai06.htm, last accessed 10 September 2006).

Waldby, C. 2000: *The Visible Human Project: Informatic Bodies and Posthuman Medicine* (New York: Routledge).

Zeese, Z., 2006: "Return of the Petri Dish Warriors," *Counterpunch*, 1 June.

19 Transnational transgenes: the political ecology of maize in Mexico

Joel Wainwright and Kristin L. Mercer

Introduction

Few environmental issues today are as hotly debated as those surrounding genetically-modified (or "GM") food. Although genetic modification remains limited to major crops (particularly cotton, soy, and maize), the total area in the world planted with GM seeds has been increasing steadily since the mid-1990s. By 2007, ~7 percent of the world's acreage (in 22 countries) was planted with GM seeds (BIO 2006; Hindo 2007), a percentage that continues to increase. While the USA alone accounts for the majority of the world's GM agriculture (measured by area), other countries are reducing barriers to planting GM crops, including China, India, and Brazil (Miller 2006; Hindo 2007). For many environmentalists, the rapid spread of GM agriculture is a sign of a great failure, a sweeping loss of territory in a global struggle for a more organic, sustainable way of life.

At the forefront of this global trend is the humble maize plant (also known as corn). Most of the maize grown in the US today is transgenic, and the many products from its harvest – animal feed, corn starch, corn oil, high-fructose syrup – are so widely distributed throughout the US food supply that, if you live in the US, it is a near certainty that you will eat GM maize today.

Although the "corn belt" of the US has long been the world's leading center of maize production, it is not the center of origin. Maize was domesticated in Mexico, where farming families have grown it as a main component of their diets for millennia. It should be little surprise, then, that the GM debate has taken root in the maize-fields of Mexico. From Oaxaca and Chiapas east to Quintana Roo and north to the US-Mexico border, Mexican farmers and consumers look anxiously toward their maize-fields and tortillas, wondering if transgenes have made their way from the USA into Mexican maize. The Government of Mexico placed a moratorium on planting GM maize in 1999 (CEC 2004; Reuters 2007), but there are transgenes in Mexican maize-fields today – perhaps the world's most famous *transnational transgenes*. This chapter tells their story.

***

When the North American Free Trade Agreement (NAFTA) came into effect in January 1994, the long-protected Mexican agricultural market was opened to US exports, which increased sharply (see Bello 2009). The combination of increased maize imports and the suspicion that GM maize could cause problems generated great concern in Mexico. Thus in 1999, the Government of Mexico introduced a de facto moratorium on the production of GM maize – but its importation for consumption from the USA continued (González Aguirre and Aguilar Muñoz 2006). In response, ~150 non-governmental organizations called for a complete ban on the importation of GM maize in 2000; that same year, the Government signed the Cartagena Protocol on biosafety that mandates adherence to the precautionary principle while making regulatory decisions (González Aguirre and Aguilar Muñoz 2006).

So in the following year when Quist and Chapela (2001: 542) claimed to demonstrate "a high level of gene flow from industrially produced maize towards populations of progenitor landraces," the results launched a vigorous debate and spurred several other studies. Farmers and anti-GM activists erupted in protest throughout Mexico. For example in March 2004, the Commission for Environmental Cooperation (CEC), an organization that monitors transnational environmental issues among the NAFTA partners, met in Oaxaca to discuss their report on Mexican maize, transgenes, and biodiversity. Hundreds of Oaxacan farmers entered their meeting, faced the scientists and policy makers, and called for an end to imported GM maize (see McAfee 2008). The farmers brought local maize tortillas to offer the visitors.

Though much is contested about the presence of GM maize in Mexico, most agree about the stakes. Measured by volume, more maize is grown globally than any other crop (FAOSTAT 2006). The ability of maize farmers to adapt to challenges in the future (from pests and climate change for instance) is partly dependent upon the continued production and evolution of the 59 distinct, local landraces of maize that are grown in small regions of Mexico (Wellhausen *et al.* 1952; Sanchez *et al.* 2000). Mexico's maize landraces thus comprise a genetic resource of tremendous planetary importance (CEC 2004; Turrent-Fernández *et al.* 2009). Landrace maize is grown by > 80 percent of Mexico's maize farmers (Aquino *et al.* 2001), chiefly in the highlands of Mexico's rural south (Louette *et al.* 1997; Mann 2004), the poorest and most indigenous region in Mexico. The ways that these farmers select and share maize seeds constitute evolutionary processes that shape the persistence and evolution of landrace populations (Bellon and Berthaud 2006; Mercer *et al.* 2008). Thanks to their labor and the vagaries of agroecological processes, maize landraces in Mexico change continuously through human and natural selection (CEC 2004: 18).

So we could summarize the situation in Mexico by saying that one of the world's major food crops has likely experienced transgene introgression in its center of origin and cultural hearth – in violation of both the wishes of most local peasant farmers and the national moratorium on growing GM maize. In the roiling debate over GM agro-food, this situation – what many call the "transgenic contamination of Mexican maize" – is exhibit A for those who would argue in favor of strict regulation of GM agriculture (e.g. Greenpeace 2003). As one of the defining cases in the debates over GM agriculture, we can be sure that the boundaries and outcomes of the struggle over transgenic maize in Mexico will have repercussions around the world.

What's wrong with transgenes?

What exactly is the problem with GM agriculture? Even for those of us who are "GM critics," it can be difficult to answer this question. Perhaps the difficulty arises in part from the very *plurality* of criticisms of transgenic agriculture and food. We can discern six substantive criticisms of GM agriculture which, while inter-related, focus on fundamentally different scales and problems (compare Dawkins 1997; de la Perrière and Seuret 2001). Here we outline these criticisms, organized by geographical scale (see Figure 19.1). Our intention is to point out the range of the criticisms of GM agriculture – a range that is important for two reasons. First, it helps to explain the different responses to GM food in, for instance, the US and Europe (see Gaskell *et al.* 1999; Toke 2004). Second, it clarifies the implications of recent shifts in the *emphasis* of the GM debate.

***

At the smallest scale, that of the *gene*, we find the argument that transgenes from GM crops may move into populations of wild relatives or non-GM crops and produce unwanted effects by influencing the function of natural or managed ecosystems (cf. Ellstrand and Hoffman 1990; Tiedje *et al.* 1989; Snow and Moran Palma 1997; Marra 2001; Andow and Zwahlen 2006; Soleri *et al.* 2006a). Transgenes could introgress into closely-related wild species, potentially creating intractable weeds. For instance, a transgene in wild rice could create herbicide-resistant weed populations (Gealy 2005). Or transgenes could introgress into neighboring fields planted to the same species. This has been especially contentious in crop centers of origin or where organic farms border GM fields. Transgene introgression in landrace populations may alter patterns of genetic diversity (Gepts 2005; Soleri *et al.* 2006a) if advantageous traits such as effective disease resistance result in selective sweeps (Ellstrand 2003). Transgene introgression into landrace populations is often regarded as a problem in itself because the landrace population may be seen as "contaminated" (Greenpeace 2003). Farmers found to have trans-

	Scale of criticism	Perceived site of negative effects of GM agriculture
1	Gene	Gene frequencies; plant evolutionary processes
2	Organism	Human, plant, or animal health
3	Producer household	Farmers' livelihoods
4	Landscape	Genetic diversity within the agroecosystem
5	Cultural region	Cultural practices
6	Natural order	Morality

Figure 19.1 Principal criticisms of GM agriculture. GM agriculture has produced a wide array of criticisms. While inter-related, they nevertheless focus on fundamentally different problems and scales.

genes in their field without having paid for GM seed may face legal persecution from seed companies: consider the case of Monsanto versus Percy Schmeiser (see Prudham 2007). Also, because organic markets generally do not allow GM foods, transgene flow could cause organic farmers to lose market premiums.

A second criticism concerns the non-target effects of transgenes on *organisms* within the agroecosystem or on the end users (Haslberger 2003; NRC 2004; Gepts 2005). Critics cite concerns of the effects of transgenes on fauna found in agricultural environments (Andow and Zwahlen 2006), as well as fears of unexpected allergic reactions brought on by transgenic proteins in GM food. Yet many early critics have come to accept that GM food has been widely consumed in a number of countries without major health consequences. This is not to deny, of course, that there *may* be undetected or long-term negative effects. The total number of transgenic proteins that are widely consumed is still modest, and more will enter the market in the future. Nonetheless, the absence of recognized human effects has dampened this criticism (yet see Séralini *et al.* 2007).

The third line of criticism addresses the social relations of GM technology for *producer households*. A common thrust is that, by shifting political-economic power (or market share) toward a small number of large agro-food firms, the expansion of GM technologies will adversely affect the livelihoods of marginal farmers (Buttel 1989; Ojarasca 2002). In this view, GM technology is not scale-neutral, since the benefits accrue mainly to those who manage large, capital-intensive operations. Arguably the major proven advantage of GM technology has been to improve the effectiveness of commercial herbicides (sold, like the seeds, by only a few corporations), making certain herbicides more efficient. Because global agribusiness is highly concentrated around a few large seed and grain corporations (Kloppenburg 2004; Bello 2009), critics contend that the extension of GM agriculture may deepen inequalities in the control and profitability of agriculture in favor of the largest farmers and companies. For instance, only three companies (Cargill, ADM, and Zen Noh) export 82 percent of the world's maize (Patel and Memarsadeghi 2003: 4). Of course, many advocates of GM agriculture have claimed that GM seeds help small farmers. As Gardiner (2006: 56) explains, "one argument often presented [in favor of GM agriculture] claims that [GM seeds] offer a solution to the problem of world hunger." By contrast, GM opponents "often respond by claiming that such technology is neither necessary for, nor particularly likely to bring about, a solution to global food security." (The recent report from the International Assessment of Agricultural Knowledge, Science and Technology for Development (IAASTD 2008) supports the opponents' view.)

The fourth criticism of GM agriculture concerns evolutionary dynamics, albeit here at the scale of the *landscape*: the argument is that GM agriculture directly reduces agro-biodiversity. Since the creation of transgenic crop varieties is a time- and capital-intensive process, few different varieties are initially available. An array of conventional (non-GM) varieties may then be replaced by only a few GM varieties, thereby reducing the total genetic variation on the landscape (Gepts and Papa 2003). Conditions in the US corn belt, presently the world's epicenter of

GM agriculture, seems to support this claim (Kimbrell 2002). However, in recent years, more GM counterparts of conventional varieties have been produced.

The fifth line of criticism concerns *cultural practices*. In this view, landraces are not just natural objects found within the environment; they are equally social, the result of an immense collective labor, a grand collaboration with natural processes – and therefore important socionatural things to be respected and protected on cultural grounds (Brush 2004; CEC 2004). As one anti-GM activist explains, "contaminating" maize with transgenes is nothing less than "a crime against all indigenous peoples and farming communities who have safeguarded maize over millennia for the benefit of humankind" (Mendoza, cited in ETC Group 2003a: 2–3; also see ETC Group 2003b; Dawkins 1997: 44–46; Pilcher 2006). Nationalism may play a role here. Mexican cultural practices – from literature to cooking, music and farming – have long emphasized the foundational role of maize in defining national identity (or *Mexicanidad*), as though Mexico itself is rooted in a maize-field. Consider Figure 19.2, a painting by Diego Rivera from the *Palacio Nacional*, at the center of Mexico's central government and Mexico City. As with Rivera, the prevailing view in Mexico today is that its indigenous farmers have given the world a gift by creating and sustaining maize, and the world should return the favor by respecting its genetic integrity.

The sixth set of criticisms concerns nature as a metaphysical totality, i.e. nature *qua* natural order or Creation. Some have argued that creating transgenic organisms amounts to intervening into nature in ways that are contrary to religious values. Take, for instance, the position of the Catholic Church. Pope John Paul II spoke of transgenic technologies as "challenges to the natural order" (cited in CWNews 2004). Similarly, the director of the National Catholic Rural Life Conference argues that "agricultural biotechnology brings humankind extraordinarily close to upsetting the intricate order of biological and ecological relationships upon which life and health depend" (Andrews 2005). Of course, the Church is not alone in adopting this position, but its views are especially important in Mexico, which has more avowed Catholics than any country except Brazil.

***

A few observations are warranted on the relations among these six criticisms. Although they are often combined in different ways, these criticisms of GM agriculture do not necessarily follow from one another. They are analytically distinct, but not mutually exclusive. In Mexico and elsewhere, the boundaries between these six criticisms have blurred as arguments over GM maize recombine these claims in often confounding ways. For instance, farmers in Mexico often indicate that maize landraces should be maintained in a "pure" state, or at least that farmers should be able to choose whether or not they plant GM seeds. Likewise, when Greenpeace-Mexico activists scaled the Independence Column on Mexico City's Paseo de la Reforma in October 2009 in protest, their banners carried the slogan, "*Nuestro maíz es primero, ¡traidores!* (our maize is first, traitors!)" (Greenpeace 2009). Where does this argument fall in Figure 19.1? It could, in fact, touch upon all six criticisms. If farmers who have maintained traditional maize landraces come to perceive their seeds to be contaminated – and scientists cannot

Figure 19.2 Painting by Diego Rivera (*c.* 1929) from the *Palacio Nacional* in Mexico City, portraying an indigenous (ostensibly proto-Mexican) maize farmer, with seeds, planting stick, and maize. Below the farmer the text lists crops that "El mundo debe a México" – the world owes to Mexico. The first is "el maíz."

prove otherwise – then cultural, *in situ,* conservation practices may erode (Bellon and Berthaud 2006). Interviews with farmers show that changes brought on by the use of GM crops – such as in yield and/or seed networks – tend to make farmers less interested in planting new types of maize (Soleri *et al.* 2006a). The social relations of production are thus fundamental to the unfolding evolutionary drama that sustains maize landraces.

Analysis of these distinct criticisms is especially urgent today because the GM critics are rapidly losing ground. Clearly, the increasing number of farmers planting GM seeds coupled with the growing global flows of GM agro-foods have shifted the discourse over GM agriculture (cf. Hindo 2007). The prevailing debate is no longer whether GM agriculture is good or bad or whether it should be allowed. Today the question is *how to regulate* GM agriculture's *ecological* effects, specifically transgene introgression (criticism one in our table), which is where recent debate in the WTO and other international bodies has focused (WTO 2006a, b). Evolutionary ecology has become the hegemonic terrain where today's debates over GM agriculture play out.

***

In rural southern Mexico, crop fields tend to be evolving polycultural ecosystems. While farmers conserve crop genetic diversity on their farms, their plants are always exposed to evolutionary forces, such as natural and artificial selection, drift, mutation, and gene flow – all of which facilitate crop evolution (Alvarez *et al.* 2005; Bellon and Berthaud 2006). Interest in natural hybridization and its role in shaping invasiveness, species boundaries, and introgression has led to well-developed theories of gene flow dynamics in natural ecosystems (Rieseberg *et al.* 1996; Ellstrand and Schierenbeck 2000; Lenormand 2002; Geber and Eckhart 2005). Ecologists have used these concepts to investigate whether transgenes could introgress into local landraces (Belchera *et al.* 2005; Gepts 2005; Arriaga *et al.* 2006).

Transgenes can move among crop populations in two ways: through *seed exchange* and/or *cross pollination* among adjacent fields. Mexican farmers exchange seed to revitalize their populations or initiate new seed lots (Louette *et al.* 1997; Rice *et al.* 1998; Bellon and Risopoulos 2001). They also occasionally plant seed from DICONSA (a Mexican state institution that distributes subsidized grain), which imported some maize from the USA prior to 2003 (SEDESOL 2003). Transgenic seeds may have been inadvertently introduced by farmers in this way.

Cross-pollination among fields also results in the mixing of crop gene pools. Although most maize pollen does not move far (on the order of ~15m; Louette 1997), border plants can receive pollen from adjacent fields (Ortiz Torres 1993). Because maize fields in southern Mexico are typically small and irregularly shaped, they have comparatively large edges relative to their area and can be expected to produce hybrid seed resulting from cross-pollination between neighboring varieties. The degrees to which seed exchange and cross-pollination act as sources of gene flow in a given population depend on field size and geography as well as local seed-exchange practices.

Seed-mediated gene flow can occur over longer distances, as when bulk maize is transported and seed is lost *en route* or sold at its destination. Individuals can also move quantities of seeds long distances – even internationally. Local movement of seed occurs when farmers exchange seed within and between their communities. Farmers expose their maize landraces to artificial selection by saving seed to plant the following year and by management practices, such as herbicide application. Environmental conditions likewise select for fitness-enhancing traits. If transgenes are advantageous under these conditions, they would likely introgress into landrace populations.

Are there transgenes in Mexican maize landraces?

The burning question is thus: are there transgenes in Mexico's maize-fields, and if so, how many and where? This was not the question that Quist and Chapela (2001) set out to answer when their research launched the debate. Their aim was rather to establish a baseline for landrace genetics prior to the arrival of transgenes in rural Oaxaca (Ezcurra *et al.* 2001). Yet they found that approximately 1 percent of kernels in all four landrace ears collected from two communities contained transgenes (though their small sample size reduces the strength of this conclusion). Other claims made by the authors, such as the evidence of repeated introductions of transgenes into landrace genomes, were later discredited (Metz and Futterer 2002; Kaplinsky *et al.* 2002; see reply, Quist and Chapela 2002), but the primary claim – that transgenes were present – was not. As we will see, this finding has been repeated by others, both in and outside Oaxaca (see Serratos-Hernández *et al.* 2007 regarding the discovery of transgenes in maize fields south of Mexico City).

Quist and Chapela's finding inspired a string of studies conducted by numerous, diverse groups, each employing different seed sampling methods, varying the communities, fields, and ears. Several of these studies have found transgenes, albeit at low levels. Fully 95 percent of the 21 landrace maize fields sampled in Oaxaca and Puebla in 2001 had variable, low levels of transgenes (Ezcurra *et al.* 2001). (The methodology was later found to yield false positives, and subsequent unpublished results appear to indicate lower levels of transgenes). The Centro Internacional de Mejoramiento de Maíz y Trigo (2002) tested *ex situ* germplasm bank accessions from Oaxaca collected in 1997 and 1999 and found all to be free of detectable levels of transgenes.

As part of an effort to broaden the geographical reach of the testing and to increase popular involvement in the research, in 2003 and 2004 several NGOs tested maize landraces from nine states using kits by Agdia (a USA-based testing-equipment company). After conducting tests "to determine the presence or absence of five types of proteins that are present in GM organisms," they found at least one of three transgenes in many of the communities sampled (ETC Group 2003a: 4). Unfortunately, details on their methods and results are vague.

One recent study (Ortiz-García *et al.* 2005a, 2005b) sampled the Sierra de Juárez, the same region of Oaxaca where Quist and Chapela (2001) found transgenes. They sampled 4–5 ears from 1–5 fields within a total of 18 municipalities in 2003 and 2004, for a total of 153,746 seeds from 870 plants. Half of the ~306 seeds from each ear were sent to one laboratory for genetic analysis, and half went to another. They found no transgenes.

In response to this unexpected result, Ortiz-García *et al.* note that common evolutionary forces "may have prevented [transgenes] from persisting at detectable frequencies in the sampled seed" (Ortiz-García *et al.* 2005a, p. 12342). They argue that the amount of transgenic seed entering the region (i.e. gene flow) may have declined for two reasons: first, DICONSA ceased importing grain by 2003 (SEDESOL 2003); second, increased GM awareness may have led farmers to avoid planting DICONSA or other imported seed. Moreover, Ortiz-García *et al.* (2005a) posit that repeated backcrossing of transgenic plants with non-transgenic plants may have considerably reduced the frequency of transgenes since the sampling by Quist and Chapela (2001). They also suggest that genes at low frequencies can be randomly lost due to genetic drift. Finally, if transgenic plants produce relatively fewer seeds or pollen (i.e. are less fit), selection would act to reduce the frequency of transgenes in landraces.

The variation between Quist and Chapela (2001) and Ortiz-Garcia *et al.* (2005a) could be explained by differences in methodology and sampling error. Yet it is also possible that these results could be accurate without contradiction. Between the two sample periods – Quist and Chapela in 2000, Ortiz-García *et al.* in 2003 and 2004 – the levels of transgenes may have fallen to undetectable levels. Such rapid evolution is not unheard of (Reznick and Ghalambor 2001), particularly where selection and population bottlenecks are extreme. Since the sampling for both studies was confined to one region where transgenic maize would not necessarily be expected (Cleveland *et al.* 2005), their results do not confirm the absence of transgenes in landraces in other parts of Oaxaca or Mexico. By contrast, the ETC Group (2003a) may have sampled in higher-risk areas, suggesting that monitoring is needed beyond Oaxaca.

New issues surrounding sampling protocols emerged when Cleveland *et al.* (2005) responded to Ortiz-García *et al.* (2005a). Their criticism centers on the sampling scheme, the joint interpretation of tests across locations, and subsequent conclusions concerning the putative rarity of transgenes. More narrowly, they fault Ortiz-García *et al.* (2005a) for overestimating the precision of their sample (and other technical errors). They conclude that Ortiz-García *et al.* (2005a) should not have made any claims about the presence of transgenes in the region studied; rather, they should have limited themselves to claims about *individual* locations.

The consensus emerging between Cleveland *et al.* (2005) and Ortiz-García *et al.* (2005c) is that studies should focus on testing fewer seeds taken from many ears and many fields (Ortiz-García *et al.* 2006; Soleri *et al.* 2006b). Still, the amount of sampling and testing needed to bring the probability of *failing* to detect transgenes down to a reasonable level is very high. And such an approach is more complex and expensive, particularly given the enormous scale of Mexico's maize acreage. This debate highlights the difficulty in finding a way to identify regions with maize landraces in need of "decontamination" – as many farmers in Mexico have demanded (Wainwright and Mercer 2009). In effect, the debate could continue indefinitely because the acceptable level of "contamination" cannot be determined. As Ortiz-García *et al.* (2005c: 6) note:

Because it is impossible to prove that transgenes are absent in a given region, discussions about the *consequences* of undetected transgenic plants should acknowledge that even extremely low frequencies could result in biological and/or socioeconomic effects, depending on the transgenes in question and how they are viewed by local farmers.

Herein lies the dilemma of "decontamination": farmers and conservationists need precise answers that science cannot presently provide. In demanding the impossible, the call for "decontamination" exposes this fundamental dilemma.

***

A recent paper by Piñeyro-Nelson *et al.* (2009a, b) has introduced a new chapter in the debate over the presence of transgenes in Mexican landrace populations (see also Snow 2009). They collected maize over the course of three years of maize from 25 communities in Puebla and Oaxaca (one household was sampled per community), focusing primarily on the Sierra Juárez de Oaxaca where Quist and Chapela (2001) collected their positive samples in 2000. They tested these maize landrace collections with multiple robust molecular techniques.

Their study makes two key contributions to the debate. First, they confirm the presence of transgenes in maize seeds from three localities, of which at least two were collected from the Sierra Juárez in 2001. Despite a lack of positive results from the 2002 samples, more extensive resampling of two of the positive 2001 communities in 2004 confirmed the persistence of transgenes in these areas. This resampling of 30 fields in each locality yielded 11 with evidence of transgenes. This is the first study which followed up on previous positives and found that transgenes had persisted. Such persistence implies introgression.

Second, this study shows that the uneven frequencies of transgenes across landrace populations greatly reduces our ability to detect them. Transgene introgression can be expected to follow a predictable geographical pattern: the introduction of a transgenic variety into a given area will raise the frequency of transgenes in surrounding fields before affecting fields that are further away, producing an aggregation of maize fields with transgenes. Consequently, to maximize detection probabilities across a landscape, it is crucial to avoid "over-sampling of individual fields when frequencies differ strongly among them" (2009a: 10).

Notwithstanding the clarity and rigor of this study, Piñeyro-Nelson *et al.* (2009a) received a sharp critique written by the founder/CEO and VP of Genetic ID, a company hired by the authors of Piñeyro-Nelson *et al.* to conduct part of their genetic analysis. Two new arguments surfaced. First, in their critique, Schoel and Fagan (2009) contend that "Piñeyro-Nelson *et al.* (2009a) essentially *came up with* negative results in their survey of Oaxaca for transgenic maize" (4144, our italics). In other words, the critics assert that Piñeyro-Nelson *et al.* fudged the data. They explain:

Although sample 5 appears to be positive, it is hard to conclude from the provided data whether this is a true positive result . . . The two sets of authors essentially disagree about how to interpret evidence that levels of transgene presence are 'above detection level',

but weak. Schoel and Fagan's technique is to call these results negative, whereas Piñeyro-Nelson *et al.* prefer to examine those samples more carefully, using multiple methods.

In a lengthy reply, Piñeyro-Nelson *et al.* focus on methodological details that go beyond the scope of this chapter. Here is the core of their rebuttal:

> It is impossible to test [Schoel and Fagan's] statements using the standard methods of science because they have not provided full, transparent disclosure of many details in their methods because of intellectual property claims by their company, Genetic ID. Even with limited information, we have shown. . .that [they] base their statements on a system . . . which is overly permissive of false negatives. [Schoel and Fagan] seem honestly convinced of the infallibility of their proprietary system at Genetic ID, and do not see the need to account for it transparently. However, the value of a scientific procedure lies not in its infallibility but rather in the qualities that make it falsifiable by independent confirmation (Piñeyro-Nelson *et al.* 2009b: 4149).

Thus, the first debate turns on the means of confirming results in a rigorous and transparent fashion.

The second issue concerns, again, sampling protocols and maize diversity. Schoel and Fagan conclude: "the sample number was too small in both [studies] (Ortiz-García *et al.* and Piñeyro-Nelson *et al.*) and that sampling was not representative of the total Oaxacan maize population" (4144). Thus, the data fail to show that there are transgenes in Oaxaca.

Piñeyro-Nelson *et al.* (2009b) reply by appealing to the strengths of the broader literature, suggesting that their results are bolstered by the pattern in the data: "An overwhelming body of evidence has accumulated from many experienced researchers, using a diversity of independent methods and often published under the most stringent. . .standards to state that transgenic DNA sequences are present in collections of Mexican maize landraces.. . .[Schoel and Fagan] provide nothing but the face value of their statements to support their position" (2009b: 4149). In conclusion, they reassert their claim that, given the genetic diversity of Mexico's maize landraces, acceptable detection methods may perform differently:

> We learn from the experience of this discussion that such monitoring is extremely difficult under the conditions of high variability [that are] characteristic of the most delicate environmental situations. In the case of maize, we see that methods that may be acceptable in highly homogenous situations should not be expected to work in the diverse conditions found, for example, in the centres of origin and ongoing diversification of crops.
>
> (Ibid.)

Conclusions

The ongoing debate surrounding GM maize raises many important questions for political ecologists. We conclude by considering three.

The first stems from Piñeyro-Nelson *et al.*'s (2009b) conclusion that the "monitoring of transgenic DNA in. . .living organisms is of great environmental importance." Are transgenes-out-of-place really an "environmental conservation" issue? If so, how do we know this? On what basis could there be a "political ecology of transgenes"? As we have seen, critics of GM agriculture argue that the incorporation of genetic elements with unknown effects may complicate future efforts at corn breeding and production (CEC 2004) and cause the loss of important agricultural practices (Mann 2004). In this view, transgenic introgression into maize landraces in Mexico *is* a serious environmental issue, not to mention a political one. By contrast, many advocates of GM agriculture – including especially US-based global agribusiness corporations, and their supporters – typically regard such criticisms as subjective and unscientific. For one thing, incorporation of transgenes into maize landraces has, paradoxically perhaps, *increased* the genetic diversity of those landraces. This sort of fundamental disagreement filters through the language of the GM dispute, and each side regards certain terms as legitimate and some as illegitimate (cf. Jefferson 2001). Although scientific research is typically seen as the foundation of good environmental thinking, the case of transgenes in Mexican maize demonstrates some of the complexities – and limitations – of scientific practices for addressing environmental issues. What is at stake in the way we answer the first question, in short, is the framework in which to analyze the scientific debate over the presence of transgenes.

The second question follows from Piñeyro-Nelson *et al.*'s recent finding that, in the maize-fields of Oaxaca, transgenes "were present in our sample at a frequency of 0.011 based on PCR, and at a frequency of 0.0089 based on S[outhern] B[lot] hybridization" (2009a, 758). Just for the sake of argument, let's assume that (a) their sample frequency provided a reliable estimate of transgene frequencies of all Mexico's maize-fields (a claim they emphatically do not make). Moreover, assume (b) the scientific debate were to end with their paper (as we have seen, it won't). Let's assume, in other words, that a consensus were to emerge that transgenes are present in Mexico's maize landraces at a frequency of ~0.011, or ~1 percent. *Is this quantity problematic?* On what basis could we reply, "yes"? Reflecting upon the six criticisms of transgenes: exactly how many transgenes are "too many"? The key point to recognize here is that the source of authority in the GM debate – the scientists studying these issues (particularly ecologists) – *cannot answer these questions for us.* That is because the questions that arise from all six criticisms of GM agriculture are difficult (or impossible) to quantify, to measure scientifically. Thus we are left in a paradoxical situation: the political debate over transgenes relies upon scientific research, yet the scientific practices cannot resolve the political debate. One thing is clear: there is no sure path to reducing the uncertainty about GM agriculture's ecological effects, and the difficulty in detecting transgenes only underscores the importance of those practices that may prevent introgression from occurring in the first place – such as the moratorium on planting GM maize.

This brings us to our third question, concerning geographical responsibilities in a world of cross-border transgenic traffic. How should we conceptualize responsibility in a world where transgenes may cross national boundaries? Should regulatory agencies from the US – like the USDA's Biotechnology Regulatory Service (see BRS 2009) – have responsibility to regulate transgenes from US laboratories and maize fields that stray across borders? Presently, they do not. Neither does the CEC. In other words, there is no transnational government agency charged with regulating the flow of transgenes from the US into Mexico. Nor is there an international body with overarching regulatory power governing transnational transgenes. Should there be? It is easy to imagine that fear of transgenes could empower states to increase their claim to the regulation of life itself – that transnational transgenes could stimulate new forms of biopolitical regulation. The conditions for a alternative response to the challenge of transnational transgenes remain to be created.

***

The GM maize dispute has captured the attention of many people. Yet it is far from the only threat to the landraces and livelihoods of maize farmers today. Indeed, Mexican farmers face a wicked trinity of challenges. First: declining prices for their produce. The average price received by Mexican farmers selling maize in their domestic market collapsed in the 1990s, falling by more than 50 percent between 1990 and 2000 (Rosset 2006). Second: neoliberalism in the countryside, meaning declining subsidies for production and a shake-out of *ejido* lands (Barros-Nock 2000; Richard 2009; Bello 2009). Third: worsening natural conditions of production and the looming specter of rapid climate change. Climate models predict higher average temperatures and lower precipitation expected for central and southern Mexico (McSweeney *et al.* 2009). Taken together, the future for the subaltern social class that sustains Mexico's maize landraces looks bleak. This is not to deny that this class, Mexico's peasant-workers, will resist; indeed it will. As Walden Bello reminds us,

> One should never dismiss the toughness and resilience of the Mexican peasantry. Especially today, as the neoliberal model collapses, export markets dry up, low-skill jobs in the United States disappear, and industry spirals into depression owing to the financial crisis, the countryside is likely to see a return of hundreds of thousands of peasant-workers, seeking salvation in the land.

(2009: 52)

No doubt Mexico's peasant-worker communities will persist, and with them some maize landrace varieties. But a major qualification is needed: their livelihoods must be sustained under the increasingly difficult conditions of a declining traditional agrarian society, characterized by functional dualism (de Janvry 1981). Thus the future will only bring more challenges for these rural communities that sustain diverse maize landraces. Transgenic introgression is but one of these, and it is probably not, in our view, the most threatening. In this light, what is perhaps most remarkable about the GM dispute is that it has crystallized the spirit of resistance to the multiple processes that threaten the landrace maize *milpa*.

Acknowledgments

Earlier versions of subsections of this chapter were published in *Agriculture, Ecosystems and Environment* 123 and *Geoforum* 40 (3). We thank Elsevier for permission to republish these subsections. Mercer thanks the International Institute for Education, COMEXUS, and a Fulbright-García Robles grant for funding her research in Mexico in 2005. Wainwright conducted research for this paper as a Killam Fellow in the Department of Geography at the University of British Columbia. Our paper also benefited from conversations with L. Campbell, W. Jones, K. McSweeney, H. Perales, S. Prudham, and A. Snow. We are solely responsible for errors.

References

Alvarez, N., Garine, E., Khasah, C., Dounias, E., Hossaert-McKey, M., and McKey, D., 2005. Farmers' practices, metapopulation dynamics, and conservation of agricultural biodiversity on-farm: A case study of sorghum among the Duupa in sub-sahelian Cameroon. *Biological Conservation* 121: 533–543.

Andow, D. A., and Zwahlen, C., 2006. Assessing environmental risks of transgenic plants. *Ecology Letters* 9, 196–214.

Andrews, D., 2005. "Questioning the moral imperative of biotechnology for feeding the world." Unpublished April 19, 2005 lecture at Vincentian Center for Church and Society, Saint John's University (New York). Avaliable online at http://www.ncrlc.com/Saint-John's-University.html (accessed 21 April 2006).

Aquino, P., Carrión, F., Calvo, R., and Flores, D., 2001. Selected maize statistics. In: P. Pingali, ed. *CIMMYT 1999–2000 World Maize Facts and Trends: Meeting World Maize Needs: Technological Opportunities and Priorities for the Public Sector*. CIMMYT, Mexico, D.F., pp. 45–57.

Arriaga, L., Huerta, E., Lira-Saade, R., Moreno, E., and Alarcón, J., 2006. Assessing the risk of releasing transgenic *Cucurbita* spp. in Mexico. *Agriculture, Ecosystems, and Environment* 112: 291–299.

Barros-Nock, M., 2000. "The Mexican peasantry and the *ejido* in the neoliberal period," in D. Bryceson, C. Kay, and J. Moois., eds. *Disappearing Peasantries: Rural Labor in Africa, Asia, and Latin America*. London: Intermediate Technology Publications.

Belchera, K., Nolana, J., and Phillips, P., 2005. Genetically modified crops and agricultural landscapes: spatial patterns of contamination. *Ecological Economics* 53: 387–401.

Bello, W., 2009. *The Food Wars*. New York: Verso.

Bellon, M. and Berthaud, J., 2006. Traditional Mexican agricultural systems and the potential impacts of transgenic varieties on maize diversity. *Agriculture and Human Values* 23, 3–14.

Bellon, M. and Risopoulos, J., 2001. Small-scale farmers expand the benefits of improved maize germplasm: A case study from Chiapas, Mexico. *World Development* 29, 799–811.

(BIO) Biotechnology Industry Organization, 2006. "2006 Global Biotech Crop Acreage Increases by 13 Percent Over 2005." Available online at http://www.bio.org/ataglance/acreage.asp (accessed 3 March 2007).

(BRS) United States Department of Agriculture – Biotechnology Regulatory Service. 2009. Available online at http://www.aphis.usda.gov/biotechnology/brs_main.shtml (accessed 15 October 2009).

Brush, S., 2004. *Farmers' Bounty: Locating Crop Diversity in the Contemporary World.* New Haven, CT: Yale University Press.

Buttel, F., 1989. Social science research on biotechnology and agriculture: A critique. *Rural Sociologist* 9: 5–15.

(CEC) Commission for Environmental Cooperation, 2004. *Maize and Biodiversity: The Effects of Transgenic Maize in Mexico.* Montréal, Canada: CEC.

(CIMMYT) Centro Internacional de Mejoramiento de Maíz y Trigo, 2002. Los últimos escrutinios de maíz criollo en México no revelan la presencia del promotor relacionado con los transgenes. CIMMYT, Mexico, D.F.

Cleveland, D.A., Soleri, D., Aragon Cuevas, F., Crossa, J., and Gepts, P., 2005. Detecting (trans)gene flow to landraces in centers of crop origin: Lessons from the case of maize in Mexico. *Environmental Biosafety Research* 4, 197–208.

CWNews, 2004. Pope urges preserving diversity in food production. Rome, Oct. 15, 2004. Available online at http://www.cwnews.com/news/viewstory.cfm?recnum=32803 (accessed 21 April 2006).

Dawkins, K., 1997. *Gene wars: The Politics of Biotechnology.* New York: Seven Stories/Open Media.

De Janvry, A., 1981. *The Agrarian Question and Reform in Latin America.* Baltimore, MD: Johns Hopkins University Press.

De la Perrière, R.A., and Seuret, F., 2001. *Brave New Seeds: The Threat of GM Crops to Farmers.* London: Zed.

Ellstrand, N.C., 2003. *Dangerous Liaisons? When Cultivated Plants Mate with Their Wild Relatives.* Baltimore, MD: Johns Hopkins University Press.

Ellstrand, N.C., and Hoffman, C.A., 1990. Hybridization as An Avenue of Escape for Engineered Genes. *BioScience* 40: 438–442.

Ellstrand, N.C., and Schierenbeck, K.A., 2000. Hybridization as a stimulus for the evolution of invasiveness in plants? *Proceedings of the National Academy of Sciences* 97, 7043–7050.

(ETC Group) Action Group on Erosion, Technology and Concentration, 2003a. Contamination of Bt genetically modified maize in Mexico much worse than feared. www.etcgroup.org, Mexico City, Mexico.

—— 2003b. Open letter from international civil society organizations on transgenic contamination in the centers of origin and diversity. www.etcgroup.org, Mexico City, Mexico.

Ezcurra, E., Ortiz, S., and Soberón Mainero, J., 2001. "Evidence of gene flow from transgenic maize to local varieties in Mexico" in *LMOs and the Environment: Proceedings of an International Conference.* Durham, NC: OECD, pp. 289–295.

FAOSTAT, 2006. Accessed June 16, 2006, at http://faostat.fao.org.

Gardiner, S., 2006. A core precautionary principle. *The Journal of Political Philosophy* 14(1), 33–60.

Gaskell, G., Bauer, M., Durant, J., and Allum, N., 1999. Worlds apart? The reception of genetically modified foods in Europe and the US. *Science* 285: 384–387. [Published July 1999 and retracted June 2000.]

Gealy, D.R. 2005. "Gene movement between rice (*Oryza sativa*) and weedy rice (*Oryza sativa*) – a U.S. temperate rice perspective" in J. Gressel (ed.) *Crop Ferality and Volunteerism.* Boca Raton, FL: CRC Press, Inc.

Geber, M. A., and Eckhart, V.M., 2005. Experimental studies of adaptation in *Clarkia xantiana.* II. Fitness variation across a subspecies border. *Evolution* 59: 521–531.

Gepts, P., 2005. "Introduction of transgenic crops in centers of origin and domestication" in D.L. Kleinman, A.J. Kinchy, and J. Handelsman, *Controversies in Science and Technology: From Maize to Menopause*. Madison: University of Wisconsin Press.

Gepts, P., and Papa, R., 2003. Possible effects of (trans)gene flow from crops on the genetic diversity of landraces and wild relatives. *Environmental Biosafety Research* 2: 89–103.

González Aguirre, R.L., and Aguilar Muñoz, E.J., 2006. Bioseguridad y actores sociales. Presentation at the 2006 Latin American Studies Association Conference (San Juan, Puerto Rico).

Greenpeace, 2003. Maize Under Threat: GE Maize Contamination in Mexico. Available online at http://www.greenpeace.org/raw/content/international/press/reports/maize-under-threat-ge-maize.pdf (accessed 9 October 2007).

—— 2009. Nuestro maíz es primero, ¡traidores! Available online at http://www.greenpeace.org/mexico/news/traicion (accessed 22 October 2009).

Haslberger, A.G. 2003. Codex guidelines for GM foods include the analysis of unintended effects. *Nature Biotechnology* 21: 739–741.

Hindo, B., 2007. Monsanto: Winning the ground war. *Business Week*. December 17, 2007: 34–41.

(IAASTD) International Assessment of Agricultural Knowledge, Science and Technology for Development, 2008. "International assessment of agricultural knowledge, science and technology for development global summary for decision makers." Available online at http://www.agassessment.org/index.cfm?Page=About_IAASTD&ItemID=2 (accessed 23 May 2008).

Jefferson, R., 2001. "Transcending transgenics: Are there "babies in the bathwater," or is that a dorsal fin?" in P. Pardey (ed.) *The Future of Food: Biotechnology Markets in an International Setting*. Washington, D.C.: International Food Policy Research Institute and Johns Hopkins University Press, 75–92.

Kaplinsky, N., Braun, D., Lisch, D., Hay, A., Hake, S., and Freeling, M., 2002. Maize transgene results in Mexico are artifacts. *Nature* 416, 601.

Kloppenburg, J., 2004. *First the Seed: The Political Economy of Plant Biotechnology*. 2nd ed. Madison: University of Wisconsin Press.

Kimbrell, A., (ed.), 2002. *Fatal Harvest: The Tragedy of Industrial Agriculture*. Sausalito, CA: Foundation for Deep Ecology.

Lenormand T., 2002. Gene flow and the limits to natural selection. *Trends in Ecology & Evolution* 17: 183–189.

Louette, D. 1997. "Seed exchange among farmers and gene flow among maize varieties in traditional agricultural systems" in Serratos, J.A., Wilcox, M.C., and Castillo González, F., (Eds.), *Gene Flow Among Maize Landraces, Improved Maize Varieties, and Teosinte: Implications for Transgenic Maize*. Mexico, D.F.: CIMMYT, pp. 56–66.

Louette, D., Charrier, A., and Berthaud, J., 1997. In situ conservation of maize in Mexico: Genetic diversity and maize seed management in a traditional community. *Economic Botany* 51, 20–38.

McAfee, K., 2008. Beyond techno-science: Transgenic maize in the fight over Mexico's future. *Geoforum* 39: 148–260.

McSweeney, C., New, M. and Lizcano, G. 2009. UNDP Climate change country profile: Mexico. Available online at http://country-profiles.geog.ox.ac.uk/ (accessed 29 October 2009).

Mann, C., 2004. Diversity on the farm: How traditional crops around the world help to feed us all, and why we should reward the people who grow them. Amherst: University of

Massachusetts. Available online at http://www.umass.edu/peri/crop.html (accessed 3 March 2007).

Marra, M., 2001. "Agricultural biotechnology: A critical review of the impact evidence to date" in P. Pardey, ed. *The Future of Food: Biotechnology Markets in an International Setting.* Washington, DC: International Food Policy Research Institute and Johns Hopkins University Press: 155–184.

Mercer, K., Martínez-Vásquez, A., and Perales Rivera, H., (2008). Asymmetrical local adaptation of maize landraces along an altitudinal gradient. *Evolutionary Applications* 1(3), 489–500.

Metz, M., and Fütterer, J., 2002. Suspect evidence of transgenic contamination. *Nature* 416, 600–601.

Miller, J., 2006. Stark raving mad: French farmers, activists battle over rise in genetically altered corn. *The Wall Street Journal*, 12 October 2006: B1.

(NRC) National Research Council, 2004. *Biological Confinement of Genetically Engineered Organisms.* Washington, DC: The National Academies Press,.

Ojarasca. 2002. *La Jornada* (Mexico) 58, February 2002. Cited in Ramon Vera Herrera. 2004. Available online at http://americas.irc-online.org/reports/2004/sp_0408maiz.html (accessed October 2009).

Ortiz-García, S., Ezcurra, E., Schoel, B., Acevedo, F., Soberón, J., Snow, A.A., 2005a. Absence of detectable transgenes in local landraces of maize in Oaxaca, Mexico (2003–4). *Proceedings of the National Academy of Sciences*, USA 102, 12338–12343.

—— 2005b. Correction. *Proceedings of the National Academy of Sciences*, USA 102, 18242.

—— 2005c. Reply to Cleveland *et al.*'s critique of "Absence of detectable transgenes in local landraces of maize in Oaxaca, Mexico (2003–4)." *Environmental Biosafety Research* 4, 209–215.

—— 2006. Transgenic maize in Mexico. *BioScience* 56, 709.

Ortiz Torres, E., 1993. Aislamiento y dispersión de polen en la producción semilla de maíz. MS Thesis. College of Postgraduates, Montecillo, México. 81 pp.

Patel, R., and Memarsadeghi, S. 2003. *Agricultural Restructuring and Concentration in the United States: Who Wins? Who Loses?* Institute for Food and Development Policy (Food First), Policy Brief No. 6.

Pilcher, J., 2006. "Taco Bell, Maseca, and slow food: A postmodern apocalypse for Mexico's peasant cuisine?" in R. Wilk, ed. *Fast Food/Slow Food: The Cultural Economy of the Global Food System.* Lanham, MD: AltaMira Press, 69–81.

Piñeyro-Nelson, A., Van Heerwaarden, J., Perales, H., Serratos-Hernandez, A., Rangel, A., Hufford, M., Gepts, P., Garay-Arroyo, A., Rivera-Bustamante, R., and Álvarez-Buylla, E. 2009a. Transgenes in Mexican maize: molecular evidence and methodological considerations for GMO detection in landrace populations. *Molecular Ecology* 18(4), 750–761.

—— 2009b. Resolution of the Mexican transgene detection controversy: Error sources and scientific practice in commercial and ecological contexts. *Molecular Ecology* 18, 4145–4150.

Prudham, S. 2007. The fictions of autonomous invention: Accumulation by dispossession, commodification and life patents in Canada. *Antipode* 39(3), 406–429.

Quist, D. and Chapela, I., 2001. Transgenic DNA introgressed into traditional maize landraces in Oaxaca, Mexico. *Nature* 414, 541–543.

—— 2002. Quist and Chapela reply. *Nature* 416, 602.

Reuters, 2007. Tortilla-hungry Mexico setting rules on GMO corn. Reuters, July 17, 2007. Available online at www.reuters.com (accessed 9 October 2007).

Reznick, D.N., and Ghalambor, C.K., 2001. The population ecology of contemporary adaptations: What empirical studies reveal about the conditions that promote adaptive evolution. *Genetica* 112–113, 183–198.

Rice, E., Smale, M., and Blanco, J.L., 1998. Farmers' use of improved seed selection practices in Mexican maize: Evidence and issues from the Sierra de Santa Marta. *World Deveopment.* 26, 1625–1640.

Richard, A., 2009. Withered milpas: Governmental disaster and the Mexican countryside. *Journal of Latin American and Caribbean Anthropology* 13: 2, 387–413.

Rieseberg, L. H., Sinervo, B., Linder, C.R., Ungerer, M.C., and Arias, D.M., 1996. Role of gene interactions in hybrid speciation: Evidence from ancient and experimental hybrids. *Science* (Washington) 272: 741–745.

Rosset, P., 2006. *Food is Different: Why We Must Get the WTO Out of Agriculture.* New York: Zed Books.

Sanchez, G., Goodman, M., and Stuber, C.W., 2000. Isozymatic and morphological diversity in the races of maize of Mexico. *Economic Botany* 54, 43–59.

Schoel, B., and Fagan, J. 2009. Comment: Insufficient evidence for the discovery of transgenes in Mexican landraces. *Molecular Ecology* 18, 4143–4144.

(SEDESOL) Secretaría de Desarrollo Social, 2003. Apoya SEDESOL al campo Mexicano; ni un sólo grano de maíz comprado al exterior este año. Available online at http://www.sedesol.gob.mx/prensa/comunicados/c_074_2003.htm (accessed 15June 2006).

Séralini, G., Cellier, D., and Spiroux de Vendomois, J., 2007. New analysis of a rat feeding study with a genetically modified maize reveals signs of hepatorenal toxicity. *Archives of Environmental Contamination and Toxicology* 52: 596–602.

Serratos-Hernández, J., Gómez-Olivares, J., Salinas-Arreortua, N., Buendía-Rodríguez, E., Islas-Gutiérrez, F., and de-Ita, A. 2007. Transgenic proteins in maize in the soil conservation area of the Federal District. *Frontiers in Ecology and the Environment* 5, 247–552.

Snow, A. 2009. Unwanted transgenes re-discovered in Oaxacan maize. *Molecular Ecology* 18(4), 569–571.

Snow, A.A. and Moran-Palma, P., 1997. Commercialization of transgenic plants: potential ecological risks. *Bioscience* 47: 86–96.

Soleri, D., Cleveland, D., Aragón Cuevas, F., 2006a. Transgenic crops and crop varietal diversity: the case of maize in Mexico. *BioScience.* 56, 503–513.

—— 2006b. Response from Soleri and colleagues. *BioScience.* 56, 709–710.

Tiedje, J.M., Colwell, R.K. , Grossman, Y.L., Hodson, R.E., Lenski, R.E., Mack, R.N., and Regal, P.J., 1989. The planned introduction of genetically engineered organisms – ecological considerations and recommendations. *Ecology* 70: 298–315.

Toke, D., 2004. *The Politics of GM Food: A Comparative Study of the UK, USA, and EU.* New York: Routledge.

Turrent-Fernández, A., Serratos-Hernández, J., Mejía-Andrade, H., and Espinosa-Calderón, A. 2009. Evaluation proposal of possible impact of transgene accumulation in Mexican maize landraces. *Agrociencia* 43: 257–265.

Wainwright, J., and K. Mercer, 2009. "The dilemma of decontamination: A Gramscian analysis of the Mexican transgenic maize dispute," *Geoforum* 40 (3) 345–354.

Wellhausen, E., Roberts, J., Roberts, L. M., and Hernández, E., 1952. *Races of maize in Mexico: Their origin, characteristics, and distribution.* Cambridge, MA: Harvard University Press.

(WTO) World Trade Organization, 2006a. Reports out on biotech disputes. Available online at http://www.wto.org/english/news_e/news06_e/291r_e.htm (accessed 3 March 2007).

—— 2006b. European Communities – Measures affecting the approval and marketing of biotech products: reports of the panel, 29 September 2006. Available online at http://www.wto.org/english/tratop_e/dispu_e/cases_e/ds291_e.htm (accessed 3 March 2007).

Index

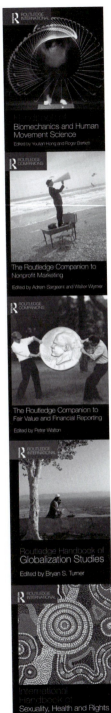

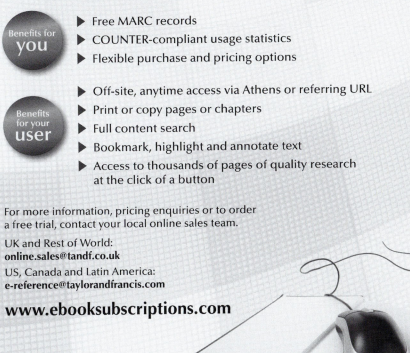